Third Edition

ABSTRACT ALGEBRA

John A. Beachy
William D. Blair
Northern Illinois University

WAVELAND

PRESS, INC.

Long Grove, Illinois

For information about this book, contact:
Waveland Press, Inc.
4180 IL Route 83, Suite 101
Long Grove, IL 60047-9580
(847) 634-0081
info@waveland.com
www.waveland.com

10-digit ISBN 1-57766-443-4
13-digit ISBN 978-1-57766-443-7

Contents

PREFACE

This edition would probably not have been written without the impetus from George Bergman, of the University of California, Berkeley. After using the book, on more than one occasion he sent us a large number of detailed suggestions on how to improve the presentation. Many of these were in response to questions from his students, so we owe an enormous debt of gratitude to his students, as well as to Professor Bergman. We believe that our responses to his suggestions and corrections have measurably improved the book.

We would also like to acknowledge important corrections and suggestions that we received from Marie Vitulli, of the University of Oregon, and from David Doster, of Choate Rosemary Hall. We have also benefitted over the years from numerous comments from our own students and from a variety of colleagues. We would like to add Doug Bowman, Dave Rusin, and Jeff Thunder to the list of colleagues given in the preface to the second edition.

In this edition we have added a number of exercises, we have added 1 to all rings, and we have done our best to weed out various errors and misprints.

We use the book in a linear fashion, but there are some alternatives to that approach. With students who already have some acquaintance with the material in Chapters 1 and 2, it would be possible to begin with Chapter 3, on groups, using the first two chapters for a reference. We view these chapters as studying cyclic groups and permutation groups, respectively. Since Chapter 7 continues the development of group theory, it is possible to go directly from Chapter 3 to Chapter 7.

Chapter 5 contains basic facts about commutative rings, and contains many examples which depend on a knowledge of polynomial rings from Chapter 4. Chapter 5 also depends on Chapter 3, since we make use of facts about groups in the development of ring theory, particularly in Section 5.3 on factor rings. After covering Chapter 5, it is possible to go directly to Chapter 9, which has more ring theory and some applications to number theory.

Our development of Galois theory in Chapter 8 depends on results from Chapters 5 and 6. Section 8.4, on solvability by radicals, requires a significant amount of material from Chapter 7.

Rather than outlining a large number of possible paths through various parts of the text, we have to ask the instructor to read ahead and use a great deal of caution in choosing any paths other than the ones we have suggested above.

We would like to point out to both students and instructors that there is some supplementary material available on the book's website:

$www.math.niu.edu/\sim beachy/abstract\_algebra/$.

Finally, we would like to thank our publisher, Neil Rowe, for his continued support of our writing.

DeKalb, Illinois   *John A. Beachy*
                                                                                                        William D. Blair
                                                                                                        September 1, 2005

PREFACE TO THE SECOND EDITION

An abstract algebra course at the junior/senior level, whether for one or two semesters, has been a well-established part of the curriculum for mathematics majors for over a generation. Our book is intended for this course, and has grown directly out of our experience in teaching the course at Northern Illinois University.

As a prerequisite to the abstract algebra course, our students are required to have taken a sophomore level course in linear algebra that is largely computational, although they have been introduced to proofs to some extent. Our classes include students preparing to teach high school, but almost no computer science or engineering students. We certainly do not assume that all of our students will go on to graduate school in pure mathematics.

In searching for appropriate text books, we have found several texts that start at about the same level as we do, but most of these stay at that level, and they do not teach nearly as much mathematics as we desire. On the other hand, there are several fine books that start and finish at the level of our Chapters 3 through 6, but these books tend to begin immediately with the abstract notion of group (or ring), and then leave the average student at the starting gate. We have in the past used such books, supplemented by our Chapter 1.

Historically the subject of abstract algebra arose from concrete problems, and it is our feeling that by beginning with such concrete problems we will be able to generate the student's interest in the subject and at the same time build on the foundation with which the student feels comfortable.

Although the book starts in a very concrete fashion, we increase the level of sophistication as the book progresses, and, by the end of Chapter 6, all of the topics taught in our course have been covered. It is our conviction that the level of sophistication should increase, slowly at first, as the students become familiar with the subject. We think our ordering of the topics speaks directly to this assertion.

Recently there has been a tendency to yield to demands of "relevancy," and to include "applications" in this course. It is our feeling that such inclusions often tend to be superficial. In order to make room for the inclusion of applications, some important mathematical concepts have to be sacrificed. It is clear that one must have substantial experience with abstract algebra before any genuine applications can be treated. For this reason we feel that the most honest introduction concentrates on the algebra. One of the reasons frequently given for treating applications is that they motivate the student. We prefer to motivate the subject with concrete problems from areas that the students have previously encountered, namely, the integers and polynomials over the real numbers.

One problem with most treatments of abstract algebra, whether they begin with group theory or ring theory, is that the students simultaneously encounter for the first time both abstract mathematics and the requirement that they produce proofs of their own devising. By taking a more concrete approach than is usual, we hope to separate these two initiations.

In three of the first four chapters of our book we discuss familiar concrete mathematics: number theory, functions and permutations, and polynomials. Although the objects of study are concrete, and most are familiar, we cover quite a few nontrivial ideas and at the same time introduce the student to the subtle ideas of mathematical proof. (At Northern Illinois University, this course and Advanced Calculus are the traditional places for students to learn how to write proofs.) After studying Chapters 1 and 2, the students have at their disposal some of the most important examples of groups—permutation groups, the group of integers modulo n, and certain matrix groups. In Chapter 3 the abstract definition of a group is introduced, and the students encounter the notion of a group armed with a variety of concrete examples.

Probably the most difficult notion in elementary group theory is that of a factor group. Again this is a case where the difficulty arises because there are, in fact, two new ideas encountered together. We have tried to separate these by treating the notions of equivalence relation and partition in Chapter 2 in the context of sets and functions. We consider there the concept of factoring a function into "better" functions, and show how the notion of a partition arises in this context. These ideas are related to the integers modulo n, studied in Chapter 1. When factor groups are introduced in Chapter 3, we have partitions and equivalence relations at our disposal, and we are able to concentrate on the group structure introduced on the equivalence classes.

In Chapter 4 we return to a more concrete subject when we derive some important properties of polynomials. Here we draw heavily on the students' familiarity with polynomials from high school algebra and on the parallel between the properties of the integers studied in Chapter 1 and the polynomials. Chapter 5 then introduces the abstract definition of a ring after we have already encountered several important examples of rings: the integers, the integers modulo n, and the ring of polynomials with coefficients in any field.

From this point on our book looks more like a traditional abstract algebra textbook. After rings we consider fields, and we include a discussion of root adjunction as well as the three problems from antiquity: squaring the circle, duplicating the cube, and trisecting an angle. We also discuss splitting fields and finite fields here. We feel that the first six chapters represent the most that students at institutions such as ours can reasonably absorb in a year.

Chapter 7 returns to group theory to consider several more sophisticated ideas including those needed for Galois theory, which is the subject matter of Chapter 8.

In Chapter 9 we return to a study of rings, and consider questions of unique factorization. As a number theoretic application, we present a proof of Fermat's last theorem for the exponent 3. In fact, this is the last of a thread of number theoretic applications that run through the text, including a proof of the quadratic reciprocity law in Section 6.7 and a study of primitive roots modulo p in Section 7.5. The applications to number theory provide topics suitable for honors students.

The last three chapters are intended to make the book suitable for an honors course or for classes of especially talented or well-prepared students. In these chapters the writing style is rather terse and demanding. Proofs are included for the Sylow theorems, the structure theorem for finite abelian groups, theorems on the simplicity of the alternating group and the special linear group over a finite field, the fundamental theorem of Galois theory, Abel's theorem on the insolvability of the quintic, and the theorem that a polynomial ring over a unique factorization domain is again a unique factorization domain.

The only prerequisite for our text is a sophomore level course in linear algebra. We do not assume that the student has been required to write, or even read, proofs before taking our course. We do use examples from matrix algebra in our discussion of group theory, and we draw on the computational techniques learned in the linear algebra course—see, for example, our treatment of the Euclidean algorithm in Chapter 1.

We have included a number of appendices to which the student may be referred for background material. The appendices on induction and on the complex numbers might be appropriate to cover in class, and so they include some exercises.

In our classes we usually intend to cover Chapters 1, 2 and 3 in the first semester, and most of Chapters 4, 5 and 6 in the second semester. In practice, we usually begin the second semester with group homomorphisms and factor groups, and end with geometric constructions. We have rarely had time to cover splitting fields and finite fields. For students with better preparation, Chapters 1 and 2 could be covered more quickly. The development is arranged so that Chapter 7 on the structure of groups can be covered immediately after Chapter 3. On the other hand, the material from Chapter 7 is not really needed until Section 8.4, at which point we need results on solvable groups.

We have included answers to some of the odd numbered computational exercises. In the exercise sets, the problems for which answers are given in the answer key are marked by the symbol †.

ACKNOWLEDGMENTS

To list all of the many sources from which we have learned is almost impossible. Perhaps because we are ring theorists ourselves, we have been attracted to and influenced by the work of two ring theorists—I. N. Herstein in *Topics in Algebra* and N. Jacobson in *Basic Algebra I, II*. In most cases our conventions, notation, and symbols are consistent with those used by Jacobson. We certainly need to mention the legacy of E. Noether, which we have met via the classic text *Algebra* by B. L. van der Waerden. Our treatment of Galois theory is influenced by the writing of E. Artin. In many ways our approach to abstract concepts via concrete examples is similar in flavor to that of Birkhoff and Mac Lane in *A Survey of Modern Algebra*, although we have chosen to take a naive approach to the development of the number systems and have omitted any discussion of ordered fields. We have also been influenced by the historical approaches and choice of material in *Abstract Algebra: A First Course* by L. Goldstein and *Introduction to Abstract Algebra* by L. Shapiro.

Many colleagues have taught from preliminary versions of parts of this book. We would like to thank a number of them for their comments: Harvey Blau, Harald Ellers, John Ewell, Tac Kambayashi, Henry Leonard, John Lindsey, Martin Lorenz, Donald McAlister, Robert McFadden, Gunnar Sigurdsson, George Seelinger, Doug Weakley, and John Wolfskill. Various students have offered comments and pointed out errors. We are particularly indebted to Svetlana Butler, Penny Fuller, Lauren Grubb, Michelle Mace, and Susan Talarico for giving us lists of misprints. We would like to thank all of the reviewers of our previous version, including: Victor Camillo, The University of Iowa; John C. Higgins, Brigham Young University; I. Martin Isaacs, University of Wisconsin, Madison; Paul G. Kumpel, State University of New York at Stony Brook; and Mark L. Teply, University of Wisconsin, Milwaukee. We would also like to thank Neil Rowe of Waveland Press for keeping our book in print at a reasonable price.

This seems to be an appropriate place to record our thanks to Goro Azumaya and Lance Small (respectively) for their inspiration, influence, and contributions to our mathematical development. Finally, we would like to thank our families: Marcia, Gwendolyn, Elizabeth, and Hannah Beachy and Kathy, Carla, and Stephanie Blair.

John A. Beachy
William D. Blair

TO THE STUDENT

This book has grown out of our experiences in teaching abstract algebra over a considerable period of time. Our students have generally had three semesters of calculus, followed by a semester of linear algebra, and so we assume only this much background. This has meant that our students have had some familiarity with the abstract concepts of vector spaces and linear transformations. They have even had to write out a few short proofs in previous courses. But they have not usually been prepared for the depth in our course, where we require that almost everything be proved quite carefully. Learning to write proofs has always been a major stumbling block. The best advice we can give in this regard is to urge you to talk to your teacher. Each ten minutes of help in the early going will save hours later.

Don't be discouraged if you can't solve all of the exercises. Do the ones you are assigned, try lots of others, and come back to the ones you can't do on the first try. From time to time there will be "misplaced" exercises. By this we mean exercises for which you have sufficient tools to solve the problem as it appears, but which have easier solutions after better techniques have been introduced at a later stage. Simply attempting one of these problems (even if the attempt ends in failure) can help you understand certain ideas when they are introduced later. For this reason we urge you to keep coming back to exercises that you cannot solve on the first try.

We urge any student who feels in need of a pep talk to reread this part of the preface. The same general comment applies to the introductions to each chapter, and to the notes at the ends of chapters. When first read, these introductory comments are meant to motivate the material that follows by indicating why it is interesting or important and at the same time relating this new material to things from the student's background. They are also intended to tie together various concepts to be introduced in the chapter, and so some parts will make more sense after the relevant part of the chapter has been covered in detail. Not only will the introductions themselves make more sense on rereading, but the way in which they tie the subject matter of the chapter into the broader picture should be easier to understand.

We often hear comments or questions similar to the ones we have listed below. We hope that our responses will be helpful to you as you begin studying our book.

"I have to read the text several times before I begin to understand it."

Yes, you should probably expect to have to do this. In fact, you might benefit from a "slow reading" rather than a "speed reading" course. There aren't many pages in a section, so you can afford to read them line by line. You should make sure that you can supply any reasons that may have been left out. We have written the book with the intention of gradually raising the level as it progresses. That simply means that we take more for granted, that we leave out more details. We hope to force you to become more sophisticated as you go along, so that you can supply more and more of the details on your own.

"I understand the definitions and theorems, but I can't do the problems."

Please forgive us for being skeptical of this statement. Often it just simply isn't the case. How do you really understand a definition or a theorem? Being able to write down a definition constitutes the first step. But to put it into context you need a good variety of examples, which should allow you to relate a new definition to facts you already know. With each definition you should associate several examples, simple enough to understand thoroughly, but complex enough to illustrate the properties inherent in the definition. In writing the book we have tried to provide good examples for each definition, but you may need to come up with your own examples based on your particular background and interests.

Understanding a proof is similar. If you can follow every step in the proof, and even write it out by yourself, then you have one degree of understanding. A complementary aspect of really understanding the proof is to be able to show exactly what it means for some simple examples that you can easily grasp. Sometimes it is helpful to take an example you understand well and follow through each step of the proof, applying it to the example you have in mind.

Trying to use "lateral thinking" is often important in solving a problem. It is easy to get stuck in one approach to a problem. You need to keep asking yourself if there are other ways to view the problem. Time to reflect is important. You need to do the groundwork in trying to understand the question and in reviewing relevant definitions and theorems from the text. Then you may benefit from simply taking a break and allowing your subconscious mind to sort some things out and make some connections. If you do the preparation well, you will find that a solution or method of attack may occur to you at quite unlikely times, when you are completely relaxed or even absorbed in something unrelated.

After emphasizing the use of examples, we need to discuss the next complaint, which is a standard one.

"I need more examples."

This is probably true, since even though we have tried to supply a good variety of examples, we may not have included the ones that best tie into your previous experience. We can't overemphasize the importance of examples in providing motivation, as well as in understanding definitions and theorems. Keep in mind that the exercises at the end of each section provide a source for more examples.

In fact, each time you learn a new definition you should choose several examples to help you remember the definition. We have tried to present the examples that we feel would be the most helpful in understanding definitions and theorems. This approach is rather different from that in a calculus text, where many examples are simply intended to be "model exercises." That approach doesn't work, anyway, in an abstract algebra text, since many exercises by their very nature involve "one of a kind" proofs.

For real understanding you must learn to construct your own examples. A good example should be simple enough for you to grasp, but not so simple that it doesn't illuminate the relevant points. We hope that you will learn to construct good examples for yourself while reading our book. We have drawn our examples from areas that we hope are familiar. We use ordinary addition and multiplication of various sets of numbers, composition of functions, and multiplication of matrices to illustrate the basic algebraic concepts that we want to study.

"When I try to find a proof, I don't know where to start."

Partly, this is just a matter of experience. Just as in any area, it takes some practice before you will feel able to use the various ideas with some ease. It is also probably a matter of some "math anxiety," because actually doing a proof can seem a little mysterious. How is it possible to come up with all the right ideas, in just the right order?

It is true that there are certain approaches that an experienced mathematician would know to try first while attempting to solve a problem. In the text we will try to alert you to these. In fact, sometimes we have suggested a few techniques to keep in mind while attacking the assigned problems. If you get stuck, see what happens in some simple examples. If all else fails, make a list of all of the results in the text that have the hypothesis of your proof as their hypothesis. Also make a list of all those results which have the conclusion of your proof as their conclusion. Then you can use these results to help you narrow the gap between the hypothesis and conclusion of the proof you are working on.

"I'm no good at writing proofs."

There are really several parts to proving something: understanding the problem, finding a solution, and writing it down in a logical fashion.

What is involved in writing a proof? Isn't it just an explanation? Of course, it has to include enough detail to be convincing, but it shouldn't include unnecessary details which might only obscure the real reasons why things work as they do. One way to test this is to see if your proof will convince another student in the class. You should even ask yourself whether or not it will convince you when you read it while studying for the final exam.

Constructing a proof is like building a bridge. Construction begins at both ends and continues until it is possible to put in the final span that links both sides. In the same way, in actually constructing a proof, it is often necessary to simplify or rewrite or expand both the hypothesis and the conclusion. You need to try to make the gap between the two as small as possible, so that you can finally see the steps that link them.

The bridge is designed to be used by people who simply start at one side and move across to the other. In writing down a proof you should have the same goal, so that a reader can start at the hypothesis and move straight ahead to the conclusion. Writing a clear proof is like any writing—it will probably take several revisions, even after all of the key ideas are in place. (We want to make sure that you don't suffer from writer's block because you believe that a proof should appear on your paper, line after line, in perfect order.)

Of course, we can't avoid the real problem. Sometimes the proofs are quite difficult and require a genuine idea. In doing your calculus homework, you may have followed the time-honored technique used by most students. If you couldn't do a problem, you would look for an example of exactly the same type, reading the text only until you found one. That technique often is good enough to solve routine computational problems, but in a course such as this you should not expect to find models for all of the problems that you are asked to solve as exercises. These problems may very well be unique. The only way to prepare to do them is to read the text in detail.

"I keep trying, but I don't seem to be making any progress."

We can only encourage you to keep trying. Sometimes it seems a bit like learning to ride a bicycle. There is a lot of struggling and effort, trial and error, and it can be really discouraging to see your friends all of a sudden riding pretty well, while you keep falling over. Then one day it just seems to happen—you can do it, and you never really forget how.

WRITING PROOFS

Logic is the glue that holds together the proofs that you will be writing. Logical connectives such as "and," "or," "if ... then ...," and "not" are used to build compound statements out of simpler ones. We assume that you are more or less familiar with these terms, but we need to make a few comments because they are used in mathematics in a precise fashion.

Let $\mathcal{P}$ and $\mathcal{Q}$ be statements, that is, declarative sentences which are either true or false. The word "or" can be ambiguous in ordinary English usage. It may mean "$\mathcal{P}$ or $\mathcal{Q}$, but not both," which we call the *exclusive* "or," or it may mean "$\mathcal{P}$ or $\mathcal{Q}$, or possibly both," which we call the *inclusive* "or." In mathematics, it is generally agreed to use "or" only in the inclusive sense. That is, the compound statement "$\mathcal{P}$ or $\mathcal{Q}$" is true precisely when one of the following occurs: (i) $\mathcal{P}$ is true and $\mathcal{Q}$ is false; (ii) $\mathcal{Q}$ is true and $\mathcal{P}$ is false; (iii) $\mathcal{P}$ is true and $\mathcal{Q}$ is true.

The expression "if $\mathcal{P}$ then $\mathcal{Q}$" is called a **conditional expression**, and is the single most important form that we will use. Here $\mathcal{P}$ is the **hypothesis** and $\mathcal{Q}$ is the **conclusion**. By definition, this expression is true in all cases except when $\mathcal{P}$ is true and $\mathcal{Q}$ is false. In fact, "if $\mathcal{P}$ then $\mathcal{Q}$" has the same meaning as "$\mathcal{Q}$ or not $\mathcal{P}$." There are several equivalent ways to say "if $\mathcal{P}$ then $\mathcal{Q}$." We can say "$\mathcal{P}$ implies $\mathcal{Q}$," or "$\mathcal{P}$ is sufficient for $\mathcal{Q}$," or "$\mathcal{Q}$ if $\mathcal{P}$," or "$\mathcal{Q}$ necessarily follows from $\mathcal{P}$."

Two expressions related to "if $\mathcal{P}$ then $\mathcal{Q}$" are its **contrapositive** "if not $\mathcal{Q}$ then not $\mathcal{P}$" and its **converse** "if $\mathcal{Q}$ then $\mathcal{P}$." The expression "$\mathcal{P}$ implies $\mathcal{Q}$" is logically equivalent to its contrapositive "not $\mathcal{Q}$ implies not $\mathcal{P}$," but is not logically equivalent to its converse "$\mathcal{Q}$ implies $\mathcal{P}$."

For example, the most intuitive way to define a one-to-one function f from a set S into a set T is to require that the following condition holds for all $x_1, x_2 \in S$: if $x_1 \neq x_2$ in S, then $f(x_1) \neq f(x_2)$. In practice, though, it is easier to work with equalities, and so the definition is usually reformulated using the contrapositive of the stated condition: if $f(x_1) = f(x_2)$, then $x_1 = x_2$.

The **biconditional** is the statement "$\mathcal{P}$ if and only if $\mathcal{Q}$." We can also say "$\mathcal{P}$ is equivalent to $\mathcal{Q}$," or "$\mathcal{P}$ is necessary and sufficient for $\mathcal{Q}$."

The precision of our mathematical language is abused at one point. Definitions are usually stated in a form such as "a number is said to be even if it is divisible by 2." It must be understood that the biconditional is being used, since the statement is clearly labeled as a definition, and so the meaning of the definition is "a number is said to be even if and only if it is divisible by 2."

We are now ready to illustrate some techniques of proof: **direct proof**, and **indirect proof**. In a direct proof that a statement $\mathcal{P}$ implies a statement $\mathcal{Q}$, the proof should begin with the hypothesis that $\mathcal{P}$ is true and end with the conclusion that $\mathcal{Q}$ is true. In an indirect proof we actually prove the contrapositive of the desired result, so the proof should begin with the hypothesis that "not $\mathcal{Q}$" is true and end with the conclusion that "not $\mathcal{P}$" is true. In a **proof by contradiction**, the proof should begin with the assumption that the conclusion of the theorem is false, and end with a contradiction, in which some statement is shown to be both true and false.

We begin with a direct proof of the well-known fact that the square of an even integer is even. To give a convincing proof, we need something concrete to work with, like an equation. We will use the condition that defines a number to be even if it is a multiple of 2. An equivalent condition is that the number can be factored, with 2 as one of the factors.

Proposition. *If an integer is even, then its square is also even.*

Proof. Assume that n is an even integer. Then since n has 2 as a factor, we can write

$$n = 2m ,$$

for some integer m. We can square both sides of this equation, to get

$$n^2 = (2m)^2 = 2(2m^2) \ .$$

The new equation shows that the square of n has 2 as a factor, and so n^2 is an even integer. □

You can see that in the proof we began with the hypothesis that n is an even integer, and we ended with the conclusion that n^2 is an even integer. To fill in the necessary steps to get from the hypothesis to the conclusion we needed to work with the definition of the terms involved in the statement. We suggest that as a first step you should try, whenever possible, to use definitions and theorems that return you to the more familiar world of high school algebra, with concrete equations involving numbers. The next step is to become familiar with equations that hold in a more general context, say for matrices. Some of the familiar rules will still hold, but some may fail. As one example, contrast this statement from high school algebra: "If both sides of an equation are multiplied by the same number, then the equation is still valid," with the corresponding statement from matrix theory: "If both sides of a matrix equation are multiplied on the left by the same matrix, then the equation is still valid." The second statement is similar to the first, but greater care must be taken with matrices, because matrix multiplication does not in general satisfy the commutative law.

We next give an example of an indirect proof, in which we prove the contrapositive of the desired result. We will use the fact that adding one to an integer changes its parity, so that it changes from even to odd, or from odd to even. This means that one way to describe all odd integers is to say that they can be expressed as one plus an even integer, so they have the form $2m + 1$, for some integer m. This also shows that every integer is either even or odd, but not both.

Proposition. *If the square of an integer is even, then the integer itself must be even.*

Proof. Suppose that the desired conclusion is false. Then the integer in question, say n, must be odd, so it has the form

$$n = 2m + 1 \ .$$

But then n^2 has the form

$$n^2 = (2m + 1)^2 = 4m^2 + 4m + 1 = 2(2m^2 + 2m) + 1 \ ,$$

which shows that n^2 must be odd. Thus the hypothesis that the square is even must be false. We have proved the contrapositive of the desired statement, and this completes the proof. □

If you have already studied Chapter 1, then you will surely have realized that it is possible to give a direct proof of the previous proposition, based on Lemma 1.2.5.

It is called Euclid's lemma, and states that if p is a prime number that is a factor of the product ab of two integers a and b, then p must be a factor either of a or of b. If we take the special case $p = 2$, $a = n$, and $b = n$, then the lemma reduces to the statement that if p is a factor of n^2, then p must be a factor of n itself.

We next give an example of a proof by contradiction. In this method of proof, we assume that the conclusion of the theorem is false, and attempt to arrive at a contradiction. The form of the contradiction should be that some statement is both true and false. We will prove that $\sqrt{2}$ is an irrational numbers, that is, that it cannot be expressed as a quotient of integers, of the form m/n.

Proposition. *The square root of 2 is an irrational number.*

Proof. Suppose that the conclusion of the theorem is false, in other words, that $\sqrt{2}$ is a rational number. Then we can write

$$\sqrt{2} = \frac{m}{n}$$

for some integers m and n, where n is nonzero. Furthermore, we can cancel common factors of m and n until there are no such common factors left, so we can assume that the fraction m/n has been reduced to lowest terms.

Multiplying both sides of the above equation by n, and then squaring both sides, yields the equation

$$2n^2 = m^2 .$$

This shows that m^2 is an even integer, so by our previous proposition, the number m itself must be even. This means that we can factor 2 out of m, so we can write $m = 2k$ for some integer k. Making this substitution gives

$$2n^2 = (2k)^2 = 4k^2 ,$$

and then we can divide both sides of the equation to obtain

$$n^2 = 2k^2 .$$

As before, this shows that n^2 is even, and it follows that n is even. We have now reached a contradiction to the assumption that $\sqrt{2}$ can be written as a fraction m/n in lowest terms, since the numerator and denominator both have 2 as a factor. □

Our final proof illustrates that in some cases a great deal of interesting information can be obtained by looking at something from two different points of view. We recall that if n is a positive integer, then the symbol $n!$ (read n factorial) is defined by $n! = n(n-1) \cdots 2 \cdot 1$. The **binomial coefficient** $\binom{n}{i}$ (pronounced n choose i) is defined by

$$\binom{n}{i} = \frac{n!}{i!(n-i)!} .$$

With this notation, the binomial theorem states that

$$(a + b)^n = \sum_{i=0}^{n} \binom{n}{i} a^i b^{n-i} .$$

The binomial coefficients can be generated recursively. We show the first few by giving part of **Pascal's triangle**.

```
                          1
                      1       1
                  1       2       1
              1       3       3       1
          1       4       6       4       1
      1       5      10      10       5       1
  1       6      15      20      15       6       1
1     7     21      35      35      21       7      1
1   8    28     56      70      56      28     8     1
```

Proposition. *For any positive integer n, we have* $\sum_{i=0}^{n} \binom{n}{i} = 2^n$.

Proof. Let S be a set with n elements. We will count the number of subsets of S in two different ways.

First, to construct a subset with i elements, we must choose i of the n elements of S, and this can be done in $\binom{n}{i}$ ways. Adding $\binom{n}{i}$ from $i = 0$ through n counts all subsets of S. Thus $\sum_{i=0}^{n} \binom{n}{i}$ is the number of subsets of S.

On the other hand, when constructing a subset of S, for each of the n elements of S we must choose whether or not to include that element. This gives us a total of 2^n choices, and so we conclude that S has 2^n subsets. Hence the desired equality holds. □

So far we have discussed the construction of the proof of an individual theorem or proposition. Theorems don't exist in isolation; they are part of a body of results. While studying such a body of results, it is important to step back from time to time to get a global picture. It is of course necessary to note which definitions, theorems, and propositions are used to obtain each result, but it is also important when reviewing a topic to note which theorems are obtained by applying a given result. In order to understand a result's place in the full scheme of things you should note not only its ancestors, but also its descendents.

As you read through a chapter, think of the collection of results as a tapestry woven from individual strands. The true value of each individual theorem only emerges as you see parts of the whole tapestry.

HISTORICAL BACKGROUND

The word "algebra" entered the mathematical vocabulary from Arabic over one thousand years ago, and for almost all of that time it has meant the study of equations. The "algebra" of equations is at a higher level of abstraction than arithmetic, in that symbols may be used to represent unknown numbers. "Modern algebra" or "abstract algebra" dates from the nineteenth century, when problems from number theory and the theory of equations led to the study of abstract mathematical models. In these models, symbols might represent numbers, polynomials, permutations, or elements of various other sets in which arithmetic operations could be defined. Mathematicians attempted to identify the relevant underlying principles, and to determine their logical consequences in very general settings.

One of the problems that has motivated a great deal of work in algebra has been the problem of solving equations by radicals. We begin our discussion with the familiar quadratic formula

$$x = \frac{-b \pm \sqrt{b^2 - 4ac}}{2a}.$$

This formula gives a solution of the equation $ax^2 + bx + c = 0$, where $a \neq 0$, expressed in terms of its coefficients, and using a square root. More generally, we say that an equation

$$a_n x^n + \ldots + a_1 x + a_0 = 0$$

is **solvable by radicals** if the solutions can be given in a form that involves sums, differences, products, or quotients of the coefficients $a_n, \ldots, a_1, a_0$, together with square roots, cube roots, etc., of such combinations of the coefficients.

Quadratic and even cubic equations were studied as early as Babylonian times. In the second half of the eleventh century, Omar Khayyam (1048–1131) wrote a book on algebra, which contained a systematic study of cubic equations. His approach was mainly geometric, and he found the roots of the equations as intersections of conic sections. A general method for solving cubic equations numerically eluded the Greeks and later oriental mathematicians. The solution of the cubic equation represented for the Western world the first advance beyond classical mathematics.

General cubic equations were reduced to the form $x^3 + px + q = 0$. In the early sixteenth century, a mathematician by the name of Scipione del Ferro (1465–1526) solved one particular case of the cubic. He did not publish his solution, but word of the discovery became known, and several others were also successful in solving the equations. The solutions were published in a textbook by Girolamo Cardano (1501–1576) in 1545. This caused a bitter dispute with another mathematician, who had independently discovered the formulas, and claimed to have given them to Cardano under a pledge of secrecy. (For additional details see the notes at the end of Chapter 4.) The solution of the equation $x^3 + px = q$ was given by Cardano in the form

$$x = \sqrt[3]{\frac{q}{2} + \sqrt{\frac{p^3}{27} + \frac{q^2}{4}}} - \sqrt[3]{-\frac{q}{2} + \sqrt{\frac{p^3}{27} + \frac{q^2}{4}}}\ .$$

A solution to the general quartic equation was also given, in which the solution could be expressed in terms of radicals involving the coefficients. (See Section A.6 of the appendix.)

Subsequently, attempts were made to find similar solutions to the general quintic equation, but without success. The development of calculus led to methods for approximating roots, and the theory of equations became analytic. One result of this approach was the discovery by Jean le Rond D'Alembert (1717–1783) in 1746 that every algebraic equation of degree n has n roots in the set of complex numbers. Although it was not until 1801 that a correct proof was published, by Carl F. Gauss (1777-1855), this discovery changed the emphasis of the question from the existence of roots to whether equations of degree 5 or greater could be solved by radicals.

In 1798, Paolo Ruffini (1765–1822) published a proof claiming to show that the quintic could not be solved by radicals. The proof was not complete, although the general idea was correct. A full proof was finally given by Niels Abel (1802–1829) in 1826. A complete answer to the question of which equations are solvable by radicals was found by Evariste Galois (1811-1832). Galois was killed in a duel, and did not live to see the remarkable consequences of the papers he submitted to the French Academy. (See the introduction to Chapter 8 for further details.) Galois considered certain permutations of the roots of a polynomial—those that leave the coefficients fixed—and showed that the polynomial is solvable by radicals if and only if the associated group of permutations has certain properties. (See the notes at the end of Chapter 3 and the introduction to Chapter 8). This theory, named after Galois, contains deep and very beautiful results, and is the subject of Chapter 8. Although it is not always possible to cover that material in a beginning course in abstract algebra, it is toward this goal that many of the results in this book were originally directed.

The final chapter of the book studies the question of unique factorization, in certain subsets of the set of complex numbers, and for polynomials in several variables. Much of the original investigation was motivated by attempts to prove Fermat's last theorem. (See the introduction to Chapter 9.)

The mathematics necessary to answer the question of solvability by radicals and the question of unique factorization includes the development of a good deal of the theory of groups, rings, and fields, which has subsequently been applied in many other areas, including physics and computer science. In studying these areas we have used a modern, axiomatic approach rather than an historical one.

Chapter 1

INTEGERS

In this chapter we will develop some of the properties of the set of integers

$$\mathbf{Z} = \{\ldots, -2, -1, 0, 1, 2, \ldots\}$$

that are needed in our later work. The use of $\mathbf{Z}$ for the integers reflects the strong German influence on the modern development of algebra; $\mathbf{Z}$ comes from the German word for numbers, "Zahlen." Some of the computational techniques we study here will reappear numerous times in later chapters. Furthermore, we will construct some concrete examples that will serve as important building blocks for later work on groups, rings, and fields.

To give a simple illustration of how we will use elementary number theory, consider the matrix $A = \begin{bmatrix} 0 & 1 \\ -1 & 0 \end{bmatrix}$. The powers of A are $A^2 = \begin{bmatrix} -1 & 0 \\ 0 & -1 \end{bmatrix}$, $A^3 = \begin{bmatrix} 0 & -1 \\ 1 & 0 \end{bmatrix}$, $A^4 = \begin{bmatrix} 1 & 0 \\ 0 & 1 \end{bmatrix}$, $A^5 = \begin{bmatrix} 0 & 1 \\ -1 & 0 \end{bmatrix}$, etc. Since A^4 is the identity matrix I, the powers begin to repeat at A^5, as we can see by writing

$$\begin{aligned} A^5 &= A^4 A = IA = A, \\ A^6 &= A^4 A^2 = IA^2 = A^2, \\ A^7 &= A^4 A^3 = IA^3 = A^3, \qquad \text{etc.} \end{aligned}$$

. How can we find A^{231}, for example? If we divide 231 by 4, we get 57, with remainder 3, so $231 = 4 \cdot 57 + 3$. This provides our answer, since

$$A^{231} = A^{4\cdot57+3} = A^{4\cdot57} A^3 = (A^4)^{57} A^3 = I^{57} A^3 = IA^3 = A^3.$$

We can see that two powers A^j and A^k are equal precisely when j and k differ by a multiple of 4. Altogether there are only the following four powers:

$$\begin{bmatrix} 1 & 0 \\ 0 & 1 \end{bmatrix}, \quad \begin{bmatrix} 0 & 1 \\ -1 & 0 \end{bmatrix}, \quad \begin{bmatrix} -1 & 0 \\ 0 & -1 \end{bmatrix}, \quad \begin{bmatrix} 0 & -1 \\ 1 & 0 \end{bmatrix}.$$

A very similar situation occurs when we analyze the positive powers of the complex number i. We have $i^1 = i$, $i^2 = -1$, $i^3 = -i$, and $i^4 = 1$. As before, we see that $i^j = i^k$ if and only if j and k differ by a multiple of 4.

As a slightly different example, consider the positive powers of the complex number

$$\omega = -\frac{1}{2} + \frac{\sqrt{3}}{2}i \ .$$

There are only three distinct powers of ω, as shown below:

$$\omega \ = \ -\frac{1}{2} + \frac{\sqrt{3}}{2}i \ ,$$

$$\omega^2 \ = \ \left(-\frac{1}{2} + \frac{\sqrt{3}}{2}i\right)\left(-\frac{1}{2} + \frac{\sqrt{3}}{2}i\right) \ = \ \frac{1}{4} - \frac{2\sqrt{3}}{4}i + \frac{3}{4}i^2 \ = \ -\frac{1}{2} - \frac{\sqrt{3}}{2}i \ ,$$

$$\omega^3 \ = \ \omega^2\omega \ = \ \left(-\frac{1}{2} - \frac{\sqrt{3}}{2}i\right)\left(-\frac{1}{2} + \frac{\sqrt{3}}{2}i\right) \ = \ \frac{1}{4} - \frac{3}{4}i^2 \ = \ 1 \ .$$

From this point on, the positive powers begin to repeat, and $\omega^j = \omega^k$ if and only if j and k differ by a multiple of 3.

To give a unified approach to situations analogous to the ones above, in which we need to consider numbers that exhibit similar behavior when they differ by a multiple of a number n, we will develop the notion of congruence modulo n. The notion of a congruence class will enable us to think of the collection of numbers that behave in the same way as a single entity. The simplest example is congruence modulo 2. When we consider two numbers to be similar if they differ by a multiple of 2, we are just saying that the two numbers are similar if they have the same parity (both are even, or both are odd). Another familiar situation of this type occurs when telling time, since on a clock we do not distinguish between times that differ by a multiple of 12 (or 24 if you are in Europe or the military).

In this chapter we will develop only enough number theory to be of use in later chapters, when we study groups, rings, and fields. Historically, almost all civilizations have developed the integers (at least the positive ones) for use in agriculture, commerce, etc. After the elementary operations (addition, subtraction, multiplication, and division) have been understood, human curiosity has taken over and individuals have begun to look for deeper properties that the integers may possess.

Nonmathematicians are often surprised that research is currently being done in mathematics. They seem to believe that all possible questions have already been answered. At this point an analogy may be useful. Think of all that is known as being contained in a ball. Adding knowledge enlarges the ball, and this means that the surface of the ball—the interface between known and unknown where research occurs—also becomes larger. In short, the more we know, the more questions there are to ask. In number theory, perhaps more than in any other branch of mathematics, there are still many unanswered questions that can easily be posed. In fact, it seems that often the simplest sounding questions require the deepest tools to resolve.

One aspect of number theory that has particular applications in algebra is the one that concerns itself with questions of divisibility and primality. Fortunately for our study of algebra, this part of number theory is easily accessible, and it is with these properties of integers that we will deal in this chapter. Number theory got its start with Euclid and much of what we do in the first two sections appears in his book *Elements*.

Our approach to number theory will be to study it as a tool for later use. In the notes at the end of this chapter, we mention several important problems with which number theorists are concerned. You can read the notes at this point, before studying the material in the chapter. In fact, we suggest that you read them now, because we hope to indicate why number theory is so interesting in its own right.

1.1 Divisors

Obviously, at the beginning of the book we must decide where to start mathematically. We would like to give a careful mathematical development, including proofs of virtually everything we cover. However, that would take us farther into the foundations of mathematics than we believe is profitable in a beginning course in abstract algebra. As a compromise, we have chosen to assume a knowledge of basic set theory and some familiarity with the set of integers.

For the student who is concerned about how the integers can be described formally and how the basic properties of the integers can be deduced, we have provided some very sketchy information in the appendix. Even there we have taken a naive approach, rather than formally treating the basic notions of set theory as undefined terms and giving the axioms that relate them. We have included a list of the Peano postulates, which use concepts and axioms of set theory to characterize the natural numbers. We then give an outline of the logical development of the set of integers, and larger sets of numbers.

In the beginning sections of this chapter we will assume some familiarity with the set of integers, and we will simply take for granted some of the basic arithmetic and order properties of the integers. (These properties should be familiar from elementary school arithmetic. They are listed in detail in Section A.3 of the appendix.) The set $\{0, \pm 1, \pm 2, \ldots\}$ of **integers** will be denoted by $\mathbf{Z}$ throughout the text, while we will use $\mathbf{N}$ for the set $\{0, 1, 2, \ldots\}$ of **natural numbers**.

Our first task is to study divisibility. We will then develop a theory of prime numbers based on our work with greatest common divisors. The fact that exact division is not always possible within the set of integers should not be regarded as a deficiency. Rather, it is one source of the richness of the subject of number theory and leads to many interesting and fundamental propositions about the integers.

1.1.1 Definition. *An integer a is called a* ***multiple*** *of an integer b if a = bq for some integer q. In this case we also say that b is a* ***divisor*** *of a, and we use the notation b | a.*

In the above case we can also say that b is a **factor** of a, or that a is **divisible** by b. If b is not a divisor of a, meaning that $a \neq bq$ for any $q \in \mathbf{Z}$, then we write $b \nmid a$. The set of all multiples of an integer a will be denoted by $a\mathbf{Z}$.

Be careful when you use the notation $b \mid a$. It describes a relationship between integers a and b and does *not* represent a fraction. Furthermore, a handwritten vertical line | can easily be confused with the symbol /. The statement $2 \mid 6$ is a true statement; $6 \mid 2$ is a statement that is false. On the other hand, the equation $6/2 = 3$ is written correctly, since the fraction $6/2$ *does* represent the number 3. We have at least three different uses for a vertical line: for "such that" in the "set-builder" notation { | }, when talking about the absolute value of a number, and to indicate that one integer is a divisor of another.

We note some elementary facts about divisors. If $a \neq 0$ and $b \mid a$, then $|b| \leq |a|$ since $|b| \leq |b||q| = |a|$ for some nonzero integer q. It follows from this observation that if $b \mid a$ and $a \mid b$, then $|b| = |a|$ and so $b = \pm a$. Therefore, if $b \mid 1$, then since it is always true that $1 \mid b$, we must have $b = \pm 1$.

Note that the only multiple of 0 is 0 itself. On the other hand, for any integer a we have $0 = a \cdot 0$, and thus 0 is a multiple of any integer. With the notation we have introduced, the set of all multiples of 3 is $3\mathbf{Z} = \{0, \pm 3, \pm 6, \pm 9, \ldots\}$. To describe $a\mathbf{Z}$ precisely, we can write

$$a\mathbf{Z} = \{m \in \mathbf{Z} \mid m = aq \ \text{for some} \ q \in \mathbf{Z}\}.$$

Suppose that a is a multiple of b. Then every multiple of a is also a multiple of b, and in fact we can say that a is a multiple of b if and only if every multiple of a is also a multiple of b. In symbols we can write $b \mid a$ if and only if $a\mathbf{Z} \subseteq b\mathbf{Z}$. Exercise 15 asks for a more detailed proof of this statement.

Before we study divisors and multiples of a fixed integer, we need to state an important property of the set of natural numbers, which we will take as an axiom.

1.1.2 Axiom (Well-Ordering Principle). *Every nonempty set of natural numbers contains a smallest element.*

The well-ordering principle is often used in arguments by contradiction. If we want to show that all natural numbers have some property, we argue that if the set of natural numbers not having the property were nonempty, it would have a least member, and then we deduce a contradiction from this, using the particular facts of the situation. The theory of mathematical induction (see Appendix A.4) formalizes that sort of argument.

Let S be a nonempty set of integers that has a lower bound. That is, there is an integer b such that $b \leq n$ for all $n \in S$. If $b \geq 0$, then S is actually a set of natural numbers, so it contains a smallest element by the well-ordering principle. If $b < 0$, then adding $|b|$ to each integer in S produces a new set T of natural numbers, since $n + |b| \geq 0$ for all $n \in S$. The set T must contain a smallest element, say t, and it is easy to see that $t - |b|$ is the smallest element of S. This allows us to use, if necessary, a somewhat stronger version of the well-ordering principle: every set of integers that is bounded below contains a smallest element.

The first application of the well-ordering principle will be to prove the division algorithm. In familiar terms, the division algorithm states that dividing an integer a by a positive integer b gives a quotient q and nonnegative remainder r, such that r is less than b. You could write this as

$$\frac{a}{b} = q + \frac{r}{b},$$

but since we are studying properties of the set of integers, we will avoid fractions and write this equation in the form

$$a = bq + r.$$

For example, if $a = 29$ and $b = 8$, then

$$29 = 8 \cdot 3 + 5,$$

so the quotient q is 3 and the remainder r is 5. You must be careful when a is a negative number, since the remainder must be nonnegative. Simply changing signs in the previous equation, we have

$$-29 = (8)(-3) + (-5),$$

which does not give an appropriate remainder. Rewriting this in the form

$$-29 = (8)(-4) + 3$$

gives the correct quotient $q = -4$ and remainder $r = 3$.

Solving for r in the equation $a = bq + r$ shows that $r = a - bq$, and that r must be the smallest nonnegative integer that can be written in this form, since $0 \leq r < b$. This observation clarifies the relationship between the quotient and remainder, and forms the basis of our proof that the division algorithm can be deduced from the well-ordering principle. Another way to see this relationship is to notice that you could find the remainder and quotient by repeatedly subtracting b from a and noting that you have the remainder in the required form when you obtain a nonnegative integer less than b.

The next theorem on "long division with remainder" has traditionally been called the "division algorithm".

1.1.3 Theorem (Division Algorithm). *For any integers a and b, with $b > 0$, there exist unique integers q (the **quotient**) and r (the **remainder**) such that*

$$a = bq + r, \text{ with } 0 \le r < b.$$

Proof. Consider the set $R = \{a - bq : q \in \mathbf{Z}\}$. The elements of R are the potential remainders, and among these we need to find the smallest nonnegative one. We want to apply the well-ordering principle to the set R^+ of nonnegative integers in R, so we must first show that R^+ is nonempty. Since $b \ge 1$, the number $a - b(-|a|) = a + b \cdot |a|$ is nonnegative and belongs to R^+, so R^+ is nonempty.

Now by the well-ordering principle, R^+ has a smallest element, and we will call this element r. We will show that $a = bq + r$, with $0 \le r$ and $r < b$. By definition, $r \ge 0$, and since $r \in R^+$, we must have $r = a - bq$ for some integer q. We cannot have $r \ge b$, since if we let $s = r - b$ we would have $s \ge 0$ and $s = a - b(q + 1) \in R^+$. Since $s < r$, this would contradict the way r was defined, and therefore we must have $r < b$. We have now proved the existence of r and q satisfying the conditions $a = bq + r$ and $0 \le r < b$.

To show that q and r are unique, suppose that we can also write $a = bp + s$ for integers p and s with $0 \le s < b$. We have $0 \le r < b$ and $0 \le s < b$, and this implies that $|s - r| < b$. But $bp + s = bq + r$ and so $s - r = b(q - p)$, which shows that $b \mid (s - r)$. The only way that b can be a divisor of a number with smaller absolute value is if that number is 0, and so we must have $s - r = 0$, or $s = r$. Then $bp = bq$, which implies that $p = q$ since $b > 0$. Thus the quotient and remainder are unique, and we have completed the proof of the theorem. $\square$

Given integers a and b, with $b > 0$, we can use the division algorithm to write $a = bq + r$, with $0 \le r < b$. Since $b \mid a$ if and only if there exists $q \in \mathbf{Z}$ such that $a = bq$, we see that $b \mid a$ if and only if $r = 0$. This simple observation gives us a useful tool in doing number theoretic proofs. To show that $b \mid a$ we can use the division algorithm to write $a = bq + r$ and then show that $r = 0$. This technique makes its first appearance in the proof of Theorem 1.1.4.

A set of multiples $a\mathbf{Z}$ has the property that the sum or difference of two integers in the set is again in the set, since $aq_1 \pm aq_2 = a(q_1 \pm q_2)$. We say that the set $a\mathbf{Z}$ is **closed under addition and subtraction**. This will prove to be a very important property in our later work. The next theorem shows that this property characterizes sets of multiples, since a nonempty set of integers is closed under addition and subtraction if and only if it is a set of the form $a\mathbf{Z}$, for some nonnegative integer a.

1.1.4 Theorem. *Let I be a nonempty set of integers that is closed under addition and subtraction. Then I either consists of zero alone or else contains a smallest positive element, in which case I consists of all multiples of its smallest positive element.*

$a = bq + r$

Proof. Since I is nonempty, either it consists of 0 alone, or else it contains a nonzero integer a. In the first case we are done. In the second case, if I contains the nonzero integer a, then it must contain the difference $a - a = 0$, and hence the difference $0 - a = -a$, since I is assumed to be closed under subtraction. Now either a or $-a$ is positive, so I contains at least one positive integer. Having shown that the set of positive integers in I is nonempty, we can apply the well-ordering principle to guarantee that it contains a smallest member, say b.

Next we want to show that I is equal to the set $b\mathbf{Z}$ of all multiples of b. To show that $I = b\mathbf{Z}$, we will first show that $b\mathbf{Z} \subseteq I$, and then show that $I \subseteq b\mathbf{Z}$.

Any nonzero multiple of b is given by just adding b (or $-b$) to itself a finite number of times, so since I is closed under addition, it must contain all multiples of b. Thus $b\mathbf{Z} \subseteq I$.

On the other hand, to show that $I \subseteq b\mathbf{Z}$ we must take any element c in I and show that it is a multiple of b, or equivalently, that $b \mid c$. (Now comes the one crucial idea in the proof.) Using the division algorithm we can write $c = bq + r$, for some integers q and r with $0 \leq r < b$. Since I contains bq and is closed under subtraction, it must also contain $r = c - bq$. But this is a contradiction unless $r = 0$, because b was chosen to be the smallest positive integer in I and yet $r < b$ by the division algorithm. We conclude that $r = 0$, and therefore $c = bq$, so $b \mid c$ and we have shown that $I \subseteq b\mathbf{Z}$.

This completes the proof that $I = b\mathbf{Z}$. $\square$

One of the main goals of Chapter 1 is to develop some properties of prime numbers, which we will do in Section 1.2. Before discussing prime numbers themselves, we will introduce the notion of relatively prime numbers, and this definition in turn depends on the notion of the greatest common divisor of two numbers. Our definition of the greatest common divisor is given in terms of divisibility, rather than in terms of size, since it is this form that is most useful in writing proofs. Exercise 20 gives an equivalent formulation that focuses on size.

1.1.5 Definition. *Let a and b be integers, not both zero. A positive integer d is called the **greatest common divisor** of a and b if*

  (i) *d is a divisor of both a and b, and*

  (ii) *any divisor of both a and b is also a divisor of d.*

The greatest common divisor of a and b will be denoted by $\gcd(a, b)$ *or* (a, b).

Our first observation is that $\gcd(0, 0)$ is undefined, but if a is any nonzero integer, then $\gcd(a, 0)$ is defined and equal to $|a|$. The definition of the greatest common divisor can be shortened by using our notation for divisors. If a and b are integers, not both zero, and d is a positive integer, then $d = \gcd(a, b)$ if

  (i) $d \mid a$ and $d \mid b$, and

  (ii) if $c \mid a$ and $c \mid b$, then $c \mid d$.

The fact that we have written down a definition of the greatest common divisor does not guarantee that there is such a number. Furthermore, the use of the word "the" has to be justified, since it implies that there can be only one greatest common divisor. The next theorem will guarantee the existence of the greatest common divisor, and the question of uniqueness is easily answered: if d_1 and d_2 are greatest common divisors of a and b, then the definition requires that $d_1 \mid d_2$ and $d_2 \mid d_1$, so $d_1 = \pm d_2$. Since both d_1 and d_2 are positive, we have $d_1 = d_2$.

If a and b are integers, then we will refer to any integer of the form $ma + nb$, where $m, n \in \mathbf{Z}$, as a **linear combination** of a and b. The next theorem gives a very useful connection between greatest common divisors and linear combinations.

1.1.6 Theorem. *Let a and b be integers, not both zero. Then a and b have a greatest common divisor, which can be expressed as the smallest positive linear combination of a and b.*

Moreover, an integer is a linear combination of a and b if and only if it is a multiple of their greatest common divisor.

Proof. Let I be the set of all linear combinations of a and b, that is,

$$I = \{x \in \mathbf{Z} \mid x = ma + nb \text{ for some } m, n \in \mathbf{Z}\}.$$

The set I is nonempty since it contains $a = 1 \cdot a + 0 \cdot b$ and $b = 0 \cdot a + 1 \cdot b$. It is closed under addition and subtraction since if $k_1, k_2 \in I$, then $k_1 = m_1 a + n_1 b$ and $k_2 = m_2 a + n_2 b$ for some integers m_1, m_2, n_1, n_2. Thus

$$k_1 \pm k_2 = (m_1 a + n_1 b) \pm (m_2 a + n_2 b) = (m_1 \pm m_2)a + (n_1 \pm n_2)b$$

also belong to I. By Theorem 1.1.4, the set I consists of all multiples of the smallest positive integer it contains, say d. Since $d \in I$, $d = ma + nb$ for some integers m and n.

Since we already know that d is positive, to show that $d = (a, b)$ we must show that (i) $d \mid a$ and $d \mid b$ and (ii) if $c \mid a$ and $c \mid b$, then $c \mid d$. First, d is a divisor of every element in I, so $d \mid a$ and $d \mid b$ since $a, b \in I$. Secondly, if $c \mid a$ and $c \mid b$, say $a = cq_1$ and $b = cq_2$, then

$$d = ma + nb = m(cq_1) + n(cq_2) = c(mq_1 + nq_2),$$

which shows that $c \mid d$.

The second assertion follows from the fact that I, the set of all linear combinations of a and b, is equal to $d\mathbf{Z}$, the set of all multiples of d. $\square$

You are probably used to finding the greatest common divisor of a and b by first finding their prime factorizations. This is an effective technique for small numbers, but we must postpone a discussion of this method until after we have studied prime factorizations in Section 1.2. In practice, for large numbers it can be very difficult to find prime factors, whereas the greatest common divisor can be found in many fewer steps by using the method we discuss next.

The greatest common divisor of two numbers can be computed by using a procedure known as the *Euclidean algorithm*. (Our proof of the existence of the greatest common divisor did not include an explicit method for finding it.) Before discussing the Euclidean algorithm, we need to note some properties of the greatest common divisor. First, if a and b are not both zero, then it is not difficult to see that $\gcd(a, b) = \gcd(|a|, |b|)$. Furthermore, if $b > 0$ and $b \mid a$, then $(a, b) = b$.

The next observation provides the basis for the Euclidean algorithm. If $b \neq 0$ and $a = bq + r$, then $(a, b) = (b, r)$. This can be shown by noting first that a is a multiple of (b, r) since it is a linear combination of b and r. Then $(b, r) \mid (a, b)$ since b is also a multiple of (b, r). A similar argument using the equality $r = a - bq$ shows that $(a, b) \mid (b, r)$, and it follows that $(a, b) = (b, r)$.

Given integers $a > b > 0$, the **Euclidean algorithm** uses the division algorithm repeatedly to obtain

$$
\begin{array}{llll}
a &= bq_1 + r_1 & \text{with} & 0 \leq r_1 < b \\
b &= r_1 q_2 + r_2 & \text{with} & 0 \leq r_2 < r_1 \\
r_1 &= r_2 q_3 + r_3 & \text{with} & 0 \leq r_3 < r_2 \\
& & \text{etc.}
\end{array}
$$

If $r_1 = 0$, then $b \mid a$, and so $(a, b) = b$. Since $r_1 > r_2 > \ldots$, the remainders get smaller and smaller, and after a finite number of steps we obtain a remainder $r_{n+1} = 0$. The algorithm ends with the equation

$$ r_{n-1} = r_n q_{n+1} + 0. $$

This gives us the greatest common divisor:

$$ (a, b) = (b, r_1) = (r_1, r_2) = \ldots = (r_{n-1}, r_n) = (r_n, 0) = r_n . $$

Example 1.1.1.

In showing that $(24, 18) = 6$, we have $(24, 18) = (18, 6)$ since $24 = 18 \cdot 1 + 6$, and $(18, 6) = 6$ since $6 \mid 18$. Thus $(24, 18) = (18, 6) = 6$. □

Example 1.1.2.

To show that $(126, 35) = 7$, we first have $(126, 35) = (35, 21)$ since $126 = 35 \cdot 3 + 21$. Then $(35, 21) = (21, 14)$ since $35 = 21 \cdot 1 + 14$, and $(21, 14) = (14, 7)$ since $21 = 14 \cdot 1 + 7$. Finally, $(14, 7) = 7$ since $14 = 7 \cdot 2$. Thus $(126, 35) = (35, 21) = (21, 14) = (14, 7) = 7$. □

Example 1.1.3.

In finding $(83, 38)$, we can arrange the work in the following manner:

$$
\begin{array}{rclcrcl}
83 & = & 38 \cdot 2 + 7 & \qquad & (83, 38) & = & (38, 7) \\
38 & = & 7 \cdot 5 + 3 & & (38, 7) & = & (7, 3) \\
7 & = & 3 \cdot 2 + 1 & & (7, 3) & = & (3, 1) \\
3 & = & 3 \cdot 1 & & (3, 1) & = & 1 \; .
\end{array}
$$

If you only need to find the greatest common divisor, stop as soon as you can compute it in your head. In showing that $(83, 38) = 1$, note that since 7 has no positive divisors except 1 and 7 and is not a divisor of 38, it is clear immediately that $(38, 7) = 1$. $\square$

Example 1.1.4.

Sometimes it is necessary to find the linear combination of a and b that gives (a, b). In finding $(126, 35)$ in Example 1.1.2 we had the following equations:

$$
\begin{array}{rclcrcl}
a & = & bq_1 + r_1 & \qquad & 126 & = & 35 \cdot 3 + 21 \\
b & = & r_1 q_2 + r_2 & & 35 & = & 21 \cdot 1 + 14 \\
r_1 & = & r_2 q_3 + d & & 21 & = & 14 \cdot 1 + 7 \\
r_2 & = & dq_4 + 0 & & 14 & = & 7 \cdot 2 + 0 \; .
\end{array}
$$

The next step is to solve for the nonzero remainder in each of the equations (omitting the last equation):

$$
\begin{array}{rclcrcl}
r_1 & = & a + (-q_1)b & \qquad & 21 & = & 1 \cdot 126 + (-3) \cdot 35 \\
r_2 & = & b + (-q_2)r_1 & & 14 & = & 1 \cdot 35 + (-1) \cdot 21 \\
d & = & r_1 + (-q_3)r_2 & & 7 & = & 1 \cdot 21 + (-1) \cdot 14 \; .
\end{array}
$$

We then work with the last equation $d = r_1 + (-q_3)r_2$, which contains the greatest common divisor, as desired, but may not be a linear combination of the original integers a and b. We can obtain the desired linear combination by substituting for the intermediate remainders, one at a time. Our first equation is

$$
7 \; = \; 1 \cdot 21 + (-1) \cdot 14 \; .
$$

We next substitute for the previous remainder 14, using the equation $14 = 1 \cdot 35 + (-1) \cdot 21$. This gives the following equation, involving a linear combination of 35 and 21:

$$
\begin{aligned}
7 & = \; 1 \cdot 21 + (-1) \cdot [1 \cdot 35 + (-1) \cdot 21] \\
& = \; (-1) \cdot 35 + 2 \cdot 21 \; .
\end{aligned}
$$

Finally, we use the first equation $21 = 1 \cdot 126 + (-3) \cdot 35$ to substitute for the remainder 21. This allows us to represent the greatest common divisor 7 as a linear combination of 126 and 35:

$$
\begin{aligned}
7 &= (-1) \cdot 35 + 2 \cdot [1 \cdot 126 + (-3) \cdot 35] \\
&= 2 \cdot 126 + (-7) \cdot 35 \, . \qquad \square
\end{aligned}
$$

The technique introduced in the previous example can easily be extended to the general situation in which it is desired to express (a, b) as a linear combination of a and b. After solving for the remainder in each of the relevant equations, we obtain

$$
\begin{aligned}
r_1 &= a + (-q_1)b \\
r_2 &= b + (-q_2)r_1 \\
r_3 &= r_1 + (-q_3)r_2 \\
r_4 &= r_2 + (-q_4)r_3 \\
&\ \ \vdots
\end{aligned}
$$

At each step, the expression for the remainder depends upon the previous two remainders. By substituting into the successive equations and then rearranging terms, it is possible to express each remainder (in turn) as a linear combination of a and b. The final step is to express (a, b) as a linear combination of a and b.

The Euclidean algorithm can be put into a convenient matrix format that keeps track of the remainders and linear combinations at the same time. To find (a, b), the idea is to start with the following system of equations:

$$
\begin{aligned}
x &\quad\ = a \\
y &= b
\end{aligned}
$$

and find, by using elementary row operations, an equivalent system of the following form:

$$
\begin{aligned}
m_1 x &+ n_1 y = (a, b) \\
m_2 x &+ n_2 y = 0 \, .
\end{aligned}
$$

Beginning with the matrix

$$
\begin{bmatrix} 1 & 0 & a \\ 0 & 1 & b \end{bmatrix},
$$

we use the division algorithm to write $a = bq_1 + r_1$. We then subtract q_1 times the bottom row from the top row, to get

$$
\begin{bmatrix} 1 & -q_1 & r_1 \\ 0 & 1 & b \end{bmatrix}.
$$

We next write $b = r_1 q_2 + r_2$, and subtract q_2 times the top row from the bottom row. This gives the matrix

$$\begin{bmatrix} 1 & -q_1 & r_1 \\ -q_2 & 1 + q_1 q_2 & r_2 \end{bmatrix}$$

and it can be checked that this algorithm produces rows in the matrix that give each successive remainder, together with the coefficients of the appropriate linear combination of a and b. The procedure is continued until one of the entries in the right-hand column is zero. Then the other entry in this column is the greatest common divisor, and its row contains the coefficients of the desired linear combination.

Example 1.1.5.

In using the matrix form of the Euclidean algorithm to compute $(126, 35)$ we begin with the equations $x = 126$ and $y = 35$. We have the following matrices:

$$\begin{bmatrix} 1 & 0 & 126 \\ 0 & 1 & 35 \end{bmatrix} \rightsquigarrow \begin{bmatrix} 1 & -3 & 21 \\ 0 & 1 & 35 \end{bmatrix} \rightsquigarrow \begin{bmatrix} 1 & -3 & 21 \\ -1 & 4 & 14 \end{bmatrix} \rightsquigarrow$$

$$\begin{bmatrix} 2 & -7 & 7 \\ -1 & 4 & 14 \end{bmatrix} \rightsquigarrow \begin{bmatrix} 2 & -7 & 7 \\ -5 & 18 & 0 \end{bmatrix},$$

ending with the equations $2x - 7y = 7$ and $-5x + 18y = 0$. Thus $(126, 35) = 7$, and substituting $x = 126$ and $y = 35$ in the equation $2x - 7y = 7$ gives us a linear combination $7 = 2 \cdot 126 + (-7) \cdot 35$.

Substituting into the second equation $-5x + 81y = 0$ also gives us some interesting information. Any multiple of the linear combination $0 = (-5) \cdot 126 + 18 \cdot 35$ can be added to the above representation of the greatest common divisor. Thus, for example, we also have $7 = (-3) \cdot 126 + 11 \cdot 35$ and $7 = (-8) \cdot 126 + 29 \cdot 35$. □

Example 1.1.6.

In matrix form, the solution for $(83, 38)$ is the following:

$$\begin{bmatrix} 1 & 0 & 83 \\ 0 & 1 & 38 \end{bmatrix} \rightsquigarrow \begin{bmatrix} 1 & -2 & 7 \\ 0 & 1 & 38 \end{bmatrix} \rightsquigarrow \begin{bmatrix} 1 & -2 & 7 \\ -5 & 11 & 3 \end{bmatrix} \rightsquigarrow$$

$$\begin{bmatrix} 11 & -24 & 1 \\ -5 & 11 & 3 \end{bmatrix} \rightsquigarrow \begin{bmatrix} 11 & -24 & 1 \\ -38 & 83 & 0 \end{bmatrix}.$$

Thus $(83, 38) = 1$ and $(11)(83) + (-24)(38) = 1$. □

The number (a, b) can be written in many different ways as a linear combination of a and b. The matrix method gives a linear combination with $0 = m_1a + n_1b$, so if $(a, b) = ma + nb$, then adding the previous equation gives $(a, b) = (m + m_1)a + (n + n_1)b$. In fact, any multiple of the equation $0 = m_1a + n_1b$ could have been added, so there are infinitely many linear combinations of a and b that give (a, b).

EXERCISES: SECTION 1.1

Before working on the exercises, you must make sure that you are familiar with all of the definitions and theorems of this section. You also need to be familiar with the techniques of proof that have been used in the theorems and examples in the text. As a reminder, we take this opportunity to list several useful approaches.

—When working questions involving divisibility you may find it useful to go back to the definition. If you rewrite $b \mid a$ as $a = bq$ for some $q \in \mathbf{Z}$, then you have an equation involving integers, something concrete and familiar to work with.

—To show that $b \mid a$, try to write down an expression for a that has b as a factor.

—Another approach to proving that $b \mid a$ is to use the division algorithm to write $a = bq + r$, where $0 \le r < b$, and show that $r = 0$.

—Theorem 1.1.6 is extremely useful in questions involving greatest common divisors. Remember that finding *some* linear combination of a and b is not necessarily good enough to determine $\gcd(a, b)$. You must show that the linear combination you believe is equal to $\gcd(a, b)$ is actually the *smallest* positive linear combination of a and b.

Exercises for which a solution is given in the answer key are marked by the symbol †.

1. A number n is called **perfect** if it is equal to the sum of its proper positive divisors (those divisors different from n). The first perfect number is 6 since $1 + 2 + 3 = 6$. For each number between 6 and the next perfect number, make a list containing the number, its proper divisors, and their sum.

   Note: If you reach 40, you have missed the next perfect number.

2. Find the quotient and remainder when a is divided by b.

   (a) $a = 99$, $b = 17$

   (b) $a = -99$, $b = 17$

   (c) $a = 17$, $b = 99$

   (d) $a = -1017$, $b = 99$

3. Use the Euclidean algorithm to find the following greatest common divisors.

   †(a) $(35, 14)$

   (b) $(15, 11)$

   †(c) $(252, 180)$

   (d) $(513, 187)$

   †(e) $(7655, 1001)$

4. Use the Euclidean algorithm to find the following greatest common divisors.

   (a) $(6643, 2873)$

   (b) $(7684, 4148)$

   (c) $(26460, 12600)$

   (d) $(6540, 1206)$

   (e) $(12091, 8439)$

5.†For each part of Exercise 3, find integers m and n such that (a, b) is expressed in the form $ma + nb$.

6. For each part of Exercise 4, find integers m and n such that (a, b) is expressed in the form $ma + nb$.

7. Let a, b, c be integers. Give a proof for these facts about divisors:

   (a) If $b \mid a$, then $b \mid ac$.

   (b) If $b \mid a$ and $c \mid b$, then $c \mid a$.

   (c) If $c \mid a$ and $c \mid b$, then $c \mid (ma + nb)$ for any integers m, n.

8. Let a, b, c be integers such that $a + b + c = 0$. Show that if n is an integer which is a divisor of two of the three integers, then it is also a divisor of the third.

9. Let a, b, c be integers.

   (a) Show that if $b \mid a$ and $b \mid (a + c)$, then $b \mid c$.

   (b) Show that if $b \mid a$ and $b \nmid c$, then $b \nmid (a + c)$.

10. Let a, b, c be integers, with $c \neq 0$. Show that $bc \mid ac$ if and only if $b \mid a$.

11. Show that if $a > 0$, then $(ab, ac) = a(b, c)$.

12. Show that if n is any integer, then $(10n + 3, 5n + 2) = 1$.

13. Show that if n is any integer, then $(a + nb, b) = (a, b)$.

14. For what positive integers n is it true that $(n, n + 2) = 2$? Prove your claim.

15. Give a detailed proof of the statement in the text that if a and b are integers, then $b \mid a$ if and only if $a\mathbf{Z} \subseteq b\mathbf{Z}$.

16. Let a, b, c be integers, with $b > 0, c > 0$, and let q be the quotient and r the remainder when a is divided by b.

    (a) Show that q is the quotient and rc is the remainder when ac is divided by bc.

    (b) Show that if q' is the quotient when q is divided by c, then q' is the quotient when a is divided by bc. (Do not assume that the remainders are zero.)

17. Let a, b, n be integers with $n > 1$. Suppose that $a = nq_1 + r_1$ with $0 \leq r_1 < n$ and $b = nq_2 + r_2$ with $0 \leq r_2 < n$. Prove that $n \mid (a - b)$ if and only if $r_1 = r_2$.

18. Show that any nonempty set of integers that is closed under subtraction must also be closed under addition. (Thus part of the hypothesis of Theorem 1.1.4 is redundant.)

19. Let a, b, q, r be integers such that $b \neq 0$ and $a = bq + r$. Prove that $(a, b) = (b, r)$ by showing that (b, r) satisfies the definition of the greatest common divisor of a and b.

20. Perhaps a more natural definition of the greatest common divisor is the following: Let a and b be integers, not both zero. An integer d is called the greatest common divisor of the nonzero integers a and b if (i) $d \mid a$ and $d \mid b$, and (ii) $c \mid a$ and $c \mid b$ implies $d \geq c$. Show that this definition is equivalent to Definition 1.1.5.

21. Prove that the sum of the cubes of any three consecutive positive integers is divisible by 3.

22.†Find all integers x such that $3x + 7$ is divisible by 11.

23. Develop a theory of integer solutions x, y of equations of the form $ax + by = c$, where a, b, c are integers. That is, when can an equation of this form be solved, and if it can be solved, how can all solutions be found? Test your theory on these equations:

$$60x + 36y = 12, \qquad 35x + 6y = 8, \qquad 12x + 18y = 11.$$

Finally, give conditions on a and b under which $ax + by = c$ has solutions for every integer c.

24. Formulate a definition of the greatest common divisor of three integers a, b, c (not all zero). With the appropriate definition you should be able to prove that the greatest common divisor is a linear combination of a, b and c.

1.2 Primes

The main focus of this section is on prime numbers. Our method will be to investigate the notion of two integers which are relatively prime, that is, those which have no common divisors except ± 1. Using some facts which we will prove about them, we will be able to prove the prime factorization theorem, which states that every nonzero integer can be expressed as a product of primes. Finally, we will be able to use prime factorizations to learn more about greatest common divisors and least common multiples.

1.2.1 Definition. *The nonzero integers a and b are said to be **relatively prime** if $(a, b) = 1$.*

1.2.2 Proposition. *Let a, b be nonzero integers. Then $(a, b) = 1$ if and only if there exist integers m, n such that $ma + nb = 1$.*

Proof. If a and b are relatively prime, then by Theorem 1.1.6 integers m and n can be found for which $ma + nb = 1$. To prove the converse, we only need to note that if there exist integers m and n with $ma + nb = 1$, then 1 must be the smallest positive linear combination of a and b, and thus $(a, b) = 1$, again by Theorem 1.1.6. $\square$

Proposition 1.2.2 will be used repeatedly in the proof of the next result. A word of caution—it is often tempting to jump from the equation $d = ma + nb$ to the conclusion that $d = (a, b)$. For example, $16 = 2 \cdot 5 + 3 \cdot 2$, but obviously $(5, 2) \neq 16$. The most that it is possible to say (using Theorem 1.1.6) is that d is a multiple of (a, b). Of course, if $ma + nb = 1$, then Proposition 1.2.2 implies that $(a, b) = 1$.

1.2.3 Proposition. *Let a, b, c be integers, where $a \neq 0$ or $b \neq 0$.*
    (a) *If $b \mid ac$, then $b \mid (a, b) \cdot c$.*
    (b) *If $b \mid ac$ and $(a, b) = 1$, then $b \mid c$.*
    (c) *If $b \mid a$, $c \mid a$ and $(b, c) = 1$, then $bc \mid a$.*
    (d) *$(a, bc) = 1$ if and only if $(a, b) = 1$ and $(a, c) = 1$.*

Proof. (a) Assume that $b \mid ac$. To show that $b \mid (a, b) \cdot c$, we will try to find an expression for $(a, b) \cdot c$ that has b as an obvious factor. We can write $(a, b) = ma + nb$ for some $m, n \in \mathbf{Z}$, and then multiplying by c gives

$$(a, b) \cdot c = mac + nbc .$$

Now b is certainly a factor of nbc, and by assumption it is also a factor of ac, so it is a factor of mac and therefore of the sum $mac + nbc$. Thus $b \mid (a, b) \cdot c$.
    (b) Simply letting $(a, b) = 1$ in part (a) gives the result immediately.
    (c) If $b \mid a$, then $a = bq$ for some integer q. If $c \mid a$, then $c \mid bq$, so if $(b, c) = 1$, it follows from part (b) that $c \mid q$, say with $q = cq_1$. Substituting for q in the equation $a = bq$ gives $a = bcq_1$, and thus $bc \mid a$.
    (d) Suppose that $(a, bc) = 1$. Then $ma + n(bc) = 1$ for some integers m and n, and by viewing this equation as $ma + (nc)b = 1$ and $ma + (nb)c = 1$ we can see that $(a, b) = 1$ and $(a, c) = 1$.
    Conversely, suppose that $(a, b) = 1$ and $(a, c) = 1$. Then $m_1 a + n_1 b = 1$ for some integers m_1 and n_1, and $m_2 a + n_2 c = 1$ for some integers m_2 and n_2. Multiplying these two equations gives

$$(m_1 m_2 a + m_1 n_2 c + m_2 n_1 b)a + (n_1 n_2)bc = 1,$$

which shows that $(a, bc) = 1$. $\square$

1.2.4 Definition. *An integer $p > 1$ is called a **prime number** if its only divisors are ± 1 and $\pm p$. An integer $a > 1$ is called **composite** if it is not prime.*

To determine whether or not a given integer $n > 1$ is prime, we could just try to divide n by each positive integer less than n. This method of trial division is very inefficient, and for this reason various sophisticated methods of "primality testing" have been developed. The need for efficient tests has become particularly apparent recently, because of applications to computer security that make use of cryptographic algorithms. To determine the complete list of all primes up to some bound, there is a useful procedure handed down from antiquity.

Example 1.2.1 (Sieve of Eratosthenes).

The primes less than a fixed positive integer a can be found by the following procedure. List all positive integers less than a (except 1), and cross off every even number except 2. Then go to the first number that has not been crossed off, which will be 3, and cross off all higher multiples of 3. Continue this process to find all primes less than a. You can stop after you have crossed off all proper multiples of primes p for which $p < \sqrt{a}$, since you will have crossed off every number less than a that has a proper factor. (If b is composite, say $b = b_1 b_2$, then either $b_1 \leq \sqrt{b}$ or $b_2 \leq \sqrt{b}$.) For example, we can find all primes less than 20 by just crossing off all multiples of 2 and 3, since $5 > \sqrt{20}$:

$$2 \quad 3 \quad \cancel{4} \quad 5 \quad \cancel{6} \quad 7 \quad \cancel{8} \quad \cancel{9} \quad \cancel{10}$$
$$11 \quad \cancel{12} \quad 13 \quad \cancel{14} \quad \cancel{15} \quad \cancel{16} \quad 17 \quad \cancel{18} \quad 19 \quad .$$

This method is attributed to the Greek mathematician Eratosthenes, and is called the **sieve of Eratosthenes**.

Similarly, the integers less than a and relatively prime to a can be found by crossing off the prime factors of a and all of their multiples. For example, the prime divisors of 36 are 2 and 3, and so the positive integers less than 36 and relatively prime to it can be found as follows:

$$1 \quad \cancel{2} \quad \cancel{3} \quad \cancel{4} \quad 5 \quad \cancel{6} \quad 7 \quad \cancel{8} \quad \cancel{9} \quad \cancel{10} \quad 11 \quad \cancel{12}$$
$$13 \quad \cancel{14} \quad \cancel{15} \quad \cancel{16} \quad 17 \quad \cancel{18} \quad 19 \quad \cancel{20} \quad \cancel{21} \quad \cancel{22} \quad 23 \quad \cancel{24}$$
$$25 \quad \cancel{26} \quad \cancel{27} \quad \cancel{28} \quad 29 \quad \cancel{30} \quad 31 \quad \cancel{32} \quad \cancel{33} \quad \cancel{34} \quad 35 \quad . \qquad \square$$

Euclid's lemma, the next step in our development of the fundamental theorem of arithmetic, is the one that requires our work on relatively prime numbers. We will use Proposition 1.2.3 (b) in a crucial way.

1.2.5 Lemma (Euclid). *An integer $p > 1$ is prime if and only if it satisfies the following property: for all integers a and b, if $p \mid ab$, then either $p \mid a$ or $p \mid b$.*

Proof. Suppose that p is prime and $p \mid ab$. We know that either $(p, a) = p$ or $(p, a) = 1$, since (p, a) is always a divisor of p and p is prime. In the first case $p \mid a$ and we are done. In the second case, since $(p, a) = 1$, we can apply Proposition 1.2.3 (b) to show that $p \mid ab$ implies $p \mid b$. Thus we have shown that if $p \mid ab$, then either $p \mid a$ or $p \mid b$.

Conversely, suppose that p satisfies that given condition. If p were composite, then we could write $p = ab$ for some positive integers smaller than p. The condition would imply that either $p \mid a$ or $p \mid b$, which would be an obvious contradiction. □

The following corollary extends Euclid's lemma to the product of more than two integers. In the proof we will use mathematical induction, which we hope is familiar to you. If you do not remember how to use induction, you should read the discussion in Appendix A.4.

1.2.6 Corollary. *If p is a prime number, and $p \mid a_1 a_2 \cdots a_n$ for integers a_1, a_2, ..., a_n, then $p \mid a_i$ for some i with $1 \leq i \leq n$.*

Proof. In order to use the principle of mathematical induction, let P_n be the following statement: if $p \mid a_1 a_2 \cdots a_n$, then $p \mid a_i$ for some $1 \leq i \leq n$. The statement P_1 is clearly true. Next, assume that the statement P_k is true, that is, if $p \mid a_1 a_2 \cdots a_k$, then $p \mid a_i$ for some $1 \leq i \leq k$. If $p \mid a_1 a_2 \cdots a_k a_{k+1}$, for integers $a_1, a_2, \ldots, a_k, a_{k+1}$, then applying Euclid's lemma to $a = a_1 a_2 \cdots a_k$ and $b = a_{k+1}$ yields that $p \mid a_1 a_2 \cdots a_k$ or $p \mid a_{k+1}$. In case $p \mid a_1 a_2 \cdots a_k$, the truth of the statement P_k implies that $p \mid a_i$ for some $1 \leq i \leq k$. Thus, in either case, $p \mid a_i$ for some $1 \leq i \leq k + 1$, and hence the statement P_{k+1} is true. By the principle of mathematical induction (as stated in Theorem A.4.2 of Appendix A.4), the statement P_n holds for all positive integers n. □

The next theorem, on prime factorization, is sometimes called the fundamental theorem of arithmetic. The naive way to prove that an integer a can be written as a product of primes is to note that either a is prime and we are done, or else a is composite, say $a = bc$. Then the same argument can be applied to b and c, and continued until a has been broken up into a product of primes. (This process must stop after a finite number of steps because of the well-ordering principle.) We also need to prove that any two factorizations of a number are in reality the same. The idea of the proof is to use Euclid's lemma to pair the primes in one factorization with those in the other. In fact, the proof of the uniqueness of the factorization requires a more delicate argument than the proof of the existence of the factorization.

1.2.7 Theorem (Fundamental Theorem of Arithmetic). *Any integer $a > 1$ can be factored uniquely as a product of prime numbers, in the form*

$$a = p_1^{\alpha_1} p_2^{\alpha_2} \cdots p_n^{\alpha_n},$$

where $p_1 < p_2 < \ldots < p_n$ and the exponents $\alpha_1, \alpha_2, \ldots, \alpha_n$ are all positive.

Proof. Suppose that there is some integer that cannot be written as a product of primes. Then the set of all integers $a > 1$ that have no prime factorization must be nonempty, so as a consequence of the well-ordering principle it must have a smallest member, say b. Now b cannot itself be a prime number since then it would have a prime factorization. Thus b is composite, and we can write $b = cd$ for positive integers c, d that are smaller than b. Since b was assumed to be the smallest positive integer not having a factorization into primes, and c and d are smaller, then both c and d must have factorizations into products of primes. This shows that b also has such a factorization, which is a contradiction. Since multiplication is commutative, the prime factors can be ordered in the desired manner.

If there exists an integer > 1 for which the factorization is not unique, then by the well-ordering principle there exists a smallest such integer, say a. Assume that a has two factorizations $a = p_1^{\alpha_1} p_2^{\alpha_2} \cdots p_n^{\alpha_n}$ and $a = q_1^{\beta_1} q_2^{\beta_2} \cdots q_m^{\beta_m}$, where $p_1 < p_2 < \ldots < p_n$, and $q_1 < q_2 < \ldots < q_m$, with $\alpha_i > 0$ for $i = 1, \ldots, n$, and $\beta_i > 0$ for $i = 1, \ldots, m$. By Corollary 1.2.6 of Euclid's lemma, $q_1 \mid p_k$ for some k with $1 \leq k \leq n$ and $p_1 \mid q_j$ for some j with $1 \leq j \leq m$. Since all of the numbers p_i and q_i are prime, we must have $q_1 = p_k$ and $p_1 = q_j$. Then $p_1 = q_1$ since $q_1 \leq q_j = p_1 \leq p_k = q_1$. Hence we can let

$$s = \frac{a}{p_1} = \frac{a}{q_1} = p_1^{\alpha_1 - 1} p_2^{\alpha_2} \cdots p_n^{\alpha_n} = q_1^{\beta_1 - 1} q_2^{\beta_2} \cdots q_m^{\beta_m} .$$

If $s = 1$ then $a = p_1$ has a unique factorization, contrary to the choice of a. If $s > 1$, then since $s < a$ and s has two factorizations, we again have a contradiction to the choice of a. $\square$

If the prime factorization of an integer is known, then it is easy to list all of its divisors. If $a = p_1^{\alpha_1} p_2^{\alpha_2} \cdots p_n^{\alpha_n}$, then b is a divisor of a if and only if $b = p_1^{\beta_1} p_2^{\beta_2} \cdots p_n^{\beta_n}$, where $\beta_i \leq \alpha_i$ for all i. Thus we can list all possible divisors of a by systematically decreasing the exponents of each of its prime divisors.

Example 1.2.2.

The positive divisors of 12 are $1, 2, 3, 4, 6, 12$; the positive divisors of 8 are $1, 2, 4, 8$; and the positive divisors of 36 are $1, 2, 3, 4, 6, 9, 12, 18, 36$. In Figure 1.2.1, we have arranged the divisors so as to show the divisibility relations among them. There is a path (moving upward only) from a to b if and only if $a \mid b$.

In constructing the first diagram in Figure 1.2.1, it is easiest to use the prime factorization of 12. Since $12 = 2^2 3$, we first divide 12 by 2 to get 6 and then divide again by 2 to get 3. This gives the first side of the diagram, and to construct the opposite side of the diagram we divide each number by 3.

If the number has three different prime factors, then we would need a three-dimensional diagram. (Visualize the factors as if on the edges of a box.) With more than three distinct prime factors, the diagrams lose their clarity.   □

Figure 1.2.1:

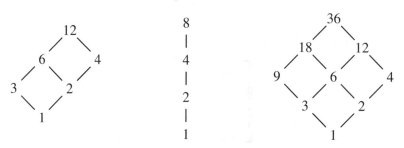

The following proof, although easy to follow, is an excellent example of the austere beauty of mathematics.

1.2.8 Theorem (Euclid). *There exist infinitely many prime numbers.*

Proof. Suppose that there were only finitely many prime numbers, say p_1, p_2, . . . , p_n. Then consider the number $a = p_1 p_2 \cdots p_n + 1$. By Theorem 1.2.7, the number a has a prime divisor, say p. Now p must be one of the primes we listed, so $p \mid (p_1 p_2 \cdots p_n)$, and since $p \mid a$, it follows that $p \mid (a - p_1 p_2 \cdots p_n)$. This is a contradiction since p cannot be a divisor of 1.   □

Example 1.2.3.

Consider the numbers $2^2 - 1 = 3$, $2^3 - 1 = 7$, $2^4 - 1 = 15$, $2^5 - 1 = 31$, and $2^6 - 1 = 63$. The prime exponents each give rise to a prime, while the composite exponents each give a composite number. Is this true in general? Continuing to investigate prime exponents gives $2^7 - 1 = 127$, which is prime, but $2^{11} - 1 = 2047 = 23 \cdot 89$. Thus a prime exponent may or may not yield a prime number.

On the other hand, it is always true that a composite exponent yields a composite number. To prove this, let n be composite, say $n = qm$ (where q and m are integers greater than 1), and consider $2^n - 1 = 2^{qm} - 1$. We need to find a nontrivial factorization of $2^{qm} - 1 = (2^q)^m - 1$. We can look at this as $x^m - 1$, and then we have the familiar factorization

$$x^m - 1 = (x - 1)(x^{m-1} + x^{m-2} + \ldots + x^2 + x + 1) .$$

Substituting $x = 2^q$ shows that $2^q - 1$ is a factor of $2^n - 1$. Now $1 < 2^q - 1 < 2^n - 1$ since both q and m are greater than 1, and so we have found a nontrivial factorization of $2^n - 1$. $\square$

The final concept we study in this section is the least common multiple of two integers. Its definition is parallel to that of the greatest common divisor. We can characterize it in terms of the prime factorizations of the two numbers, or by the fact that the product of two numbers is equal to the product of their least common multiple and greatest common divisor.

1.2.9 Definition. *A positive integer m is called the **least common multiple** of the nonzero integers a and b if*

   (i) *m is a multiple of both a and b, and*

   (ii) *any multiple of both a and b is also a multiple of m.*

We will use the notation $\text{lcm}[a, b]$ *or* $[a, b]$ *for the least common multiple of a and b.*

When written out in symbols, the definition of the least common multiple looks like this: $m = \text{lcm}[a, b]$ if (i) $a \mid m$ and $b \mid m$, and (ii) if $a \mid c$ and $b \mid c$, then $m \mid c$.

There are times, as in next proposition, when it is convenient to allow the prime factorization of a number to include primes with exponent 0. This leads to a representation that is no longer unique, but it is particularly useful to be able to write the prime factorizations of two different integers in terms of the *same* primes.

1.2.10 Proposition. *Let a and b be positive integers with prime factorizations $a = p_1^{\alpha_1} p_2^{\alpha_2} \cdots p_n^{\alpha_n}$ and $b = p_1^{\beta_1} p_2^{\beta_2} \cdots p_n^{\beta_n}$, where $\alpha_i \geq 0$ and $\beta_i \geq 0$ for all i.*

   (a) *Then $a \mid b$ if and only if $\alpha_i \leq \beta_i$ for $i = 1, 2, \ldots, n$.*

   (b) *For each i, let $\delta_i = \min\{\alpha_i, \beta_i\}$ and $\mu_i = \max\{\alpha_i, \beta_i\}$. Then*

$$\gcd(a, b) = p_1^{\delta_1} p_2^{\delta_2} \cdots p_n^{\delta_n} \quad and \quad \text{lcm}[a, b] = p_1^{\mu_1} p_2^{\mu_2} \cdots p_n^{\mu_n} .$$

Proof. (a) Suppose that $\alpha_i \leq \beta_i$ for $i = 1, 2, \ldots, n$. Let $\gamma_i = \beta_i - \alpha_i$, for $i = 1, 2, \ldots, n$, and set $c = p_1^{\gamma_1} p_2^{\gamma_2} \cdots p_n^{\gamma_n}$ (note that $\gamma_i \geq 0$ for $i = 1, 2, \ldots, n$). Then

$$
\begin{aligned}
ac &= p_1^{\alpha_1} p_2^{\alpha_2} \cdots p_n^{\alpha_n} p_1^{\gamma_1} p_2^{\gamma_2} \cdots p_n^{\gamma_n} = p_1^{\alpha_1 + \gamma_1} p_2^{\alpha_2 + \gamma_2} \cdots p_n^{\alpha_n + \gamma_n} \\
&= p_1^{\beta_1} p_2^{\beta_2} \cdots p_n^{\beta_n} = b .
\end{aligned}
$$

Since $b = ac$, we have $a \mid b$.

Conversely, suppose that $a \mid b$. Then there exists $c \in \mathbf{Z}$ such that $b = ac$. For any prime p such that $p \mid c$, we have $p \mid b$, and so $p = p_j$ for some j with $1 \leq j \leq n$.

Thus c has a factorization $c = p_1^{\gamma_1} p_2^{\gamma_2} \cdots p_n^{\gamma_n}$, where $\gamma_i \geq 0$ for $i = 1, 2, \ldots, n$. Since $b = ac$, we have

$$p_1^{\beta_1} p_2^{\beta_2} \cdots p_n^{\beta_n} = p_1^{\alpha_1} p_2^{\alpha_2} \cdots p_n^{\alpha_n} p_1^{\gamma_1} p_2^{\gamma_2} \cdots p_n^{\gamma_n} = p_1^{\alpha_1+\gamma_1} p_2^{\alpha_2+\gamma_2} \cdots p_n^{\alpha_n+\gamma_n} \,,$$

where $\beta_i = \alpha_i + \gamma_i$ for $i = 1, 2, \ldots, n$. Because $\gamma_i \geq 0$, we have $\alpha_i \leq \beta_i$ for $i = 1, 2, \ldots, n$.

(b) The proof follows immediately from part (a) and the definitions of the least common multiple and greatest common divisor. $\square$

As a corollary of Proposition 1.2.10, it is clear that

$$\gcd(a, b) \cdot \mathrm{lcm}[a, b] = ab \,.$$

This can also be shown directly from the definitions, as we have noted in Exercise 15.

For small numbers it is probably easiest to use their prime factorizations to find their greatest common divisor and least common multiple. It takes a great deal of work to find the prime factors of a large number, even on a computer making use of sophisticated algorithms. In contrast, the Euclidean algorithm is much faster, so its use is more efficient for finding the greatest common divisor of large numbers.

Example 1.2.4.

In the previous section we computed $(126, 35)$. To do this using Proposition 1.2.10 we need the factorizations $126 = 2^1 \cdot 3^2 \cdot 7^1$ and $35 = 5^1 \cdot 7^1$. We then add terms so that we have the same primes in each case, to get $126 = 2^1 \cdot 3^2 \cdot 5^0 \cdot 7^1$ and $35 = 2^0 \cdot 3^0 \cdot 5^1 \cdot 7^1$. Thus we obtain $(126, 35) = 2^0 \cdot 3^0 \cdot 5^0 \cdot 7^1 = 7$ and $[126, 35] = 2^1 \cdot 3^2 \cdot 5^1 \cdot 7^1 = 630$. $\square$

EXERCISES: SECTION 1.2

When proving results in these exercises, we recommend that you first try to use Proposition 1.2.2, Proposition 1.2.3, or Lemma 1.2.5, before trying to use the very powerful Fundamental Theorem of Arithmetic.

1. Find the prime factorizations of each of the following numbers, and use them to compute the greatest common divisor and least common multiple of the given pairs of numbers.

   †(a) 35, 14

   (b) 15, 11

   †(c) 252, 180

   (d) 7684, 4148

   †(e) 6643, 2873

2. Use the sieve of Eratosthenes to find all prime numbers less than 200.

3.†For each composite number a, with $4 \leq a \leq 20$, find all positive numbers less than a that are relatively prime to a.

4. Find all positive integers less than 60 and relatively prime to 60.

   Hint: Use techniques similar to the sieve of Eratosthenes.

5.†For each of the numbers 9, 15, 20, 24 and 100, give a diagram of all divisors of the number, showing the divisibility relationships. (See Example 1.2.2.)

6. For each of the following numbers, give a diagram of all divisors of the number, showing the divisibility relationships.

   (a) 60

   (b) 1575

7. Let m and n be positive integers such that $m + n = 57$ and $[m, n] = 680$. Find m and n.

8. Let a, b be positive integers, and let $d = (a, b)$. Since $d \mid a$ and $d \mid b$, there exist integers h, k such that $a = dh$ and $b = dk$. Show that $(h, k) = 1$.

9. Let a, b, c be positive integers, and let $d = (a, b)$. Since $d \mid a$, there exists an integer h with $a = dh$. Show that if $a \mid bc$, then $h \mid c$.

10. Show that $a\mathbf{Z} \cap b\mathbf{Z} = [a, b]\mathbf{Z}$.

11. Let a, b be nonzero integers, and let p be a prime. Show that if $p \mid [a, b]$, then either $p \mid a$ or $p \mid b$.

12. Let a, b, c be nonzero integers. Show that $(a, b) = 1$ and $(a, c) = 1$ if and only if $(a, [b, c]) = 1$.

13. Let a, b be nonzero integers. Prove that $(a, b) = 1$ if and only if $(a + b, ab) = 1$.

14. Let a, b be nonzero integers with $(a, b) = 1$. Compute $(a + b, a - b)$.

15. Let a and b be positive integers, and let m be an integer such that $ab = m(a, b)$. Without using the prime factorization theorem, prove that $(a, b)[a, b] = ab$ by verifying that m satisfies the necessary properties of $[a, b]$.

16. A positive integer a is called a **square** if $a = n^2$ for some $n \in \mathbf{Z}$. Show that the integer $a > 1$ is a square if and only if every exponent in its prime factorization is even.

17. Show that if the positive integer a is not a square, then $a \neq b^2/c^2$ for integers b, c. Thus any positive integer that is not a square must have an irrational square root.

   Hint: Use Exercise 16 to show that $ac^2 \neq b^2$.

18. Show that if a, b are positive integers such that $(a, b) = 1$ and ab is a square, then a and b are also squares.

19. Let p and q be prime numbers. Prove that $pq + 1$ is a square if and only if p and q are twin primes.

20. A positive integer is called **square-free** if it is a product of distinct primes. Prove that every positive integer can be written uniquely as a product of a square and a square-free integer.

21. Prove that if $a > 1$, then there is a prime p with $a < p \leq a! + 1$.

22. Show that for any $n > 0$, there are n consecutive composite numbers.

23. Show that if n is a positive integer such that $2^n + 1$ is prime, then n is a power of 2.

24. Show that $\log 2 / \log 3$ is not a rational number.

25. If a, b, c are positive integers such that $a^2 + b^2 = c^2$, then (a, b, c) is called a **Pythagorean triple**. For example, $(3, 4, 5)$ and $(5, 12, 13)$ are Pythagorean triples. Assume that (a, b, c) is a Pythagorean triple in which the only common divisors of a, b, c are ± 1.

    (a) Show that a and b cannot both be odd.

    (b) Assume that a is even. Show that there exist relatively prime integers m and n such that $a = 2mn$, $b = m^2 - n^2$, and $c = m^2 + n^2$.

    Hint: Factor $a^2 = c^2 - b^2$ after showing that $(c + b, c - b) = 2$.

1.3   Congruences

For many problems involving integers, all of the relevant information is contained in the remainders obtained by dividing by some fixed integer n. Since only n different remainders are possible $(0, 1, \ldots, n - 1)$, having only a finite number of cases to deal with can lead to considerable simplifications. For small values of n it even becomes feasible to use trial-and-error methods.

Example 1.3.1.

A famous theorem of Lagrange states that every positive integer can be written as sum of four squares. (See the notes at the end of this chapter for a short discussion of this problem.) To illustrate the use of remainders in solving a number theoretic problem, we will show that any positive integer whose remainder is 7 when divided by 8 cannot be written as the sum of three squares. Therefore this theorem of Lagrange is as sharp as possible.

If $n = a^2 + b^2 + c^2$, then when both sides are divided by 8, the remainders must be the same. It will follow from Proposition 1.3.3 that we can compute the remainder of $n = a^2 + b^2 + c^2$ by adding the remainders of a^2, b^2, and c^2 (and subtracting a multiple of 8 if necessary). By the same proposition, we can compute the remainders of a^2, b^2, and c^2 by squaring the remainders of a, b, and c (and subtracting a multiple of 8 if necessary). The possible remainders for a, b, and c are $0, 1, \ldots, 7$, and squaring and taking remainders yields only the values 0, 1, and 4. To check the possible remainders for $a^2 + b^2 + c^2$ we only need to add together three such terms. (If we get a sum larger than 7 we subtract 8.) A careful analysis of all of the cases shows that we cannot obtain 7 as a remainder for $a^2 + b^2 + c^2$. Thus we cannot express any integer n whose remainder is 7 when divided by 8 in the form $n = a^2 + b^2 + c^2$. □

Trial and error techniques similar to those of Example 1.3.1 can sometimes be used to show that a polynomial equation has no integer solution. For example, if $x = c$ is a solution of the equation $a_k x^k + \ldots + a_1 x + a_0 = 0$, then $a_k c^k + \ldots + a_1 c + a_0$ must be divisible by every integer n. If some n can be found for which $a_k x^k + \ldots + a_1 x + a_0$ is never divisible by n, then this can be used to prove that the equation has no integer solutions. For example, $x^3 + x + 1 = 0$ has no integer solutions since $c^3 + c + 1$ is odd for all integers c, and thus is never divisible by 2.

A more familiar situation in which we carry out arithmetic after dividing by a fixed integer is the addition of hours on a clock (where the fixed integer is 12). Another example is given by the familiar rules "even plus even is even," "even times even is even," etc., which are useful in other circumstances (where the fixed integer is 2). Gauss introduced the following congruence notation, which simplifies computations of this sort.

1.3.1 Definition. *Let n be a positive integer. Integers a and b are said to be **congruent modulo** n if they have the same remainder when divided by n. This is denoted by writing $a \equiv b$ (mod n).*

If we use the division algorithm to write $a = nq + r$, where $0 \leq r < n$, then $r = n \cdot 0 + r$. It follows immediately from the previous definition that $a \equiv r$ (mod n). In particular, any integer is congruent modulo n to one of the integers $0, 1, 2, \ldots, n-1$.

We feel that the definition we have given provides the best intuitive understanding of the notion of congruence, but in almost all proofs it will be easiest to use the characterization given by the next proposition. Using this characterization makes it possible to utilize the facts about divisibility that we have developed in the preceding sections of this chapter.

1.3.2 Proposition. *Let a, b, and $n > 0$ be integers. Then $a \equiv b$ (mod n) if and only if $n \mid (a - b)$.*

Proof. If $a \equiv b$ (mod n), then a and b have the same remainder when divided by n, so the division algorithm gives $a = nq_1 + r$ and $b = nq_2 + r$. Solving for the common remainder gives $a - nq_1 = b - nq_2$. Thus $a - b = n(q_1 - q_2)$, and so $n \mid (a - b)$.

To prove the converse, assume that $n \mid (a - b)$. Then there exists $k \in \mathbf{Z}$ with $a - b = nk$, and hence $b = a - nk$. If upon applying the division algorithm we have $a = nq + r$, with $0 \leq r < n$, then $b = a - nk = (nq + r) - nk = n(q - k) + r$. Since $0 \leq r < n$, division of b by n also yields the remainder r. Hence $a \equiv b \pmod{n}$.   □

When working with congruence modulo n, the integer n is called the **modulus**. By the preceding proposition, $a \equiv b \pmod{n}$ if and only if $a - b = nq$ for some integer q. We can write this in the form $a = b + nq$, for some integer q. This observation gives a very useful method of replacing a congruence with an equation (over $\mathbf{Z}$). On the other hand, Proposition 1.3.3 shows that any equation can be converted to a congruence modulo n by simply changing the $=$ sign to $\equiv$. In doing so, any term congruent to 0 can simply be omitted. Thus the equation $a = b + nq$ would be converted back to $a \equiv b \pmod{n}$.

Congruence behaves in many ways like equality. The following properties, which are obvious from the definition of congruence modulo n, are a case in point. Let a, b, c be integers. Then

(i) $a \equiv a \pmod{n}$;

(ii) if $a \equiv b \pmod{n}$, then $b \equiv a \pmod{n}$;

(iii) if $a \equiv b \pmod{n}$ and $b \equiv c \pmod{n}$, then $a \equiv c \pmod{n}$.

The following theorem carries this analogy even further. Perhaps its most important consequence is that when adding, subtracting, or multiplying congruences you may substitute any congruent integer. For example, to show that $99^2 \equiv 1 \pmod{100}$, it is easier to substitute -1 for 99 and just show that $(-1)^2 = 1$.

1.3.3 Proposition. *Let $n > 0$ be an integer. Then the following conditions hold for all integers a, b, c, d:*

(a) *If $a \equiv c \pmod{n}$ and $b \equiv d \pmod{n}$, then $a \pm b \equiv c \pm d \pmod{n}$, and $ab \equiv cd \pmod{n}$.*

(b) *If $a + c \equiv a + d \pmod{n}$, then $c \equiv d \pmod{n}$. If $ac \equiv ad \pmod{n}$ and $(a, n) = 1$, then $c \equiv d \pmod{n}$.*

Proof. (a) If $a \equiv c \pmod{n}$ and $b \equiv d \pmod{n}$, then $n \mid (a - c)$ and $n \mid (b - d)$. Adding shows that $n \mid ((a + b) - (c + d))$, and subtracting shows that $n \mid ((a - b) - (c - d))$. Thus $a \pm b \equiv c \pm d \pmod{n}$.

Since $n \mid (a - c)$, we have $n \mid (ab - cb)$, and since $n \mid (b - d)$, we have $n \mid (cb - cd)$. Adding shows that $n \mid (ab - cd)$ and thus $ab \equiv cd \pmod{n}$.

(b) If $a + c \equiv a + d \pmod{n}$, then $n \mid ((a + c) - (a + d))$. Thus $n \mid (c - d)$ and so $c \equiv d \pmod{n}$.

If $ac \equiv ad \pmod{n}$, then $n \mid (ac - ad)$, and since $(n, a) = 1$, it follows from Proposition 1.2.3 (b) that $n \mid (c - d)$. Thus $c \equiv d \pmod{n}$.   □

The consequences of Proposition 1.3.3 can be summarized as follows.

(i) For any number in the congruence, you can substitute any congruent integer.

(ii) You can add or subtract the same integer on both sides of a congruence.

(iii) You can multiply both sides of a congruence by the same integer.

(iv) Canceling, or dividing both sides of a congruence by the same integer, must be done very carefully. You may divide both sides of a congruence by an integer a only if $(a, n) = 1$. For example, $30 \equiv 6 \pmod 8$, but dividing both sides by 6 gives $5 \equiv 1 \pmod 8$, which is certainly false. On the other hand, since 3 is relatively prime to 8, we may divide both sides by 3 to get $10 \equiv 2 \pmod 8$.

Proposition 1.3.3 shows that the remainder upon division by n of $a+b$ or ab can be found by adding or multiplying the remainders of a and b when divided by n and then dividing by n again if necessary. For example, if $n = 8$, then 101 has remainder 5 and 142 has remainder 6 when divided by 8. Thus $101 \cdot 142 = 14,342$ has the same remainder as 30 (namely, 6) when divided by 8. Formally, $101 \equiv 5 \pmod 8$ and $142 \equiv 6 \pmod 8$, so it follows that $101 \cdot 142 \equiv 5 \cdot 6 \equiv 6 \pmod 8$.

As a further example, we compute the powers of 2 modulo 7. Rather than computing each power and then dividing by 7, we reduce modulo 7 at each stage of the computations:

$$2^2 \equiv 4 \pmod 7,$$
$$2^3 \equiv 2^2 2 \equiv 4 \cdot 2 \equiv 1 \pmod 7,$$
$$2^4 \equiv 2^3 2 \equiv 1 \cdot 2 \equiv 2 \pmod 7,$$
$$2^5 \equiv 2^4 2 \equiv 2 \cdot 2 \equiv 4 \pmod 7.$$

From the way in which we have done the computations, it is clear that the powers will repeat. In fact, since there are only finitely many remainders modulo n, the powers of any integer will eventually begin repeating modulo n.

1.3.4 Proposition. *Let a and $n > 1$ be integers. There exists an integer b such that $ab \equiv 1 \pmod n$ if and only if $(a, n) = 1$.*

Proof. If there exists an integer b such that $ab \equiv 1 \pmod n$, then we have $ab = 1 + qn$ for some integer q. This can be rewritten to give a linear combination of a and n equal to 1, and so $(a, n) = 1$.

Conversely, if $(a, n) = 1$, then there exist integers s, t such that $sa + tn = 1$. Letting $b = s$ and reducing the equation to a congruence modulo n gives $ab \equiv 1 \pmod n$. $\square$

We are now ready to present a systematic study of linear congruences that involve unknowns. The previous proposition shows that the congruence

$$ax \equiv 1 \pmod n$$

has a solution if and only if $(a, n) = 1$. In fact, the proof of the proposition shows that the solution can be obtained by using the Euclidean algorithm to write $1 = ab + nq$ for some $b, q \in \mathbf{Z}$, since then $1 \equiv ab \pmod n$.

The next theorem determines all solutions of a linear congruence of the form

$$ax \equiv b \ (\mathrm{mod}\ n) \ .$$

Of course, if the numbers involved are small, it may be simplest just to use trial and error. For example, to solve $3x \equiv 2 \ (\mathrm{mod}\ 5)$, we only need to substitute $x = 0, 1, 2, 3, 4$. Thus by trial and error we can find the solution $x \equiv 4 \ (\mathrm{mod}\ 5)$.

In many ways, solving congruences is like solving equations. There are a few important differences, however. A linear equation over the integers (an equation of the form $ax = b$, where $a \neq 0$) has at most one solution. On the other hand, the linear congruence $2x \equiv 2 \ (\mathrm{mod}\ 4)$ has the two solutions $x \equiv 1 \ (\mathrm{mod}\ 4)$ and $x \equiv 3 \ (\mathrm{mod}\ 4)$.

For linear equations, it may happen that there is no solution. The same is true for linear congruences. For example, trial and error shows that the congruence $3x \equiv 2 \ (\mathrm{mod}\ 6)$ has no solution. Thus the first step in solving a linear congruence is to use the theorem to determine whether or not a solution exists.

We say that two solutions r and s to the congruence $ax \equiv b \ (\mathrm{mod}\ n)$ are **distinct solutions modulo** n if r and s are not congruent modulo n. Thus in the next theorem the statement "d distinct solutions modulo n" means that there are d solutions $s_1, s_2, \ldots, s_d$ such that if $i \neq j$, then s_i and s_j are not congruent modulo n. This terminology is necessary in order to understand what we mean by "solving" the congruence $ax \equiv b \ (\mathrm{mod}\ n)$. In the next section, we will introduce the concept of a "congruence class" to clarify the situation.

1.3.5 Theorem. *Let a, b and $n > 1$ be integers. The congruence $ax \equiv b$ (mod n) has a solution if and only if b is divisible by d, where $d = (a, n)$. If $d \mid b$, then there are d distinct solutions modulo n, and these solutions are congruent modulo n/d.*

Proof. To prove the first statement, observe that $ax \equiv b \ (\mathrm{mod}\ n)$ has a solution if and only if there exist integers s and q such that $as = b + nq$, or, equivalently, $as + (-q)n = b$. Thus there is a solution if and only if b can be expressed as a linear combination of a and n. By Theorem 1.1.6 the linear combinations of a and n are precisely the multiples of d, so there is a solution if and only if $d \mid b$.

To prove the second statement, assume that $d \mid b$, and let $m = n/d$. Suppose that x_1 and x_2 are solutions of the congruence $ax \equiv b \ (\mathrm{mod}\ n)$, giving $ax_1 \equiv ax_2 \ (\mathrm{mod}\ n)$. Then $n \mid a(x_1 - x_2)$, and so it follows from Proposition 1.2.3 (a) that $n \mid d(x_1 - x_2)$. Thus $m \mid (x_1 - x_2)$, and so $x_1 \equiv x_2 \ (\mathrm{mod}\ m)$. On the other hand, if $x_1 \equiv x_2 \ (\mathrm{mod}\ m)$, then $m \mid (x_1 - x_2)$, and so $n \mid d(x_1 - x_2)$ since $n = dm$. Then since $d \mid a$ we can conclude that $n \mid a(x_1 - x_2)$, and so $ax_1 \equiv ax_2 \ (\mathrm{mod}\ n)$.

We can choose the distinct solutions from among the n remainders $0, 1, \ldots, n{-}1$. Given one such solution, we can find all others in the set by adding multiples of n/d, giving a total of d distinct solutions. $\square$

We now describe an algorithm for solving linear congruences of the form

$$ax \equiv b \pmod{n} .$$

We first compute $d = (a, n)$, and if $d \mid b$, then we write the congruence $ax \equiv b \pmod{n}$ as an equation $ax = b + qn$. Since d is a common divisor of a, b, and n, we can write $a = da_1$, $b = db_1$, and $n = dm$. Thus we get $a_1 x = b_1 + qm$, which yields the congruence

$$a_1 x \equiv b_1 \pmod{m} ,$$

where $a_1 = a/d$, $b_1 = b/d$, and $m = n/d$.

It follows immediately from Proposition 1.2.10 that since $d = (a, n)$, the numbers a_1 and m must be relatively prime. Thus by Proposition 1.3.4 we can apply the Euclidean algorithm to find an integer c such that $ca_1 \equiv 1 \pmod{m}$. Multiplying both sides of the congruence $a_1 x \equiv b_1 \pmod{m}$ by c gives the solution

$$x \equiv cb_1 \pmod{m} .$$

Finally, since the original congruence was given modulo n, we should give our answer modulo n instead of modulo m. The congruence $x \equiv cb_1 \pmod{m}$ can be converted to the equation $x = cb_1 + mk$, which yields the solution $x \equiv cb_1 + mk \pmod{n}$. The solution modulo m determines d distinct solutions modulo n. The solutions have the form $s_0 + km$, where s_0 is any particular solution of $x \equiv b_1 c \pmod{m}$ and k is any integer.

Example 1.3.2 (Homogeneous linear congruences).

In this example we consider the special case of a linear homogeneous congruence

$$ax \equiv 0 \pmod{n} .$$

In this case there always exists a solution, namely $x \equiv 0 \pmod{n}$, but this may not be the only solution modulo n.

As the first step in the solution we obtain $a_1 x \equiv 0 \pmod{n_1}$, where $a = da_1$ and $n = dn_1$. Since a_1 and n_1 are relatively prime, by part (b) of Proposition 1.3.3 we can cancel a_1, to obtain

$$x \equiv 0 \pmod{n_1} , \qquad \text{with } n_1 = \frac{n}{\gcd(a, n)} .$$

We have d distinct solutions modulo n.

For example, $28x \equiv 0 \pmod{48}$ reduces to $x \equiv 0 \pmod{12}$, and $x \equiv 0, 12, 24, 36$ are the four distinct solutions modulo 48. $\square$

Example 1.3.3.

To solve the congruence

$$60x \equiv 90 \ (\text{mod } 105) \ ,$$

we first note that $(60, 105) = 15$, and then check that $15 \mid 90$, so that there will indeed be a solution. Dividing the corresponding equation $60x = 90 + 105q$ by 15, we obtain the equation $4x = 6 + 7q$, which reduces to the congruence

$$4x \equiv 6 \ (\text{mod } 7) \ .$$

To solve this congruence, we need an integer c with $c \cdot 4 \equiv 1 \ (\text{mod } 7)$, so in effect we must solve another congruence, $4z \equiv 1 \ (\text{mod } 7)$. We could use the Euclidean algorithm, but with such a small modulus, trial and error is quicker, and it is easy to see that $c = 2$ will work.

We now multiply both sides of the congruence $4x \equiv 6 \ (\text{mod } 7)$ by 2, to obtain $8x \equiv 12 \ (\text{mod } 7)$, which reduces to

$$x \equiv 5 \ (\text{mod } 7) \ .$$

Writing the solution in the form of an equation, we have $x = 5 + 7k$, so $x \equiv 5 + 7k \ (\text{mod } 105)$. By adding multiples of 7 to the particular solution $x_0 = 5$, we obtain the solutions $\ldots, -2, 5, 12, 19, \ldots$. There are 15 distinct solutions modulo 105, so we have

$$x \equiv 5, 12, 19, 26, 33, 40, 47, 54, 61, 68, 75, 82, 89, 96, 103 \ (\text{mod } 105) \ . \ \square$$

In the next theorem we show how to solve two simultaneous congruences over moduli that are relatively prime. The motivation for the proof of the next theorem is as follows. Assume that the congruences $x \equiv a \ (\text{mod } n)$ and $x \equiv b \ (\text{mod } m)$ are given. If we can find integers y and z with

$$y \equiv 1 \ (\text{mod } n) \qquad\qquad y \equiv 0 \ (\text{mod } m)$$

$$z \equiv 0 \ (\text{mod } n) \qquad\qquad z \equiv 1 \ (\text{mod } m)$$

then $x = ay + bz$ will be a solution to the pair of simultaneous congruences $x \equiv a \ (\text{mod } n)$ and $x \equiv b \ (\text{mod } m)$. This can be seen by reducing modulo n and then modulo m.

1.3.6 Theorem (Chinese Remainder Theorem). *Let n and m be positive integers, with* $(n, m) = 1$. *Then the system of congruences*

$$x \equiv a \ (mod \ n) \qquad x \equiv b \ (mod \ m)$$

has a solution. Moreover, any two solutions are congruent modulo mn.

Proof. Since $(n, m) = 1$, there exist integers r and s such that $rm + sn = 1$. Then $rm \equiv 1$ (mod n) and $sn \equiv 1$ (mod m). Following the suggestion in the preceding paragraph, we let $x = arm + bsn$. Then a direct computation verifies that $x \equiv arm \equiv a$ (mod n) and $x \equiv bsn \equiv b$ (mod m).

If x is a solution, then adding any multiple of mn is obviously still a solution. Conversely, if x_1 and x_2 are two solutions of the given system of congruences, then they must be congruent modulo n and modulo m. Thus $x_1 - x_2$ is divisible by both n and m, so it is divisible by mn since by assumption $(n, m) = 1$. Therefore $x_1 \equiv x_2$ (mod mn). $\square$

Example 1.3.4.

> The proof of Theorem 1.3.6 actually shows how to solve the given system of congruences. For example, if we wish to solve the system
>
> $$x \equiv 7 \ (\text{mod } 8) \qquad x \equiv 3 \ (\text{mod } 5)$$
>
> we first use the Euclidean algorithm to write $2 \cdot 8 - 3 \cdot 5 = 1$. Then $x = 7(-3)(5) + 3(2)(8) = -57$ is a solution, and the general solution is $x = -57 + 40t$. The smallest nonnegative solution is therefore 23, so we have
>
> $$x \equiv 23 \ (\text{mod } 40) \ . \ \square$$

Another proof of the existence of a solution in Theorem 1.3.6 can be given as follows. In some respects this method of solution is more intuitive and provides a convenient algorithm for solving the congruences. Given the congruences

$$x \equiv a \ (\text{mod } n) \qquad x \equiv b \ (\text{mod } m)$$

we can rewrite the first congruence as an equation in the form $x = a + qn$ for some $q \in \mathbf{Z}$. To find a simultaneous solution, we only need to substitute this expression for x in the second congruence, giving $a + qn \equiv b$ (mod m), or

$$qn \equiv b - a \ (\text{mod } m) \ .$$

Since $(n, m) = 1$, we can solve the congruence $nz \equiv 1$ (mod m), and using this solution we can solve for q in the congruence $qn \equiv b - a$ (mod m).

Recall that we converted the first congruence $x \equiv a$ (mod m) to the equation $x = a + qn$. Now that we have a value for q, we can substitute, and so this gives the simultaneous solutions to the two congruences in the form $x = a + qn$. We can choose as a particular solution the smallest positive integer in this form. The general solution is obtained by adding multiples of mn.

Example 1.3.5.

To illustrate the second method of solution, again consider the system

$$x \equiv 7 \ (\text{mod } 8) \qquad x \equiv 3 \ (\text{mod } 5) \ .$$

The first congruence gives us the equation $x = 7 + 8q$, and then substituting we obtain $7 + 8q \equiv 3$ (mod 5), or equivalently, $3q \equiv -4$ (mod 5). Multiplying by 2, since $2 \cdot 3 \equiv 1$ (mod 5), gives $q \equiv -8$ (mod 5) or $q \equiv 2$ (mod 5). This yields the particular solution $x = 7 + 2 \cdot 8 = 23$.   □

EXERCISES: SECTION 1.3

1. Solve the following congruences.

   †(a) $4x \equiv 1$ (mod 7)

   (b) $2x \equiv 1$ (mod 9)

   †(c) $5x \equiv 1$ (mod 32)

   (d) $19x \equiv 1$ (mod 36)

2. Write n as a sum of four squares for $1 \le n \le 20$.

3. Solve the following congruences.

   †(a) $10x \equiv 5$ (mod 21)

   (b) $10x \equiv 5$ (mod 15)

   †(c) $10x \equiv 4$ (mod 15)

   (d) $10x \equiv 4$ (mod 14)

4. Solve the following congruence. $20x \equiv 12$ (mod 72)

5.† Solve the following congruence. $25x \equiv 45$ (mod 60)

6. Find all integers x such that $3x + 7$ is divisible by 11.

   (New techniques are available for this problem, which was Exercise 22 in Section 1.1)

7. The smallest positive solution of the congruence $ax \equiv 0$ (mod n) is called the **additive order** of a modulo n. Find the additive orders of each of the following elements, by solving the appropriate congruences.

   †(a) 8 modulo 12

   (b) 7 modulo 12

   †(c) 21 modulo 28

   (d) 12 modulo 18

8. Prove that if p is a prime number and a is any integer such that $p \nmid a$, then the additive order of a modulo p is equal to p.

9. Prove that if $n > 1$ and $a > 0$ are integers and $d = (a, n)$, then the additive order of a modulo n is n/d.

10. Let a, b, n be positive integers. Prove that if $a \equiv b$ (mod n), then $(a, n) = (b, n)$.

11. Show that 7 is a divisor of $(6! + 1)$, 11 is a divisor of $(10! + 1)$, and 19 is a divisor of $(18! + 1)$.

12. Show that $4 \cdot (n^2 + 1)$ is never divisible by 11.

13. Prove that the sum of the cubes of any three consecutive positive integers is divisible by 9. (Compare Exercise 21 of Section 1.1.)

14. Find the units digit of $3^{29} + 11^{12} + 15$.

    Hint: Choose an appropriate modulus n, and then reduce modulo n.

15. Solve the following congruences by trial and error.

    †(a) $x^2 \equiv 1$ (mod 16)

    (b) $x^3 \equiv 1$ (mod 16)

    †(c) $x^4 \equiv 1$ (mod 16)

    (d) $x^8 \equiv 1$ (mod 16)

16. Solve the following congruences by trial and error.

    (a) $x^3 + 2x + 2 \equiv 0$ (mod 5)

    (b) $x^4 + x^3 + x^2 + x + 1 \equiv 0$ (mod 2)

    (c) $x^4 + x^3 + 2x^2 + 2x \equiv 0$ (mod 3)

17. List and solve all quadratic congruences modulo 3. That is, list and solve all congruences of the form $ax^2 + bx + c \equiv 0$ (mod 3). The only coefficients you need to consider are 0, 1, 2.

18. Solve the following system of congruences.

$$x \equiv 15 \ (\text{mod } 27) \qquad x \equiv 16 \ (\text{mod } 20)$$

19.†Solve the following system of congruences.

$$x \equiv 11 \;(\text{mod } 16) \qquad x \equiv 18 \;(\text{mod } 25)$$

20. Solve the following system of congruences.

$$2x \equiv 5 \;(\text{mod } 7) \qquad 3x \equiv 4 \;(\text{mod } 8)$$

   Hint: First reduce to the usual form.

21. Solve the following system of congruences.

$$x \equiv a \;(\text{mod } n) \qquad x \equiv b \;(\text{mod } n + 1)$$

22. Extend the techniques of the Chinese remainder theorem to solve the following system of congruences.

$$2x \equiv 3 \;(\text{mod } 7) \qquad x \equiv 4 \;(\text{mod } 6) \qquad 5x \equiv 50 \;(\text{mod } 55)$$

23. This exercise extends the Chinese remainder theorem. Let m, n be positive integers, with $(m, n) = d$ and $[m, n] = k$. Prove that the system of congruences

$$x \equiv a \;(\text{mod } n) \qquad x \equiv b \;(\text{mod } m)$$

   has a solution if and only if $a \equiv b \;(\text{mod } d)$, and in this case any two solutions are congruent modulo k.

24. (Casting out nines) Show that the remainder of an integer n when divided by 9 is the same as the remainder of the sum of its digits when divided by 9.

   Hint: For example, $7862 \equiv 7 + 8 + 6 + 2 \;(\text{mod } 9)$. How you can use the digits of 7862 to express it in terms of powers of 10?

   Note: "Casting out nines" is a traditional method for checking a sum of a long column of large numbers by reducing each of the numbers modulo 9 and checking the sum modulo 9. This exercise shows that the method is practical, because it provides a quick algorithm for reducing an integer modulo 9.

25. Find a result similar to casting out nines for the integer 11.

26. Let p be a prime number and let a, b be any integers. Prove that

$$(a + b)^p \equiv a^p + b^p \;(\text{mod } p) .$$

27. Prove that in any Pythagorean triple (a, b, c), either a or b is divisible by 3, and one of a, b, c is divisible by 5.

28. Prove that there exist infinitely many prime numbers of the form $4m + 3$ (where m is an integer).

1.4 Integers Modulo *n*

In working with congruences, we have established that in computations involving addition, subtraction, and multiplication, we can consider congruent numbers to be interchangeable. In this section we will formalize this point of view. We will now consider entire congruence classes as individual entities, and we will work with these entities much as we do with ordinary numbers. The point of introducing the notation given below is to allow us to use our experience with ordinary numbers as a guide to working with congruence classes. Most of the laws of integer arithmetic hold for the arithmetic of congruence classes. The notable exception is that the product of two nonzero congruence classes may be zero.

1.4.1 Definition. *Let a and n > 0 be integers. The set of all integers which have the same remainder as a when divided by n is called the **congruence class of** a **modulo** n, and is denoted by $[a]_n$, where*

$$[a]_n = \{x \in \mathbf{Z} \mid x \equiv a \ (mod \ n)\} \ .$$

*The collection of all congruence classes modulo n is called the **set of integers modulo** n, denoted by $\mathbf{Z}_n$.*

Note that $[a]_n = [b]_n$ if and only if $a \equiv b \pmod{n}$. When the modulus is clearly understood from the context, the subscript n can be omitted and $[a]_n$ can be written simply as $[a]$.

A given congruence class can be denoted in many ways. For example, $x \equiv 5 \pmod 3$ if and only if $x \equiv 8 \pmod 3$, since $5 \equiv 8 \pmod 3$. This shows that $[5]_3 = [8]_3$. We sometimes say that an element of $[a]_n$ is a **representative of the congruence class**. Each congruence class $[a]_n$ has a unique nonnegative representative that is smaller than n, namely, the remainder when a is divided by n. This shows that there are exactly n distinct congruence classes modulo n. For example, the congruence classes modulo 3 can be represented by 0, 1, and 2.

$$
\begin{aligned}
[0]_3 &= \{\ldots, -9, -6, -3, 0, 3, 6, 9, \ldots\} \\
[1]_3 &= \{\ldots, -8, -5, -2, 1, 4, 7, 10, \ldots\} \\
[2]_3 &= \{\ldots, -7, -4, -1, 2, 5, 8, 11, \ldots\}
\end{aligned}
$$

Each integer belongs to exactly one congruence class modulo 3, since the remainder on division by 3 is unique. In general, each integer belongs to a unique congruence class modulo n. Hence we have

$$\mathbf{Z}_n = \{[0]_n, [1]_n, \ldots, [n-1]_n\} \ .$$

The set $\mathbf{Z}_2$ consists of $[0]_2$ and $[1]_2$, where $[0]_2$ is the set of even numbers and $[1]_2$ is the set of odd numbers. With the new notation, the familiar rules "even + even = even,"   "odd + even = odd,"   "odd + odd = even"

can be expressed as

$$[0]_2 + [0]_2 = [0]_2 , \qquad [1]_2 + [0]_2 = [1]_2 , \qquad [1]_2 + [1]_2 = [0]_2 .$$

Similarly,

"even × even = even,"   "even × odd = even,"   "odd × odd = odd"

can be expressed as

$$[0]_2 \cdot [0]_2 = [0]_2 , \qquad [0]_2 \cdot [1]_2 = [0]_2 , \qquad [1]_2 \cdot [1]_2 = [1]_2 .$$

These rules can be summarized by giving an addition table and a multiplication table (Table 1.4.1).

Table 1.4.1: Addition and Multiplication in $\mathbf{Z}_2$

+	[0]	[1]		·	[0]	[1]
[0]	[0]	[1]		[0]	[0]	[0]
[1]	[1]	[0]		[1]	[0]	[1]

To use the addition table, select an element a from the first column, and an element b from the top row. Read from left to right in the row to which a belongs, until reaching the column to which b belongs. The corresponding entry in the table is $a + b$. In this table, as we will sometimes do elsewhere, we have simplified our notation for congruence classes by omitting the subscript in $[a]_n$.

A similar addition and multiplication can be introduced in $\mathbf{Z}_n$, for any n. Given congruence classes in $\mathbf{Z}_n$, we add (or multiply) them by picking representatives of each congruence class. We then add (or multiply) the representatives, and find the congruence class to which the result belongs. This can be written formally as follows.

Addition: $[a]_n + [b]_n = [a + b]_n$

Multiplication: $[a]_n \cdot [b]_n = [ab]_n$

In $\mathbf{Z}_{12}$, for example, we have $[8]_{12} = [20]_{12}$ and $[10]_{12} = [34]_{12}$. Adding congruence classes gives the same answer, no matter which representatives we use: $[8]_{12} + [10]_{12} = [18]_{12} = [6]_{12}$ and also $[20]_{12} + [34]_{12} = [54]_{12} = [6]_{12}$.

1.4.2 Proposition. *Let n be a positive integer, and let a, b be any integers. Then the addition and multiplication of congruence classes given below are well-defined:*

$$[a]_n + [b]_n = [a+b]_n \ , \qquad\qquad [a]_n \cdot [b]_n = [ab]_n \ .$$

Proof. We must show that the given formulas do not depend on the integers a and b which have been chosen to represent the congruence classes with which we are concerned. Suppose that x and y are any other representatives of the congruence classes $[a]_n$ and $[b]_n$, respectively. Then $x \equiv a \pmod{n}$ and $y \equiv b \pmod{n}$, and so we can apply Proposition 1.3.3. It follows from that proposition that $x + y \equiv a + b \pmod{n}$ and $xy \equiv ab \pmod{n}$, and thus we have $[x]_n + [y]_n = [a+b]_n$ and $[x]_n \cdot [y]_n = [ab]_n$. Since the formulas we have given do not depend on the particular representatives chosen, we say that addition and multiplication are "well-defined." $\square$

The familiar rules for addition and multiplication carry over from the addition and multiplication of integers. A complete discussion of these rules will be given in Chapter 5, when we study ring theory. If $[a]_n, [b]_n \in \mathbf{Z}_n$ and $[a]_n + [b]_n = [0]_n$, then $[b]_n$ is called an **additive inverse** of $[a]_n$. By Proposition 1.3.3 (b), additive inverses are unique. We will denote the additive inverse of $[a]_n$ by $-[a]_n$. It is easy to see that $-[a]_n$ is in fact equal to $[-a]_n$, since $[a]_n + [-a]_n = [a-a]_n = [0]_n$.

For any elements $[a]_n, [b]_n, [c]_n$ in $\mathbf{Z}_n$, the following laws hold.

Associativity:
$$([a]_n + [b]_n) + [c]_n = [a]_n + ([b]_n + [c]_n)$$
$$([a]_n \cdot [b]_n) \cdot [c]_n = [a]_n \cdot ([b]_n \cdot [c]_n)$$

Commutativity:
$$[a]_n + [b]_n = [b]_n + [a]_n$$
$$[a]_n \cdot [b]_n = [b]_n \cdot [a]_n$$

Distributivity:
$$[a]_n \cdot ([b]_n + [c]_n) = [a]_n \cdot [b]_n + [a]_n \cdot [c]_n$$

Identities:
$$[a]_n + [0]_n = [a]_n$$
$$[a]_n \cdot [1]_n = [a]_n$$

Additive inverses:
$$[a]_n + [-a]_n = [0]_n$$

We will give a proof of the distributive law and leave the proofs of the remaining properties as an exercise. If $a, b, c \in \mathbf{Z}$, then

$$\begin{aligned}
[a]_n \cdot ([b]_n + [c]_n) &= [a]_n \cdot ([b+c]_n) = [a(b+c)]_n \\
&= [ab+ac]_n = [ab]_n + [ac]_n \\
&= [a]_n \cdot [b]_n + [a]_n \cdot [c]_n \ .
\end{aligned}$$

The steps in the proof depend on the definitions of addition and multiplication and the equality $a(b + c) = ab + ac$, which is the distributive law for $\mathbf{Z}$.

In doing computations in $\mathbf{Z}_n$, the one point at which particular care must be taken is the cancellation law, which no longer holds in general. Otherwise, in almost all cases your experience with integer arithmetic can be trusted when working with congruence classes. A quick computation shows that $[6]_8 \cdot [5]_8 = [6]_8 \cdot [1]_8$, but $[5]_8 \neq [1]_8$. It can also happen that the product of nonzero classes is equal to zero. For example, $[6]_8 \cdot [4]_8 = [0]_8$.

1.4.3 Definition. *If $[a]_n$ belongs to $\mathbf{Z}_n$, and $[a]_n[b]_n = [0]_n$ for some nonzero congruence class $[b]_n$, then $[a]_n$ is called a **divisor of zero**.*

If $[a]_n$ is not a divisor of zero, then in the equation $[a]_n[b]_n = [a]_n[c]_n$ we may cancel $[a]_n$, to get $[b]_n = [c]_n$. To see this, if $[a]_n[b]_n = [a]_n[c]_n$, then $[a]_n([b]_n - [c]_n) = [a]_n[b - c]_n = [0]_n$, and so $[b]_n - [c]_n$ must be zero since $[a]_n$ is not a divisor of zero. This shows that $[b]_n = [c]_n$.

1.4.4 Definition. *If $[a]_n$ belongs to $\mathbf{Z}_n$, and $[a]_n[b]_n = [1]_n$, for some congruence class $[b]_n$, then $[b]_n$ is called a **multiplicative inverse** of $[a]_n$ and is denoted by $[a]_n^{-1}$.*

*In this case, we say that $[a]_n$ is an **invertible** element of $\mathbf{Z}_n$, or a **unit** of $\mathbf{Z}_n$.*

The next proposition (which is just a restatement of Proposition 1.3.4) shows that a has a multiplicative inverse modulo n if and only if $(a, n) = 1$. When a satisfies this condition, it follows from Proposition 1.3.3 (b) that any two solutions to $ax \equiv 1 \pmod{n}$ are congruent modulo n, and so we are justified in referring to *the* multiplicative inverse of $[a]_n$, whenever it exists.

In $\mathbf{Z}_7$, each nonzero congruence class contains representatives which are relatively prime to 7, and so each nonzero congruence class has a multiplicative inverse. We can list them as $[1]_7^{-1} = [1]_7$, $[2]_7^{-1} = [4]_7$, $[3]_7^{-1} = [5]_7$, and $[6]_7^{-1} = [6]_7$. We did not need to list $[4]_7^{-1}$ and $[5]_7^{-1}$ since, in general, if $[a]_n^{-1} = [b]_n$, then $[b]_n^{-1} = [a]_n$.

From this point on, if the meaning is clear from the context we will omit the subscript on congruence classes. Using this convention in $\mathbf{Z}_n$, we note that if $[a]$ has a multiplicative inverse, then it cannot be a divisor of zero, since $[a][b] = [0]$ implies $[b] = [a]^{-1}([a][b]) = [a]^{-1}[0] = [0]$.

1.4.5 Proposition. *Let n be a positive integer.*

   (a) *The congruence class $[a]_n$ has a multiplicative inverse in $\mathbf{Z}_n$ if and only if $(a, n) = 1$.*

   (b) *A nonzero element of $\mathbf{Z}_n$ either has a multiplicative inverse or is a divisor of zero.*

Proof. (a) If $[a]$ has a multiplicative inverse, say $[a]^{-1} = [b]$, then $[a][b] = [1]$. Therefore $ab \equiv 1 \pmod{n}$, which implies that $ab = 1 + qn$ for some integer q. Thus $ab + (-q)n = 1$, and so $(a, n) = 1$.

Conversely, if $(a, n) = 1$, then there exist integers b and q such that $ab + qn = 1$. Reducing modulo n shows that $ab \equiv 1 \pmod{n}$, and so $[b] = [a]^{-1}$.

(b) Assume that a represents a nonzero congruence class, so that $n \nmid a$. If $(a, n) = 1$, then $[a]$ has a multiplicative inverse. If not, then $(a, n) = d$, where $1 < d < n$. In this case, since $d \mid n$ and $d \mid a$, we can find integers k, b with $n = kd$ and $a = bd$. Then $[k]$ is a nonzero element of $\mathbf{Z}_n$, but

$$[a][k] = [ak] = [bdk] = [bn] = [0],$$

which shows that $[a]$ is a divisor of zero. $\square$

1.4.6 Corollary. *The following conditions on the modulus $n > 0$ are equivalent.*

(1) *The number n is prime.*

(2) $\mathbf{Z}_n$ *has no divisors of zero, except $[0]_n$.*

(3) *Every nonzero element of $\mathbf{Z}_n$ has a multiplicative inverse.*

Proof. Since n is prime if and only if every positive integer less than n is relatively prime to n, Corollary 1.4.6 follows from Proposition 1.4.5. $\square$

The proof of Proposition 1.4.5 (a) shows that if $(a, n) = 1$, then the multiplicative inverse of $[a]$ can be computed by using the Euclidean algorithm.

Example 1.4.1.

For example, to find $[11]^{-1}$ in $\mathbf{Z}_{16}$ using the matrix form of the Euclidean algorithm (see the discussion preceding Example 1.1.5) we have the following computation:

$$\begin{bmatrix} 1 & 0 & 16 \\ 0 & 1 & 11 \end{bmatrix} \rightsquigarrow \begin{bmatrix} 1 & -1 & 5 \\ 0 & 1 & 11 \end{bmatrix}$$

$$\begin{bmatrix} 1 & -1 & 5 \\ -2 & 3 & 1 \end{bmatrix} \rightsquigarrow \begin{bmatrix} 11 & -16 & 0 \\ -2 & 3 & 1 \end{bmatrix}.$$

Thus $16(-2) + 11 \cdot 3 = 1$, which shows that $[11]_{16}^{-1} = [3]_{16}$.

When the numbers are small, as in this case, it is often easier to use trial and error. The positive integers less than 16 and relatively prime to 16 are $1, 3, 5, 7, 9, 11, 13, 15$. It is easier to use the representatives $\pm 1, \pm 3, \pm 5, \pm 7$ since if $[a][b] = [1]$, then $[-a][-b] = [1]$, and so $[-a]^{-1} = -[a]^{-1}$. Now we observe that $3 \cdot 5 = 15 \equiv -1 \pmod{16}$, so $3(-5) \equiv 1 \pmod{16}$. Thus $[3]_{16}^{-1} = [-5]_{16} = [11]_{16}$ and $[-3]_{16}^{-1} = [5]_{16}$. Finally, $7 \cdot 7 \equiv 1 \pmod{16}$, so $[7]_{16}^{-1} = [7]_{16}$ and $[-7]_{16}^{-1} = [-7]_{16} = [9]_{16}$. $\square$

Another way to find the inverse of an element $[a] \in \mathbf{Z}_n$ is to take successive powers of $[a]$. If $(a, n) = 1$, then $[a]$ is not a zero divisor, and so no power of $[a]$ can be zero. We let $[a]^0 = [1]$. The set of powers $[1], [a], [a]^2, [a]^3, \ldots$ must contain fewer than n distinct elements, so after some point there must be a repetition. Suppose that the first repetition occurs for the exponent m, say $[a]^m = [a]^k$, with $k < m$. Then $[a]^{m-k} = [a]^0 = [1]$ since we can cancel $[a]$ from both sides a total of k times. This shows that for the first repetition we must have had $k = 0$, so actually $[a]^m = [1]$. From this we can see that $[a]^{-1} = [a]^{m-1}$.

Example 1.4.2.

To find $[11]_{16}^{-1}$, we can list the powers of $[11]_{16}$. We have $[11]^2 = [-5]^2 = [9]$, $[11]^3 = [11]^2[11] = [99] = [3]$, and $[11]^4 = [11]^3[11] = [33] = [1]$. Thus again we see that $[11]_{16}^{-1} = [3]_{16}$. □

We are now ready to continue our study of equations in $\mathbf{Z}_n$. A linear congruence of the form $ax \equiv b \pmod{n}$ can be viewed as a linear equation $[a]_n[x]_n = [b]_n$ in $\mathbf{Z}_n$. If $[a]_n$ has a multiplicative inverse, then there is a unique congruence class $[x]_n = [a]_n^{-1}[b]_n$ that is the solution to the equation. Without the notation for congruence classes we would need to modify the statement regarding uniqueness to say that if x_0 is a solution of $ax \equiv b \pmod{n}$, then so is $x_0 + qn$, for any integer q.

It is considerably harder to solve nonlinear congruences of the form $a_k x^k + \ldots + a_1 x + a_0 \equiv 0 \pmod{n}$, where $a_k, \ldots, a_0 \in \mathbf{Z}$. It can be shown that in solving congruences modulo n of degree greater than or equal to 1, the problem reduces to solving congruences modulo p^α for the prime factors of n. This question is usually addressed in a course on elementary number theory, where the Chinese remainder theorem is used to show how to determine the solutions modulo a prime power p^α (for integers $\alpha \geq 2$) from the solutions modulo p. Then to determine the solutions modulo p we can proceed by trial and error, simply substituting each of $0, 1, \ldots, p - 1$ into the congruence. Fermat's theorem (Corollary 1.4.12) can be used to reduce the problem to considering polynomials of degree at most $p - 1$.

We will prove this theorem of Fermat as a special case of a more general theorem due to Euler. Another proof will also be given in Section 3.2, which takes advantage of the concepts we will have developed by then. The statement of Euler's theorem involves a function of paramount importance in number theory and algebra, which we now introduce.

1.4.7 Definition. *Let n be a positive integer. The number of positive integers less than or equal to n which are relatively prime to n will be denoted by $\varphi(n)$. This function is called* **Euler's φ-function,** *or the* **totient function.**

In Section 1.2 we gave a procedure for listing the positive integers less than n and relatively prime to n. However, in many cases we only need to determine the numerical value of $\varphi(n)$, without actually listing the numbers themselves. With the formula in Proposition 1.4.8, $\varphi(n)$ can be given in terms of the prime factorization of n. Note that $\varphi(1) = 1$.

1.4.8 Proposition. *If the prime factorization of n is $n = p_1^{\alpha_1} p_2^{\alpha_2} \cdots p_k^{\alpha_k}$, where $\alpha_i > 0$ for $1 \leq i \leq k$, then*

$$\varphi(n) = n \left(1 - \frac{1}{p_1}\right) \left(1 - \frac{1}{p_2}\right) \cdots \left(1 - \frac{1}{p_k}\right).$$

Proof. See Exercises 17, 29, and 30. A proof of this result will also be presented in Section 3.5. $\square$

Example 1.4.3.

Using the formula in Proposition 1.4.8, we have

$$\varphi(10) = 10 \left(\frac{1}{2}\right) \left(\frac{4}{5}\right) = 4 \quad \text{and} \quad \varphi(36) = 36 \left(\frac{1}{2}\right) \left(\frac{2}{3}\right) = 12 . \ \square$$

1.4.9 Definition. *The set of units of $\mathbf{Z}_n$, the congruence classes $[a]$ such that $(a, n) = 1$, will be denoted by $\mathbf{Z}_n^{\times}$.*

1.4.10 Proposition. *The set $\mathbf{Z}_n^{\times}$ of units of $\mathbf{Z}_n$ is closed under multiplication.*

Proof. This can be shown either by using Proposition 1.2.3 (d) or by using the formula $([a][b])^{-1} = [b]^{-1}[a]^{-1}$. $\square$

The number of elements of $\mathbf{Z}_n^{\times}$ is given by $\varphi(n)$. The next theorem should be viewed as a result on powers of elements in $\mathbf{Z}_n^{\times}$, although it is phrased in the more familiar congruence notation.

1.4.11 Theorem (Euler). *If $(a, n) = 1$, then $a^{\varphi(n)} \equiv 1 \ (mod \ n)$.*

Proof. In the set $\mathbf{Z}_n$, there are $\varphi(n)$ congruence classes which are represented by an integer relatively prime to n. Let these representatives be $\{a_1, \ldots, a_{\varphi(n)}\}$. For the given integer a, consider the congruence classes represented by the products $\{aa_1, \ldots, aa_{\varphi(n)}\}$. By Proposition 1.3.3 (b) these are all distinct because $(a, n) = 1$.

Since each of the products is still relatively prime to n, we must have a representative from each of the $\varphi(n)$ congruence classes we started with. Therefore

$$a_1 a_2 \cdots a_{\varphi(n)} \equiv (aa_1)(aa_2) \cdots (aa_{\varphi(n)}) \equiv a^{\varphi(n)} a_1 a_2 \cdots a_{\varphi(n)} \pmod{n}.$$

Since the product $a_1 \cdots a_{\varphi(n)}$ is relatively prime to n, we can cancel it in the congruence

$$a_1 a_2 \cdots a_{\varphi(n)} \equiv a^{\varphi(n)} a_1 a_2 \cdots a_{\varphi(n)} \pmod{n},$$

and so we have $a^{\varphi(n)} \equiv 1 \pmod{n}$. $\square$

1.4.12 Corollary (Fermat). *If p is a prime number, then for any integer a we have* $a^p \equiv a \pmod{p}$.

Proof. If $p \mid a$, then trivially $a^p \equiv a \equiv 0 \pmod{p}$. If $p \nmid a$, then $(a, p) = 1$ and Euler's theorem gives $a^{\varphi(p)} \equiv 1 \pmod{p}$. Then since $\varphi(p) = p - 1$, we have $a^p \equiv a \pmod{p}$. $\square$

It is instructive to include another proof of Fermat's "little" theorem, one that does not depend on Euler's theorem. Expanding $(a + b)^p$ we obtain

$$(a + b)^p = a^p + pa^{p-1}b + \frac{p(p - 1)}{1 \cdot 2} a^{p-2}b^2 + \ldots + pab^{p-1} + b^p.$$

For $k \neq 0, k \neq p$, each of the coefficients

$$\frac{p!}{k!(p - k)!}$$

is an integer and has p as a factor, since p is a divisor of the numerator but not the denominator. Therefore

$$(a + b)^p \equiv a^p + b^p \pmod{p}.$$

Using induction, this can be extended to more terms, giving $(a + b + c)^p \equiv a^p + b^p + c^p \pmod{p}$, etc. Writing a as $(1 + 1 + \ldots + 1)$ shows that

$$a^p = (1 + 1 + \ldots + 1)^p \equiv 1^p + \ldots + 1^p \equiv a \pmod{p}.$$

As a final remark we note that if $(a, n) = 1$, then the multiplicative inverse of $[a]_n$ can be given explicitly as $[a]_n^{\varphi(n)-1}$, since by Euler's theorem, $a \cdot a^{\varphi(n)-1} \equiv 1 \pmod{n}$. Note also that for a given n the exponent $\varphi(n)$ in Euler's theorem may not be the smallest exponent possible. For example, in $\mathbf{Z}_8$ the integers $\pm 1, \pm 3$, are relatively prime to 8, and Euler's theorem states that $a^4 \equiv 1 \pmod{8}$ for each of these integers. In fact, $a^2 \equiv 1 \pmod{8}$ for $a = \pm 1, \pm 3$.

EXERCISES: SECTION 1.4

1. Make addition and multiplication tables for the following sets.

    (a) $\mathbf{Z}_3$

    (b) $\mathbf{Z}_4$

    †(c) $\mathbf{Z}_{12}$

2. Make multiplication tables for the following sets.

    (a) $\mathbf{Z}_6$

    (b) $\mathbf{Z}_7$

    (c) $\mathbf{Z}_8$

3. Find the multiplicative inverses of the given elements (if possible).

    †(a) [14] in $\mathbf{Z}_{15}$

    (b) [38] in $\mathbf{Z}_{83}$

    †(c) [351] in $\mathbf{Z}_{6669}$

    (d) [91] in $\mathbf{Z}_{2565}$

4. Let a and b be integers.

    (a) Prove that $[a]_n = [b]_n$ if and only if $a \equiv b \pmod{n}$.

    (b) Prove that either $[a]_n \cap [b]_n = \emptyset$ or $[a]_n = [b]_n$.

5. Prove that each congruence class $[a]_n$ in $\mathbf{Z}_n$ has a unique representative r that satisfies $0 \le r < n$.

6. Let m and n be positive integers such that $m \mid n$. Show that for any integer a, the congruence class $[a]_m$ is the union of the congruence classes $[a]_n$, $[a+m]_n$, $[a+2m]_n$, ..., $[a + n - m]_n$.

7. Prove that the associative and commutative laws hold for addition and multiplication of congruence classes, as defined in Proposition 1.4.2.

8. Use Proposition 1.3.3 (b) to show that if $[b]$ and $[c]$ are both multiplicative inverses of $[a]$ in $\mathbf{Z}_n$, then $b \equiv c \pmod{n}$.

9. Let $(a, n) = 1$. The smallest positive integer k such that $a^k \equiv 1 \pmod{n}$ is called the **multiplicative order** of $[a]$ in $\mathbf{Z}_n^\times$.

    †(a) Find the multiplicative orders of [5] and [7] in $\mathbf{Z}_{16}^\times$.

    (b) Find the multiplicative orders of [2] and [5] in $\mathbf{Z}_{17}^\times$.

10. Let $(a, n) = 1$. If $[a]$ has multiplicative order k in $\mathbf{Z}_n^\times$, show that $k \mid \varphi(n)$.

11.† In $\mathbf{Z}_9^\times$ each element is equal to a power of [2]. (Verify this.) Can you find a congruence class in $\mathbf{Z}_8^\times$ such that each element of $\mathbf{Z}_8^\times$ is equal to some power of that class? Answer the same question for $\mathbf{Z}_7^\times$.

12. Generalizing Exercise 11, we say that the set of units $\mathbf{Z}_n^{\times}$ of $\mathbf{Z}_n$ is **cyclic** if it has an element of multiplicative order $\varphi(n)$. Show that $\mathbf{Z}_{10}^{\times}$ and $\mathbf{Z}_{11}^{\times}$ are cyclic, but $\mathbf{Z}_{12}^{\times}$ is not.

13. An element $[a]$ of $\mathbf{Z}_n$ is said to be **idempotent** if $[a]^2 = [a]$.

    †(a) Find all idempotent elements of $\mathbf{Z}_6$ and $\mathbf{Z}_{12}$.

    (b) Find all idempotent elements of $\mathbf{Z}_{10}$ and $\mathbf{Z}_{30}$.

14. If p is a prime number, show that $[0]$ and $[1]$ are the only idempotent elements in $\mathbf{Z}_p$.

15. If n is not a prime power, show that $\mathbf{Z}_n$ has an idempotent element different from $[0]$ and $[1]$.

    Hint: Suppose that $n = bc$, with $(b, c) = 1$. Solve the simultaneous congruences $x \equiv 1 \pmod{b}$ and $x \equiv 0 \pmod{c}$.

16. An element $[a]$ of $\mathbf{Z}_n$ is said to be **nilpotent** if $[a]^k = [0]$ for some k. Show that $\mathbf{Z}_n$ has no nonzero nilpotent elements if and only if n has no factor that is a square (except 1).

17. Using the formula for $\varphi(n)$, compute $\varphi(27)$, $\varphi(81)$, and $\varphi(p^\alpha)$, where p is a prime number. Give a proof that the formula for $\varphi(n)$ is valid when $n = p^\alpha$, where p is a prime number.

18. Show that if a and b are positive integers such that $a \mid b$, then $\varphi(a) \mid \varphi(b)$.

19. Find all integers $n > 1$ such that $\varphi(n) = 2$.

20. Show that $\varphi(1) + \varphi(p) + \ldots + \varphi(p^\alpha) = p^\alpha$ for any prime number p and any positive integer α.

21. Show that if $n > 2$, then $\varphi(n)$ is even.

22. For $n = 12$ show that $\sum_{d \mid n} \varphi(d) = n$. Do the same for $n = 18$.

23. Show that if $n > 1$, then the sum of all positive integers less than n and relatively prime to n is $n\varphi(n)/2$. That is, $\sum_{0 < a < n,\ (a,n)=1} a = n\varphi(n)/2$.

24. Show that if p is a prime number, then the congruence $x^2 \equiv 1 \pmod{p}$ has only the solutions $x \equiv 1$ and $x \equiv -1$.

25. Let a, b be integers, and let p be a prime number of the form $p = 2k + 1$. Show that if $p \nmid a$ and $a \equiv b^2 \pmod{p}$, then $a^k \equiv 1 \pmod{p}$.

26. Let $p = 2k + 1$ be a prime number. Show that if a is an integer such that $p \nmid a$, then either $a^k \equiv 1 \pmod{p}$ or $a^k \equiv -1 \pmod{p}$.

27. Prove Wilson's theorem, which states that if p is a prime number, then $(p - 1)! \equiv -1 \pmod{p}$.

    Hint: $(p-1)!$ is the product of all elements of $\mathbf{Z}_p^{\times}$. Pair each element with its inverse, and use Exercise 24. For three special cases see Exercise 11 in Section 1.3.

28. Prove that if $(m, n) = 1$, then $n^{\varphi(m)} + m^{\varphi(n)} \equiv 1 \pmod{mn}$.

29. Prove that if m, n are positive integers with $(m, n) = 1$, then $\varphi(mn) = \varphi(m)\varphi(n)$.

    Hint: Use the Chinese remainder theorem to show that each pair of elements $[a]_m$ and $[b]_n$ (in $\mathbf{Z}_m$ and $\mathbf{Z}_n$ respectively) corresponds to a unique element $[x]_{mn}$ in $\mathbf{Z}_{mn}$. Then show that under this correspondence, $[a]$ and $[b]$ are units if and only if $[x]$ is a unit.

30. Use Exercise 17 and Exercise 29 to prove Proposition 1.4.8.

Notes

The prime numbers are the basic the basic building blocks in number theory, since every positive integer can be written (essentially uniquely) as a product of prime numbers. (If you are reading this before studying the chapter, perhaps we need to remind you that an integer $p > 1$ is called prime if its only positive divisors are 1 and p.) Euclid considered primes and proved that there are infinitely many. When we look at the sequence of primes

$$2, \ 3, \ 5, \ 7, \ 11, \ 13, \ 17, \ 19, \ 23, \ 29, \ 31, \ \ldots$$

we observe that except for 2, all primes are odd. Any two odd primes on the list must differ by at least 2, but certain pairs of "twin primes" that differ by the minimal amount 2 do appear, for example,

$$(3, 5), \ (5, 7), \ (11, 13), \ (17, 19), \ (29, 31), \ (41, 43), \ldots.$$

Are there infinitely many "twin prime" pairs? The answer to this innocent question is unknown.

Although any positive integer is a product of primes, what about sums? Another open question is attributed to Christian Goldbach (1690–1764). He asked whether every even integer greater than 2 can be written as the sum of two primes. (Since the sum of two odd primes is even, the only way to write an odd integer as a sum of two primes is to use an odd prime added to 2. That means that the only odd primes that can be represented as a sum of two primes are the ones that occur as the larger prime in a pair of "twin primes.") We invite you to experiment in writing some even integers as sums of two primes.

A beautiful theorem proved by Joseph Louis Lagrange (1736–1813) in 1770 states that every positive integer can be written as the sum of four squares (where an integer of the form n^2 is called a square). Could we get by with fewer than four squares? The answer is no; try representing 7 as a sum of three squares. This naturally leads to the question of which positive integers can be written as the sum of three squares. The answer is that n can be written as a sum of three squares if and only if n is not of the form $4^m(8k+7)$, where m, k are any nonnegative integers. This theorem was first correctly proved by Gauss and appears in his famous book *Disquisitiones Arithmeticae* (1801).

This raises the question of which positive integers can be written as the sum of two squares. The answer in this case is slightly more complicated. It is that n can be written as the sum of two squares if and only if when we factor n as a product of primes, all those primes that give a remainder of 3 upon division by 4 have even exponents. The first published proof of this fact (dating from 1749) is due to Leonhard Euler (1707–1783). Around 1640 Pierre de Fermat (1601–1665) had stated, without proof, all three of these theorems on the representation of n as a sum of squares.

Our fourth and final topic deals with another statement of Fermat, usually known as "Fermat's last theorem." The ancient Greeks (the Pythagoreans, in particular) knew that certain triples (x, y, z) of nonzero integers can satisfy the equation

$$x^2 + y^2 = z^2 ,$$

for example,

$$(3, 4, 5), \ \ (5, 12, 13), \ \ (8, 15, 17), \ \ (7, 24, 25), \ldots .$$

(See Exercise 25 of Section 1.2.) Fermat considered a generalization of this equation, and asked whether for any integer $n > 2$ there exists a triple (x, y, z) such that

$$x^n + y^n = z^n .$$

In the margin of his copy of a number theory text he stated that he had a wonderful proof that there exists no such triple for $n \geq 3$, but he went on to say that the margin was not wide enough to write it out. His assertion dates from 1637, and mathematicians have spent the last 350 years searching for a proof! Finally, in 1993, Andrew Wiles announced that he had completed the proof of "Fermat's last theorem." A gap was found in his initial proof, but within a year, Wiles, with the assistance of Richard Taylor, found a way to complete the proof. A long paper by Wiles, together with a shorter one by Taylor and Wiles, fill the May, 1995 issue of the *Annals of Mathematics*. This proof will stand as one of the major accomplishments of our time.

Fermat is clearly the first truly modern number theorist, and he deserves much of the credit for the subject as we know it today. Another important milestone in modern number theory is Gauss's *Disquisitiones Arithmeticae*, which changed number theory from a "hodge-podge" of results into a coherent subject. The material on congruences in Section 1.3 first appeared there, and contributed much to the systematic organization of number theory.

Chapter 2

FUNCTIONS

In studying mathematical objects, we need to develop ways of classifying them, and to do this, we must have various methods for comparing them. Since we will be studying algebraic objects which usually consist of a set together with additional structure, functions provide the most important means of comparison. In this chapter we will study functions in preparation for their later use, when we will utilize functions that preserve the relevant algebraic structure. Recall that in linear algebra the appropriate functions to work with are those that preserve scalar multiplication and vector addition, namely, linear transformations.

To give an example of such a comparison, we point out that the multiplicative structure of the set of all positive real numbers is algebraically similar to the additive structure of the set of all real numbers. The justification for this statement lies in the existence of the log and exponential functions, which provide one-to-one correspondences between the two sets and convert multiplication to addition (and back again).

When we need to show that two structures are essentially the same, we will need to use one-to-one correspondences. As a second example, consider the set

$$\mathbf{Z}_4 = \{[0]_4, [1]_4, [2]_4, [3]_4\}$$

using addition of congruence classes, and the set

$$\mathbf{Z}_5^\times = \{[1]_5, [2]_5, [3]_5, [4]_5\}$$

using multiplication of congruence classes. Although we are using addition in one case and multiplication in the other, there are clear similarities. In $\mathbf{Z}_4$ each congruence class can be written as a sum of $[1]_4$'s; in $\mathbf{Z}_5^\times$ each congruence class can be written as a product of $[2]_5$'s. We will see in Chapter 3 that this makes the two algebraic structures essentially the same.

To make the idea of similarity precise, we will work with one-to-one correspondences between the relevant structures. We will restrict ourselves to one-to-one correspondences that preserve the algebraic aspects of the structures. In the above example, if $[1]_4$ corresponds to $[2]_5$ and sums correspond to products, then we must have $[1]_4 \leftrightarrow [2]_5$, $[2]_4 \leftrightarrow [4]_5$, $[3]_4 \leftrightarrow [3]_5$ and $[0]_4 \leftrightarrow [1]_5$.

We can also obtain useful information from functions that are not necessarily one-to-one correspondences. In the introduction to Chapter 1 we discussed the problem of investigating the powers of the matrix $A = \begin{bmatrix} 0 & 1 \\ -1 & 0 \end{bmatrix}$. It is useful to consider the exponential function $f(n) = A^n$ that assigns to each integer n the corresponding power of the matrix A. The rules for exponents of matrices involve the relationship between addition of integers and taking powers of the matrix and show that the function respects the inherent algebraic structure. If we collect together the exponents that yield equal powers of A, we obtain the congruence classes of integers modulo 4. In fact, this function determines a natural one-to-one correspondence between $\mathbf{Z}_4$ and the powers of A.

In working with particular mathematical objects it is often useful to consider distinct objects to be essentially the same, just for the immediate purpose. In the previous example, we can consider two exponents to be the same if the corresponding powers of A are equal. In studying the Euclidean plane it is useful to consider all triangles that are congruent to each other to be "essentially the same." In Chapter 1 we studied the notion of congruence modulo n, where we did not differentiate between integers that had the same remainder on division by n. That led us to construct sets $\mathbf{Z}_n$ from the set of integers $\mathbf{Z}$.

This idea of collecting together similar objects leads to the important notion of a partition, or what amounts to the same thing—an equivalence relation. We partition a set into subsets consisting of those objects that we do not wish to distinguish from each other, and then we say that two objects belonging to the same subset of the partition are equivalent. Thus equivalence relations enable us to study objects by making distinctions between them that are no finer than those needed for the purpose at hand. Of course, for some other purpose we might need to preserve some other distinction that is lost in the equivalence relation we are using. In that case we would resort to a different equivalence relation, in which the subsets in the partition were smaller.

The notion of a one-to-one correspondence, when expressed in the concept of a permutation, can be used to describe symmetry of geometric objects and also symmetry in other situations. For example, the symmetry inherent in a square or equilateral triangle can be expressed in terms of various permutations of its vertices. Our study of permutations in this chapter will provide the motivation for a number of important concepts in later chapters.

We should emphasize that in this chapter we are developing important parts of our mathematical language which will be used throughout the remainder of the book. The reader needs to be thoroughly familiar with these ideas.

2.1 Functions

The concept of a function should already be familiar from the calculus and linear algebra courses which we assume as a prerequisite for this book. At that level, it is standard to define a function f from a set S of real numbers into a set T of real numbers to be a "rule" that assigns to each real number x in S a unique real number y in T. An example would be the function $f(x)$ given by the rule $f(x) = x^2 + 3$, where f assigns to 1 the value 4, to 2 the value 7, and so on.

The graph of the function $f(x) = x^2 + 3$ is the set of points in the real plane described by $\{(x, y) \mid y = x^2 + 3\}$. Using the "rule" definition of a function recalled in the previous paragraph, a set of points in the plane is the graph of a function from the set of all real numbers into the set of all real numbers if and only if for each real number x there is a unique number y such that (x, y) belongs to the set. In a calculus course this is often expressed by saying that a set of points in the plane is the graph of a function if and only if every vertical line intersects the set in exactly one point.

In our development, we have chosen to take the concepts of set and element of a set as primitive (undefined) ideas. See Section A.1 of the appendix for a quick review of some basic set theory. The approach we will take is to define functions in terms of sets, and so we will do this by identifying a function with its graph.

It is convenient to introduce some notation for the sets we will be using. The symbol **R** will be used to denote the set of all real numbers. We will leave the precise development of the real numbers to a course in advanced calculus and simply view them as the set of all decimal numbers. They can be viewed as coordinates of points on a straight line, as in an introductory calculus course.

We will use the symbol **Q** to denote the set of ratios of integers, or **rational numbers**; that is

$$\mathbf{Q} = \left\{ \frac{m}{n} \,\middle|\, m, n \in \mathbf{Z} \text{ and } n \neq 0 \right\}$$

where we must agree that m/n and p/q represent the same ratio if $mq = np$. Of course, we can view the set of integers **Z** as a subset of **Q** by identifying the integer m with the fraction $m/1$. The rational numbers can be viewed as a subset of **R**, since fractions correspond to either terminating or repeating decimals.

The set $\mathbf{C} = \{a + bi \mid a, b \in \mathbf{R} \text{ and } i^2 = -1\}$ is called the set of **complex numbers**. Addition and multiplication of complex numbers are defined as follows:

$$(a + bi) + (c + di) = (a + c) + (b + d)i \,,$$

$$(a + bi)(c + di) = (ac - bd) + (ad + bc)i \,.$$

Note that $a + bi = c + di$ if and only if $a = c$ and $b = d$. See Section A.5 of the appendix for more details on the properties of complex numbers, which will be developed from a more rigorous point of view in Section 4.3.

We now return to the definition of a function. To describe the graph of a function, we need to consider ordered pairs. Let A and B be any sets. The **Cartesian product** of A and B is formed from ordered pairs of elements of A and B. Formally, we define the Cartesian product of A and B as

$$A \times B = \{(a, b) \mid a \in A \text{ and } b \in B\}.$$

In this set, ordered pairs (a_1, b_1) and (a_2, b_2) are equal if and only if $a_1 = a_2$ and $b_1 = b_2$. For example, if $A = \{1, 2, 3\}$ and $B = \{4, 5, 6\}$, then

$$A \times B = \{(1, 4), (1, 5), (1, 6), (2, 4), (2, 5), (2, 6), (3, 4), (3, 5), (3, 6)\}.$$

Since the use of ordered pairs of elements should be familiar from calculus and linear algebra courses, we will not go into further detail on Cartesian products at this point. (If there is any possibility of confusing the greatest common divisor of the integers m and n with an ordered pair, we will write $\gcd(m, n)$ for the greatest common divisor.)

2.1.1 Definition. *Let S and T be sets. A **function** from S into T is a subset F of $S \times T$ such that for each element $x \in S$ there is exactly one element $y \in T$ such that $(x, y) \in F$.*

*The set S is called the **domain** of the function, and the set T is called the **codomain** of the function. The subset*

$$\{y \in T \mid (x, y) \in F \text{ for some } x \in S\}$$

*of the codomain is called the **image** of the function.*

Many authors prefer to use the word **range** for what we have called the codomain of a function. The word range is also sometimes used for what we call the image of the function. For this reason, we will try to avoid using the word range.

Example 2.1.1.

> Let $S = \{1, 2, 3\}$ and $T = \{4, 5, 6\}$. The subsets $F_1 = \{(1, 4), (2, 5), (3, 6)\}$ and $F_2 = \{(1, 4), (2, 4), (3, 4)\}$ of $S \times T$ both define functions since in both cases each element of S occurs exactly once among the ordered pairs. For each of F_1 and F_2 the domain is S and the codomain is T. The image of F_1 is the set $\{4, 5, 6\} = T$, while the image of F_2 is the proper subset $\{4\} \subset T$.
>
> As above, let $S = \{1, 2, 3\}$ and $T = \{4, 5, 6\}$. The subset $F_3 = \{(1, 4), (3, 6)\}$ of $S \times T$ does not define a function with domain S because the element $2 \in S$ does not appear as the first component of any ordered pair. Note that F_3 is a function if the domain is changed to the set $\{1, 3\}$. Unlike the conventions used in calculus, the domain and codomain must be specified as well as the "rule of correspondence" (list of pairs) when you are presenting a function. The

subset $F_4 = \{(1, 4), (2, 4), (2, 5), (3, 6)\}$ of $S \times T$ does not define a function since 2 appears as the first component of two ordered pairs. When a candidate such as F_4 fails to be a function in this way, we say that the collection of ordered pairs does not make a "well-defined" assignment to each element of the domain S. □

If $F \subseteq S \times T$ defines a function, then we will simply write $y = f(x)$ whenever $(x, y) \in F$. Thus f determines a "rule" that assigns to $x \in S$ the unique element $y \in T$, and we will call F the **graph** of f. We will also use the familiar notation $f : S \to T$. We continue to emphasize the importance of the domain and codomain. In particular, we even distinguish between the functions $f : S \to T$ and $g : S \to T'$ where $T \subset T'$ and $\{(x, f(x)) \mid x \in S\} = \{(x, g(x)) \mid x \in S\}$.

When using the notation $f : S \to T$ for a function, the image of f is usually written $f(S)$. More generally, if $A \subseteq S$, then

$$f(A) = \{y \in T \mid y = f(a) \text{ for some } a \in A\}$$

is called the **image of A under f**.

We think it will be useful to rewrite Example 2.1.1 using the "rule" definition of a function.

Example 2.1.1. *(continued)*

Recall that $S = \{1, 2, 3\}$ and $T = \{4, 5, 6\}$. Instead of defining the functions F_1 and F_2 as subsets of $S \times T$, we can use the more familiar notation to their definitions as follows. We define the function $f_1 : S \to T$ by using the rule $f_1(1) = 4$, $f_1(2) = 5$, and $f_1(3) = 6$, and we define the function $f_2 : S \to T$ by using the rule $f_2(1) = 4$, $f_2(2) = 4$, and $f_2(3) = 4$. The images of these functions are $f_1(S) = \{4, 5, 6\} = T$, and $f_2(S) = \{4\}$, and their graphs are the subsets F_1 and F_2 (defined previously).

The formula $f_3(1) = 4$ and $f_3(3) = 6$ does not define a function with domain S because the rule does not assign any element of T to the element $2 \in S$.

The formula $f_4(1) = 4$, $f_4(2) = 4$, $f_4(2) = 5$, and $f_4(3) = 6$ does not define a function since the element $2 \in S$ has two elements of T assigned to it, and thus $f_4(2)$ is not uniquely defined. □

Example 2.1.2 (Inclusion function).

If A is a subset of the set T, we define the **inclusion function** $\iota : A \to T$ by setting $\iota(x) = x$, for all $x \in A$. The graph of ι is

$$I = \{(x, x) \in A \times T \mid x \in A\},$$

and it is easily shown that I defines a function from A to T. The image of ι is just $\iota(A) = A$. □

Although the definition we have given provides a rigorous definition of a function, in the language of set theory, the more familiar definition and notation are usually easier to work with. But it is important to understand the precise definition. If you feel unsure of a concept involving functions, it is worth your time to try to rephrase it in terms of the formal definition using graphs. Becoming comfortable with both the formal and informal definitions will give you two ways to think about a function, so that you can use whichever one is appropriate.

For example, if we attempt to use the square root to define a function $f : \mathbf{R} \to \mathbf{R}$, we immediately run into a problem: the square root of a negative number cannot exist in the set of real numbers. There are two natural ways in which this can be remedied. We can restrict the domain to the set $\mathbf{R}^+$ of all positive real numbers, in which case the formula $f(x) = \sqrt{x}$ yields a function $f : \mathbf{R}^+ \to \mathbf{R}$. On the other hand, we can enlarge the codomain to the set $\mathbf{C}$ of all complex numbers, in which case the formula $f(x) = \sqrt{x}$ yields a function $f : \mathbf{R} \to \mathbf{C}$. Note that we are following the convention that if x is a nonnegative real number, then $\sqrt{x}$ is the nonnegative real number whose square is x, and if x is a negative real number, then $\sqrt{x}$ is $i\sqrt{|x|}$.

With the familiar notation, a function $f : S \to T$ must be determined by a rule or formula that assigns to each element $x \in S$ a unique element $f(x) \in T$. It is often the case that the uniqueness of $f(x)$ is in question. When checking that for each $x \in S$ the corresponding element $f(x) \in T$ is uniquely determined, we will say that we are checking that the function f is **well-defined.** Problems arise when the element x can be described in more than one way, and the rule or formula for $f(x)$ depends on how x is written.

Example 2.1.3.

Consider the formula $f(m/n) = m$, which might be thought to define a function $f : \mathbf{Q} \to \mathbf{Z}$. The difficulty is that a fraction has many equivalent representations, and the formula depends on one particular choice. For example, $f(1/2) = 1$, according to the formula, while $f(3/6) = 3$. Since we know that $1/2 = 3/6$ in the set of rational numbers, we are forced to conclude that the formula does not define a function from the set of rational numbers into the set of integers since it is not well-defined.

On the other hand, the formula $f(m/n) = 2m/3n$ does define a function $f : \mathbf{Q} \to \mathbf{Q}$. To show that f is well-defined, suppose that $m/n = p/q$. Then $mq = np$, and so multiplying both sides by 6 gives $2m \cdot 3q = 3n \cdot 2p$, which implies that $2m/3n = 2p/3q$, and thus $f(m/n) = f(p/q)$.   □

Example 2.1.4.

In defining functions on $\mathbf{Z}_n$ it is necessary to be very careful that the given formula is independent of the numbers chosen to represent each congruence class.

On the one hand, the formula $f([x]_4) = [x]_6$ does not define a function from $\mathbf{Z}_4$ into $\mathbf{Z}_6$. To see this, we only need to note that although $[0]_4 = [4]_4$, the formula specifies that $f([0]_4) = [0]_6$, whereas $f([4]_4) = [4]_6$, giving two different values in $\mathbf{Z}_6$, since $[0]_6 \neq [4]_6$.

On the other hand, the formula $f([x]_4) = [3x]_6$ does define a function from $\mathbf{Z}_4$ into $\mathbf{Z}_6$. If $a \equiv b \pmod{4}$, then $4 \mid (a - b)$, and so multiplying by 3 shows that $12 \mid 3(a - b)$. But then of course $6 \mid 3(a - b)$, which implies that $3a \equiv 3b \pmod{6}$. Thus $[a]_4 = [b]_4$ implies $f([a]_4) = f([b]_4)$, proving that f is well-defined. $\quad\square$

Two subsets F and G of $S \times T$ define the same function from S into T if and only if the subsets are equal. In terms of the associated functions $f : S \to T$ and $g : S \to T$, this happens if and only if $f(x) = g(x)$ for all $x \in S$, giving the familiar condition for equality of functions, since f and g have the same domain and codomain. To see this, note that if $x \in S$, then $(x, f(x)) \in F$, and so we also have $(x, f(x)) \in G$, which means that $f(x) = g(x)$, since a given first component has exactly one possible second component in the graph of the function. On the other hand, if $f(x) = g(x)$ for all $x \in S$, then the ordered pairs $(x, f(x))$ and $(x, g(x))$ are equal for all $x \in S$, so we have $F = G$.

If $f : S \to T$ and $g : T \to U$ are functions, then the composite of f and g is defined by the formula $(g \circ f)(x) = g(f(x))$ for all $x \in S$. Its graph is the subset of $S \times U$ defined by

$$\{(x, z) \mid (x, y) \in F \text{ and } (y, z) \in G \text{ for some } y \in T\},$$

where F and G are the graphs of f and g, respectively. We note that $g \circ f$ is indeed a function, for given $x \in S$, there is a unique element $y \in T$ such that $y = f(x)$, and then there is a unique element $z \in U$ such that $z = g(y)$. Thus $z = g(f(x))$ is uniquely determined by x. This justifies the following formal definition.

2.1.2 Definition. *Let $f : S \to T$ and $g : T \to U$ be functions. The **composite** $g \circ f$ of f and g is the function from S to U defined by the formula $(g \circ f)(x) = g(f(x))$ for all $x \in S$.*

The composite of two functions is defined only when the codomain of the first function is the same as the domain of the second function. Thus the composite of two functions may be defined in one order but not in the opposite order. (In calculus books, some authors allow the composite of two functions, $g \circ f$, to be defined when the codomain of f is merely a subset of the domain of g.)

The following analogy comparing a function to a deterministic computer program may be useful to students with some background in computer science. If $f : S \to T$ is a function, then we can think of the set S as consisting of the possible values which can be used as input for the program, and the set T as consisting of the potential output values. The conditions defining a function ensure that (i) any element of S can be used as an input and (ii) the output is uniquely determined by the input. If $g : T \to U$, then the composite function $g \circ f$ corresponds to a new program obtained by linking the two given programs, taking the output from the first program and using it as the input for the second program.

Example 2.1.5.

Let $f : \mathbf{R} \to \mathbf{R}$ be given by $f(x) = x^2$ and $g : \mathbf{R} \to \mathbf{R}$ be given by $g(x) = x + 1$, for all $x \in \mathbf{R}$. Then $g \circ f$ is given by

$$(g \circ f)(x) = g(f(x)) = g(x^2) = x^2 + 1$$

and $f \circ g$ is given by

$$(f \circ g)(x) = f(g(x)) = f(x + 1) = (x + 1)^2 = x^2 + 2x + 1 .$$

This example shows that you should not expect $g \circ f$ and $f \circ g$ to be equal.   □

Example 2.1.6.

As in calculus, one use of composite functions is to start with a given function and try to write it as the composite of two simpler functions. Here is one such example. If $h : \mathbf{R} \to \mathbf{R}$ is defined by $h(x) = e^{x^2}$, for all $x \in R$, then we can write $h = g \circ f$, where $f : \mathbf{R} \to \mathbf{R}$ is defined by $f(x) = x^2$, for all $x \in R$, and $g : \mathbf{R} \to \mathbf{R}$ is defined by $g(x) = e^x$, for all $x \in \mathbf{R}$.   □

Suppose that we are given three functions $f : S \to T$, $g : T \to U$, and $h : U \to V$. If we wish to compose these functions to obtain a function from S into V, then there are two ways to proceed. We could first form $g \circ f : S \to U$ and then compose with h to get $h \circ (g \circ f) : S \to V$, or we could first form $h \circ g : T \to V$ and then compose with f to get $(h \circ g) \circ f : S \to V$. These procedures define the same function, since in both cases the definition of the composition of functions leads to the expression $h(g(f(x)))$, for all $x \in S$. Thus we can say that composition of functions is **associative**, and write $h \circ (g \circ f) = (h \circ g) \circ f$. In practice, this allows us to ignore the use of parentheses, so that we can just write hgf for the composite function.

2.1.3 Proposition. *Composition of functions is associative.*

Proof. Let $f : S \to T$, $g : T \to U$, and $h : U \to V$ be functions. Then for each $x \in S$, we have

$$(h \circ (g \circ f))(x) = (h(g \circ f))(x) = h(g(f(x)))$$

and also

$$((h \circ g) \circ f)(x) = (h \circ g)(f(x)) = h(g(f(x))) \ .$$

This shows that $h \circ (g \circ f)$ and $(h \circ g) \circ f$ are equal as functions. $\quad\square$

In the definition of a function $f : S \to T$, every element of the domain S must appear as the first entry of some ordered pair in the graph of f, but nothing is said about the necessity of each element of the codomain T appearing as the second entry of some ordered pair. For example, in the function $f : \mathbf{R} \to \mathbf{R}$ defined by $f(x) = x^2$ for all $x \in \mathbf{R}$, no negative number appears as the second coordinate of any point on the graph of f. This particular example also points out the fact that two different elements of the domain may have assigned to them the same element of the codomain. Functions which avoid either or both of these forms of behavior are important enough to warrant the following definition.

2.1.4 Definition. *Let $f : S \to T$ be a function. Then f is said to map S **onto** T if for each element $y \in T$ there exists an element $x \in S$ with $f(x) = y$. If f maps S onto T, then we say that f is an **onto** function.*

*If $f(x_1) = f(x_2)$ implies $x_1 = x_2$ for all elements $x_1, x_2 \in S$, then f is said to be a **one-to-one** function.*

*If f is both one-to-one and onto, then it is called a **one-to-one correspondence** from S to T.*

Some other terminology is also in common use. An onto function is said to be *surjective* or is called a *surjection*. Similarly, a one-to-one function is said to be *injective* or is called an *injection*. In this terminology, which comes from the French, a one-to-one correspondence is a *bijection*. This terminology has the advantage of avoiding the unfortunate use of the preposition "onto" as an adjective. We have decided to continue to use the word "onto" since it its use is so common in the mathematical literature. When you use it, remember that it has a technical meaning, so that the phrase "f maps S onto T" is different from the phrase "f maps S into T."

It may be helpful to think of onto functions in the following terms. If $f : S \to T$, then f is an onto function if and only if for each $y \in T$ the equation $y = f(x)$ has a solution $x \in S$. Thus f is an onto function if and only if the image $f(S)$ is equal to the codomain T. With this point of view, to show that $f : \mathbf{R} \to \mathbf{R}$ defined by $f(x) = x^3 + 1$ is onto, we need to show that for each $y \in \mathbf{R}$ we can solve the equation $y = x^3 + 1$, for x in terms of y. All we have to do is give the solution explicitly: $x = \sqrt[3]{y - 1}$.

The definition we have given for a one-to-one function is perhaps not the most intuitive one. For $f : S \to T$ to be one-to-one we want to know that $f(x_1) \neq f(x_2)$ whenever $x_1 \neq x_2$ (for $x_1, x_2 \in S$). However, if we try to apply this definition to show that the function $f(x) = x^3 + 1$ of the previous paragraph is one-to-one, we would need to work with inequalities. Working with equalities is much more familiar, so it is useful to reformulate the definition. Rewording the phrase "$f(x_1) \neq f(x_2)$ whenever $x_1 \neq x_2$" to read "if $x_1 \neq x_2$, then $f(x_1) \neq f(x_2)$" allows us to pass to the logically equivalent contrapositive statement "if $f(x_1) = f(x_2)$, then $x_1 = x_2$." Using the second statement, which we have taken as the definition, to show that f is one-to-one we only need to show that if $(x_1)^3 + 1 = (x_2)^3 + 1$, then $x_1 = x_2$. This is easy to do by just subtracting 1 from both sides and taking the cube root, which yields a unique value.

Example 2.1.7.

Let $S = \{1, 2, 3\}$ and $T = \{4, 5, 6\}$. The function $f_1 : S \to T$ defined by $f_1(1) = 4$, $f_1(2) = 5$, and $f_1(3) = 6$ is a one-to-one correspondence because it is both one-to-one and onto. The function $f_2 : S \to T$ defined by $f_2(1) = 4$, $f_2(2) = 4$, and $f_2(3) = 4$ is not one-to-one since $f_2(2) = f_2(3)$, and it is not onto since its image is $f_2(S) = \{4\} \neq T$.

With these sets we cannot give an example that is one-to-one but not onto, and we cannot give an example that is onto but not one-to-one. Let $S = \{1, 2, 3\}$ and $U = \{7, 8\}$. The function $f_3 : S \to U$ defined by $f_3(1) = 7$, $f_3(2) = 8$, and $f_3(3) = 8$ is onto since its image is U, but it is not one-to-one since $f_3(2) = f_3(3)$. The function $f_4 : U \to S$ defined by $f_4(7) = 1$ and $f_4(8) = 2$ is one-to-one, but it is not onto since its image is $\{1, 2\} \neq S$. $\square$

Example 2.1.8.

Let $f : S \to T$ be a function. Define $\widehat{f} : S \to f(S)$ by $\widehat{f}(x) = f(x)$, for all $x \in S$. Then by definition $\widehat{f}$ is an onto function. If $\iota : f(S) \to T$ is the inclusion function, then $f = i \circ \widehat{f}$, and we have written f as the composite of an onto function and a one-to-one function. $\square$

2.1.5 Proposition. *Let $f : S \to T$ and $g : T \to U$ be functions.*

(a) *If f and g are one-to-one, then $g \circ f$ is one-to-one.*

(b) *If f and g are onto, then $g \circ f$ is onto.*

Proof. (a) Assume that f and g are one-to-one functions. Let $x_1, x_2 \in S$. If $(g \circ f)(x_1) = (g \circ f)(x_2)$, then $g(f(x_1)) = g(f(x_2))$, and so $f(x_1) = f(x_2)$ since g is one-to-one. Then since f is one-to-one, we have $x_1 = x_2$. This shows that $g \circ f$ is one-to-one.

(b) Assume that f and g are onto functions, and let $z \in U$. Since g is onto, there exists $y \in T$ such that $g(y) = z$. Since f is onto, there exists $x \in S$ such that $f(x) = y$. Hence $(g \circ f)(x) = g(f(x)) = g(y) = z$, and this shows that $g \circ f$ is onto. □

We now want to study one-to-one correspondences in more detail. We first note that Proposition 2.1.5 implies that the composite of two one-to-one correspondences is a one-to-one correspondence. One of the most important properties that we want to show is that any one-to-one correspondence $f : S \rightarrow T$ is "reversible."

Let us return to our earlier analogy between functions and programs. If we are given a function $f : S \rightarrow T$, which we think of as a program, then we want to be able to construct another program (a new function $g : T \rightarrow S$) that is capable of taking any output from the program f and retrieving the corresponding input. Another way to express this is to say that if the two programs are linked via composition, they would end up doing nothing: that is, $g(f(x)) = x$ for all $x \in S$ and $f(g(y)) = y$ for all $y \in T$.

We need some formal notation at this point.

2.1.6 Definition. *Let S be a set. The **identity** function $1_S : S \rightarrow S$ is defined by the formula $1_S(x) = x$ for all $x \in S$.*

*If $f : S \rightarrow T$ is a function, then a function $g : T \rightarrow S$ is called an **inverse** for f if $g \circ f = 1_S$ and $f \circ g = 1_T$.*

An identity function has the following important property. If $f : S \rightarrow T$, then for all $x \in S$ we have $f(1_S(x)) = f(x)$, showing that $f \circ 1_S = f$. Similarly, we have $1_T \circ f = f$. In particular, if $f : S \rightarrow S$, then $f \circ 1_S = f$ and $1_S \circ f = f$. Thus the identity function 1_S plays the same role for composition of functions as the number 1 does for multiplication.

If g is an inverse for f, then the definition shows immediately that f is an inverse for g. The next proposition shows that a function has an inverse precisely when it is one-to-one and onto. It also implies that the inverse function is again one-to-one and onto.

Suppose that $g, h : T \rightarrow S$ are both inverses for $f : S \rightarrow T$. Then on the one hand, $(g \circ f) \circ h = 1_S \circ h = h$, while on the other hand, $g \circ (f \circ h) = g \circ 1_T = g$. Since composition of functions is associative, the two expressions must be equal, so $h = g$. This shows that inverses are unique, and so the use of the notation f^{-1} for the inverse of f is justified.

2.1.7 Proposition. *Let* $f : S \to T$ *be a function. If* f *has an inverse, then it must be one-to-one and onto. Conversely, if* f *is one-to-one and onto, then it has a unique inverse.*

Proof.   First assume that f has an inverse $g : T \to S$ such that $g \circ f = 1_S$ and $f \circ g = 1_T$. Given any element $y \in T$, we have $y = 1_T(y) = f(g(y))$, and so f maps $g(y)$ onto y, showing that f is onto. If $x_1, x_2 \in S$ with $f(x_1) = f(x_2)$, then applying g gives $g(f(x_1)) = g(f(x_2))$, and so we must have $x_1 = x_2$ since $g \circ f = 1_S$. Thus f is one-to-one.

Conversely, assume that f is one-to-one and onto. We will define a function $g : T \to S$ as follows. For each $y \in T$, there exists an element $x \in S$ with $f(x) = y$ since f is onto. Furthermore, there is only one such $x \in S$ since f is one-to-one. This allows us to define $g(y) = x$, and it follows immediately from this definition that $g(f(x)) = x$ for all $x \in S$. For any $y \in T$, we have $g(y) = x$ for the element $x \in S$ for which $f(x) = y$. Thus $f(g(y)) = f(x) = y$ for all $y \in T$, showing that g is an inverse for f.

As in the remarks preceding the proposition, suppose that $h : T \to S$ is also an inverse for f. Then

$$h = h \circ 1_T = h \circ (f \circ g) = (h \circ f) \circ g = 1_S \circ g = g$$

and the uniqueness is established. $\square$

It is instructive to prove that a one-to-one and onto function has an inverse by using the graph of the function. Assume that $f : S \to T$ is one-to-one and onto, and let F denote the graph of f. We will define an inverse $g : T \to S$ by giving its graph G. Let $G = \{(y, x) \mid (x, y) \in F\}$. Then we have clearly defined a subset of $T \times S$. Since f is onto, for each $y \in T$ there exists $x \in S$ such that $(x, y) \in F$, and so for each $y \in T$, there exists $(y, x) \in G$. Furthermore, the element x is uniquely determined by y, since f is one-to-one, showing that for each $y \in T$ there is only one ordered pair in G with first component y. This shows that G does in fact define a function. The graph of $g \circ f$ in $S \times S$ is $\{(x, x) \mid x \in S\}$, and the graph of $f \circ g$ in $T \times T$ is $\{(y, y) \mid y \in T\}$, and so g is the inverse of f.

The final result in the section applies only to finite sets. If S and T are finite sets, and $f : S \to T$ is a one-to-one correspondence, then three things occur: f is one-to-one, f is onto, and S and T have the same number of elements. As a consequence of the next proposition, any two of these conditions imply the third.

2.1.8 Proposition. *Let* $f : S \to T$ *be a function, and assume that* S *and* T *are finite sets with the same number of elements. Then* f *is a one-to-one correspondence if either* f *is one-to-one or* f *is onto.*

Proof. Suppose that S and T both have n elements.

First assume that f is one-to-one. Let $S = \{x_1, x_2, \ldots, x_n\}$, and consider the subset

$$B = \{f(x_1), f(x_2), \ldots, f(x_n)\} \subseteq T .$$

Since f is one-to-one, the elements $f(x_i)$ are distinct, for $i = 1, 2, \ldots, n$, and so B contains n elements. Since B is a subset of T and T also has n elements, we must have $B = T$, showing that f is onto. Thus f is a one-to-one correspondence.

On the other hand, assume that f is onto. Suppose that $T = \{y_1, y_2, \ldots, y_n\}$ and $f(z) = y_i$ and $f(z') = y_i$ for some $z \neq z'$ in S and some i with $1 \leq i \leq n$. Since f is onto, for each $j \neq i$ (with $1 \leq j \leq n$) there exists an element z_j such that $f(z_j) = y_j$. Consider the subset

$$A = \{z, z', z_1, \ldots, z_{i-1}, z_{i+1}, \ldots, z_n\} \subseteq S .$$

The elements z_j are distinct since f is a function. Thus A is a subset that has $n + 1$ elements, which is impossible. Hence having $f(z) = f(z')$ for distinct elements z and z' is impossible. We conclude that f is one-to-one, and so f is a one-to-one correspondence. $\square$

Example 2.1.9.

> Define $f : \mathbf{Z} \to \mathbf{Z}$ by $f(n) = 2n$, for all $n \in \mathbf{Z}$. Then f is one-to-one but not onto. On the other hand, if $g : \mathbf{Z} \to \mathbf{Z}$ is defined by letting $g(n) = n$ if n is odd and $g(n) = n/2$ if n is even, then g is onto but not one-to-one. $\square$

Example 2.1.9 shows that it is possible to have a set S and a function $f : S \to S$ such that f is one-to-one but not onto. It can also happen that f is onto but not one-to-one. Proposition 2.1.8 shows that this can only happen if S is an infinite set. In fact, it can be proved that this characterizes infinite sets: a set S is infinite if and only if there exists a one-to-one correspondence between S and a proper subset of S.

EXERCISES: SECTION 2.1

We use $\mathbf{R}^+$ to denote the set of positive real numbers.

1. In each of the following parts, determine whether the given function is one-to-one and whether it is onto.

   †(a) $f : \mathbf{R} \to \mathbf{R}$; $f(x) = x + 3$

   (b) $f : \mathbf{C} \to \mathbf{C}$; $f(x) = x^2 + 2x + 1$

   †(c) $f : \mathbf{Z}_n \to \mathbf{Z}_n$; $f([x]_n) = [mx + b]_n$, where $m, b \in \mathbf{Z}$

   (d) $f : \mathbf{R}^+ \to \mathbf{R}$; $f(x) = \ln x$

2. In each of the following parts, determine whether the given function is one-to-one and whether it is onto.

   (a) $f : \mathbf{R} \to \mathbf{R}$; $f(x) = x^2$

   (b) $f : \mathbf{C} \to \mathbf{C}$; $f(x) = x^2$

   (c) $f : \mathbf{R}^+ \to \mathbf{R}^+$; $f(x) = x^2$

   (d) $f : \mathbf{R}^+ \to \mathbf{R}^+$; $f(x) = \begin{cases} x & \text{if } x \text{ is rational} \\ x^2 & \text{if } x \text{ is irrational} \end{cases}$

3.† For each one-to-one and onto function in Exercise 1, find the inverse of the function.

   Hint: It might not hurt to review the section on inverse functions in your calculus book.

4. For each one-to-one and onto function in Exercise 2, find the inverse of the function.

5. In each of the following parts, determine whether the given function is one-to-one and whether it is onto. If the function is both one-to-one and onto, find the inverse of the function.

   (a) $f : \mathbf{R}^2 \to \mathbf{R}^2$; $f(x, y) = (x + y, y)$

   (b) $f : \mathbf{R}^2 \to \mathbf{R}^2$; $f(x, y) = (x + y, x + y)$

   (c) $f : \mathbf{R}^2 \to \mathbf{R}^2$; $f(x, y) = (2x + y, x + y)$

6. Let $S = \{1, 2, 3\}$ and $T = \{4, 5\}$.

   †(a) How many functions are there from S into T? from T into S?

   (b) How many of the functions from S into T are one-to-one? How many are onto?

   (c) How many of the functions from T into S are one-to-one? How many are onto?

7. (a) Does the formula $f(x) = 1/(x^2 + 1)$ define a function $f : \mathbf{R} \to \mathbf{R}$?

   (b) Does the formula given in part (a) define a function $f : \mathbf{C} \to \mathbf{C}$?

8. Which of the following formulas define functions from the set of rational numbers into itself? (Assume in each case that n, m are integers and that n is nonzero.)

   †(a) $f\left(\dfrac{m}{n}\right) = \dfrac{m + 1}{n + 1}$

   (b) $g\left(\dfrac{m}{n}\right) = \dfrac{2m}{3n}$

   †(c) $h\left(\dfrac{m}{n}\right) = \dfrac{m + n}{n^2}$

   (d) $k\left(\dfrac{m}{n}\right) = \dfrac{(m - n)^2}{n^2}$

   †(e) $p\left(\dfrac{m}{n}\right) = \dfrac{4m^2}{7n^2} - \dfrac{m}{n}$

   (f) $q\left(\dfrac{m}{n}\right) = \dfrac{m + 1}{m}$

9. Show that each of the following formulas yields a well-defined function.

    (a) $f : \mathbf{Z}_8 \to \mathbf{Z}_8$ defined by $f([x]_8) = [mx]_8$, for any $m \in \mathbf{Z}$

    (b) $g : \mathbf{Z}_8 \to \mathbf{Z}_{12}$ defined by $g([x]_8) = [6x]_{12}$

    (c) $h : \mathbf{Z}_{12} \to \mathbf{Z}_4$ defined by $h([x]_{12}) = [x]_4$

    (d) $p : \mathbf{Z}_{10} \to \mathbf{Z}_5$ defined by $p([x]_{10}) = [x^2 + 2x - 1]_5$

    (e) $q : \mathbf{Z}_4 \to \mathbf{Z}_{12}$ defined by $q([x]_4) = [9x]_{12}$

10. In each of the following cases, give an example to show that the formula does not define a function.

    †(a) $f : \mathbf{Z}_8 \to \mathbf{Z}_{10}$ defined by $f([x]_8) = [6x]_{10}$

    (b) $g : \mathbf{Z}_2 \to \mathbf{Z}_5$ defined by $g([x]_2) = [x]_5$

    †(c) $h : \mathbf{Z}_4 \to \mathbf{Z}_{12}$ defined by $h([x]_4) = [x]_{12}$

    (d) $p : \mathbf{Z}_{12} \to \mathbf{Z}_5$ defined by $p([x]_{12}) = [2x]_5$

11. Let k and n be positive integers. For a fixed $m \in \mathbf{Z}$, define the formula $f : \mathbf{Z}_n \to \mathbf{Z}_k$ by $f([x]_n) = [mx]_k$, for $x \in \mathbf{Z}$. Show that f defines a function if and only if $k|mn$.

12. Let k, m, n be positive integers such that $k|mn$. Show that the function $f : \mathbf{Z}_n \to \mathbf{Z}_k$ defined in Exercise 11 by $f([x]_n) = [mx]_k$ is a one-to-one correspondence if and only if $k = n$ and $(m, n) = 1$.

13. Let $f : A \to B$ be a function, and let $f(A) = \{f(a) \mid a \in A\}$ be the image of f. Show that f is onto if and only if $f(A) = B$.

14. Let $f : A \to B$ and $g : B \to C$ be one-to-one and onto. Show that $(g \circ f)^{-1}$ exists and that $(g \circ f)^{-1} = f^{-1} \circ g^{-1}$.

15. Let $f : A \to B$ and $g : B \to C$ be functions. Prove that if $g \circ f$ is one-to-one, then f is one-to-one, and that if $g \circ f$ is onto, then g is onto.

16. Let $f : A \to B$ be a function. Prove that f is onto if and only if there exists a function $g : B \to A$ such that $f \circ g = 1_B$.

17. Let $f : A \to B$ be a function. Prove that f is onto if and only if $h \circ f = k \circ f$ implies $h = k$, for every set C and all choices of functions $h : B \to C$ and $k : B \to C$.

18. Let A be a nonempty set, and let $f : A \to B$ be a function. Prove that f is one-to-one if and only if there exists a function $g : B \to A$ such that $g \circ f = 1_A$.

19. Let $f : A \to B$ be a function. Prove that f is one-to-one if and only if $f \circ h = f \circ k$ implies $h = k$, for every set C and all choices of functions $h : C \to A$ and $k : C \to A$.

20. Define $f : \mathbf{Z}_{mn} \to \mathbf{Z}_m \times \mathbf{Z}_n$ by $f([x]_{mn}) = ([x]_m, [x]_n)$. Show that f is a function and that f is onto if and only if $\gcd(m, n) = 1$.

2.2   Equivalence Relations

It is very useful to introduce the notion of an equivalence relation. The basic idea of an equivalence relation is to collect together elements that, even though they differ, all behave in the same fashion with respect to some property of interest. We then treat this collection of similar elements as a single entity. This approach can be taken in many different situations and is fundamental in abstract algebra.

We have already used this idea in Chapter 1, when we collected together integers congruent modulo n and used these sets to construct elements $[a]_n$ of the new object $\mathbf{Z}_n$. In this section we will adopt the point of view that studying equivalence relations is simply a continuation of our study of functions.

Suppose that we are working with a function $f : S \to T$. If f is one-to-one, then for elements $x_1, x_2 \in S$ we know that $f(x_1) = f(x_2)$ if and only if $x_1 = x_2$. Of course, this is not true in general. For example, if $f : \mathbf{Z} \to \mathbf{Z}$ is the function that assigns to each integer its remainder when divided by 10, then infinitely many integers are mapped to each of the possible remainders $0, 1, \ldots, 9$. In this example we might want to consider two integers as equivalent whenever the function maps them to the same value. Note that two integers are treated the same by our function if they are congruent modulo 10. In general, for a function $f : S \to T$, we can say that two elements $x_1, x_2 \in S$ are equivalent with respect to f if $f(x_1) = f(x_2)$.

As with the definition of a function, for the sake of precision we will use set theory to give our formal definition of an equivalence relation. We then immediately give an equivalent definition which is more useful and, we hope, more intuitive.

2.2.1 Definition. *Let S be a set. A subset R of $S \times S$ is called an **equivalence relation** on S if*

(i) *for all $a \in S$, $(a, a) \in R$;*

(ii) *for all $a, b \in S$, if $(a, b) \in R$ then $(b, a) \in R$;*

(iii) *for all $a, b, c \in S$, if $(a, b) \in R$ and $(b, c) \in R$, then $(a, c) \in R$.*

We will write $a \sim b$ to denote the fact that $(a, b) \in R$.

The symbol $\sim$ is called **tilde**. For elements a, b the relation $a \sim b$ is usually read "a is equivalent to b" or "a tilde b."

Using the above definition and notation, it is clear that R is an equivalence relation if and only if for all $a, b, c \in S$ we have

(i) [Reflexive law] $a \sim a$;

(ii) [Symmetric law] if $a \sim b$, then $b \sim a$;

(iii) [Transitive law] if $a \sim b$ and $b \sim c$, then $a \sim c$.

We will usually use these conditions rather than the formal definition.

The most fundamental equivalence relation is given by simple equality of elements; that is, for $a, b \in S$, define $a \sim b$ if $a = b$. For this relation the reflexive, symmetric, and transitive laws are clear. Under this relation, for any $a \in S$ the only element equivalent to a is a itself. In fact, equality is the only equivalence relation for which each element is related only to itself. For other equivalence relations we are interested in the set of all elements related to a given element. In reading the following definition, keep in mind the example of congruence classes modulo n.

2.2.2 Definition. *Let $\sim$ be an equivalence relation on the set S. For a given element $a \in S$, we define the **equivalence class** of a to be the set of all elements of S that are equivalent to a. We will use the notation $[a]$. In symbols,*

$$[a] = \{x \in S \mid x \sim a\}.$$

*The notation $S/\sim$ will be used for the collection of equivalence classes of S defined by the equivalence relation $\sim$. We say that $S/\sim$ is the **factor set** of the relation $\sim$.*

The elements of the factor set $S/\sim$ are the equivalence classes under the relation $\sim$, and are constructed as sets, but once the factor set has been constructed we no longer picture the equivalence classes as sets. Intuitively, the factor set $S/\sim$ is thought of as a set of objects which are obtained by "gluing together" certain elements of S to get the elements $[a]$ of the new set $S/\sim$.

Example 2.2.1 (Congruence modulo n).

Let n be a positive integer. For integers a, b we define $a \sim b$ if $n \mid (a - b)$. This, of course, is logically equivalent to the definition of congruence modulo n. It is not difficult to check that congruence modulo n defines an equivalence relation on the set $\mathbf{Z}$ of integers. First, the reflexive property holds since for any integer a we certainly have $a \sim a$ since $n \mid (a - a)$. Next, the symmetric property holds since if $a \sim b$, then $n \mid (a - b)$ and so $n \mid (b - a)$, showing that $b \sim a$. Finally, the transitive property holds since if $a \sim b$ and $b \sim c$, then $n \mid (a - b)$ and $n \mid (b - c)$, so adding shows that $n \mid (a - c)$, and thus $a \sim c$. The equivalence classes of this relation are the familiar congruence classes $[a]_n$.

As an alternate proof, we could use our original definition of congruence modulo n, in which we said that integers a and b are congruent modulo n if we obtain equal remainders when using the division algorithm to divide both a and b by n. Then the reflexive, symmetric, and transitive properties really follow from the identical properties for equality. $\square$

Because of our work with congruences in Chapter 1, the previous example is one of the most familiar nontrivial equivalence relations. The next example is probably the most basic one. In fact, later in this section we will show that every equivalence relation arises in this way from a function.

Example 2.2.2 (S/f).

Let $f : S \to T$ be any function. For $x_1, x_2 \in S$ we define $x_1 \sim_f x_2$ if $f(x_1) = f(x_2)$. Then for all $x_1, x_2, x_3 \in S$ we have (i) $f(x_1) = f(x_1)$; (ii) if $f(x_1) = f(x_2)$, then $f(x_2) = f(x_1)$; and (iii) if $f(x_1) = f(x_2)$ and $f(x_2) = f(x_3)$, then $f(x_1) = f(x_3)$. This shows that $\sim_f$ defines an equivalence relation on the set S. The proof of this is easy because the equivalence relation is defined in terms of equality of the images $f(x)$, and equality is the most elementary equivalence relation. The collection of all equivalence classes of S under $\sim_f$ will be denoted by S/f. We say that S/f is the **factor set of S determined by** f.

Later in this section we will show that any function $f : S \to T$ naturally induces a one-to-one function $\overline{f} : S/f \to T$. Thus by introducing the factor set S/f we can study f by studying the equivalence relation it defines on S and the corresponding one-to-one function $\overline{f}$. $\quad\square$

Example 2.2.3 (Equivalence of rational numbers).

Let A be the set of all integers and let B be the set of all nonzero integers. On the set $S = A \times B$ of ordered pairs, define $(m, n) \sim (p, q)$ if $mq = np$. This defines an equivalence relation, which we can show as follows. Certainly $(m, n) \sim (m, n)$, since $mn = nm$. If $(m, n) \sim (p, q)$, then $mq = np$ implies $pn = qm$, and so $(p, q) \sim (m, n)$. Finally, suppose that $(m, n) \sim (p, q)$ and $(p, q) \sim (s, t)$. Then $mq = np$ and $pt = qs$, and multiplying the first equation by t and the second by n gives $mqt = npt$ and $npt = nqs$. After equating mqt and nqs and cancelling q (which is nonzero by assumption), we obtain $mt = ns$, and so $(m, n) \sim (s, t)$. Thus we have verified the reflexive, symmetric, and transitive laws, showing that we have in fact defined an equivalence relation.

The equivalence class of $(m, n) \in S$ is usually denoted by m/n, and this equivalence relation is the basis for the standard rule for equality of fractions. The equivalence classes of this equivalence relation form the set $\mathbf{Q}$ of rational numbers.

As a passing remark, we note that in this example using the ordered pair formulation of the notion of an equivalence relation would lead to considering ordered pairs of ordered pairs. $\quad\square$

Example 2.2.4.

Consider the set of all differentiable functions from $\mathbf{R}$ into $\mathbf{R}$. For two such functions $f(x)$ and $g(x)$ we define $f \sim g$ if the derivatives $f'(x)$ and $g'(x)$

are equal. It can easily be checked directly that the properties defining an equivalence relation hold in this case. Furthermore, the equivalence class of the function f is the set of all functions of the form $f(x) + C$, for a constant C. Using a somewhat more sophisticated point of view, we can consider differentiation as a function, whose domain is the set of all differentiable functions and whose codomain is the set of all functions. Then the equivalence relation we have defined arises as in the previous Example 2.2.2. □

Example 2.2.5.

Let S be the set $\mathbf{R}^2$ of points in the Euclidean plane; that is, $S = \{(x, y) \mid x, y \in \mathbf{R}\}$. Define $(x_1, y_1) \sim (x_2, y_2)$ if $x_1 = x_2$. There are several ways to check that we have defined an equivalence relation. First, the reflexive, symmetric, and transitive laws are easy to check since $\sim$ is defined in terms of equality of the first components. As a second method we could use Example 2.2.2. If we define $f : \mathbf{R}^2 \to \mathbf{R}$ by $f(x, y) = x$, then $f(x_1, y_1) = f(x_2, y_2)$ if and only if $x_1 = x_2$. Thus the relation $\sim_f$ defined by f is the same as $\sim$, and it follows that $\sim$ is an equivalence relation.

Now let us find the equivalence classes of the equivalence relation we have defined. The equivalence class of (a, b) consists of all points in the plane that have the same first coordinate; that is, $[(a, b)]$ is just the line $x = a$, which we will denote by L_a. Then since distinct vertical lines are parallel, we see that $L_a \cap L_b = \emptyset$ for $a \neq b$. We can summarize by saying that each point in the plane belongs to exactly one of the equivalence classes L_a. □

The following fact, observed at the end of Example 2.2.5, holds for all equivalence relations.

2.2.3 Proposition. *Let S be a set, and let $\sim$ be an equivalence relation on S. Then each element of S belongs to exactly one of the equivalence classes of S determined by the relation $\sim$.*

Proof. If $a \in S$, then $a \sim a$, and so $a \in [a]$. If $a \in [b]$ for some $b \in S$, then we claim that $[a] = [b]$. To show this, we must check that each element of $[a]$ belongs to $[b]$, and also that each element of $[b]$ must belong to $[a]$. If $x \in [a]$, then $x \sim a$ by definition. We have assumed that $a \in [b]$, so $a \sim b$, and then it follows from the transitive law that $x \sim b$, and thus $x \in [b]$. On the other hand, if $x \in [b]$, then $x \sim b$. By the symmetric law, $a \sim b$ implies $b \sim a$, and so this gives us $x \sim a$, showing that $x \in [a]$. We have thus shown that each element $a \in S$ belongs to exactly one of the equivalence classes determined by $\sim$. □

An alternate point of view with regard to equivalence relations is afforded by the notion of a partition of a set. In Proposition 2.2.5 we will see that any factor

set arising from an equivalence relation turns out to be a partition, and that every
partition gives rise to an equivalence relation for which it is the factor set.

2.2.4 Definition. *Let S be a set. A collection $\mathcal{P}$ of nonempty subsets of S is called
a **partition** of S if each element of S belongs to exactly one of the members of $\mathcal{P}$.*

In this terminology, Proposition 2.2.3 implies that any equivalence relation on
a nonempty set S determines a partition of the set. We will show that the converse
is also true. We first give some additional examples of partitions.

Example 2.2.6.

The equivalence relation that we considered in Example 2.2.5 partitions the
Euclidean plane $\mathbf{R}^2$ by using the collection of all vertical lines.

Another interesting partition of $\mathbf{R}^2$ uses lines radiating from the origin. First,
let S_0 be the x-axis, $y = 0$. Since the origin $(0, 0)$ belongs to S_0, it cannot
belong to any other set in the partition, so for any nonzero real number a, let
S_a be the line $y = ax$ with the origin removed. Let S_∞ be the y-axis with the
origin removed. Then each point in the plane (except the origin) determines
a unique line through the origin, and so each point belongs to exactly one of
the sets we have defined.   □

Example 2.2.7.

It is obvious that the collection of all circles with center at the origin forms
a partition of the plane $\mathbf{R}^2$. (We must allow the origin itself to be considered
a degenerate circle.) There is a corresponding equivalence relation, which
can be described algebraically by defining $(x_1, y_1) \sim (x_2, y_2)$ if $x_1^2 + y_1^2 = x_2^2 + y_2^2$.   □

2.2.5 Proposition. *Any partition $\mathcal{P}$ of the set S determines a unique equivalence
relation $\sim$ on S such that $\mathcal{P}$ is the factor set $S/\sim$.*

*Conversely, if $\sim$ is any equivalence relation on S, then the factor set $S/\sim$ is a
partition of S that determines the equivalence relation $\sim$.*

Proof. Assume that we are given a partition $\mathcal{P}$ of the set S. Then the given partition
yields an equivalence relation on S by defining $a \sim b$ when a and b belong to
the same element of $\mathcal{P}$. To prove that we have defined an equivalence relation,
we proceed as follows. For all $a \in S$ we have $a \sim a$ since a belongs to exactly
one subset in the partition, and thus the reflexive law holds. It is obvious from
the definition that the relation is symmetric. Finally, for $a, b, c \in S$ suppose that

$a \sim b$ and $b \sim c$. Then a and b both belong to some subset of the partition, say P_1. Similarly, b and c both belong to some subset, say P_2. Since we are given a partition, the element b can belong to only one subset, so we have $P_1 = P_2$, and then this implies that a and c belong to the same subset, so we have $a \sim c$, and the transitive law holds.

Let P_α be an element of $\mathcal{P}$. Then P_α is nonempty, and so it contains some element $a \in S$. We claim that $P_\alpha = [a]$. To see this, let $x \in P_\alpha$. Then $x \sim a$, so $x \in [a]$, and thus $P_\alpha \subseteq [a]$. If $x \in [a]$, then $x \sim a$, and so x belongs to the same element of $\mathcal{P}$ as a, namely $x \in P_\alpha$. Thus $[a] \subseteq P_\alpha$, and so $[a] = P_\alpha$. We have now shown that $\mathcal{P} \subseteq S/\sim$.

To show the reverse inclusion, let $[a] \in S/\sim$. Let P_α be the unique element of $\mathcal{P}$ for which $a \in P_\alpha$. We will show that $P_\alpha = [a]$. Let $x \in P_\alpha$. Then $x \sim a$ since $x, a \in P_\alpha$, so $x \in [a]$, and thus $P_\alpha \subseteq [a]$. If $x \in [a]$ then $x \sim a$, and so x and a belong to the same element of $\mathcal{P}$. Since $a \in P_\alpha$ we must have $x \in P_\alpha$, and thus $[a] \subseteq P_\alpha$, showing that $[a] = P_\alpha \in \mathcal{P}$. This completes the proof that $\mathcal{P} = S/\sim$.

Conversely, if $\sim$ is an equivalence relation on S, it follows immediately from Proposition 2.2.3 that the factor set $S/\sim$ is a partition of S. We need to show that the factor set determines the equivalence relation $\sim$.

Let $\approx$ be the new equivalence relation on S determined by the partition $S/\sim$. If $a, b \in S$ and $a \sim b$, then a and b belong to $[a] \in S/\sim$, and since a, b belong to the same set in the partition $S/\sim$, we have $a \approx b$. Conversely, if $a \approx b$, then a, b belong to the same element of $S/\sim$, say $[x]$. Then $a \sim x$ and $b \sim x$, and so $a \sim b$. Thus for all $a, b \in S$ we have $a \sim b$ if and only if $a \approx b$, and so $\sim$ and $\approx$ are the same equivalence relation. $\square$

From our intuitive informal point of view, the way we think of factor sets and partitions of S are diametrically opposite. The sets that make up the partition are the collections of elements which are to be "glued together," and the elements of the factor set are what we get after the "gluing" process.

We recall from Example 2.2.2 that when an equivalence relation $\sim$ on a set S is defined by a function $f : S \rightarrow T$, we use the notation S/f for the factor set, instead of the notation $S/\sim$.

Example 2.2.8.

Let S be any set, and let $\sim$ be an equivalence relation on S. Define a function $\pi : S \rightarrow S/\sim$ by $\pi(x) = [x]$, for all $x \in S$. Proposition 2.2.3 shows that each element $x \in S$ belongs to a unique equivalence class, and so this shows that π is a function.

For any $x_1, x_2 \in S$ we have $x_1 \sim x_2$ if and only if $\pi(x_1) = \pi(x_2)$. This shows that π defines the original equivalence relation $\sim$, so that the factor sets $S/\sim$ and S/π are identical. Thus we have just verified our earlier remark that every equivalence relation can be realized as the equivalence relation defined in a natural way by a function. $\square$

2.2.6 Definition. *Let S be a set, and let $\sim$ be an equivalence relation on S. The function $\pi : S \to S/\sim$ defined by $\pi(x) = [x]$, for all $x \in S$, is called the **natural projection** from S onto the factor set $S/\sim$.*

2.2.7 Theorem. *If $f : S \to T$ is any function, and $\sim_f$ is the equivalence relation defined on S by letting $x_1 \sim_f x_2$ if $f(x_1) = f(x_2)$, for all $x_1, x_2 \in S$, then there is a one-to-one correspondence between the elements of the image $f(S)$ of S under f and the equivalence classes in the factor set S/f of the relation $\sim_f$.*

Proof. Define $\overline{f} : S/f \to f(S)$ by $\overline{f}([x]) = f(x)$, for all $x \in S$. The function $\overline{f}$ is well-defined since if x_1 and x_2 belong to the same equivalence class in the factor set S/f, then by definition of the equivalence relation we must have $f(x_1) = f(x_2)$. Furthermore, $\overline{f}$ is onto since if $y \in f(S)$, then $y = f(x)$ for some $x \in S$, and thus we have $y = \overline{f}([x])$. Finally, $\overline{f}$ is one-to-one since if $\overline{f}([x_1]) = \overline{f}([x_2])$, then $f(x_1) = f(x_2)$, and so by definition $x_1 \sim_f x_2$, which implies that $[x_1] = [x_2]$. Thus we have shown that $\overline{f}$ is a one-to-one correspondence. $\square$

If $f : S \to T$ is a function, then we can use the function $\overline{f}$ defined in Theorem 2.2.7 to write f as a composite of better-behaved functions. Let $\iota : f(S) \to T$ be the inclusion mapping, defined by $\iota(y) = y$, for all $y \in f(S)$. Let $\pi : S \to S/f$ be the natural projection, defined by $\pi(x) = [x]$, for all $x \in S$. Then for each $x \in X$ we have

$$\iota \circ \overline{f} \circ \pi(x) = \iota(\overline{f}(\pi(x))) = \iota(\overline{f}([x])) = \iota(f(x)) = f(x) ,$$

which shows that $f = \iota \circ \overline{f} \circ \pi$. This composite function is illustrated in Figure 2.2.1.

Figure 2.2.1:

$$S \xrightarrow{\ \pi\ } S/f \xrightarrow{\ \overline{f}\ } f(S) \xrightarrow{\ \iota\ } T$$

Figure 2.2.2 illustrates the equality $f = \iota \circ \overline{f} \circ \pi$ more vividly. In summary, the point is to express f in such a way that the first function is onto, the second function is both one-to-one and onto, and the last function is one-to-one.

We end this section with another very useful concept from set theory.

2.2.8 Definition. *Let $S \to T$ be a function. If $B \subseteq T$, then the set*

$$\{x \in S \mid f(x) \in B\}$$

*is called the **inverse image** of B under f.*

Figure 2.2.2:

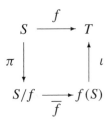

If $f : S \to T$ is a function and $B \subseteq T$, we will use the notation $f^{-1}(B)$ for the inverse image of B. You must be very careful when using this notation, since it does not imply that f has an inverse function. If $y \in f(S)$, it is customary to write $f^{-1}(y)$ rather than $f^{-1}(\{y\})$, even though this is an abuse of our notation.

The notion of an inverse image of an element is closely connected to the results we have been discussing. If $f : S \to T$ is any function, then the inverse image of an element of $f(S)$ is just the corresponding equivalence class in the factor set $S/\!\sim_f$. In fact, we can write

$$S/f = \{f^{-1}(y) \mid y \in f(S)\} .$$

If f is a one-to-one and onto function, then it has an inverse, and for each $y \in T$ the inverse image $f^{-1}(y)$ consists of a single element. This element is the image of y under the inverse function, and so in this case the new notation $f^{-1}(y)$ coincides with the notation we have already been using for an inverse function.

EXERCISES: SECTION 2.2

1. It is shown in Theorem 2.2.7 that if $f : S \to T$ is a function, then there is a one-to-one correspondence between the elements of $f(S)$ and the equivalence classes of S/f. For each of the following functions, find $f(S)$ and S/f and exhibit the one-to-one correspondence between them.

   †(a) $f : \mathbf{Z} \to \mathbf{C}$ given by $f(n) = i^n$ for all $n \in \mathbf{Z}$

   (b) $g : \mathbf{Z} \to \mathbf{Z}_{12}$ given by $g(n) = [8n]_{12}$ for all $n \in \mathbf{Z}$

   †(c) $h : \mathbf{Z}_{12} \to \mathbf{Z}_{12}$ defined by $h([x]_{12}) = [9x]_{12}$

   (d) $k : \mathbf{Z}_{12} \to \mathbf{Z}_{12}$ defined by $k([x]_{12}) = [5x]_{12}$

2. Repeat Exercise 1 for each of the following functions.

   (a) $f : \mathbf{Z}_{12} \to \mathbf{Z}_{12}$ defined by $f([x]_{12}) = [4x]_{12}$

(b) $g : \mathbf{Z}_{24} \to \mathbf{Z}_{24}$ defined by $g([x]_{24}) = [4x]_{24}$

(c) $p : \mathbf{Z}_{12} \to \mathbf{Z}_{24}$ defined by $p([x]_{12}) = [4x]_{24}$

(d) $q : \mathbf{Z}_{24} \to \mathbf{Z}_{12}$ defined by $q([x]_{24}) = [4x]_{12}$

3. For each of the following relations on $\mathbf{R}$, determine which of the three conditions of Definition 2.2.1 hold.

   (a) For $a, b \in \mathbf{R}$, define $a \sim b$ if $a \le b$.

   (b) For $a, b \in \mathbf{R}$, define $a \sim b$ if $a - b \in \mathbf{Q}$.

   (c) For $a, b \in \mathbf{R}$, define $a \sim b$ if $|a - b| \le 1$.

4. Let S be the set of all ordered pairs (m, n) of positive integers. For $(a_1, a_2) \in S$ and $(b_1, b_2) \in S$, define $(a_1, a_2) \sim (b_1, b_2)$ if $a_1 + b_2 = a_2 + b_1$. Show that $\sim$ is an equivalence relation.

5. On $\mathbf{R}^2$, define $(a_1, a_2) \sim (b_1, b_2)$ if $a_1^2 + a_2^2 = b_1^2 + b_2^2$. Check that this defines an equivalence relation. What are the equivalence classes?

6.†In $\mathbf{R}^3$, consider the standard (x, y, z)-coordinate system. We can define a partition of $\mathbf{R}^3$ by using planes parallel to the (x, y)-plane. Describe the corresponding equivalence relation by giving conditions on the coordinates x, y, z.

7. Define an equivalence relation on the set $\mathbf{R}$ that partitions the real line into subsets of length 1.

8. For integers m, n, define $m \sim n$ if and only if $n \mid m^k$ and $m \mid n^j$ for some positive integers k and j.

   (a) Show that $\sim$ is an equivalence relation on $\mathbf{Z}$.

   †(b) Determine the equivalence classes [1], [2], [6] and [12].

   (c) Give a characterization of the equivalence class $[m]$.

9. Let S be a set. A subset $R \subseteq S \times S$ is called a **circular relation** if (i) for each $a \in S$, $(a, a) \in R$ and (ii) for each $a, b, c \in S$, if $(a, b) \in R$ and $(b, c) \in R$, then $(c, a) \in R$. Show that any circular relation must be an equivalence relation.

10. Let S be a set and let $2^S = \{A \mid A \subseteq S\}$ be the collection of all subsets of S. Define $\sim$ on 2^S by letting $A \sim B$ if and only if there exists a one-to-one correspondence from A to B.

    (a) Show that $\sim$ is an equivalence relation on 2^S.

    (b) If $S = \{1, 2, 3, 4\}$, list the elements of 2^S and find each equivalence class determined by $\sim$.

11. Let W be a subspace of a vector space V over $\mathbf{R}$, (that is, the scalars are assumed to be real numbers). We say that two vectors $\mathbf{u}, \mathbf{v} \in V$ are congruent modulo W if $\mathbf{u} - \mathbf{v} \in W$, written $\mathbf{u} \equiv \mathbf{v} \pmod{W}$.

    (a) Show that $\equiv$ is an equivalence relation.

(b) Show that if r, s are scalars and $\mathbf{u}_1, \mathbf{u}_2, \mathbf{v}_1, \mathbf{v}_2$ are vectors in V such that $\mathbf{u}_1 \equiv \mathbf{v}_1 \pmod{W}$ and $\mathbf{u}_2 \equiv \mathbf{v}_2 \pmod{W}$, then $r\mathbf{u}_1 + s\mathbf{u}_2 \equiv r\mathbf{v}_1 + s\mathbf{v}_2 \pmod{W}$.

(c) Let $[\mathbf{u}]_W$ denote the equivalence class of the vector $\mathbf{u}$. Set $U = \{[\mathbf{u}]_W \mid \mathbf{u} \in V\}$. Define $+$ and $\cdot$ on U by $[\mathbf{u}]_W + [\mathbf{v}]_W = [\mathbf{u} + \mathbf{v}]_W$ and $r \cdot [\mathbf{u}]_W = [r\mathbf{u}]_W$ for all $\mathbf{u}, \mathbf{v} \in V$ and $r \in R$. Show that U is a vector space with respect to these operations.

(d) Let $V = \mathbf{R}^2$, and let $W = \{(x, 0) \mid x \in \mathbf{R}\}$. Describe the equivalence class $[(x, y)]_W$ geometrically. Show that $T : \mathbf{R} \to U$ defined by $T(y) = [(0, y)]_W$ is a linear transformation that is one-to-one and onto.

12. Let $T = \{(x, y, z) \in \mathbf{R}^3 \mid (x, y, z) \neq (0, 0, 0)\}$. Define $\sim$ on T by $(x_1, y_1, z_1) \sim (x_2, y_2, z_2)$ if there exists a nonzero real number λ such that $x_1 = \lambda x_2$, $y_1 = \lambda y_2$, and $z_1 = \lambda z_2$.

(a) Show that $\sim$ is an equivalence relation on T.

(b) Give a geometric description of the equivalence class of (x, y, z).

The set $T / \sim$ is called the **real projective plane**, and is denoted by $\mathbf{P}^2$. The class of (x, y, z) is denoted by $[x, y, z]$, and is called a **point**.

(c) Let $(a, b, c) \in T$, and suppose that $(x_1, y_1, z_1) \sim (x_2, y_2, z_2)$. Show that if $ax_1 + by_1 + cz_1 = 0$, then $ax_2 + by_2 + cz_2 = 0$. Conclude that

$$L = \{[x, y, z] \in \mathbf{P}^2 \mid ax + by + cz = 0\}$$

is a well-defined subset of $\mathbf{P}^2$. Such sets L are called **lines**.

(d) Show that the triples $(a_1, b_1, c_1) \in T$ and $(a_2, b_2, c_2) \in T$ determine the same line if and only if $(a_1, b_1, c_1) \sim (a_2, b_2, c_2)$.

(e) Given two distinct points of $\mathbf{P}^2$, show that there exists exactly one line that contains both points.

(f) Given two distinct lines, show that there exists exactly one point that belongs to both lines.

(g) Show that the function $f : \mathbf{R}^2 \to \mathbf{P}^2$ defined by $f(x, y) = [x, y, 1]$ is a one-to-one function. This is one possible embedding of the "affine plane" into the projective plane. We sometimes say that $\mathbf{P}^2$ is the "completion" of $\mathbf{R}^2$.

(h) Show that the embedding of part (g) takes lines to "lines."

(i) If two lines intersect in $\mathbf{R}^2$, show that the image of their intersection is the intersection of their images (under the embedding defined in part (g)).

(j) If two lines are parallel in $\mathbf{R}^2$, what happens to their images under the embedding into $\mathbf{P}^2$?

2.3   Permutations

We will now study one-to-one correspondences in more detail, particularly for finite sets. Our emphasis in this section will be on computations with such functions. We need to develop some notation that will make it easier to work with such functions, especially when finding the composite of two functions. We will change our notation slightly, using Greek letters for permutations, and instead of writing $\sigma \circ \tau$ for the composite of two permutations, we will simply write $\sigma\tau$, and refer to this as the **product** of the two permutations.

2.3.1 Definition. *Let S be a set. A function* $\sigma : S \rightarrow S$ *is called a* **permutation** *of S if* σ *is one-to-one and onto.*

   The set of all permutations of S will be denoted by $\mathrm{Sym}(S)$.

   The set of all permutations of the set $\{1, 2, \ldots, n\}$ *will be denoted by* S_n.

Proposition 2.1.5 shows that the composite of two permutations in $\mathrm{Sym}(S)$ is again a permutation. It is obvious that the identity function on S is one-to-one and onto. Proposition 2.1.7 shows that any permutation in $\mathrm{Sym}(S)$ has an inverse function that is also one-to-one and onto. We can summarize these important properties as follows:

   (i) if $\sigma, \tau \in \mathrm{Sym}(S)$, then $\tau\sigma \in \mathrm{Sym}(S)$;

   (ii) $1_S \in \mathrm{Sym}(S)$;

   (iii) if $\sigma \in \mathrm{Sym}(S)$, then $\sigma^{-1} \in \mathrm{Sym}(S)$.

   We also mention that the composition of permutations is associative, by Proposition 2.1.3.

   We need to develop some notation for working with permutations in S_n. Given $\sigma \in S_n$, note that σ is completely determined as soon as we know $\sigma(1), \sigma(2), \ldots, \sigma(n)$, and so we introduce the notation

$$\sigma = \left(\begin{array}{cccc} 1 & 2 & \cdots & n \\ \sigma(1) & \sigma(2) & \cdots & \sigma(n) \end{array} \right),$$

where under each integer i we write the image of i.

   For example, if $S = \{1, 2, 3\}$ and $\sigma : S \rightarrow S$ is given by $\sigma(1) = 2$, $\sigma(2) = 3$, $\sigma(3) = 1$, then we would write $\sigma = \left(\begin{array}{ccc} 1 & 2 & 3 \\ 2 & 3 & 1 \end{array} \right)$.

   Since any element σ in S_n is one-to-one and onto, in the above notation for σ each element of S must appear once and only once in the second row. Thus an element $\sigma \in S_n$ is completely determined once we know the order in which the elements of S appear in the second row.

   To count the number of elements of S_n we only need to count the number of possible second rows. Since there are n elements in $S = \{1, 2, \ldots, n\}$, there are n choices for the first element $\sigma(1)$ of the second row. Since the element that is assigned to $\sigma(1)$ cannot be used again, we have $n - 1$ choices when we wish to assign a value to $\sigma(2)$, and thus there are $n \cdot (n-1)$ ways to assign values to both $\sigma(1)$ and $\sigma(2)$. Now there are $n-2$ choices for $\sigma(3)$ and a total of $n(n-1)(n-2)$ ways to assign values to $\sigma(1)$, $\sigma(2)$, and $\sigma(3)$. Continuing in this fashion (there is an induction argument here), we have a total of $n!$ ways to assign values to $\sigma(1)$, $\sigma(2), \ldots, \sigma(n)$. Thus S_n has $n!$ elements.

The notation we have introduced is useful for computing in S_n. Suppose that

$$\sigma = \begin{pmatrix} 1 & 2 & \cdots & n \\ \sigma(1) & \sigma(2) & \cdots & \sigma(n) \end{pmatrix} \text{ and } \tau = \begin{pmatrix} 1 & 2 & \cdots & n \\ \tau(1) & \tau(2) & \cdots & \tau(n) \end{pmatrix}.$$

Then to compute the composition

$$\sigma\tau = \begin{pmatrix} 1 & 2 & \cdots & n \\ \sigma(\tau(1)) & \sigma(\tau(2)) & \cdots & \sigma(\tau(n)) \end{pmatrix}$$

we proceed as follows. To find $\sigma(\tau(i))$ we first look under i in τ to get $j = \tau(i)$, and then we find $\sigma\tau(i) = \sigma(j)$ by looking under j in σ.

Example 2.3.1.

Let $\sigma = \begin{pmatrix} 1 & 2 & 3 & 4 \\ 4 & 3 & 1 & 2 \end{pmatrix}$ and $\tau = \begin{pmatrix} 1 & 2 & 3 & 4 \\ 2 & 3 & 4 & 1 \end{pmatrix}$. To compute $\sigma\tau$ we have $\tau(1) = 2$ and then $\sigma(2) = 3$, giving $\sigma\tau(1) = 3$. Next we have $\tau(2) = 3$ and $\sigma(3) = 1$, giving $\sigma\tau(2) = 1$. Continuing this procedure we obtain $\sigma\tau = \begin{pmatrix} 1 & 2 & 3 & 4 \\ 3 & 1 & 2 & 4 \end{pmatrix}$. A similar computation gives $\tau\sigma = \begin{pmatrix} 1 & 2 & 3 & 4 \\ 1 & 4 & 2 & 3 \end{pmatrix}$. $\square$

Given $\sigma = \begin{pmatrix} 1 & 2 & \cdots & n \\ \sigma(1) & \sigma(2) & \cdots & \sigma(n) \end{pmatrix}$ in S_n, it is easy to compute σ^{-1}. To find $\sigma^{-1}(j)$ we find j in the second row of σ, say $j = \sigma(i)$. The inverse of σ must reverse this assignment, and so under j we write i, giving $\sigma^{-1}(j) = i$. This can be accomplished easily by simply turning the two rows of σ upside down and then rearranging terms.

Example 2.3.2.

If $\sigma = \begin{pmatrix} 1 & 2 & 3 & 4 \\ 4 & 3 & 1 & 2 \end{pmatrix}$, then

$$\sigma^{-1} = \begin{pmatrix} 4 & 3 & 1 & 2 \\ 1 & 2 & 3 & 4 \end{pmatrix} = \begin{pmatrix} 1 & 2 & 3 & 4 \\ 3 & 4 & 2 & 1 \end{pmatrix}. \quad \square$$

The double-row notation that we have introduced for permutations is transparent but rather cumbersome. We now want to introduce another notation that is more compact and also helps to convey certain information about the permutation. Consider the permutation $\sigma = \begin{pmatrix} 1 & 2 & 3 & 4 & 5 \\ 3 & 1 & 4 & 2 & 5 \end{pmatrix}$. We do not necessarily have to write the first row in numerical order, and in this case it is informative to change the order as follows. The first column expresses the fact that $\sigma(1) = 3$, and using this we interchange the second and third columns. Now $\sigma(3) = 4$, and so as the third column we choose the one with 4 in the first row. Continuing gives us $\sigma = \begin{pmatrix} 1 & 3 & 4 & 2 & 5 \\ 3 & 4 & 2 & 1 & 5 \end{pmatrix}$. Now writing $(1, 3, 4, 2)$ would give us all of the necessary information to describe σ, since $\sigma(1) = 3, \sigma(3) = 4, \sigma(4) = 2$, and $\sigma(2) = 1$. In the new notation we do not need to mention $\sigma(5)$ since $\sigma(5) = 5$.

2.3.2 Definition. *Let S be a set, and let $\sigma \in \mathrm{Sym}(S)$. Then σ is called a **cycle of length** k if there exist elements $a_1, a_2, \ldots, a_k \in S$ such that $\sigma(a_1) = a_2, \sigma(a_2) = a_3,$ $\ldots, \sigma(a_{k-1}) = a_k, \sigma(a_k) = a_1,$ and $\sigma(x) = x$ for all other elements $x \in S$ with $x \neq a_i$ for $i = 1, 2, \ldots, k$.*

In this case we write $\sigma = (a_1, a_2, \ldots, a_k)$.

We can also write $\sigma = (a_2, a_3, \ldots, a_k, a_1)$ or $\sigma = (a_3, \ldots, a_k, a_1, a_2)$, etc. The notation for a cycle of length $k \geq 2$ can thus be written in k different ways, depending on the starting point. We will use (1) to denote the identity permutation. This seems to be the most natural choice, although, in fact, $(1) = (a)$ for any cycle (a) of length 1. Of course, if you are working in S_n and it is clear that you are referring to the set $S = \{1, 2, \ldots, n\}$, you can use the notation 1_S.

Example 2.3.3.

In S_5 the permutation $\begin{pmatrix} 1 & 2 & 3 & 4 & 5 \\ 3 & 2 & 4 & 1 & 5 \end{pmatrix}$ is a cycle of length 3, written $(1, 3, 4)$. The permutation $\begin{pmatrix} 1 & 2 & 3 & 4 & 5 \\ 3 & 5 & 4 & 1 & 2 \end{pmatrix}$ is not a cycle, since

$$\begin{pmatrix} 1 & 2 & 3 & 4 & 5 \\ 3 & 5 & 4 & 1 & 2 \end{pmatrix} = \begin{pmatrix} 1 & 2 & 3 & 4 & 5 \\ 3 & 2 & 4 & 1 & 5 \end{pmatrix} \begin{pmatrix} 1 & 2 & 3 & 4 & 5 \\ 1 & 5 & 3 & 4 & 2 \end{pmatrix}$$

$$= (1, 3, 4)(2, 5)$$

is the product of two cycles. $\quad \square$

Example 2.3.4.

For any permutation of a finite set there is an associated diagram, found by representing the elements of the set as points and joining two points with an arrow if the permutation maps one to the other. Since any permutation is a one-to-one and onto function, at each point of the set there is one and only one incoming arrow and one and only one outgoing arrow. To find the diagram of a permutation such as

$$\begin{pmatrix} 1 & 2 & 3 & 4 & 5 & 6 & 7 & 8 & 9 & 10 & 11 & 12 \\ 8 & 2 & 10 & 11 & 5 & 9 & 4 & 6 & 1 & 3 & 12 & 7 \end{pmatrix}$$

we first rearrange the columns to give the following form:

$$\begin{pmatrix} 1 & 8 & 6 & 9 & 2 & 3 & 10 & 4 & 11 & 12 & 7 & 5 \\ 8 & 6 & 9 & 1 & 2 & 10 & 3 & 11 & 12 & 7 & 4 & 5 \end{pmatrix}.$$

The associated diagram is given in Figure 2.3.1.

Figure 2.3.1:

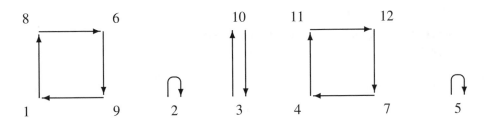

The diagram of $\sigma = \begin{pmatrix} 1 & 2 & 3 & 4 & 5 \\ 3 & 2 & 4 & 1 & 5 \end{pmatrix}$ is given in Figure 2.3.2.

Note that the diagram of a cycle of length k would consist of a connected component with k vertices, while all other components of it would contain only one element. This diagram would clearly illustrate why such a permutation is called a cycle. □

Example 2.3.5.

Let $(1, 4, 2, 5)$ and $(2, 6, 3)$ be cycles in S_6. Then

$$(1, 4, 2, 5) = \begin{pmatrix} 1 & 2 & 3 & 4 & 5 & 6 \\ 4 & 5 & 3 & 2 & 1 & 6 \end{pmatrix}$$

Figure 2.3.2:

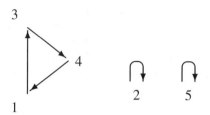

and

$$(2, 6, 3) = \begin{pmatrix} 1 & 2 & 3 & 4 & 5 & 6 \\ 1 & 6 & 2 & 4 & 5 & 3 \end{pmatrix}.$$

In computing the product of these two cycles we have

$$
\begin{aligned}
(1, 4, 2, 5)(2, 6, 3) &= \begin{pmatrix} 1 & 2 & 3 & 4 & 5 & 6 \\ 4 & 5 & 3 & 2 & 1 & 6 \end{pmatrix} \begin{pmatrix} 1 & 2 & 3 & 4 & 5 & 6 \\ 1 & 6 & 2 & 4 & 5 & 3 \end{pmatrix} \\
&= \begin{pmatrix} 1 & 2 & 3 & 4 & 5 & 6 \\ 4 & 6 & 5 & 2 & 1 & 3 \end{pmatrix} \\
&= (1, 4, 2, 6, 3, 5) ,
\end{aligned}
$$

which is again a cycle.

Note that it is not true in general that the product of two cycles is again a cycle. In particular,

$$(1, 4, 2, 5)(1, 4, 2, 5) = \begin{pmatrix} 1 & 2 & 3 & 4 & 5 & 6 \\ 2 & 1 & 3 & 5 & 4 & 6 \end{pmatrix}$$

is not a cycle.   □

2.3.3 Definition. *Let* $\sigma = (a_1, a_2, \ldots, a_k)$ *and* $\tau = (b_1, b_2, \ldots, b_m)$ *be cycles in* Sym(S), *for a set* S. *Then* σ *and* τ *are said to be* **disjoint** *if* $a_i \neq b_j$ *for all* i, j.

It often happens that $\sigma\tau \neq \tau\sigma$ for two permutations σ, τ. For example, in S_3 we have

$$\begin{pmatrix} 1 & 2 & 3 \\ 2 & 1 & 3 \end{pmatrix} \begin{pmatrix} 1 & 2 & 3 \\ 3 & 2 & 1 \end{pmatrix} = \begin{pmatrix} 1 & 2 & 3 \\ 3 & 1 & 2 \end{pmatrix}$$

but on the other hand

$$\begin{pmatrix} 1 & 2 & 3 \\ 3 & 2 & 1 \end{pmatrix} \begin{pmatrix} 1 & 2 & 3 \\ 2 & 1 & 3 \end{pmatrix} = \begin{pmatrix} 1 & 2 & 3 \\ 2 & 3 & 1 \end{pmatrix}.$$

If $\sigma\tau = \tau\sigma$, then we say that σ and τ **commute**. Using this terminology, the next proposition shows that disjoint cycles always commute.

2.3.4 Proposition. *Let S be any set. If σ and τ are disjoint cycles in $\mathrm{Sym}(S)$, then $\sigma\tau = \tau\sigma$.*

Proof. Let $\sigma = (a_1, a_2, \ldots, a_k)$ and $\tau = (b_1, b_2, \ldots, b_m)$ be disjoint. If $i = a_j$ for some $j < k$, then

$$\sigma\tau(i) = \sigma(\tau(a_j)) = \sigma(a_j) = a_{j+1} = \tau(a_{j+1}) = \tau(\sigma(a_j)) = \tau\sigma(i)$$

because τ leaves $a_1, a_2, \ldots, a_k$ fixed. In case $j = k$, we use $\sigma(a_j) = a_1 = \tau(a_1)$. A similar computation can be given if $i = b_j$ for some j, since then σ leaves $b_1, b_2, \ldots, b_m$ fixed. If i appears in neither cycle, then both σ and τ leave it fixed, so $\sigma\tau(i) = \sigma(i) = i = \tau(i) = \tau\sigma(i)$. $\square$

Let σ be any permutation in $\mathrm{Sym}(S)$, for any set S. Taking the composition of σ with itself any number of times still gives us a permutation, and so for any positive integer i we define
$$\sigma^i = \sigma\sigma\cdots\sigma \qquad i \text{ times.}$$

Formally, we can define the powers of σ inductively by letting $\sigma^i = \sigma\sigma^{i-1}$ for $i \geq 2$. Then the following properties can be established by using induction. For positive integers m and n we have

$$\sigma^m\sigma^n = \sigma^{m+n} \qquad \text{and} \qquad (\sigma^m)^n = \sigma^{mn} .$$

To illustrate, we have

$$\sigma^2\sigma^3 = (\sigma\sigma)(\sigma\sigma\sigma) = \sigma^5 \text{ and } (\sigma^2)^3 = (\sigma^2)(\sigma^2)(\sigma^2) = (\sigma\sigma)(\sigma\sigma)(\sigma\sigma) = \sigma^6 .$$

To preserve these laws of exponents, we define $\sigma^0 = (1)$, where (1) is the identity element of $\mathrm{Sym}(S)$, and $\sigma^{-n} = (\sigma^n)^{-1}$. Using these definitions, it can be shown that
$$\sigma^m\sigma^n = \sigma^{m+n} \qquad \text{and} \qquad (\sigma^m)^n = \sigma^{mn}$$

for all integers m, n.

2.3.5 Theorem. *Every permutation in S_n can be written as a product of disjoint cycles. The cycles of length ≥ 2 that appear in the product are unique.*

Proof. Let $S = \{1, 2, \ldots, n\}$ and let $\sigma \in S_n = \mathrm{Sym}(S)$. If we apply successive powers of σ to 1, we have the elements $1, \sigma(1), \sigma^2(1), \sigma^3(1), \ldots$, and after some point there must be a repetition since S has only n elements. Suppose that $\sigma^m(1) = \sigma^k(1)$ is the first repetition, with $m > k \geq 0$. If $k > 0$, then applying σ^{-1} to both sides of the equation a total of k times gives $\sigma^{m-k}(1) = 1$, which contradicts the choice of m. Thus the first time a repetition occurs it is 1 that is repeated.

If we let r be the least positive exponent for which we have $\sigma^r(1) = 1$, then the elements $1, \sigma(1), \sigma^2(1), \ldots \sigma^{r-1}(1)$ are all distinct, giving us a cycle of length r:

$$(1, \sigma(1), \sigma^2(1), \ldots, \sigma^{r-1}(1)) .$$

If $r < n$, let a be the least integer not in $(1, \sigma(1), \ldots, \sigma^{r-1}(1))$ and form the cycle

$$(a, \sigma(a), \sigma^2(a), \ldots, \sigma^{s-1}(a))$$

in which s is the least positive integer such that $\sigma^s(a) = a$. (Such an exponent s exists by an argument similar to the one given above.) If $r + s < n$, then let b be the least positive integer not in the set

$$\{1, \sigma(1), \ldots, \sigma^{r-1}(1), a, \sigma(a), \ldots, \sigma^{s-1}(a)\}$$

and form the cycle beginning with b. We continue in this way until we have exhausted S. Then

$$\sigma = (1, \sigma(1), \ldots, \sigma^{r-1}(1)) \, (a, \sigma(a), \ldots, \sigma^{s-1}(a)) \, \cdots$$

and we have written σ as a product of disjoint cycles.

In fact we have given an algorithm for finding the necessary cycles. Since the cycles are disjoint, by the previous proposition the product does not depend on their order. Note that cycles of length 1 can be omitted. It is left as an exercise to show that a given permutation can be expressed as a product of disjoint cycles in only one way (if the order is disregarded). □

Since any cycle of length one in S_n is the identity function, we will usually omit cycles of length one when we write a permutation as a product of disjoint cycles. Thus if a permutation $\sigma \in S_n$ is written as a product of disjoint cycles of length greater than or equal to two, it is not hard to verify that for any i ($1 \le i \le n$) missing from these cycles we know that $\sigma(i) = i$. On the other hand, if you want each i (for $1 \le i \le n$) to appear in the notation, you may include all of the cycles of length one.

Example 2.3.6.

Let $\sigma = \begin{pmatrix} 1 & 2 & 3 & 4 & 5 & 6 & 7 & 8 \\ 5 & 2 & 7 & 6 & 3 & 8 & 1 & 4 \end{pmatrix}$. Applying the algorithm given in the proof of Theorem 2.3.5, we can write $\sigma = (1, 5, 3, 7)(2)(4, 6, 8) = (1, 5, 3, 7)(4, 6, 8)$. □

In general, if we wish to multiply (compose) two permutations in cycle notation, we do not have to return to the double-row format. We can use the algorithm in Theorem 2.3.5 to write the product (composition) as a product of disjoint cycles, remembering to work from right to left. The procedure is more difficult to describe in words than it is to carry out, but nonetheless we will now try to give a brief description of it.

Say we wish to find the composite function $(a_1, a_2, \ldots, a_k)(b_1, b_2, \ldots, b_m)$. To find the image of i we first see if i is equal to some b_s. If it is, then we know that $(b_1, b_2, \ldots, b_m)$ maps $i = b_s$ to b_{s+1}. (In case $s = m$, i is mapped to b_1.) We then look for b_{s+1} in $(a_1, a_2, \ldots, a_k)$. If it appears here, say $b_{s+1} = a_t$, then $(a_1, a_2, \ldots, a_k)$ maps b_{s+1} to a_{t+1}. (In case $t = k$, b_{s+1} is mapped to a_1.) Thus the composite function maps i to a_{t+1}. If b_{s+1} does not appear in $(a_1, a_2, \ldots, a_k)$, then $(a_1, a_2, \ldots, a_k)$ maps b_{s+1} to b_{s+1}, and so the composite function maps i to b_{s+1} in this case.

Likewise, if i does not appear in $(b_1, b_2, \ldots, b_m)$, then it leaves i fixed and we only need to look for i in $(a_1, a_2, \ldots, a_k)$. If $i = a_s$, then $(a_1, a_2, \ldots, a_k)$ maps i to a_{s+1} (in case $s = k$, i is mapped to a_1), and so the composite function maps i to a_{s+1}. If i does not appear in either $(a_1, a_2, \ldots, a_k)$ or $(b_1, b_2, \ldots, b_m)$, then i is left fixed by the composite function.

Thus if we want to write our product as a product of disjoint cycles, start with $i = 1$ and find the image of i, say j, as we have outlined above. Now, starting with j, repeat the procedure until we return to 1, completing the first cycle. Then, starting with the least integer that does not appear in that cycle, we apply the same procedure until the second cycle is complete. This is repeated until all entries of the two original cycles have been used.

Example 2.3.7.

Consider the cycles $(2, 5, 1, 4, 3)$ and $(4, 6, 2)$ in S_6. We have

$$(2, 5, 1, 4, 3)(4, 6, 2) = (1, 4, 6, 5)(2, 3)$$

and we note again that the product of two cycles need not be a cycle.   □

Since every positive power of σ must belong to S_n, while there are only finitely many elements in S_n, there must exist positive integers $i > j$ such that $\sigma^i = \sigma^j$. Taking the composition with σ^{-1} a total of j times shows that $\sigma^{i-j} = (1)$. Thus we have shown that there is a positive integer m such that $\sigma^m = (1)$.

If $\sigma = (a_1, a_2, \ldots, a_m)$ is a cycle of length m, then applying σ m times to any a_i, $i = 1, 2, \ldots, m$ gives a_i. Thus $\sigma^m = (1)$. Furthermore, m is the smallest positive power of σ that equals the identity, since $\sigma^k(a_1) = a_{k+1}$ for $1 \leq k < m$. In terms of the following definition, we have just shown that a cycle of length m has order m.

2.3.6 Definition. *Let $\sigma \in S_n$. The least positive integer m such that $\sigma^m = (1)$ is called the **order** of σ.*

2.3.7 Proposition. *Let $\sigma \in S_n$ have order m. Then for all integers i, j we have $\sigma^i = \sigma^j$ if and only if $i \equiv j \pmod{m}$.*

Proof. By assumption m is the smallest positive exponent with $\sigma^m = (1)$. If $\sigma^i = \sigma^j$, for any integers i, j, then multiplying by σ^{-j} shows that $\sigma^{i-j} = (1)$. Using the division algorithm we can write $i - j = qm + r$ for integers q, r with $0 \le r < m$. Then since

$$(1) = \sigma^{i-j} = (\sigma^m)^q \sigma^r = \sigma^r$$

we must have $r = 0$ because m is the least positive integer for which $\sigma^m = (1)$. Thus $m \mid (i - j)$ and so $i \equiv j \pmod{m}$.

Conversely, if $i \equiv j \pmod{m}$, then $i = j + mt$ for some integer t. Hence

$$\sigma^i = \sigma^{j+mt} = \sigma^j \sigma^{mt} = \sigma^j (\sigma^m)^t = \sigma^j .$$

This completes the proof. $\square$

2.3.8 Proposition. *Let $\sigma \in S_n$ be written as a product of disjoint cycles. Then the order of σ is the least common multiple of the lengths of its cycles.*

Proof. If $\sigma = (a_1, a_2, \ldots, a_m)$, then σ has order m. Furthermore, if $\sigma^k = (1)$, then $m \mid k$ by the previous proposition.

Next, if $\sigma = (a_1, a_2, \ldots, a_m)(b_1, b_2, \ldots, b_r)$ is a product of two disjoint cycles, then $\sigma^j = (a_1, a_2, \ldots, a_m)^j (b_1, b_2, \ldots, b_r)^j$ since $(a_1, a_2, \ldots, a_m)$ commutes with $(b_1, b_2, \ldots, b_r)$. If $\sigma^j = (1)$, then $(a_1, a_2, \ldots, a_m)^j = (1)$ and $(b_1, b_2, \ldots, b_r)^j = (1)$ since $(b_1, b_2, \ldots, b_r)^j$ leaves each a_i fixed and $(a_1, a_2, \ldots, a_m)^j$ leaves each b_i fixed. This happens if and only if $m \mid j$ and $r \mid j$, and then $\mathrm{lcm}[m, r]$ is a divisor of j. The smallest such j is thus $\mathrm{lcm}[m, r]$.

It should now be clear how to extend this argument to the general case. $\square$

Example 2.3.8.

The permutation $(1, 5, 3, 7)(2, 8, 4)$ has order 12 in S_8. The permutation $(1, 5, 3)(2, 8, 4, 6, 9, 7)$ has order 6 in S_9. $\square$

If we wish to compute the inverse of a cycle, then we merely reverse the order of the cycle since

$$(a_1, a_2, \ldots, a_r)(a_r, a_{r-1}, \ldots, a_1) = (1) .$$

The inverse of the product $\sigma\tau$ of two permutations is $\tau^{-1}\sigma^{-1}$ since

$$(\sigma\tau)(\tau^{-1}\sigma^{-1}) = \sigma(\tau\tau^{-1})\sigma^{-1} = \sigma(1)\sigma^{-1} = \sigma\sigma^{-1} = (1)$$

and similarly

$$(\tau^{-1}\sigma^{-1})(\sigma\tau) = (1) \ .$$

Thus we have

$$[(a_1, \ldots, a_r)(b_1, \ldots, b_m)]^{-1} = (b_m, \ldots, b_1)(a_r, \ldots, a_1) \ .$$

Note that if the cycles are disjoint, then they commute, and so the inverses do not need to be written in reverse order.

The simplest cycle, aside from (1), is one of the form (a_1, a_2). This represents an interchange of two elements. We now show that any permutation of a finite set can be obtained from a sequence of such interchanges. Whether the number of interchanges is even or odd is important in certain applications, for example in determining the signs of the elementary products used to compute the determinant of a matrix.

2.3.9 Definition. *A cycle (a_1, a_2) of length two is called a **transposition**.*

2.3.10 Proposition. *Any permutation in S_n, where $n \geq 2$, can be written as a product of transpositions.*

Proof. By Theorem 2.3.5 any permutation in S_n can be expressed as a product of cycles, and so we only need to show that any cycle can be expressed as a product of transpositions. The identity (1) can be expressed as $(1, 2)(1, 2)$. For any other permutation, we can give an explicit computation:

$$(a_1, a_2, \ldots, a_{r-1}, a_r) = (a_{r-1}, a_r)(a_{r-2}, a_r) \cdots (a_3, a_r)(a_2, a_r)(a_1, a_r) \ .$$

This completes the proof. □

In the above proof, the expression we have given for $(a_1, a_2, \ldots, a_{r-1}, a_r)$ seems to be the most natural one, given that we read composition of permutations from right to left. It is also true that

$$(a_1, a_2, \ldots, a_{r-1}, a_r) = (a_1, a_2)(a_2, a_3) \cdots (a_{r-2}, a_{r-1})(a_{r-1}, a_r) \ .$$

This expression may be easier for the student to remember. It also shows that the representation of a permutation as a product of transpositions is not unique.

Example 2.3.9.

Applying Proposition 2.3.10 to S_3 gives us the following products. We have
$(1) = (1, 2)(1, 2)$, and $(1, 2)$, $(1, 3)$, and $(2, 3)$ are already expressed as
transpositions. Finally, $(1, 2, 3) = (2, 3)(1, 3)$ and $(1, 3, 2) = (3, 2)(1, 2)$.
We could also write $(1, 2, 3) = (1, 2)(1, 3)(1, 2)(1, 3)$.

As another example, we have $(2, 5, 3, 7, 8) = (7, 8)(3, 8)(5, 8)(2, 8)$, as well
as $(2, 5, 3, 7, 8) = (2, 5)(5, 3)(3, 7)(7, 8)$, using the second method. □

In the above example we have illustrated that writing $(1, 2, 3)$ as a product of
transpositions can be done in various ways, and in fact we have written it as a
product of two transpositions in one case and four in another. In general, although
the transpositions in the product are not uniquely determined, we do have a bit of
uniqueness remaining, namely, the parity of the product. By this we mean that the
number of transpositions in the product is either always even or always odd. This
is the content of Theorem 2.3.11.

The proof of the theorem may appear to be rather complicated, so it seems
to be worthwhile to comment on the general strategy of the proof. We wish to
show that something cannot occur, and we do so by contradiction, assuming that a
counterexample exists. If so, then a counterexample of minimal length exists. After
modifying the counterexample without ever making it longer, we show that we can,
in fact, produce an even shorter counterexample. This contradicts the minimality
of the counterexample we started with. This general technique of proof goes back
to Fermat and is often quite useful. We give another proof of the same fact in
Section 3.6.

2.3.11 Theorem. *If a permutation is written as a product of transpositions in two
ways, then the number of transpositions is either even in both cases or odd in both
cases.*

Proof. We will give a proof by contradiction. Suppose that the conclusion of
the theorem is false. Then there exists a permutation σ that can be written as a
product of an even number of transpositions and as a product of an odd number of
transpositions, say

$$\sigma = \tau_1 \tau_2 \cdots \tau_{2m} = \delta_1 \delta_2 \cdots \delta_{2n+1}$$

for transpositions $\tau_1, \ldots, \tau_{2m}$ and $\delta_1, \ldots, \delta_{2n+1}$. Since $\delta_j = \delta_j^{-1}$ for $j = 1, \ldots, 2n+1$, we have $\sigma^{-1} = \delta_{2n+1} \cdots \delta_1$, and so

$$(1) = \sigma \sigma^{-1} = \tau_1 \cdots \tau_{2m} \delta_{2n+1} \cdots \delta_1 \ .$$

This shows that the identity permutation can be written as a product of an odd
number of transpositions.

Next suppose that $(1) = \rho_1 \rho_2 \cdots \rho_k$ is the shortest product of an odd number of transpositions that is equal to the identity. Note that $k \geq 3$, and suppose that $\rho_1 = (a, b)$. We observe that a must appear in at least one other transposition, say ρ_i, with $i > 1$, since otherwise $\rho_1 \cdots \rho_k(a) = b$, a contradiction. Among all products of length k that are equal to the identity, and such that a appears in the transposition on the extreme left, we assume that $\rho_1 \rho_2 \cdots \rho_k$ has the fewest number of a's.

We now show that if ρ_i is the transposition of smallest index $i > 1$ in which a occurs, then ρ_i can be moved to the left without changing the number of transpositions or the number of times that a occurs in the product. Then combining ρ_i and ρ_1 will lead to a contradiction.

Let a, u, v, and r be distinct. By computation, we see that $(u, v)(a, r) = (a, r)(u, v)$ and $(u, v)(a, v) = (a, u)(u, v)$. Hence we can move a transposition with entry a to the second position without changing the number of a's that appear, and thus we may assume that ρ_2 is the next transposition in which a occurs, say $\rho_2 = (a, c)$ for some $c \neq a$. If $c = b$, then $\rho_1 \rho_2 = (1)$, and so $(1) = \rho_3 \cdots \rho_k$ is a shorter product of an odd number of transpositions; this gives us a contradiction. If $c \neq b$, then since $(a, b)(a, c) = (a, c)(b, c)$, we see that $(1) = (a, c)(b, c)\rho_3 \cdots \rho_k$ is a product of transpositions of length k with fewer a's; this again contradicts the choice of $\rho_1, \ldots, \rho_k$. Thus we have shown that (1) cannot be written as a product of an odd number of transpositions, completing the proof. $\square$

The point of Theorem 2.3.11 is that a given permutation is either even or odd, but not both. That result makes possible the following definition.

2.3.12 Definition. *A permutation σ is called **even** if it can be written as a product of an even number of transpositions, and **odd** if it can be written as a product of an odd number of transpositions.*

Note that $(1, 2)$ is odd and $(1, 2, 3) = (2, 3)(1, 3)$ is even. In remembering the parity of a cycle, it is important to note that in this terminology a cycle of odd length is even and a cycle of even length is odd. Calling to mind the simplest case $(1, 2)$ will remind you of this.

We should also note that the identity permutation is even. Furthermore, if σ is an even permutation, then so is the inverse of σ, since given σ as a product of transpositions, we only need to reverse the order of the transpositions to write σ^{-1} as a product of transpositions.

Finally, we note that the product of two even permutations is again an even permutation, and also that the product of two odd permutations is even, while the product of an odd permutation and an even permutation is odd. This remark follows from Theorem 2.3.11 and the fact that the sum of two even integers is even, the sum of two odd integers is even, and the sum of an odd and an even integer is odd.

EXERCISES: SECTION 2.3

1. Consider the following permutations in S_7.

$$\sigma = \begin{pmatrix} 1 & 2 & 3 & 4 & 5 & 6 & 7 \\ 3 & 2 & 5 & 4 & 6 & 1 & 7 \end{pmatrix} \quad \text{and} \quad \tau = \begin{pmatrix} 1 & 2 & 3 & 4 & 5 & 6 & 7 \\ 2 & 1 & 5 & 7 & 4 & 6 & 3 \end{pmatrix}$$

Compute the following products.

†(a) $\sigma\tau$

(b) $\tau\sigma$

†(c) $\tau^2\sigma$

(d) σ^{-1}

†(e) $\sigma\tau\sigma^{-1}$

(f) $\tau^{-1}\sigma\tau$

2.† Write each of the permutations $\sigma\tau, \tau\sigma, \tau^2\sigma, \sigma^{-1}, \sigma\tau\sigma^{-1}$, and $\tau^{-1}\sigma\tau$ in Exercise 1 as a product of disjoint cycles. Write σ and τ as products of transpositions.

3. Write $\begin{pmatrix} 1 & 2 & 3 & 4 & 5 & 6 & 7 & 8 & 9 & 10 \\ 3 & 4 & 10 & 5 & 7 & 8 & 2 & 6 & 9 & 1 \end{pmatrix}$ as a product of disjoint cycles and as a product of transpositions. Construct its associated diagram, find its inverse, and find its order.

4. Find the order of each of the following permutations.

   Hint: First write each permutation as a product of disjoint cycles.

   †(a) $\begin{pmatrix} 1 & 2 & 3 & 4 & 5 & 6 \\ 6 & 4 & 5 & 3 & 2 & 1 \end{pmatrix}$

   (b) $\begin{pmatrix} 1 & 2 & 3 & 4 & 5 & 6 & 7 & 8 \\ 4 & 6 & 7 & 5 & 1 & 8 & 2 & 3 \end{pmatrix}$

   †(c) $\begin{pmatrix} 1 & 2 & 3 & 4 & 5 & 6 & 7 & 8 & 9 \\ 5 & 9 & 8 & 7 & 3 & 4 & 6 & 1 & 2 \end{pmatrix}$

   (d) $\begin{pmatrix} 1 & 2 & 3 & 4 & 5 & 6 & 7 & 8 & 9 \\ 8 & 4 & 9 & 6 & 5 & 2 & 3 & 1 & 7 \end{pmatrix}$

5. Let $3 \leq m \leq n$. Calculate $\sigma\tau^{-1}$ for the cycles $\sigma = (1, 2, \ldots, m - 1)$ and $\tau = (1, 2, \ldots, m - 1, m)$ in S_n.

6. List all of the cycles in S_4.

7.† Find the number of cycles of each possible length in S_5. Then find all possible orders of elements in S_5. (Try to do this without having to write out all 120 possible permutations.)

8. Count the permutations σ in S_6 that satisfy the conditions $\sigma(1) = 2$ and $\sigma(2) = 3$.

9. Let $\sigma, \tau \in S_n$ be permutations such that $\sigma(k) = k$ and $\tau(k) = k$ for some k with $1 \le k \le n$. Show that $\sigma^{-1}(k) = k$ and that $\rho(k) = k$, where $\rho = \sigma\tau$.

10. Let $\sigma \in S_n$, and suppose that σ is written as a product of disjoint cycles. Show that σ is even if and only if the number of cycles of even length is even. Show that σ is odd if and only if the number cycles of even length is odd.

11. Prove that in S_n, with $n \ge 3$, any even permutation is a product of cycles of length three.

    Hint: $(a, b)(b, c) = (a, b, c)$ and $(a, b)(c, d) = (a, b, c)(b, c, d)$.

12. Prove that (a, b) cannot be written as a product of two cycles of length three.

13. Let $\tau \in S_n$ be the cycle $(1, 2, \ldots, k)$ of length k, where $k \le n$.

    (a) Prove that if $\sigma \in S_n$, then $\sigma\tau\sigma^{-1} = (\sigma(1), \sigma(2), \ldots, \sigma(k))$. Thus $\sigma\tau\sigma^{-1}$ is a cycle of length k.

    (b) Let ρ be any cycle of length k. Prove that there exists a permutation $\sigma \in S_n$ such that $\sigma\tau\sigma^{-1} = \rho$.

14. Let S be any nonempty set, and let $\sigma \in \text{Sym}(S)$. For $x, y \in S$ define $x \sim y$ if $\sigma^n(x) = y$ for some $n \in \mathbf{Z}$. Show that $\sim$ defines an equivalence relation on S.

15. For $\alpha, \beta \in S_n$, let $\alpha \sim \beta$ if there exists $\sigma \in S_n$ such that $\sigma\alpha\sigma^{-1} = \beta$. Show that $\sim$ is an equivalence relation on S_n.

16. View S_3 as a subset of S_5, in the obvious way. For $\sigma, \tau \in S_5$, define $\sigma \sim \tau$ if $\sigma\tau^{-1} \in S_3$.

    (a) Show that $\sim$ is an equivalence relation on S_5.

    (b) Find the equivalence class of $(4, 5)$.

    (c) Find the equivalence class of $(1, 2, 3, 4, 5)$.

    (d) Determine the total number of equivalence classes.

Notes

Permutations are important in studying solvability by radicals. The roots of any polynomial equation with rational coefficients exist in the set of complex numbers. To determine whether these roots can be expressed in terms of the coefficients, allowing various radicals, as in the quadratic formula, it is necessary to consider the permutations, or "substitutions," of the roots that leave the basic combinations of the coefficients unchanged. In general, it is not easy to determine whether or not a particular permutation of the roots leaves all sums, differences, products, and quotients of the coefficients fixed.

In order to give a simple illustration of the effect of permuting the roots of an equation, we will consider a polynomial equation that is particularly easy to work with. We have chosen the equation

$$x^3 - 3x + 1 = 0$$

because its roots can be found from the identity $4\cos^3\theta - 3\cos\theta - \cos(3\theta) = 0$ (proved in Lemma 6.3.8) by choosing angles θ with $\cos(3\theta) = -1/2$. These roots are

$$x = 2\cos\frac{2\pi}{9}\,, \qquad x = 2\cos\frac{4\pi}{9}\,, \qquad x = 2\cos\frac{8\pi}{9}\,.$$

If we call the roots r_1, r_2, r_3, then we have the factorization

$$x^3 - 3x + 1 = (x - r_1)(x - r_2)(x - r_3)\,,$$

and expanding the right hand side shows that

$$(1) \qquad r_1 + r_2 + r_3 = 0\,, \qquad r_1 r_2 + r_2 r_3 + r_3 r_1 = -3\,, \qquad r_1 r_2 r_3 = -1\,.$$

In these equations we can permute the roots r_1, r_2, r_3. The six possibilities are obtained by letting the elements of S_3 act on the subscripts. In Figure 2.1, each element of S_3 is listed beside the corresponding permutation of the roots.

Figure 2.1:

	r_1	r_2	r_3
(1)	r_1	r_2	r_3
$(1, 2, 3)$	r_2	r_3	r_1
$(1, 3, 2)$	r_3	r_1	r_2
$(2, 3)$	r_1	r_3	r_2
$(1, 3)$	r_3	r_2	r_1
$(1, 2)$	r_2	r_1	r_3

In our study of polynomial equations, identities such as those given above in (1) will play a crucial role. We will return to this example in the notes at the end of Chapter 3, where we will show (by finding an additional identity) that only certain permutations of the roots leave all combinations of the coefficients unchanged.

Chapter 3

GROUPS

Symmetry occurs frequently and in many forms in nature. Starfish possess rotational symmetry; the human body exhibits bilateral symmetry. A third sort of symmetry appears in some wallpaper or tile patterns that can be shifted in various directions without changing their appearance.

Each coefficient of a polynomial is a symmetric function of the polynomial's roots. To see what we mean by this, consider a monic cubic polynomial $f(x)$ with roots r_1, r_2, r_3. Then

$$f(x) = (x - r_1)(x - r_2)(x - r_3) = x^3 + bx^2 + cx + d$$

where $r_1 + r_2 + r_3 = -b$, $r_1 r_2 + r_2 r_3 + r_3 r_1 = c$, and $r_1 r_2 r_3 = -d$. Notice that if we permute r_1, r_2, r_3 by (for example) replacing r_1 by r_2, r_2 by r_3, and r_3 by r_1, then the coefficients b, c, d remain unchanged. In fact, the coefficients remain unchanged under any permutation of the roots, and so we say that they are symmetric functions of the roots.

The important feature of symmetry is the way that the shapes (or roots) can be changed while the whole figure (or the coefficients) remains unchanged. Geometrically, individual points move (or, algebraically, the roots interchange) while the figure as a whole (or the polynomial) remains the same. With respect to symmetry, geometrically the important thing is not the position of the points but the operation of moving them, and similarly, with respect to considering the roots of polynomials, it is the operation of shifting the roots among themselves that is most important and not the roots themselves. This was the key insight that enabled Galois to give a complete answer to the problem of solving polynomial equations by radicals.

The mathematical idea needed for the study of symmetry is that of a group, and in this chapter we introduce this important concept. We first discuss the abstract definition of a group, and attempt to clarify it for the reader by considering a wide variety of examples in considerable detail. We include sections on permutation groups, in which we consider groups of symmetries of some geometric objects, and cyclic groups, which we have already met in the guise of $\mathbf{Z}$ and $\mathbf{Z}_n$. We also study one-to-one correspondences that preserve the group structure (you may wish to reread the discussion in the introduction to Chapter 2). The last section deals with the notion of a group formed from equivalence classes of elements of a group. The procedure by which we formed $\mathbf{Z}_n$ from $\mathbf{Z}$ can be extended to any group. Such groups, which we call "factor groups," play a crucial role in Chapter 8 in the study of Galois theory.

3.1   Definition of a Group

In describing the difference between arithmetic and algebra, it might be said that arithmetic deals exclusively with numbers, while algebra deals with letters that represent numbers. The next step in abstraction involves dealing with objects that may not even represent numbers. For example, in learning calculus it is necessary to develop an "arithmetic" for functions. To give another example, in working with matrices, it is again necessary to develop some rules for matrix operations, and these rules constitute an "arithmetic" for matrices. The common thread in these developments, from an algebraic point of view, is the idea of an operation. Thus, as operations, we have addition, multiplication, and composition of functions, together with addition and multiplication of matrices. When we write AB for a product of matrices, for example, we have created a notation that allows us to think of it as analogous to ordinary multiplication, even though it represents a more complicated computation.

The operations we will study will be *binary* operations; that is, we will consider only operations which combine two elements at a time. A useful model to use is that of a computer program that allows two inputs and combines them in some way to give a single output. If we have an operation on a particular set, then we require that combining two inputs from the set will result in an output belonging to the same set. Furthermore, the output must depend only on the inputs, so that the answer is unique (when two inputs are specified).

A binary operation $*$ on a set S is a rule that assigns to each ordered pair (a, b) of elements of S a unique element $a * b$ of S. For example, the ordinary operations of addition, subtraction, and multiplication are binary operations on the set of real numbers. The operation of division is not a binary operation on the real numbers because it is not defined for all ordered pairs of real numbers (division by zero is not defined). If we exclude zero from the set on which we are using $\div$, then for any ordered pair (a, b) of real numbers, applying the operation $\div$ we get the quotient a/b, which is uniquely defined and is again a nonzero real number, showing that we have a binary operation. Although subtraction is a binary operation on the set of all real numbers, it is not a binary operation on the set of natural numbers, since, for example, $1 - 2$ is not in the set of natural numbers.

A binary operation permits us to combine only two elements, and so *a priori* $a * b * c$ does not make sense. But $(a * b) * c$ does make sense because we first combine a and b to get $a * b$ and then combine this element with c to get $(a * b) * c$.

On the other hand, $a * (b * c)$ also makes sense because we combine b and c to get $b * c$ and then combine a with this element to get $a * (b * c)$. The point of requiring the *associative* law is that both of these options should yield the same result, giving $a * (b * c) = (a * b) * c$ for all $a, b, c \in S$. Notice that the order in which a, b, and c occur is not changed. For example, subtraction does not define an associative operation on the set $\mathbf{R}$ of real numbers, since it is not true that $a - (b - c) = (a - b) - c$ for all $a, b, c \in \mathbf{R}$ (we leave it up to the reader to find real numbers a, b, c for which the associative law fails).

3.1.1 Definition. *A **binary operation** $*$ on a set S is a function $* : S \times S \to S$ from the set $S \times S$ of all ordered pairs of elements in S into S.*

*The operation $*$ is said to be **associative** if $a * (b * c) = (a * b) * c$ for all $a, b, c \in S$.*

*An element $e \in S$ is called an **identity** element for $*$ if $a * e = a$ and $e * a = a$ for all $a \in S$.*

*If $*$ has an identity element e, and $a \in S$, then $b \in S$ is said to be an **inverse** for a if $a * b = e$ and $b * a = e$.*

To illustrate these ideas, we can look at some sets of real numbers. We have already noted that multiplication defines a binary operation on $\mathbf{R}$. The number 1 serves as an identity element, and if $a \in \mathbf{R}$ is nonzero, then $1/a$ is the inverse of a. The number 0 has no multiplicative inverse, since $0 \cdot x = 1$ has no solution in $\mathbf{R}$. If $S = \{x \in \mathbf{R} \mid x \geq 1\}$, then multiplication defines a binary operation on S, and 1 still works as an identity element. But now 1 is the only element of S that has a multiplicative inverse, since if $a > 1$, then $1/a < 1$, and thus $1/a \notin S$. If we redefine S to be $\{x \in \mathbf{R} \mid x > 1\}$, then S does not have an identity element. Finally, as an extreme example, if we choose S to be $\{x \in \mathbf{R} \mid x < 0\}$, then multiplication does not even define a binary operation on S, since it is false that the product of any two elements of S again belongs to S.

As a further illustration of the ideas in Definition 3.1.1, let S be the set of all functions from a set A into itself. If $\phi, \theta \in S$, then define $\phi * \theta$ by letting $\phi * \theta(a) = \phi(\theta(a))$ for all $a \in A$. This defines a binary operation on S, and the identity function is an identity element for the operation. Furthermore, composition of functions is associative, and the functions that have inverses are precisely the ones that are both one-to-one and onto. The set $M_n(\mathbf{R})$ of all $n \times n$ matrices with entries from the real numbers $\mathbf{R}$ provides another good example. Matrix multiplication defines a binary operation on $M_n(\mathbf{R})$, and the identity matrix serves as an identity element. The proof that associativity holds is a laborious one if done directly from the definition. The appropriate way to remember why it holds is to use the correspondence between matrices and linear transformations, under which matrix multiplication corresponds to composition of functions. Then we only need to observe that composition of functions is associative. Finally, recall that a matrix has a multiplicative inverse if and only if its determinant is nonzero. We should note that matrix multiplication does *not* define a binary operation on the set of nonzero matrices in $M_n(\mathbf{R})$, since the product of two nonzero matrices may very well be the zero matrix.

Addition of matrices defines an associative binary operation on $M_n(\mathbf{R})$, and in this case the identity element is the zero matrix. Each matrix has an inverse with respect to this operation, namely, its negative.

Since the definition of a binary operation involves a function, we sometimes face problems similar to those we have already encountered in checking that a function is well-defined. For example, consider the problem inherent in defining multiplication on the set of rational numbers

$$\mathbf{Q} = \left\{ \frac{m}{n} \,\middle|\, m, n \in \mathbf{Z} \text{ and } n \neq 0 \right\}$$

where m/n and p/q represent the same element if $mq = np$. (See Example 2.2.3 for a discussion of this equivalence relation.) If $a, b \in \mathbf{Q}$ with $a = m/n$ and $b = s/t$, then we use multiplication of integers to define $ab = ms/nt$. We must check that the product does not depend on how we choose to represent a and b. If we also have $a = p/q$ and $b = u/v$, then we must check that pu/qv is equivalent to ms/nt. Since m/n and p/q are assumed to be equivalent and s/t and u/v are assumed to be equivalent, we have $mq = np$ and $sv = tu$. Multiplying the two equations gives $(ms)(qv) = (nt)(pu)$, which shows that ms/nt is equivalent to pu/qv. This allows us to conclude that the given multiplication of rational numbers is well-defined.

3.1.2 Proposition. *Let* $*$ *be an associative binary operation on a set S.*

(a) *The operation* $*$ *has at most one identity element.*

(b) *If* $*$ *has an identity element, then any element of S has at most one inverse.*

Proof. (a) Suppose that e and e' are identity elements for $*$. Since e is an identity element, we have $e * e' = e'$, and since e' is an identity element, we have $e * e' = e$. Therefore $e = e'$.

(b) Let e be the identity element for S relative to the operation $*$. Let b and b' be inverses for the element a. Then $b * a = e$ and $a * b' = e$, and so using the fact that $*$ is associative we have

$$b' = e * b' = (b * a) * b' = b * (a * b') = b * e = b.$$

This completes the proof.   □

Part (b) of Proposition 3.1.2 justifies referring to *the* inverse of an element, whenever it exists. If $*$ is an associative binary operation on a set S, and $a \in S$ has an inverse, then we will use the notation a^{-1} to denote the **inverse** of a. Thus the equations that define an inverse (from Definition 3.1.1) can be rewritten in the following form: $a * a^{-1} = e$ and $a^{-1} * a = e$.

3.1.3 Proposition. *Let $*$ be an associative binary operation on a set S. If $*$ has an identity element and $a, b \in S$ have inverses a^{-1} and b^{-1}, respectively, then the inverse of a^{-1} exists and is equal to a, and the inverse of $a * b$ exists and is equal to $b^{-1} * a^{-1}$.*

Proof. Let e be the identity element for S relative to the operation $*$. The equations $a * a^{-1} = e$ and $a^{-1} * a = e$ that state that a^{-1} is the inverse of a also show that a is the inverse of a^{-1}. Using the associative property for $*$, the computation

$$
\begin{aligned}
(a * b) * (b^{-1} * a^{-1}) &= ((a * b) * b^{-1}) * a^{-1} \\
&= (a * (b * b^{-1})) * a^{-1} \\
&= (a * e) * a^{-1} = a * a^{-1} = e
\end{aligned}
$$

and a similar computation with $(b^{-1} * a^{-1}) * (a * b)$ shows that the inverse of $a * b$ is $b^{-1} * a^{-1}$. $\square$

The general binary operations we work with will normally be denoted multiplicatively; that is, instead of writing $a * b$ we will just write $a \cdot b$, or simply ab. Using this notation, the previous proposition shows that $(ab)^{-1} = b^{-1}a^{-1}$, provided that a and b have inverses. However, there are situations where it is natural to use a notation other than juxtaposition for the binary operation. In particular, when a binary operation $*$ satisfies the **commutative law** $a * b = b * a$, it is quite common to use additive notation for the operation.

We now come to the main goal of this section—the definition of a group. Since definitions are the basic building blocks of abstract mathematics, the student will not progress without learning all definitions very thoroughly and carefully. Learning a definition should include associating with it several examples that will immediately come to mind to illustrate the important points of the definition. Following the definition we verify some elementary properties of groups and provide some broad classes of examples: groups of numbers, with familiar operations; groups of permutations, in which the operation is composition of functions; and groups of matrices, using matrix multiplication.

3.1.4 Definition. *Let $(G, *)$ denote a nonempty set G together with a binary operation $*$ on G. That is, the following condition must be satisfied.*

(i) *Closure:* *For all $a, b \in G$, the element $a * b$ is a well-defined element of G.*

*Then G is called a **group** if the following properties hold.*

(ii) *Associativity:* *For all $a, b, c \in G$, we have $a * (b * c) = (a * b) * c$.*

(iii) *Identity:* *There exists an **identity element** $e \in G$, that is, an element $e \in G$ such that $e * a = a$ and $a * e = a$ for all $a \in G$.*

(iv) *Inverses:* *For each $a \in G$ there exists an **inverse** element $a^{-1} \in G$, that is, an element $a^{-1} \in G$ such that $a * a^{-1} = e$ and $a^{-1} * a = e$.*

Proposition 3.1.2 implies that the identity element e is unique. Proposition 3.1.3 implies that $(a^{-1})^{-1} = a$, and as an elementary consequence of this, we note that $a = b$ if and only if $a^{-1} = b^{-1}$.

In the definition of a group G we do not require commutativity. That is, we do not assume that $a * b = b * a$ for all $a, b \in G$, since we want to allow the definition to include groups in which the operation is given by composition of functions or multiplication of matrices.

In the definition, note carefully one distinction between an identity element and an inverse element: an identity element satisfies a condition for all other elements of G, whereas an inverse element is defined relative to a single element of G. The order in which the axioms (iii) and (iv) are stated is important, since it is impossible to talk about an inverse of an element until an identity element is known to exist.

In Definition 3.1.4 the axioms for a group are written out in full detail. We believe that as you are learning the definition of a group, it will help to have a "check list" of conditions that you can use to determine whether or not a set is a group under a given operation. In particular, we have listed the closure property separately, although it is actually a part of the definition of a binary operation. By doing this, we hope to make it impossible for you to forget to verify that the closure property holds when checking that a set with a given operation is a group.

After you are familiar with the definition of a group, it is convenient to have a shorter statement to remember. We next give a "compact" version of the definition. It includes all that you need to state the definition of a group; implicit in the words is a great deal of meaning.

3.1.4′ (Restatement of Definition 3.1.4) A group is a nonempty set G with an associative binary operation, such that G contains an identity element for the operation, and each element of G has an inverse in G.

If G is a group and $a \in G$, then for any positive integer n we define a^n to be the product of a with itself n times. This can also be done inductively by letting $a^n = a * a^{n-1}$. It is not difficult to show that the exponential laws

$$a^m * a^n = a^{m+n} \qquad \text{and} \qquad (a^m)^n = a^{mn}$$

must hold for all positive exponents m, n. To illustrate, we have

$$a^2 * a^3 = (a * a) * (a * a * a) = a^5$$

and

$$(a^2)^3 = (a^2) * (a^2) * (a^2) = (a * a) * (a * a) * (a * a) = a^6.$$

To extend these laws from positive exponents to all integral exponents, we define $a^0 = e$, where e is the identity element of G, and $a^{-n} = (a^n)^{-1}$. Using these definitions, the above rules for exponents extend to all integers. The general proof is left as an exercise.

To begin our examples of groups, we now consider the set $\mathbf{R}$ of all real numbers, using as an operation the standard multiplication of real numbers. It is easy to check that the first three axioms for a group are satisfied, but the fourth axiom fails because 0 can never have a multiplicative inverse ($0 \cdot x = 1$ has no solution). Thus in order to define a group using the standard multiplication, we must reduce the set we work with.

Example 3.1.1 (Multiplicative groups of numbers).

Let $\mathbf{R}^\times$ denote the set of nonzero real numbers, with the operation given by standard multiplication. The first group axiom holds since the product of any two nonzero real numbers is still nonzero. The remaining axioms are easily seen to hold, with 1 playing the role of an identity element, and $1/a$ giving the inverse of an element $a \in \mathbf{R}^\times$. Thus $\mathbf{R}^\times$ is a group under multiplication.

Similarly, we have the groups $\mathbf{Q}^\times$ of all nonzero rational numbers and $\mathbf{C}^\times$ of all nonzero complex numbers, under the operation of ordinary multiplication. If we attempt to form a multiplicative group from the integers $\mathbf{Z}$, we have to restrict ourselves to just ± 1, since these are the only integers that have multiplicative inverses in $\mathbf{Z}$.

The development of these number systems from first principles is outlined in Section A.2 of the appendix, as it is beyond the scope of this course to give a full development of them. To actually verify the group axioms in the course of such a development is a long and arduous task. We have chosen to take a naive approach, by assuming that the reader is familiar with the properties and is willing to accept that a careful development is possible. □

The previous examples exhibit some of the most familiar groups. The next proposition deals with the most basic type of group. Recall that a permutation of a set S is a one-to-one function from S onto S. We will show in Section 3.6 that groups of permutations provide the most general models of groups. For convenience, we repeat the notation established in Section 2.3. The use of the word *group* in the definition of Sym(S) will be justified by Proposition 3.1.6.

3.1.5 Definition. *The set of all permutations of a set S is denoted by* Sym(S). *The set of all permutations of the set* $\{1, 2, \ldots, n\}$ *is denoted by* S_n.

The group Sym(S) *is called the **symmetric group** on S, and S_n is called the **symmetric group of degree** n.*

3.1.6 Proposition. *If S is any nonempty set, then* Sym(S) *is a group under the operation of composition of functions.*

Proof. The closure axiom is satisfied since by Proposition 2.1.5 the composite of two one-to-one and onto functions is again one-to-one and onto. Composition of functions is associative by Proposition 2.1.3. The identity function on S serves as an identity element for $\text{Sym}(S)$. Finally, by Proposition 2.1.7 a function from S into S is one-to-one and onto if and only if it has an inverse function, and the inverse is again one-to-one and onto, so it belongs to $\text{Sym}(S)$. $\square$

To see that S_n has $n!$ elements, let $S = \{1, 2, \ldots, n\}$. To define a permutation $\sigma : S \to S$, there are n choices for $\sigma(1)$. In order to make σ one-to-one, we must have $\sigma(2) \neq \sigma(1)$, and so there are only $n - 1$ choices for $\sigma(2)$. Continuing this analysis we can see that there will be a total of $n \cdot (n - 1) \cdots 2 \cdot 1 = n!$ possible distinct permutations of S.

In Table 3.1.1 we give the multiplication table for the group S_3. We will use cycle notation for the permutations. The function that leaves all three elements $1, 2, 3$ fixed is the identity, which we will denote by (1). It is impossible for a permutation on three elements to leave exactly two elements fixed, so next we consider the functions that fix one element while interchanging the other two. The function that interchanges 1 and 2 will be denoted by $(1, 2)$. Similarly, we have $(1, 3)$ and $(2, 3)$. Finally, there are two permutations that leave no element fixed. They are denoted by $(1, 2, 3)$ for the permutation σ with $\sigma(1) = 2$, $\sigma(2) = 3$, and $\sigma(3) = 1$; and $(1, 3, 2)$ for the permutation τ with $\tau(1) = 3$, $\tau(3) = 2$, and $\tau(2) = 1$. Remember that the products represent composition of functions and must be evaluated in the usual way functions are composed, from right to left. In Table 3.1.1, to find the product $\sigma\tau$, look in the row to the right of σ for the entry in the column below τ. Note that we are using the notation $\sigma\tau$ to indicate composition of functions, as we did in Section 2.3. This is much simpler than writing $\sigma \circ \tau$ or $\sigma * \tau$.

In each row of Table 3.1.1, each element of the group occurs exactly once. The same is true in each column. This phenomenon occurs in any such group table. To explain this, suppose we look at the row corresponding to an element a. The entries in this row consist of all elements of the form ag, for $g \in G$. The next proposition, the cancellation law, implies that if $g_1 \neq g_2$, then $ag_1 \neq ag_2$. This guarantees that no group elements are repeated in the row. A similar argument applies to the column determined by a, which consists of elements of the form ga for $g \in G$.

In the next proposition, we drop the notation $a * b$ for the product of $a, b \in G$, and simply write ab instead. It is important to remember how we are using this shorthand notation. For example, addition of exponents takes on a more familiar look: $a^m a^n = a^{m+n}$. We will need to use the axioms for a group, and if we omit the symbol $*$ for the group operation, we have the following properties (see Definition 3.1.4′). We have (ii) $a(bc) = (ab)c$ for all $a, b, c \in G$, (iii) there exists $e \in G$ with $ea = a$ and $ae = a$ for all $a \in G$, and (iv) for each $a \in G$ there exists $a^{-1} \in G$ with $aa^{-1} = e$ and $a^{-1}a = e$.

Table 3.1.1: Multiplication in S_3

	(1)	(1,2,3)	(1,3,2)	(1,2)	(1,3)	(2,3)
(1)	(1)	(1,2,3)	(1,3,2)	(1,2)	(1,3)	(2,3)
(1,2,3)	(1,2,3)	(1,3,2)	(1)	(1,3)	(2,3)	(1,2)
(1,3,2)	(1,3,2)	(1)	(1,2,3)	(2,3)	(1,2)	(1,3)
(1,2)	(1,2)	(2,3)	(1,3)	(1)	(1,3,2)	(1,2,3)
(1,3)	(1,3)	(1,2)	(2,3)	(1,2,3)	(1)	(1,3,2)
(2,3)	(2,3)	(1,3)	(1,2)	(1,3,2)	(1,2,3)	(1)

3.1.7 Proposition (Cancellation Property for Groups). *Let G be a group, and let $a, b, c \in G$.*

(a) *If $ab = ac$, then $b = c$.*
(b) *If $ac = bc$, then $a = b$.*

Proof. Given $ab = ac$, multiplying on the left by a^{-1} (which exists since G is a group) gives $a^{-1}(ab) = a^{-1}(ac)$. Using the associative law, $(a^{-1}a)b = (a^{-1}a)c$. Then $eb = ec$, which shows that $b = c$. The proof of the second part of the proposition is similar. □

The next proposition provides some motivation for the study of groups. It shows that the group axioms are precisely the assumptions necessary to solve equations of the form $ax = b$ or $xa = b$.

3.1.8 Proposition. *If G is a group and $a, b \in G$, then each of the equations $ax = b$ and $xa = b$ has a unique solution.*

Conversely, if G is a nonempty set with an associative binary operation in which the equations $ax = b$ and $xa = b$ have solutions for all $a, b \in G$, then G is a group.

Proof. Let G be a group, and let $a, b \in G$. Then a has an inverse, and substituting $x = a^{-1}b$ in the equation $ax = b$ gives

$$a(a^{-1}b) = (aa^{-1})b = eb = b,$$

showing that we have found a solution. If s and t are solutions of the equation $ax = b$, then $as = b = at$, and so $s = t$ by Proposition 3.1.7, showing that we have in fact found a unique solution of $ax = b$. Similarly, $x = ba^{-1}$ can be shown to be the unique solution of the equation $xa = b$.

Conversely, suppose that G has an associative binary operation under which the equations $ax = b$ and $xa = b$ have solutions for all elements $a, b \in G$. There is at least one element $a \in G$, and so we first let e be a solution of the equation $ax = a$. Next, we will show that $be = b$ for all $b \in G$. Let $b \in G$ be given, and let c be a solution to the equation $xa = b$, so that $ca = b$. Then

$$be = (ca)e = c(ae) = ca = b.$$

Similarly, there exists an element $e' \in G$ such that $e'b = b$ for all $b \in G$. But then $e' = e'e$ and $e'e = e$, and so we conclude that $e' = e$. Thus e satisfies the conditions that show it to be an identity element for G.

Finally, given any element $b \in G$, we must find an inverse for b. Let c be a solution of the equation $bx = e$ and let d be a solution of the equation $xb = e$. Then

$$d = de = d(bc) = (db)c = ec = c$$

and so $d = c$. Thus $bc = e$ and $cb = e$, and so c is an inverse for b. This completes the proof that G is a group. $\square$

We have seen examples of groups of several different types. We already need some definitions to describe them. Groups that satisfy the commutative law are named in honor of Niels Abel, who was active in the early nineteenth century. He showed that such groups were important in the theory of equations, and gave a proof that the general fifth degree equation cannot be solved by radicals.

3.1.9 Definition. *A group G is said to be **abelian** if $ab = ba$ for all $a, b \in G$.*

In an abelian group G, the operation is very often denoted additively. With this notation, the associative law $a*(b*c) = (a*b)*c$ becomes $a+(b+c) = (a+b)+c$ for all $a, b, c \in G$. The identity element is then usually denoted by 0 and is called a **zero element**, and the equations $e * a = a$ and $a * e = a$ that define the identity e are rewritten as $0 + a = a$ and $a + 0 = a$. The additive inverse of an element a is denoted by $-a$, and satisfies the equations $a + (-a) = 0$ and $(-a) + a = 0$.

It may be useful to interpret Propositions 3.1.7 and 3.1.8 in additive notation. We begin by rewriting these propositions using our original notation $*$ for the operation. Proposition 3.1.7 says that if a, b, c are elements of a group G, then $a * b = a * c$ implies $b = c$, and $a * c = b * c$ implies $a = b$. If we use the symbol $+$ in place of $*$, we have the statement $a + b = a + c$ implies $b = c$, for all $a, b, c \in G$. Because we are assuming that the commutative law holds, the second condition that $a + c = b + c$ implies $a = b$ does not add any additional information about G. In additive notation, the equation $a * x = b$ in Proposition 3.1.8 becomes $a + x = b$, so we have the statement that in any abelian group $(G, +)$ the equation $a + x = b$ has a unique solution, for any $a, b \in G$. The solution is $x = (-a) + b$.

Let G be an abelian group, and let $a \in G$. For a positive integer n, the sum of a with itself n times will be denoted by na. In additive notation, this replaces the exponential notation a^n. It is important to remember that this is not a multiplication that takes place in G, since n is not an element of G. For instance, G might be the group $M_k(\mathbf{R})$ of $k \times k$ matrices over $\mathbf{R}$, under addition of matrices. If $n \in \mathbf{Z}$ and A is a matrix in G, it is easy to see why nA makes sense, even though the multiplication of matrices by integers does not define a binary operation on G.

In a group denoted multiplicatively, we defined $a^0 = e$. In additive notation this becomes $0 \cdot a = 0$. Notice that in the equation $0 \cdot a = 0$ the first 0 is the integer 0, while the second 0 is the identity of the group. Similarly, in multiplicative notation we defined $a^{-n} = (a^n)^{-1}$, for any positive integer n, and this becomes $(-n)a = -(na)$ in additive notation. The standard laws of exponents $a^m * a^n = a^{m+n}$ and $(a^m)^n = a^{mn}$ are then expressed as the following equations, which hold for all $a \in G$ and all $m, n \in \mathbf{Z}$:

$$ma + na = (m + n)a \qquad \text{and} \qquad m(na) = (mn)a \ .$$

Example 3.1.2 (Additive groups of numbers).

When considering groups with additive notation, the most familiar examples are found in ordinary number systems. The set of integers $\mathbf{Z}$ is probably the most basic example. Results stated in Section A.3 of the appendix show that $\mathbf{Z}$ is closed under addition, that it satisfies the associative law, that 0 serves as the additive identity element, and that any integer n has an additive inverse $-n$.

Additional additive groups can be found by considering larger sets of numbers. In particular, the set of rational numbers $\mathbf{Q}$, the set of real numbers $\mathbf{R}$, and the set of complex numbers $\mathbf{C}$ all form groups using ordinary addition. We make the convention that when we refer to $\mathbf{Z}, \mathbf{Q}, \mathbf{R}$, and $\mathbf{C}$ as groups, the operation will be understood to be ordinary addition, unless we explicitly use a different operation. In particular, this convention does not apply to variants of these symbols such as $\mathbf{R}^{\times}$. $\square$

Our next example will consist of an entire class of finite abelian groups. Before giving the example we will quickly review the notion of congruence for integers from Sections 1.3 and 1.4. We also need to introduce a notation for the size of a finite group.

3.1.10 Definition. *A group G is said to be a finite group if the set G has a finite number of elements. In this case, the number of elements is called the **order** of G, denoted by |G|. If G is not finite, it is said to be an **infinite group**.*

Let n be a positive integer, which we call the *modulus*. Then two integers a, b are *congruent modulo n*, written $a \equiv b \pmod{n}$, if a and b have the same remainder when divided by n. It can be shown that $a \equiv b \pmod{n}$ if and only if $a - b$ is divisible by n, and this condition is usually the easier one to work with. If we let $[a]_n$ denote the set of all integers that are congruent to a modulo n, then it is possible to define an addition for these *congruence classes*. Given $[a]_n$ and $[b]_n$ we define

$$[a]_n + [b]_n = [a + b]_n \ .$$

There is a question as to whether we have defined a binary operation, because the sum appears to depend on our choice of a as the representative of the congruence class $[a]_n$ and b as the representative of the congruence class $[b]_n$. Proposition 1.4.2 shows that if $a_1 \equiv a_2 \pmod{n}$ and $b_1 \equiv b_2 \pmod{n}$, then it is true that $a_1 + b_1 \equiv a_2 + b_2 \pmod{n}$, and so the sum is well-defined. The set of all congruence classes modulo n is denoted by $\mathbf{Z}_n$.

Example 3.1.3 (Group of integers modulo n).

> Let n be a positive integer. The set $\mathbf{Z}_n$ of integers modulo n is an abelian group under addition of congruence classes. The group $\mathbf{Z}_n$ is finite and $|\mathbf{Z}_n| = n$.
>
> Proposition 1.4.2 shows that addition of congruence classes defines a binary operation. The associative law holds since for all $a, b, c \in \mathbf{Z}$ we have
>
> $$\begin{aligned} [a]_n + ([b]_n + [c]_n) &= [a]_n + [b + c]_n = [a + (b + c)]_n = [(a + b) + c]_n \\ &= [a + b]_n + [c]_n = ([a]_n + [b]_n) + [c]_n \ . \end{aligned}$$
>
> The commutative law holds since for all $a, b \in \mathbf{Z}$ we have
>
> $$[a]_n + [b]_n = [a + b]_n = [b + a]_n = [b]_n + [a]_n \ .$$
>
> Because $[a]_n + [0]_n = [a + 0]_n = [a]_n$ and $[a]_n + [-a]_n = [a - a]_n = [0]_n$, all of the necessary axioms are satisfied, and $\mathbf{Z}_n$ is an abelian group.
>
> For each $a \in \mathbf{Z}$ there exists a unique integer r with $0 \le r < n$ such that $[a]_n = [r]_n$, and so $|\mathbf{Z}_n| = n$. □

As with addition of congruence classes in $\mathbf{Z}_n$, Proposition 1.4.2 implies that the multiplication defined by

$$[a]_n \cdot [b]_n = [a \cdot b]_n$$

is in fact well-defined. Furthermore, a proof utilizing the corresponding properties of multiplication of integers can be given to show that multiplication of congruence classes is associative and commutative.

If a is an integer that is relatively prime to n, then there exist integers b, m such that $ab + mn = 1$. This gives $[a]_n \cdot [b]_n + [m]_n \cdot [n]_n = [1]_n$, or simply $[a]_n \cdot [b]_n = [1]_n$, since $[n]_n = [0]_n$. Thus $[a]_n$ has a multiplicative inverse. Conversely, if $[a]_n$ has a multiplicative inverse $[b]_n$, then $ab - 1$ must be divisible by n, and this implies that a and n are relatively prime.

The set of distinct congruence classes $[a]_n$ such that $(a, n) = 1$ is denoted by $\mathbf{Z}_n^\times$. Recall that the number of elements in this set is given by the Euler φ-function.

Example 3.1.4 (Group of units modulo n).

Let n be a positive integer. The set $\mathbf{Z}_n^\times$ of units of $\mathbf{Z}_n$ is an abelian group under multiplication of congruence classes. The group $\mathbf{Z}_n^\times$ is finite and $|\mathbf{Z}_n^\times| = \varphi(n)$.

Proposition 1.4.10 shows that multiplication of congruence classes is closed on $\mathbf{Z}_n^\times$. Another way to prove this is to use Proposition 1.2.3 (d) to show that ab is relatively prime to n if and only if both a and b are relatively prime to n. We have remarked that the associative and commutative laws can easily be checked. The element $[1]_n$ serves as an identity element. The set was defined so as to include all congruence classes that have multiplicative inverses. These multiplicative inverses are again in the set, so it follows that $\mathbf{Z}_n^\times$ is a group.

As a special case, in Table 3.1.2 we include the multiplication table of $\mathbf{Z}_8^\times$. □

Table 3.1.2: Multiplication in $\mathbf{Z}_8^\times$

	[1]	[3]	[5]	[7]
[1]	[1]	[3]	[5]	[7]
[3]	[3]	[1]	[7]	[5]
[5]	[5]	[7]	[1]	[3]
[7]	[7]	[5]	[3]	[1]

We next want to consider groups of matrices. In this section we will consider only matrices with entries from the set $\mathbf{R}$ of real numbers.

Example 3.1.5 ($M_n(\mathbf{R})$ under addition).

The set of all $n \times n$ matrices with entries in $\mathbf{R}$ forms a group under matrix addition. Since addition is defined componentwise, the zero matrix is the identity of $M_n(R)$, and the additive inverse of a matrix is its negative. □

We recall how to multiply 2×2 matrices:

$$\begin{bmatrix} a_{11} & a_{12} \\ a_{21} & a_{22} \end{bmatrix} \begin{bmatrix} b_{11} & b_{12} \\ b_{21} & b_{22} \end{bmatrix} = \begin{bmatrix} a_{11}b_{11} + a_{12}b_{21} & a_{11}b_{12} + a_{12}b_{22} \\ a_{21}b_{11} + a_{22}b_{21} & a_{21}b_{12} + a_{22}b_{22} \end{bmatrix} .$$

A matrix $\begin{bmatrix} a & b \\ c & d \end{bmatrix}$ has an inverse if and only if its **determinant** $ad - bc$ is nonzero, and the inverse can be found as follows:

$$\begin{bmatrix} a & b \\ c & d \end{bmatrix}^{-1} = \frac{1}{ad - bc} \begin{bmatrix} d & -b \\ -c & a \end{bmatrix} .$$

In general, if (a_{ij}) and (b_{ij}) are $n \times n$ matrices, then the product (c_{ij}) of the two matrices is defined as the matrix whose i, j-entry is

$$c_{ij} = \sum_{k=1}^{n} a_{ik} b_{kj}$$

and this product makes sense since in **R** the two operations of addition and multiplication are well-defined.

3.1.11 Definition. *The set of all invertible $n \times n$ matrices with entries in **R** is called the **general linear group of degree** n **over the real numbers**, and is denoted by* $\mathrm{GL}_n(\mathbf{R})$.

The use of the word group in Definition 3.1.11 is justified by the next proposition.

3.1.12 Proposition. *The set $\mathrm{GL}_n(\mathbf{R})$ forms a group under matrix multiplication.*

Proof. If A and B are invertible matrices, then the formulas $(A^{-1})^{-1} = A$ and $(AB)^{-1} = B^{-1}A^{-1}$ show that $\mathrm{GL}_n(\mathbf{R})$ has inverses for each element and is closed under matrix multiplication. The identity matrix (with 1 in each entry on the main diagonal and 0 in every other entry) is the identity element of the group. We assume that you have seen the proof that matrix multiplication is associative in a previous course in linear algebra. The proof is straightforward but rather tedious. $\square$

EXERCISES: SECTION 3.1

1. Using ordinary addition of integers as the operation, show that the set of even integers is a group, but that the set of odd integers is not.

2. For each binary operation $*$ defined on a set below, determine whether or not $*$ gives a group structure on the set. If it is not a group, say which axioms fail to hold.

   †(a) Define $*$ on $\mathbf{Z}$ by $a * b = ab$.

   (b) Define $*$ on $\mathbf{Z}$ by $a * b = \max\{a, b\}$.

   †(c) Define $*$ on $\mathbf{Z}$ by $a * b = a - b$.

   (d) Define $*$ on $\mathbf{Z}$ by $a * b = |ab|$.

   †(e) Define $*$ on $\mathbf{R}^+$ by $a * b = ab$.

   (f) Define $*$ on $\mathbf{Q}$ by $a * b = ab$.

3. Let $(G, \cdot)$ be a group. Define a new binary operation $*$ on G by the formula $a * b = b \cdot a$, for all $a, b \in G$.

   (a) Show that $(G, *)$ is a group.

   (b) Give examples to show that $(G, *)$ may or may not be the same as $(G, \cdot)$.

4. Prove that multiplication of 2×2 matrices satisfies the associative law.

5. Is $GL_n(\mathbf{R})$ an abelian group? Support your answer by either giving a proof or a counterexample.

6. Write out the addition table for $\mathbf{Z}_8$.

7.† Write out the multiplication table for $\mathbf{Z}_7^\times$.

8. Write out the multiplication table for the following set of matrices over $\mathbf{Q}$:

$$\begin{bmatrix} 1 & 0 \\ 0 & 1 \end{bmatrix}, \begin{bmatrix} -1 & 0 \\ 0 & 1 \end{bmatrix}, \begin{bmatrix} 1 & 0 \\ 0 & -1 \end{bmatrix}, \begin{bmatrix} -1 & 0 \\ 0 & -1 \end{bmatrix}.$$

9. Let $G = \{x \in \mathbf{R} \mid x > 0 \text{ and } x \neq 1\}$. Define the operation $*$ on G by $a * b = a^{\ln b}$, for all $a, b \in G$. Prove that G is an abelian group under the operation $*$.

10. Show that the set $A = \{f_{m,b} : \mathbf{R} \rightarrow \mathbf{R} \mid m \neq 0 \text{ and } f_{m,b}(x) = mx + b\}$ of **affine functions** from $\mathbf{R}$ to $\mathbf{R}$ forms a group under composition of functions.

11. Show that the set of all 2×2 matrices over $\mathbf{R}$ of the form $\begin{bmatrix} m & b \\ 0 & 1 \end{bmatrix}$ with $m \neq 0$ forms a group under matrix multiplication.

12. In the group defined in Exercise 11 find all elements that commute with the element $\begin{bmatrix} 2 & 0 \\ 0 & 1 \end{bmatrix}$.

13. Let $S = \mathbf{R} - \{-1\}$. Define $*$ on S by $a * b = a + b + ab$. Show that $(S, *)$ is a group.

14. Let G be a group. We have shown that $(ab)^{-1} = b^{-1}a^{-1}$. Find a similar expression for $(abc)^{-1}$.

15. Let G be a group. If $g \in G$ and $g^2 = g$, then prove that $g = e$.

16. Show that a nonabelian group must have at least five distinct elements.

17. Let G be a group. For $a, b \in G$, prove that $(ab)^n = a^n b^n$ for all $n \in \mathbf{Z}$ if and only if $ab = ba$.

18. Let G be a group. Prove that $a^m a^n = a^{m+n}$ for all $a \in G$ and all $m, n \in \mathbf{Z}$.

19. Let G be a group. Prove that $(a^m)^n = a^{mn}$ for all $a \in G$ and all $m, n \in \mathbf{Z}$.

20. Let S be a nonempty finite set with a binary operation $*$ that satisfies the associative law. Show that S is a group if $a * b = a * c$ implies $b = c$ and $a * c = b * c$ implies $a = b$ for all $a, b, c \in S$. What can you say if S is infinite?

21. Let G be a finite set with an associative, binary operation given by a table in which each element of the set G appears exactly once in each row and column. Prove that G is a group. How do you recognize the identity element? How do you recognize the inverse of an element?

22. Let G be a group. Prove that G is abelian if and only if $(ab)^{-1} = a^{-1}b^{-1}$ for all $a, b \in G$.

23. Let G be a group. Prove that if $x^2 = e$ for all $x \in G$, then G is abelian.

    Hint: Observe that $x = x^{-1}$ for all $x \in G$.

24. Show that if G is a finite group with an even number of elements, then there must exist an element $a \in G$ with $a \neq e$ such that $a^2 = e$.

(العبارات لا تعني)

3.2 Subgroups

Verifying all of the group axioms in a particular example is often not an easy task. Many very useful examples, however, arise inside known groups. That is, we may want to consider a subset of a known group, and restrict the operation of the group to this subset. For example, it is easier to show that the set of matrices in Exercise 11 of Section 3.1 forms a group under multiplication by recognizing it as a subgroup of $GL_2(\mathbf{R})$ than it is to show that it is a group by verifying all of the properties of Definition 3.1.4.

In this section we introduce the notion of a subgroup, and present several conditions for checking that a subset of a given group is again a group. For example, we will show that if the closure axiom alone holds for a finite subset of a group, then that is enough to guarantee that the subset is a group. We will investigate one of the most important types of subgroup, that generated by a single element. Finally, we prove Lagrange's theorem, which relates the number of elements in any subgroup of a finite group to the total number of elements in the group.

If H is a subset of a group G, then for any $a, b \in H$ we can use the operation of G to define ab. This may not define a binary operation on H, unless we know that $ab \in H$ for all $a, b \in H$. In that case we say that we are using the operation on H **induced** by G.

3.2.1 Definition. *Let G be a group, and let H be a subset of G. Then H is called a **subgroup** of G if H is itself a group, under the operation induced by G.*

Example 3.2.1.

As our first examples of subgroups we consider some groups of numbers. We already know that $\mathbf{Z}$, $\mathbf{Q}$, $\mathbf{R}$, and $\mathbf{C}$ are groups under ordinary addition. Furthermore, as sets we have

$$\mathbf{Z} \subseteq \mathbf{Q} \subseteq \mathbf{R} \subseteq \mathbf{C}$$

and each group is a subgroup of the next since the given operations are consistent.

If we consider multiplicative groups of nonzero elements, we also have the subgroups

$$\mathbf{Q}^\times \subseteq \mathbf{R}^\times \subseteq \mathbf{C}^\times.$$

Note that we cannot include the set of nonzero integers in this diagram, since it is *not* a subgroup of $\mathbf{Q}^\times$. (Although the set of nonzero integers is closed under multiplication and contains an identity element, all elements except 1 and -1 fail to have multiplicative inverses.)

In the group $\mathbf{Z}$ of integers under addition consider the set of all multiples of a fixed positive integer n, denoted by

$$n\mathbf{Z} = \{x \in \mathbf{Z} \mid x = nk \text{ for some } k \in \mathbf{Z}\}.$$

Using the operation of addition induced from $\mathbf{Z}$, to show that $n\mathbf{Z}$ is a subgroup of $\mathbf{Z}$ we must check each of the axioms in the definition of a group. The closure axiom holds since if $a, b \in n\mathbf{Z}$, then we have $a = nq$ and $b = nk$ for some $q, k \in \mathbf{Z}$, and adding gives us $a + b = nq + nk = n(q + k)$. This shows that the sum of two elements in $n\mathbf{Z}$ has the correct form to belong to $n\mathbf{Z}$. The associative law holds for all elements in $\mathbf{Z}$, so in particular it holds for all elements in $n\mathbf{Z}$. The element 0 can be expressed in the form $0 = n \cdot 0$, and so it will work as an identity element for $n\mathbf{Z}$. Finally, the additive inverse of $x = nk$ has the correct form $-x = n(-k)$ to belong to $n\mathbf{Z}$, and so it also serves as an inverse in $n\mathbf{Z}$, since we have already observed that the identity elements of $\mathbf{Z}$ and $n\mathbf{Z}$ coincide.

For the final example of subgroups of sets of numbers, let $\mathbf{R}^+$ be the set of positive real numbers, and consider $\mathbf{R}^+$ as a subset of the multiplicative group $\mathbf{R}^\times$ of all nonzero real numbers. The product of two positive real numbers is again positive, so the closure axiom holds. The associative law holds for all real numbers, so in particular it holds for all real numbers in $\mathbf{R}^+$. The number 1 serves as an identity element, and the reciprocal of a positive number is still positive, so each element of $\mathbf{R}^+$ has a multiplicative inverse in $\mathbf{R}^+$. Thus we have shown that $\mathbf{R}^+$ is a subgroup of $\mathbf{R}^\times$. □

Example 3.2.2 ($\mathrm{SL}_2(\mathbf{R}) \subseteq \mathrm{GL}_2(\mathbf{R})$).

Recall our notation $\mathrm{GL}_2(\mathbf{R})$ for the set of all 2×2 invertible matrices over the real numbers $\mathbf{R}$. The set of 2×2 matrices with determinant equal to 1 is a subgroup of $\mathrm{GL}_2(\mathbf{R})$, which can be seen as follows: if $A, B \in \mathrm{GL}_2(\mathbf{R})$ with $\det(A) = 1$ and $\det(B) = 1$, then we have $\det(AB) = \det(A)\det(B) = 1$. The associative law holds for all 2×2 matrices. The identity matrix certainly has determinant equal to 1, and if $\det(A) = 1$, then $\det(A^{-1}) = 1$.

The set of all $n \times n$ matrices over $\mathbf{R}$ with determinant equal to 1 is called the **special linear group over R**, denoted by $\mathrm{SL}_n(\mathbf{R})$. Thus we have shown that $\mathrm{SL}_2(\mathbf{R})$ is a subgroup of $\mathrm{GL}_2(\mathbf{R})$. □

The associative law is always inherited by a subset of a known group, as we have noted in each of the examples of subgroups that we have considered. This means that at least one of the group axioms need not be checked. The next proposition and its corollaries are designed to give simplified conditions to use when checking that a subset is in fact a subgroup.

If a set S has a binary operation $\cdot$ defined on it, then a subset $X \subseteq S$ is said to be **closed** under the operation $\cdot$ if $a \cdot b \in X$ for all $a, b \in X$. This is the reason that the first property in the definition of a group is called the closure property. (It is actually redundant in the definition of a group since it is a part of the definition of a binary operation. However, we have included it in Definition 3.1.4 since it is very easy to forget the full implications of having a binary operation.)

The next proposition gives the conditions that are usually checked to determine whether or not a subset of a group is actually a subgroup. It can be summarized by saying that the subset must be closed under multiplication, must contain the identity, and must have inverses for each element.

3.2.2 Proposition. *Let G be a group with identity element e, and let H be a subset of G. Then H is a subgroup of G if and only if the following conditions hold:*

(i) *$ab \in H$ for all $a, b \in H$;*

(ii) *$e \in H$;*

(iii) *$a^{-1} \in H$ for all $a \in H$.*

Proof. First, assume that H is a subgroup of G. Since H is a group under the operation of G, the closure axiom guarantees that ab must belong to H whenever

a, b belong to H. There must be an identity element, say e', for H. Then considering the product in H, we have $e'e' = e'$. Now consider the same product as an element of G. Then we can write $e'e' = e'e$, and the cancellation law yields $e' = e$. Finally, if $a \in H$, then a must have an inverse b in H, with $ab = e$. But then in G we have $ab = e = aa^{-1}$, and the cancellation law implies that $a^{-1} = b$ is an element of H.

Conversely, suppose that H is a subset of G that satisfies the given conditions. Condition (i) shows that the operation of G defines a binary operation on H, and so the closure axiom holds. If $a, b, c \in H$, then in G we have the equation $a(bc) = (ab)c$, and so by considering this as an equation in H we see that H inherits the associative law. Conditions (ii) and (iii) assure that H has an identity element, and that every element of H has an inverse in H, since these elements have the same properties in H as they do when viewed as elements of G. □

Using the previous proposition, it is easy to see that for any group G, the entire set G is certainly a subgroup. At the other extreme, the set $\{e\}$ consisting of only the identity element is always a subgroup of G, called the **trivial subgroup**.

The next corollary shortens the subgroup conditions. In applying these conditions, it is crucial to show that the subset H is nonempty. Often the easiest way to do this is to show that H contains the identity element e.

3.2.3 Corollary. *Let G be a group and let H be a subset of G. Then H is a subgroup of G if and only if H is nonempty and $ab^{-1} \in H$ for all $a, b \in H$.*

Proof. First assume that H is a subgroup of G. Using condition (ii) of the previous proposition, we see that H is nonempty since $e \in H$. If $a, b \in H$, then $b^{-1} \in H$ by condition (iii) of the proposition, and so condition (i) implies that $ab^{-1} \in H$.

Conversely, suppose that H is a nonempty subset of G such that $ab^{-1} \in H$ for all $a, b \in H$. Since H is nonempty, there is at least one element a that belongs to H. Then $e \in H$ since $e = aa^{-1}$, and this product belongs to H by assumption. Next, if $a \in H$, then a^{-1} can be expressed in the form $a^{-1} = ea^{-1}$, and this product must belong to H since e and a belong to H. Finally, we must show that H is closed under products: if $a, b \in H$, then we have already shown that $b^{-1} \in H$. We can express ab in the form $a(b^{-1})^{-1}$, and then the given condition shows that ab must belong to H. □

If the subset in question known to be *finite* (and nonempty), then it is only necessary to check the closure axiom. This is a bit surprising, but very useful. The crucial step in the proof of the next corollary is to show that in this case the inverse of each element in the set can be expressed as a positive power of the element.

3.2.4 Corollary. *Let G be a group, and let H be a finite, nonempty subset of G. Then H is a subgroup of G if and only if $ab \in H$ for all $a, b \in H$.*

Proof. If H is a subgroup of G, then Proposition 3.2.2 implies that $ab \in H$ for all $a, b \in H$.

Conversely, assume that H is closed under the operation of G. We can use the previous corollary, provided we can show that $b^{-1} \in H$ whenever $b \in H$. Given $b \in H$, consider the powers $\{b, b^2, b^3, \ldots\}$ of b. These must all belong to H, by assumption, but since H is a finite set, they cannot all be distinct. There must be some repetition, say $b^n = b^m$ for positive integers $n > m$. The cancellation law then implies that $b^{n-m} = e$. Either $b = e$ or $n - m > 1$, and in the second case we then have $bb^{n-m-1} = e$, which shows that $b^{-1} = b^{n-m-1}$. Thus b^{-1} can be expressed as a positive power of b, which must belong to H. □

Example 3.2.3 (Subgroups of S_3).

In S_3, the subset $\{(1), (1, 2, 3), (1, 3, 2)\}$ is closed under multiplication. The easiest way to see this is to look at the multiplication table given in Section 3.1. Since this subset is finite, Corollary 3.2.4 shows that it is a subgroup. The subsets $\{(1), (1, 2)\}$, $\{(1), (1, 3)\}$, and $\{(1), (2, 3)\}$ can be shown in the same way to be subgroups of S_3. □

Example 3.2.4.

In the group $GL_2(\mathbf{R})$ of all invertible 2×2 matrices with real entries, let H be the following set of matrices:

$$\begin{bmatrix} 1 & 0 \\ 0 & 1 \end{bmatrix}, \quad \begin{bmatrix} -1 & 0 \\ 0 & 1 \end{bmatrix}, \quad \begin{bmatrix} 1 & 0 \\ 0 & -1 \end{bmatrix}, \quad \begin{bmatrix} -1 & 0 \\ 0 & -1 \end{bmatrix}.$$

It is easy to see that the product of any two of these matrices is a diagonal matrix with entries ± 1, which will again be in the set. Since the set is finite and closed under matrix multiplication, Corollary 3.2.4 implies that it is a subgroup of $GL_2(\mathbf{R})$. □

Example 3.2.5.

Let G be the group $GL_n(\mathbf{R})$ of all invertible $n \times n$ matrices with entries in the real numbers. Let H be the set of all diagonal matrices in G. That is, H consists of all matrices in G that have zeros in all entries except those along the main diagonal. Since the matrices in G are all invertible, the diagonal entries must all be nonzero. In showing that H is a subgroup of G, we can no longer apply Corollary 3.2.4 since H is not a finite set. It is probably easiest to just use Proposition 3.2.2. We only need to observe that the identity matrix belongs to H, that the product of two diagonal matrices with nonzero entries again has the same property, and that the inverse of a diagonal matrix with nonzero entries is again a diagonal matrix. □

If G is a group, and a is any element of G, then Proposition 3.2.6 will show that the set of all powers of a is a subgroup of G, justifying the terminology of the next definition. This is the smallest subgroup of G that contains a, and turns out to have a particularly nice structure. Groups that consist of powers of a fixed element form an important class of groups and will be studied in depth in Section 3.5.

3.2.5 Definition. *Let G be a group, and let a be any element of G.*

*The set $\langle a \rangle = \{x \in G \mid x = a^n$ for some $n \in \mathbf{Z}\}$ is called the **cyclic subgroup generated by** a.*

*The group G is called a **cyclic group** if there exists an element $a \in G$ such that $G = \langle a \rangle$. In this case a is called a **generator** of G.*

3.2.6 Proposition. *Let G be a group, and let $a \in G$.*

(a) *The set $\langle a \rangle$ is a subgroup of G.*

(b) *If K is any subgroup of G such that $a \in K$, then $\langle a \rangle \subseteq K$.*

Proof. (a) The set $\langle a \rangle$ is closed under multiplication since if $a^m, a^n \in \langle a \rangle$, then $a^m a^n = a^{m+n} \in \langle a \rangle$. Furthermore, $\langle a \rangle$ includes the identity element and includes inverses, since by definition $a^0 = e$ and $(a^n)^{-1} = a^{-n}$.

(b) If K is any subgroup that contains a, then it must contain all positive powers of a since it is closed under multiplication. It also contains $a^0 = e$, and if $n < 0$, then $a^n \in K$ since $a^n = (a^{-n})^{-1}$. Thus $\langle a \rangle \subseteq K$. $\square$

The intersection of any collection of subgroups is again a subgroup (see Exercise 17). Given any subset S of a group G, the intersection of all subgroups of G that contain S is in fact the smallest subgroup that contains S. In the case $S = \{a\}$, by the previous proposition we obtain $\langle a \rangle$. In the case of two elements a, b of a nonabelian group G, it becomes much more complicated to describe the smallest subgroup of G that contains a and b. The general problem of listing all subgroups of a given group becomes difficult very quickly as the order of the group increases.

Example 3.2.6.

In the multiplicative group $\mathbf{C}^\times$, consider the powers of i. We have $i^2 = -1$, $i^3 = -i$, and $i^4 = 1$. From this point on the powers repeat, since $i^5 = ii^4 = i$, $i^6 = i^2 i^4 = -1$, etc. For negative exponents we have $i^{-1} = -i$, $i^{-2} = -1$, and $i^{-3} = i$. Again, from this point on the powers repeat. Thus we have

$$\langle i \rangle = \{1, \, i, \, -1, \, -i\} \, .$$

The situation is quite different if we consider $\langle 2i \rangle$. In this case the powers of $2i$ are all distinct, and the subgroup generated by $2i$ is infinite:

$$\langle 2i \rangle = \left\{ \ldots, \, \frac{1}{16}, \, \frac{1}{8}i, \, -\frac{1}{4}, \, -\frac{1}{2}i, \, 1, \, 2i, \, -4, \, -8i, \, 16, \, 32i, \, \ldots \right\} . \quad \square$$

Example 3.2.7 (Z is cyclic).

Consider the group $\mathbf{Z}$, using the standard operation of addition of integers. Since the operation is denoted additively rather than multiplicatively, we must consider multiples rather than powers. Thus $\mathbf{Z}$ is cyclic if and only if there exists an integer a such that $\mathbf{Z} = \{na \mid n \in \mathbf{Z}\}$. Either $a = 1$ or $a = -1$ will satisfy the condition, so $\mathbf{Z}$ is cyclic, with generators 1 and -1. □

Example 3.2.8 ($\mathbf{Z}_n$ is cyclic).

The additive group $\mathbf{Z}_n$ of integers modulo n is also cyclic, generated by $[1]_n$, since each congruence class can be expressed as a finite sum of $[1]_n$'s. To be precise, $[k]_n = k[1]_n$.

It is very interesting to determine all possible generators of $\mathbf{Z}_n$. If $[a]_n$ is a generator of $\mathbf{Z}_n$, then in particular $[1]_n$ must be a multiple of $[a]_n$. On the other hand, if $[1]_n$ is a multiple of $[a]_n$, then certainly every other congruence class modulo n is also a multiple of $[a]_n$. Thus to determine all of the generators of $\mathbf{Z}_n$ we only need to determine the integers a such that some multiple of a is congruent to 1. These are precisely the integers that are relatively prime to n. □

Example 3.2.9 (Sometimes $\mathbf{Z}_n^\times$ is cyclic, sometimes not).

The multiplicative groups $\mathbf{Z}_n^\times$ provide many interesting examples. We first consider $\mathbf{Z}_5^\times$. We will omit the subscript on the congruence classes when it is clear from the context. We have $[2]^2 = [4]$, $[2]^3 = [3]$, and $[2]^4 = [1]$. Thus each element of $\mathbf{Z}_5^\times$ is a power of $[2]$, showing that the group is cyclic, which we write as $\mathbf{Z}_5^\times = \langle [2] \rangle$. We note that $[3]$ is also a generator, but $[4]$ is not, since $[4]^2 = [1]$, and so $\langle [4] \rangle = \{[1], [4]\} \neq \mathbf{Z}_5^\times$.

Next, consider $\mathbf{Z}_8^\times = \{[1], [3], [5], [7]\}$. The square of each element is the identity, so we have $\langle [3] \rangle = \{[1], [3]\}$, $\langle [5] \rangle = \{[1], [5]\}$, and $\langle [7] \rangle = \{[1], [7]\}$. Thus $\mathbf{Z}_8^\times$ is not cyclic. □

Example 3.2.10 (S_3 is not cyclic).

The group S_3 is not cyclic. We can list all cyclic subgroups as follows:

$\langle (1) \rangle = \{(1)\}$;

$\langle (1, 2) \rangle = \{(1), (1, 2)\}$, $\langle (1, 3) \rangle = \{(1), (1, 3)\}$, $\langle (2, 3) \rangle = \{(1), (2, 3)\}$;

$\langle (1, 2, 3) \rangle = \{(1), (1, 2, 3), (1, 3, 2)\} = \langle (1, 3, 2) \rangle$.

Since no cyclic subgroup is equal to all of S_3, it is not cyclic. That is, we have shown that there is *no* permutation σ in S_3 for which $S_3 = \langle \sigma \rangle$. □

3.2.7 Definition. *Let a be an element of the group G.*

*If there exists a positive integer n such that $a^n = e$, then a is said to have **finite order**, and the smallest such positive integer is called the **order** of a, denoted by $o(a)$.*

*If there does not exist a positive integer n such that $a^n = e$, then a is said to have **infinite order**.*

We note that the proof of Corollary 3.2.4 shows that every element of a finite group must have finite order. We have already considered the order of a permutation in Section 2.3. You may recall that problems in Sections 1.3 and 1.4 dealt with the orders of elements in $\mathbf{Z}_n$ and $\mathbf{Z}_n^\times$.

The next proposition establishes some basic facts about the order of an element. The proof of part (b) should remind you of the proof of Theorem 1.1.4. In fact, instead of writing out the details, we could have applied Theorem 1.1.4 to the set $I = \{k \in \mathbf{Z} \mid a^k = e\}$, since this set can easily be shown to be closed under addition and subtraction.

3.2.8 Proposition. *Let a be an element of the group G.*

(a) *If a has infinite order, then $a^k \neq a^m$ for all integers $k \neq m$.*

(b) *If a has finite order and $k \in \mathbf{Z}$, then $a^k = e$ if and only if $o(a) \mid k$.*

(c) *If a has finite order $o(a) = n$, then for all integers k, m, we have $a^k = a^m$ if and only if $k \equiv m \pmod{n}$. Furthermore, $|\langle a \rangle| = o(a)$.*

Proof. (a) Let a have infinite order. Suppose that $a^k = a^m$ for $k, m \in \mathbf{Z}$, with $k \geq m$. Then multiplying by $(a^m)^{-1}$ gives us $a^{k-m} = e$, and since a is not of finite order, we must have $k - m = 0$.

(b) Let $o(a) = n$, and suppose that $a^k = e$. Using the division algorithm we can write $k = nq + r$, where $0 \leq r < n$. Thus

$$a^r = a^{k-nq} = a^k a^{n(-q)} = a^k (a^n)^{-q} = e \cdot e^{-q} = e.$$

Since $0 \leq r < o(a)$ and $a^r = e$, by the definition of $o(a)$ we must have $r = 0$, and so $k = q \cdot o(a)$.

On the other hand, if $o(a) \mid k$, then $k = nq$ for some $q \in \mathbf{Z}$. But then $a^k = (a^n)^q = e^q = e$.

(c) We must show that if a has finite order, then $a^k = a^m$ if and only if $k \equiv m \pmod{n}$. We first observe that since we are working in a group, $a^k = a^m$ if and only if $a^{k-m} = e$. By part (b) this occurs if and only if $n \mid (k - m)$, which is equivalent to the statement that $k \equiv m \pmod{n}$.

Finally, using Corollary 3.2.4, it follows that the subset $S = \{e, a, \ldots, a^{n-1}\}$ is a subgroup that contains a. Hence $\langle a \rangle \subseteq S$. On the other hand, $S \subseteq \langle a \rangle$ by the definition of $\langle a \rangle$. Thus $|\langle a \rangle| = |S| = o(a)$. $\square$

If you check all of the examples that we have given of subgroups of finite groups, you will see that in every case the number of elements in the subgroup is a divisor of the order of the group. This is always true, and is a very useful result. For example, in checking whether or not a subset of a finite group is in fact a subgroup, you can immediately answer no if the number of elements in the subset is not a divisor of the order of the group. We have separated out part of the proof as a lemma.

3.2.9 Lemma. *Let H be a subgroup of the group G. For $a, b \in G$ define $a \sim b$ if $ab^{-1} \in H$. Then $\sim$ is an equivalence relation.*

Proof. The relation is reflexive since $aa^{-1} = e \in H$. It is symmetric since if $a \sim b$, then $ab^{-1} \in H$. Since H contains the inverse of each of its elements, we have $ba^{-1} = (ab^{-1})^{-1} \in H$, showing that $b \sim a$. Finally, the relation is transitive since if $a \sim b$ and $b \sim c$ for any $a, b, c \in G$, then both ab^{-1} and bc^{-1} belong to H, so since H is a subgroup, the product $ac^{-1} = (ab^{-1})(bc^{-1})$ also belongs to H, showing that $a \sim c$.   □

The equivalence relation we have used to define congruence modulo n in $\mathbf{Z}$ is a special case of Lemma 3.2.9. If the operation of G is denoted additively, then the equivalence relation of Lemma 3.2.9 is defined by setting $a \sim b$ if $a + (-b)$ belongs to H. This is usually written as $a \sim b$ if $a - b \in H$. We consider the case when $G = \mathbf{Z}$ and H is the subgroup $n\mathbf{Z}$. Since $a \equiv b \pmod{n}$ if and only if $a - b$ is a multiple of n, we see that a is congruent to b modulo n if and only if $a - b$ is an element of the subgroup $n\mathbf{Z}$. This shows how to fit our earlier work on congruences into the new terminology of group theory.

3.2.10 Theorem (Lagrange). *If H is a subgroup of the finite group G, then the order of H is a divisor of the order of G.*

Proof. Let H be a subgroup of the finite group G, with $|G| = n$ and $|H| = m$. Let $\sim$ denote the equivalence relation defined in the previous lemma.

For any $a \in G$, let $[a]$ denote the equivalence class of a. We claim that the function $\rho_a : H \to [a]$ defined by $\rho_a(x) = xa$ for all $x \in H$ is a one-to-one correspondence between H and $[a]$. We first note that the stated codomain of ρ_a is correct since if $h \in H$, then $\rho_a(h) = ha \in [a]$ because $(ha)(a^{-1}) = h \in H$. To show that ρ_a is one-to-one, suppose that $h, k \in H$ with $\rho_a(h) = \rho_a(k)$. Then $ha = ka$, and since the cancellation property holds in any group, we can conclude that $h = k$. Finally, ρ_a is onto since if $y \in G$ with $y \sim a$, then we have $ya^{-1} = h$ for some $h \in H$, and thus the equation $\rho_a(x) = y$ has the solution $x = h$ since $ha = (ya^{-1})a = y$.

Since the equivalence classes of $\sim$ partition G, each element of G belongs to precisely one of the equivalence classes. We have shown that each equivalence class has m elements, since it is in one-to-one correspondence with H. Counting the elements of G according to the distinct equivalence classes, we simply get $n = mt$, where t is the number of distinct equivalence classes. □

Example 3.2.11.

In this example we will investigate equivalence relative to two different subgroups of S_3. The proof of Lagrange's theorem shows us how to proceed in a systematic way. First, let

$$H = \langle (1, 2, 3) \rangle = \{(1), (1, 2, 3), (1, 3, 2)\} \, .$$

By definition, the elements of H are equivalent to each other and form the first equivalence class. Any other equivalence class must be disjoint from the first one and have the same number of elements, so the only possibility is that the remaining elements of G must form a second equivalence class. Therefore the equivalence relation defined by the subgroup H determines two equivalence classes:

$$\{(1), (1, 2, 3), (1, 3, 2)\} \, , \qquad\qquad \{(1, 2), (1, 3), (2, 3)\} \, .$$

Next, let us consider the subgroup K generated by $(1, 2)$. Then

$$K = \{(1), (1, 2)\}$$

so the equivalence classes must each contain two elements. The proof of Lagrange's theorem shows that we can find the equivalence class of an element a by multiplying it on the left by all elements of K. If we let $a = (1, 2, 3)$, then we have two equivalent elements $(1)(1, 2, 3) = (1, 2, 3)$ and $(1, 2)(1, 2, 3) = (2, 3)$. There are two elements remaining, and they form the third equivalence class. Thus the equivalence relation defined by the subgroup K determines three equivalence classes:

$$\{(1), (1, 2)\} \, , \qquad \{(1, 2, 3), (2, 3)\} \, , \qquad \{(1, 3, 2), (1, 3)\} \, . \ □$$

3.2.11 Corollary. *Let G be a finite group of order n.*
  (a) *For any $a \in G$, $o(a)$ is a divisor of n.*
  (b) *For any $a \in G$, $a^n = e$.*

Proof. (a) The order of a is the same as the order of $\langle a \rangle$, which by Lagrange's theorem is a divisor of the order of G.
    (b) If a has order m, then by part (a) we have $n = mq$ for some integer q. Thus $a^n = a^{mq} = (a^m)^q = e$. □

Example 3.2.12 (Euler's theorem).

We can now give a short group theoretic proof of Euler's theorem (Theorem 1.4.11). We must show that if φ denotes the Euler φ function and a is any integer relatively prime to the positive integer n, then $a^{\varphi(n)} \equiv 1 \pmod{n}$. Let $G = \mathbf{Z}_n^\times$, the group of units modulo n (see Example 3.1.4). The order of G is given by $\varphi(n)$, and so by Corollary 3.2.11, raising any congruence class to the power $\varphi(n)$ must give the identity element. The statement $[a]^{\varphi(n)} = [1]$ is equivalent to $a^{\varphi(n)} \equiv 1 \pmod{n}$. □

3.2.12 Corollary. *Any group of prime order is cyclic.*

Proof. Let G be a group of order p, where p is a prime number. Let a be an element of G different from e. Then the order of $\langle a \rangle$ is not 1, and so it must be p since it is a divisor of p. This implies that $\langle a \rangle = G$, and thus G is cyclic. □

EXERCISES: SECTION 3.2

1. In $GL_2(\mathbf{R})$, find the order of each of the following elements.

   †(a) $\begin{bmatrix} 1 & -1 \\ 1 & 0 \end{bmatrix}$

   (b) $\begin{bmatrix} 0 & 1 \\ -1 & 0 \end{bmatrix}$

   †(c) $\begin{bmatrix} 1 & 1 \\ 0 & 1 \end{bmatrix}$

   (d) $\begin{bmatrix} -1 & 1 \\ 0 & 1 \end{bmatrix}$

2. Let $A = \begin{bmatrix} 1 & -1 \\ -1 & 0 \end{bmatrix} \in GL_2(\mathbf{R})$. Show that A has infinite order by proving that
   $$A^n = \begin{bmatrix} F_{n+1} & -F_n \\ -F_n & F_{n-1} \end{bmatrix}, \text{ for } n \geq 1, \text{ where } F_0 = 0, \ F_1 = 1, \text{ and } F_{n+1} = F_n + F_{n-1}$$
   is the Fibonacci sequence.

3. Prove that the set of all rational numbers of the form m/n, where $m, n \in \mathbf{Z}$ and n is square-free, is a subgroup of $\mathbf{Q}$ (under addition).

4. Show that $\{(1), (1,2)(3,4), (1,3)(2,4), (1,4)(2,3)\}$ is a subgroup of S_4.

5. For each of the following groups, find all cyclic subgroups of the group.

   †(a) $\mathbf{Z}_6$

   (b) $\mathbf{Z}_8$

   †(c) $\mathbf{Z}_9^{\times}$

   (d) S_4

6. Let $G = GL_2(\mathbf{R})$.

   (a) Show that $T = \left\{ \begin{bmatrix} a & b \\ 0 & d \end{bmatrix} \,\middle|\, ad \neq 0 \right\}$ is a subgroup of G.

   (b) Show that $D = \left\{ \begin{bmatrix} a & 0 \\ 0 & d \end{bmatrix} \,\middle|\, ad \neq 0 \right\}$ is a subgroup of G.

7. Let $G = GL_2(\mathbf{R})$. Show that the subset S of G defined by $S = \left\{ \begin{bmatrix} a & b \\ c & d \end{bmatrix} \,\middle|\, b = c \right\}$ of symmetric 2×2 matrices does not form a subgroup of G.

8. Let $G = GL_2(\mathbf{R})$. For each of the following subsets of $M_2(\mathbf{R})$, determine whether or not the subset is a subgroup of G.

   (a) $A = \left\{ \begin{bmatrix} a & b \\ 0 & 0 \end{bmatrix} \,\middle|\, ab \neq 0 \right\}$

   (b) $B = \left\{ \begin{bmatrix} 0 & b \\ c & 0 \end{bmatrix} \,\middle|\, bc \neq 0 \right\}$

   (c) $C = \left\{ \begin{bmatrix} 1 & 0 \\ 0 & c \end{bmatrix} \,\middle|\, c \neq 0 \right\}$

9. Let $G = GL_3(\mathbf{R})$. Show that $H = \left\{ \begin{bmatrix} 1 & a & b \\ 0 & 1 & c \\ 0 & 0 & 1 \end{bmatrix} \right\}$ is a subgroup of G.

10. Let m and n be nonzero integers, with $(m, n) = d$. Show that m and n belong to $d\mathbf{Z}$, and that if H is any subgroup of $\mathbf{Z}$ that contains both m and n, then $d\mathbf{Z} \subseteq H$.

11. Let S be a set, and let a be a fixed element of S. Show that $\{\sigma \in \mathrm{Sym}(S) \mid \sigma(a) = a\}$ is a subgroup of $\mathrm{Sym}(S)$.

12.† For each of the following groups, find all elements of finite order.

   (a) $\mathbf{R}^{\times}$

   (b) $\mathbf{C}^{\times}$

13. Let G be an abelian group, such that the operation on G is denoted additively. Show that $\{a \in G \mid 2a = 0\}$ is a subgroup of G. Compute this subgroup for $G = \mathbf{Z}_{12}$.

14. Let G be an abelian group. Show that the set of all elements of G of finite order forms a subgroup of G.

15. Prove that any cyclic group is abelian.

16. Prove or disprove this statement. If G is a group in which every proper subgroup is cyclic, then G is cyclic.

17. Prove that the intersection of any collection of subgroups of a group is again a subgroup.

18. Let G be the group of rational numbers, under addition, and let H, K be subgroups of G. Prove that if $H \neq \{0\}$ and $K \neq \{0\}$, then $H \cap K \neq \{0\}$.

19. Let G be a group, and let $a \in G$. The set $C(a) = \{x \in G \mid xa = ax\}$ of all elements of G that commute with a is called the **centralizer** of a.

    (a) Show that $C(a)$ is a subgroup of G.

    (b) Show that $\langle a \rangle \subseteq C(a)$.

    (c) Compute $C(a)$ if $G = S_3$ and $a = (1, 2, 3)$.

    (d) Compute $C(a)$ if $G = S_3$ and $a = (1, 2)$.

20. Compute the centralizer in $GL_2(\mathbf{R})$ of the matrix $\begin{bmatrix} 1 & 1 \\ 0 & 1 \end{bmatrix}$.

21. Let G be a group. The set $Z(G) = \{x \in G \mid xg = gx \text{ for all } g \in G\}$ of all elements that commute with every other element of G is called the **center** of G.

    (a) Show that $Z(G)$ is a subgroup of G.

    (b) Show that $Z(G) = \cap_{a \in G} C(a)$.

    (c) Compute the center of S_3.

22. Compute the center of $GL_2(\mathbf{R})$.

23. Let G be a cyclic group, and let a, b be elements of G such that neither $a = x^2$ nor $b = x^2$ has a solution in G. Show that $ab = x^2$ does have a solution in G.

24. Let G be a group with $a, b \in G$.

    (a) Show that $o(a^{-1}) = o(a)$.

    (b) Show that $o(ab) = o(ba)$.

    (c) Show that $o(aba^{-1}) = o(b)$.

25. Let G be a finite group, let $n > 2$ be an integer, and let S be the set of elements of G that have order n. Show that S has an even number of elements.

26. Let G be a group with $a, b \in G$. Assume that $o(a)$ and $o(b)$ are finite and relatively prime, and that $ab = ba$. Show that $o(ab) = o(a)o(b)$.

27. Find an example of a group G and elements $a, b \in G$ such that a and b each have finite order but ab does not.

3.3 Constructing Examples

Being able to construct your own examples is very important. To do this you need to be familiar with a good variety of groups. In this section we first study groups with orders up to 6. Then we introduce the notion of the direct product of two groups, which can be used to construct new groups from known ones. Finally, we introduce some new matrix groups.

As a corollary of Lagrange's theorem, we proved that any group of prime order must be cyclic. This shows that any group of order 2, 3, or 5 must be cyclic. It is then easy to write down what the multiplication table of the group must look like.

Next, take the case of a group G of order 4. As another corollary of Lagrange's theorem, we proved that the order of any element of a group must be a divisor of the order of the group. Since $|G| = 4$, we can only have elements (different from the identity) of order 2 or 4. If there exists an element $a \in G$ of order 4, then its four powers e, a, a^2, and a^3 must be the only elements in G, and so G is cyclic. On the other hand, if there is no element of order 4, then every element not equal to e must have order 2. This means that in the multiplication table for G the element e must occur down the main diagonal, and then by using the fact that each element must occur exactly once in each row and column, it can be shown that there is only one possible pattern for the table.

This analysis of the table does not imply that there exists a group with such a multiplication table. In particular, there is no guarantee that the associative law holds. However, a group that serves as a model for the table is not difficult to find. The group $\mathbf{Z}_8^\times$ of units modulo 8 is easily checked to have four elements, and the square of every element is the identity $[1]_8$. Thus the multiplication table for a group of four elements has one of only two possible patterns. These are listed in Table 3.3.1.

Table 3.3.1: Multiplication Tables for Groups of Order 4

	e	a	a^2	a^3
e	e	a	a^2	a^3
a	a	a^2	a^3	e
a^2	a^2	a^3	e	a
a^3	a^3	e	a	a^2

	e	a	b	c
e	e	a	b	c
a	a	e	c	b
b	b	c	e	a
c	c	b	a	e

We know of two basic examples of groups of order 6, the group $\mathbf{Z}_6$ of integers modulo 6, which is cyclic, and the group S_3 of permutations on three elements. The multiplication table for a cyclic group of order 6 is given in Table 3.3.2.

We have described S_3 by explicitly listing the permutations that belong to it. It is possible to give another description, which is easier to work with in many cases. Let $e = (1)$, let $a = (1, 2, 3)$, and let $b = (1, 2)$. Using the multiplication table

Table 3.3.2: Multiplication Table for a Cyclic Group of Order 6

	e	a	a^2	a^3	a^4	a^5
e	e	a	a^2	a^3	a^4	a^5
a	a	a^2	a^3	a^4	a^5	e
a^2	a^2	a^3	a^4	a^5	e	a
a^3	a^3	a^4	a^5	e	a	a^2
a^4	a^4	a^5	e	a	a^2	a^3
a^5	a^5	e	a	a^2	a^3	a^4

given in Table 3.1.1, we see that $a^2 = (1, 3, 2)$ and then $a^3 = aa^2 = e$. For the element b we have $b^2 = e$. Using the convention $a^0 = e$ and $b^0 = e$, we can express each element of S_3 in a unique way in the form $a^i b^j$, for $i = 0, 1, 2$ and $j = 0, 1$. Specifically, we have $(1) = e$, $(1, 2, 3) = a$, $(1, 3, 2) = a^2$, $(1, 2) = b$, $(1, 3) = ab$, and $(2, 3) = a^2 b$. To multiply two elements in this form, we must be able to find the expression for their product in the standard form, but to do so only requires the formula $ba = a^2 b$, which holds since $(1, 2)(1, 2, 3) = (2, 3)$. This allows us to give the following description of the symmetric group on three elements:

$$S_3 = \{e, a, a^2, b, ab, a^2 b\}, \qquad \text{where } a^3 = e, \ b^2 = e, \ ba = a^2 b \, .$$

Using this notation, we can rewrite the multiplication table as shown in Table 3.3.3.

Table 3.3.3: Multiplication Table for S_3

	e	a	a^2	b	ab	a^2b
e	e	a	a^2	b	ab	a^2b
a	a	a^2	e	ab	a^2b	b
a^2	a^2	e	a	a^2b	b	ab
b	b	a^2b	ab	e	a^2	a
ab	ab	b	a^2b	a	e	a^2
a^2b	a^2b	ab	b	a^2	a	e

The multiplication table for any group of order 6 must have the form of Table 3.3.2 or that of Table 3.3.3. We have indicated, in Exercises 15–17, how this can be proved.

We now introduce a new method of combining two subgroups. We have already observed that the intersection of subgroups of a group is again a subgroup. If H

and K are subgroups of a group G, then $H \cap K$ is the largest subgroup of G that is contained in both H and K. On the other hand, if we consider the smallest subgroup that contains both H and K, then it certainly must contain all products of the form hk, where $h \in H$ and $k \in K$. In certain cases, these products do form a subgroup. In general, though, it may be necessary to include more elements.

3.3.1 Definition. *Let G be a group, and let S and T be subsets of G. Then*

$$ST = \{x \in G \mid x = st \text{ for some } s \in S, \ t \in T\}.$$

If H and K are subgroups of G, then we simply call HK the **product** of H and K. The next proposition shows that if G is abelian, then the product of any two subgroups is again a subgroup since the condition of the proposition is satisfied whenever the elements of H and K commute with each other. If the operation of G is denoted additively, then instead of HK we write $H + K$, and refer to the **sum** of H and K.

3.3.2 Proposition. *Let G be a group, and let H and K be subgroups of G. If $h^{-1}kh \in K$ for all $h \in H$ and $k \in K$, then HK is a subgroup of G.*

Proof. Suppose that $h^{-1}kh \in K$ for all $h \in H$ and $k \in K$. To show that HK is closed, let $g_1, g_2 \in HK$. Then $g_1 = h_1 k_1$ and $g_2 = h_2 k_2$ for some $h_1, h_2 \in H$ and $k_1, k_2 \in K$. By our assumption we have $h_2^{-1} k_1 h_2 \in K$, say $h_2^{-1} k_1 h_2 = k_3$. Thus $k_1 h_2 = h_2 k_3$, and so

$$g_1 g_2 = (h_1 k_1)(h_2 k_2) = h_1 (k_1 h_2) k_2 = h_1 (h_2 k_3) k_2 = (h_1 h_2)(k_3 k_2).$$

This shows that $g_1 g_2 \in HK$, since $h_1 h_2 \in H$ and $k_3 k_2 \in K$. Since the identity element belongs to both H and K, we have $e = e \cdot e \in HK$. Finally, if $g = hk$ for $h \in H$ and $k \in K$, then

$$g^{-1} = k^{-1} h^{-1} = (h^{-1} h) k^{-1} h^{-1} = (h^{-1})((h^{-1})^{-1} k^{-1} h^{-1}) \in HK$$

since $h^{-1} \in H$ and $k^{-1} \in K$ yields $(h^{-1})^{-1} k^{-1} h^{-1} \in K$ by the given condition. $\square$

Example 3.3.1.

In the group $\mathbf{Z}_{15}^{\times}$, let $H = \{[1], [11]\}$ and $K = \{[1], [4]\}$. These are subgroups since $[11]$ and $[4]$ both have order two. The condition that $h^{-1}kh \in K$ for all $h \in H$ and $k \in K$ holds since $\mathbf{Z}_{15}^{\times}$ is abelian. Computing all possible products in HK gives us

$$[1][1] = [1], \quad [1][4] = [4], \quad [11][1] = [11], \quad [11][4] = [14],$$

and so HK is a subgroup of order 4.

Let L be the cyclic subgroup $\{[1], [4], [7], [13]\}$ generated by $[7]$. Listing all of the distinct products shows that HL is all of $\mathbf{Z}_{15}^{\times}$. $\square$

Example 3.3.2 ($a\mathbf{Z} + b\mathbf{Z} = (a, b)\mathbf{Z}$).

Let G be the additive group $\mathbf{Z}$ of integers, and let $H = a\mathbf{Z}$ and $K = b\mathbf{Z}$. We claim that $H + K$ is the subgroup generated by $d = (a, b)$. Any element of $H + K$ is a linear combination of a and b, and so it is divisible by d, which shows that $H + K \subseteq d\mathbf{Z}$. On the other hand, d is a linear combination of a and b, so $d \in H + K$, which implies that $d\mathbf{Z} \subseteq H + K$. Thus $a\mathbf{Z} + b\mathbf{Z} = (a, b)\mathbf{Z}$. $\square$

Our basic examples thus far have come from ordinary sets of numbers, from the integers modulo n, and from groups of permutations and groups of matrices. We now study a construction that will allow us to give some additional examples that will prove to be very important.

3.3.3 Definition. *Let G_1 and G_2 be groups. The set of all ordered pairs (x_1, x_2) such that $x_1 \in G_1$ and $x_2 \in G_2$ is called the **direct product** of G_1 and G_2, denoted by $G_1 \times G_2$.*

3.3.4 Proposition. *Let G_1 and G_2 be groups.*
   (a) *The direct product $G_1 \times G_2$ is a group under the operation defined for all (a_1, a_2), $(b_1, b_2) \in G_1 \times G_2$ by*

$$(a_1, a_2)(b_1, b_2) = (a_1 b_1, a_2 b_2) .$$

   (b) *If $a_1 \in G_1$ and $a_2 \in G_2$ have orders n and m, respectively, then in $G_1 \times G_2$ the element (a_1, a_2) has order $\mathrm{lcm}[n, m]$.*

Proof. (a) The given operation defines a binary operation. The associative law holds since for all (a_1, a_2), (b_1, b_2), $(c_1, c_2) \in G_1 \times G_2$ we have

$$
\begin{aligned}
(a_1, a_2)((b_1, b_2)(c_1, c_2)) &= (a_1, a_2)(b_1 c_1, b_2 c_2) = (a_1(b_1 c_1), a_2(b_2 c_2)) \\
&= ((a_1 b_1)c_1, (a_2 b_2)c_2) = (a_1 b_1, a_2 b_2)(c_1, c_2) \\
&= ((a_1, a_2)(b_1, b_2))(c_1, c_2).
\end{aligned}
$$

If we use e_1 and e_2 to denote the identity elements in G_1 and G_2, respectively, then (e_1, e_2) is easily seen to be the identity element of the direct product. Finally, for any element $(a_1, a_2) \in G_1 \times G_2$ we have $(a_1, a_2)^{-1} = (a_1^{-1}, a_2^{-1})$.
   (b) Let $a_1 \in G_1$ and $a_2 \in G_2$ have orders n and m, respectively. Then in $G_1 \times G_2$, the order of (a_1, a_2) is the smallest positive power k such that $(a_1, a_2)^k = (e_1, e_2)$. Since $(a_1, a_2)^k = (a_1^k, a_2^k)$, this shows that the order is the smallest positive integer k such that $a_1^k = e_1$ and $a_2^k = e_2$. By Proposition 3.2.8 (b) this must be the smallest positive integer divisible by both n and m, so $k = \mathrm{lcm}[n, m]$. $\square$

To simplify the notation in the next three examples, we have omitted the brackets denoting congruence classes. We may do this elsewhere as well, when the notation becomes too cumbersome. In particular, when we consider matrices with entries in $\mathbf{Z}_n$, we will usually omit the brackets and subscript in the notation $[a]_n$.

Example 3.3.3 (Klein four-group).

In this example we give the addition table for $\mathbf{Z}_2 \times \mathbf{Z}_2$. The operation in the direct product uses the operations from the given groups in each component, so in this example we use addition modulo 2 in each component.

Table 3.3.4: Addition in $\mathbf{Z}_2 \times \mathbf{Z}_2$

	(0,0)	(1,0)	(0,1)	(1,1)
(0,0)	(0,0)	(1,0)	(0,1)	(1,1)
(1,0)	(1,0)	(0,0)	(1,1)	(0,1)
(0,1)	(0,1)	(1,1)	(0,0)	(1,0)
(1,1)	(1,1)	(0,1)	(1,0)	(0,0)

This group is usually called the **Klein four-group**. The pattern in Table 3.3.4 is the same as that of the second group in Table 3.3.1. This group is characterized by the fact that it has order 4 and each element except the identity has order 2. □

Example 3.3.4.

In the group $\mathbf{Z} \times \mathbf{Z}$, the subgroup generated by an element (m, n) consists of all multiples $k(m, n)$. This subgroup cannot contain both of the elements $(1, 0)$ and $(0, 1)$, and so no single element generates $\mathbf{Z} \times \mathbf{Z}$, showing that it is not cyclic. There are natural subgroups $\langle (1, 0) \rangle$ and $\langle (0, 1) \rangle$. The "diagonal" subgroup $\langle (1, 1) \rangle$ is also interesting.   □

Example 3.3.5.

The group $\mathbf{Z}_2 \times \mathbf{Z}_3$ is cyclic, since the element $(1, 1)$ must have order 6 by Proposition 3.3.4. (The order of 1 in $\mathbf{Z}_2$ is 2, while the order of 1 in $\mathbf{Z}_3$ is 3, and then we have $\mathrm{lcm}[2, 3] = 6$.)

On the other hand, the group $\mathbf{Z}_2 \times \mathbf{Z}_4$ is not cyclic, since in the first component the possible orders are 1 and 2, and in the second component the possible orders are 1, 2, and 4. The largest possible least common multiple we can have is 4, so there is no element of order 8 and the group is not cyclic.   □

We now want to introduce some more general matrix groups. We first need the definition of a field, which we will phrase in the language of groups. The sets $\mathbf{Q}$, $\mathbf{R}$, and $\mathbf{C}$ that we have worked with form abelian groups under addition, and in each case the set of nonzero elements also forms an abelian group under multiplication. This is also the case for $\mathbf{Z}_p$, when p is a prime number, and we would like to be able to work with groups of matrices that have entries in any of these sets. Fields are fundamental objects in abstract algebra and will be studied in depth in later chapters. We note that the definition of a field that is given in Definition 4.1.1 does not depend on the language of group theory.

3.3.5 Definition. *Let F be a set with two binary operations $+$ and $\cdot$ with respective identity elements 0 and 1, where $0 \neq 1$. Then F is called a **field** if*

(i) *the set of all elements of F is an abelian group under $+$;*

(ii) *the set of all nonzero elements of F is an abelian group under $\cdot$;*

(iii) *$a(b + c) = ab + ac$ and $(a + b)c = ac + bc$ for all $a, b, c \in F$.*

Axiom (iii) lists the **distributive laws,** which give a connection between addition and multiplication. The properties of matrix multiplication depend heavily on these laws. The distributive laws also imply that for any element a of a field F we have $a \cdot 0 = 0$. To see this, note that $0 + a \cdot 0 = a \cdot 0 = a \cdot (0 + 0) = a \cdot 0 + a \cdot 0$, and then since F is a group under addition, we can cancel $a \cdot 0$ from both sides of the equation, to get $0 = a \cdot 0$. A similar argument shows that $0 \cdot a = 0$ for all $a \in F$.

We next want to consider matrices that have entries in a field F. (Even if $\mathbf{R}$ and $\mathbf{Z}_p$ are the only fields F with which you feel comfortable, using them in the following matrix groups will give you interesting and instructive examples.) If (a_{ij}) and (b_{ij}) are $n \times n$ matrices, then the product (c_{ij}) of the two matrices is defined as the matrix whose i, j-entry is

$$c_{ij} = \sum_{k=1}^{n} a_{ik} b_{kj} .$$

This product makes sense because in F the two operations of addition and multiplication are well-defined.

3.3.6 Definition. *Let F be a field. The set of all invertible $n \times n$ matrices with entries in F is called the **general linear group of degree** n **over** F, and is denoted by $\mathrm{GL}_n(F)$.*

3.3.7 Proposition. *Let F be a field. Then $\mathrm{GL}_n(F)$ is a group under matrix multiplication.*

Proof. If A and B are invertible matrices, then the formulas $(A^{-1})^{-1} = A$ and $(AB)^{-1} = B^{-1}A^{-1}$ show that $\mathrm{GL}_n(F)$ has inverses for each element and is closed under matrix multiplication. The identity matrix (with 1 in each diagonal entry and 0 in every other entry) is an identity element. The proof that matrix multiplication is associative is left as an exercise. $\square$

Example 3.3.6 (GL$_2$(**Z**$_2$)).

This example gives the multiplication table for GL$_2$(**Z**$_2$). The total number of 2×2 matrices over **Z**$_2$ is $2^4 = 16$, but it can be checked that only six of the matrices are invertible, and these are listed in Table 3.3.5. We simply use 0 and 1 to denote the congruence classes $[0]_2$ and $[1]_2$. Note that the group GL$_2$(**Z**$_2$) is not abelian. $\square$

Table 3.3.5: Multiplication in GL$_2$(**Z**$_2$)

	$\begin{bmatrix} 1 & 0 \\ 0 & 1 \end{bmatrix}$	$\begin{bmatrix} 1 & 1 \\ 1 & 0 \end{bmatrix}$	$\begin{bmatrix} 0 & 1 \\ 1 & 1 \end{bmatrix}$	$\begin{bmatrix} 0 & 1 \\ 1 & 0 \end{bmatrix}$	$\begin{bmatrix} 1 & 1 \\ 0 & 1 \end{bmatrix}$	$\begin{bmatrix} 1 & 0 \\ 1 & 1 \end{bmatrix}$
$\begin{bmatrix} 1 & 0 \\ 0 & 1 \end{bmatrix}$	$\begin{bmatrix} 1 & 0 \\ 0 & 1 \end{bmatrix}$	$\begin{bmatrix} 1 & 1 \\ 1 & 0 \end{bmatrix}$	$\begin{bmatrix} 0 & 1 \\ 1 & 1 \end{bmatrix}$	$\begin{bmatrix} 0 & 1 \\ 1 & 0 \end{bmatrix}$	$\begin{bmatrix} 1 & 1 \\ 0 & 1 \end{bmatrix}$	$\begin{bmatrix} 1 & 0 \\ 1 & 1 \end{bmatrix}$
$\begin{bmatrix} 1 & 1 \\ 1 & 0 \end{bmatrix}$	$\begin{bmatrix} 1 & 1 \\ 1 & 0 \end{bmatrix}$	$\begin{bmatrix} 0 & 1 \\ 1 & 1 \end{bmatrix}$	$\begin{bmatrix} 1 & 0 \\ 0 & 1 \end{bmatrix}$	$\begin{bmatrix} 1 & 1 \\ 0 & 1 \end{bmatrix}$	$\begin{bmatrix} 1 & 0 \\ 1 & 1 \end{bmatrix}$	$\begin{bmatrix} 0 & 1 \\ 1 & 0 \end{bmatrix}$
$\begin{bmatrix} 0 & 1 \\ 1 & 1 \end{bmatrix}$	$\begin{bmatrix} 0 & 1 \\ 1 & 1 \end{bmatrix}$	$\begin{bmatrix} 1 & 0 \\ 0 & 1 \end{bmatrix}$	$\begin{bmatrix} 1 & 1 \\ 1 & 0 \end{bmatrix}$	$\begin{bmatrix} 1 & 0 \\ 1 & 1 \end{bmatrix}$	$\begin{bmatrix} 0 & 1 \\ 1 & 0 \end{bmatrix}$	$\begin{bmatrix} 1 & 1 \\ 0 & 1 \end{bmatrix}$
$\begin{bmatrix} 0 & 1 \\ 1 & 0 \end{bmatrix}$	$\begin{bmatrix} 0 & 1 \\ 1 & 0 \end{bmatrix}$	$\begin{bmatrix} 1 & 0 \\ 1 & 1 \end{bmatrix}$	$\begin{bmatrix} 1 & 1 \\ 0 & 1 \end{bmatrix}$	$\begin{bmatrix} 1 & 0 \\ 0 & 1 \end{bmatrix}$	$\begin{bmatrix} 0 & 1 \\ 1 & 1 \end{bmatrix}$	$\begin{bmatrix} 1 & 1 \\ 1 & 0 \end{bmatrix}$
$\begin{bmatrix} 1 & 1 \\ 0 & 1 \end{bmatrix}$	$\begin{bmatrix} 1 & 1 \\ 0 & 1 \end{bmatrix}$	$\begin{bmatrix} 0 & 1 \\ 1 & 0 \end{bmatrix}$	$\begin{bmatrix} 1 & 0 \\ 1 & 1 \end{bmatrix}$	$\begin{bmatrix} 1 & 1 \\ 1 & 0 \end{bmatrix}$	$\begin{bmatrix} 1 & 0 \\ 0 & 1 \end{bmatrix}$	$\begin{bmatrix} 0 & 1 \\ 1 & 1 \end{bmatrix}$
$\begin{bmatrix} 1 & 0 \\ 1 & 1 \end{bmatrix}$	$\begin{bmatrix} 1 & 0 \\ 1 & 1 \end{bmatrix}$	$\begin{bmatrix} 1 & 1 \\ 0 & 1 \end{bmatrix}$	$\begin{bmatrix} 0 & 1 \\ 1 & 0 \end{bmatrix}$	$\begin{bmatrix} 0 & 1 \\ 1 & 1 \end{bmatrix}$	$\begin{bmatrix} 1 & 1 \\ 1 & 0 \end{bmatrix}$	$\begin{bmatrix} 1 & 0 \\ 0 & 1 \end{bmatrix}$

Example 3.3.7 (Quaternion group).

Let Q be the following set of matrices in $GL_2(\mathbf{C})$:

$$\pm \begin{bmatrix} 1 & 0 \\ 0 & 1 \end{bmatrix}, \quad \pm \begin{bmatrix} i & 0 \\ 0 & -i \end{bmatrix}, \quad \pm \begin{bmatrix} 0 & 1 \\ -1 & 0 \end{bmatrix}, \quad \pm \begin{bmatrix} 0 & i \\ i & 0 \end{bmatrix}.$$

If we let

$$1 = \begin{bmatrix} 1 & 0 \\ 0 & 1 \end{bmatrix}, \quad \mathbf{i} = \begin{bmatrix} i & 0 \\ 0 & -i \end{bmatrix}, \quad \mathbf{j} = \begin{bmatrix} 0 & 1 \\ -1 & 0 \end{bmatrix}, \quad \mathbf{k} = \begin{bmatrix} 0 & i \\ i & 0 \end{bmatrix},$$

then computations show that we have the following identities:

$$\mathbf{i}^2 = \mathbf{j}^2 = \mathbf{k}^2 = -1 ;$$

$$\mathbf{ij} = \mathbf{k}, \quad \mathbf{jk} = \mathbf{i}, \quad \mathbf{ki} = \mathbf{j}; \qquad \mathbf{ji} = -\mathbf{k}, \quad \mathbf{kj} = -\mathbf{i}, \quad \mathbf{ik} = -\mathbf{j} .$$

Since they show that the set is closed under matrix multiplication, we have defined a subgroup of $GL_2(\mathbf{C})$. We can make a few observations: Q is not abelian and is not cyclic. In fact, -1 has order 2, while $\pm\mathbf{i}$, $\pm\mathbf{j}$, and $\pm\mathbf{k}$ have order 4. $\square$

We next investigate the smallest subgroup that contains a nonempty subset S of a group G. If $S \subseteq H$, where H is a subgroup of G, and $a, b, c \in S$, then all products such as $a^{-1}a^{-1}bab^{-1}c \cdots$ must belong to H. In fact, as we show below, the collection of all such products of elements of S constitutes the smallest subgroup that contains S. It is called the **subgroup generated by** S.

3.3.8 Definition. *Let S be a nonempty subset of the group G. A finite product of elements of S and their inverses is called a **word** in S. The set of all words in S is denoted by $\langle S \rangle$.*

3.3.9 Proposition. *Let S be a nonempty subset of the group G. Then $\langle S \rangle$ is a subgroup of G, and is equal to the intersection of all subgroups of G that contain S.*

Proof. Since S is nonempty, there is some $a \in S$, and then the word $aa^{-1} = e$ belongs to $\langle S \rangle$. If x and y are two words in S, then by definition their product xy is again a word in S. Finally, if x is a word in S, the so is x^{-1}, since taking inverses simply reverses the order and changes the sign of the exponent.

If $S \subseteq H$, where H is a subgroup of G, then since H is closed under the operation of G, it contains all words in S, and therefore it contains $\langle S \rangle$. It follows immediately that $\langle S \rangle$ is the intersection of all subgroups of G that contain S. $\square$

EXERCISES: SECTION 3.3

1.† Find HK in $\mathbf{Z}_{16}^{\times}$, if $H = \langle [3] \rangle$ and $K = \langle [5] \rangle$.

2. Find HK in $\mathbf{Z}_{21}^{\times}$, if $H = \{[1], [8]\}$ and $K = \{[1], [4], [10], [13], [16], [19]\}$.

3.† Find an example of two subgroups H and K of S_3 for which HK is not a subgroup.

4. Find the cyclic subgroup generated by $\begin{bmatrix} 2 & 1 \\ 0 & 2 \end{bmatrix}$ in $GL_2(\mathbf{Z}_3)$.

5. Prove that if G_1 and G_2 are abelian groups, then the direct product $G_1 \times G_2$ is abelian.

6. Construct an abelian group of order 12 that is not cyclic.

7.† Construct a group of order 12 that is not abelian.

8. Let G_1 and G_2 be groups, with subgroups H_1 and H_2, respectively. Show that $\{(x_1, x_2) \mid x_1 \in H_1, x_2 \in H_2\}$ is a subgroup of the direct product $G_1 \times G_2$.

9. This exercise concerns subgroups of $\mathbf{Z} \times \mathbf{Z}$.

    (a) Let $C_1 = \{(a, b) \in \mathbf{Z} \times \mathbf{Z} \mid a = b\}$. Show that C_1 is a subgroup of $\mathbf{Z} \times \mathbf{Z}$.

    (b) For each positive integer $n \geq 2$, let $C_n = \{(a, b) \in \mathbf{Z} \times \mathbf{Z} \mid a \equiv b \pmod{n}\}$. Show that C_n is a subgroup of $\mathbf{Z} \times \mathbf{Z}$.

    (c) Show that every subgroup of $\mathbf{Z} \times \mathbf{Z}$ that contains C_1 has the form C_n, for some positive integer n.

10. Let $n > 2$ be an integer, and let $X \subseteq S_n \times S_n$ be the set $X = \{(\sigma, \tau) \mid \sigma(1) = \tau(1)\}$. Show that X is not a subgroup of $S_n \times S_n$.

11. Let G_1 and G_2 be groups, and let G be the direct product $G_1 \times G_2$.
    Let $H = \{(x_1, x_2) \in G_1 \times G_2 \mid x_2 = e\}$ and let $K = \{(x_1, x_2) \in G_1 \times G_2 \mid x_1 = e\}$.

    (a) Show that H and K are subgroups of G.

    (b) Show that $HK = KH = G$.

    (c) Show that $H \cap K = \{(e, e)\}$.

12. (a) Generalize Definition 3.3.3 to the case of the direct product of n groups.

    (b) Generalize Proposition 3.3.4 to the case of the direct product of n groups. Prove that your generalization is true.

13. Let p, q be distinct prime numbers, and let $n = pq$. Show that $HK = \mathbf{Z}_n^{\times}$, for the subgroups $H = \{[x] \in \mathbf{Z}_n^{\times} \mid x \equiv 1 \pmod{p}\}$ and $K = \{[y] \in \mathbf{Z}_n^{\times} \mid y \equiv 1 \pmod{q}\}$ of $\mathbf{Z}_n^{\times}$.

    Hint: You can either use a counting argument to show that HK has $\varphi(n)$ elements, or use the Chinese Remainder Theorem to show that the sets are the same.

14. Let G be a finite group, and let H, K be subgroups of G. Prove that

$$|HK| = \frac{|H||K|}{|H \cap K|} .$$

15. Let G be a group of order 6. Show that G must contain an element of order 2 (see Exercise 24 of Section 3.1). Show that it cannot be true that every element different from e has order 2.

    Hint: Show that if every element had order 2 it would be possible to construct a subgroup of order 4.

16. Let G be a group of order 6, and suppose that $a, b \in G$ with a of order 3 and b of order 2. Show that either G is cyclic or $ab \neq ba$.

17. Let G be any group of order 6. Show that if G is not cyclic, then its multiplication table must look like that of S_3.

    Hint: If the group is not cyclic, use Exercises 15 and 16 to produce elements $a, b \in G$ with $a^3 = e$, $b^2 = e$ and $ba = a^2 b$.

18. Let c be a positive constant. Show that the set L of all matrices of the form

$$A(v) = \left(1 - \frac{v^2}{c^2}\right)^{-1/2} \begin{bmatrix} 1 & -v \\ -\dfrac{v}{c^2} & 1 \end{bmatrix} , \quad v \in \mathbf{R}, \ |v| < c$$

    is a subgroup of $\mathrm{GL}_2(\mathbf{R})$. Verify that $A(v_1)A(v_2) = A(v_3)$, where $v_3 = \dfrac{v_1 + v_2}{1 + \dfrac{v_1 v_2}{c^2}}$.

    Note: The set L is called the **Lorentz** group, and if c is the speed of light, then it models the addition of velocities under the theory of special relativity. Observe that in this model, if c were infinite, this formula for addition of velocities would reduce to $A(v_1)A(v_2) = A(v_1 + v_2)$.

3.4   Isomorphisms

In studying groups we are interested in their algebraic properties, and not in the particular form in which they are presented. For example, if we construct the multiplication tables for two fi nite groups and fi nd that they have the same patterns, although the elements might have different forms, then we would say that the groups have exactly the same algebraic properties.

Consider the subgroup $\{\pm 1\}$ of $\mathbf{Q}^\times$ and the group $\mathbf{Z}_2$. If you write out the group tables for these groups you will fi nd precisely the same pattern, as shown in Tables 3.4.1 and 3.4.2.

Actually, if we have any group G with two elements, say the identity element e and one other element a, then there is only one possibility for the multiplication

Table 3.4.1: Multiplication in $\{\pm 1\}$

$\times$	1	-1
1	1	-1
-1	-1	1

Table 3.4.2: Addition in $\mathbf{Z}_2$

$+$	$[0]$	$[1]$
$[0]$	$[0]$	$[1]$
$[1]$	$[1]$	$[0]$

table for G. We have already observed that Propositions 3.1.7 and 3.1.8 imply that in each row and column of a group table, each element of the group must occur exactly once. Since e is the identity element, $e \cdot e = e$, $e \cdot a = a$, and $a \cdot e = a$. Since a cannot be repeated in the last row of the table, we must have $a \cdot a = e$. This gives us Table 3.4.3, and shows that all groups with two elements must have exactly the same algebraic properties.

Table 3.4.3:

	e	a
e	e	a
a	a	e

We need a formal definition to describe when two groups have the same algebraic properties. To begin with, there should be a one-to-one correspondence between the elements of the groups. This means in essence that elements of one group could be renamed to correspond exactly to the elements of the second group. In addition, products of corresponding elements should correspond. If G_1 is a group with operation $*$ and G_2 is a group with operation $\star$, then any function $\phi : G_1 \to G_2$ that preserves products must have the property that $\phi(a * b) = \phi(a) \star \phi(b)$ for all $a, b \in G_1$. This expresses in a formula the fact that if we first multiply a and b to get $a * b$ and then find the corresponding element $\phi(a * b)$ of G_2, we should get exactly the same answer as if we find the corresponding elements $\phi(a)$ and $\phi(b)$ in G_2 and then compute their product $\phi(a) \star \phi(b)$ in G_2. It is rather cumbersome to write the two operations, so in our definition we will omit them, since it should be

clear from the context which operation is to be used. A one-to-one correspondence
that preserves products will be called an *isomorphism*. It is derived from the Greek
words *isos* meaning "equal" and *morphe* meaning "form."

3.4.1 Definition. *Let G_1 and G_2 be groups, and let $\phi : G_1 \to G_2$ be a function.
Then ϕ is said to be a **group isomorphism** if ϕ is one-to-one and onto and*

$$\phi(ab) = \phi(a)\phi(b)$$

*for all $a, b \in G_1$. In this case, G_1 is said to be **isomorphic** to G_2, and this is denoted
by $G_1 \cong G_2$.*

There are several immediate consequences of the definition of an isomorphism.
Using an induction argument it is possible to show that

$$\phi(a_1 a_2 \cdots a_n) = \phi(a_1)\phi(a_2) \cdots \phi(a_n)$$

for any isomorphism $\phi : G_1 \to G_2$. In particular, $\phi(a^n) = (\phi(a))^n$ for all positive
integers n. Let e_1 and e_2 be the identity elements of G_1 and G_2, respectively. Then

$$\phi(e_1) \cdot \phi(e_1) = \phi(e_1 \cdot e_1) = \phi(e_1) = \phi(e_1) \cdot e_2 ,$$

and then the cancellation law in G_2 implies that $\phi(e_1) = e_2$. Finally,

$$\phi(a^{-1}) \cdot \phi(a) = \phi(a^{-1} \cdot a) = \phi(e_1) = e_2 ,$$

so $(\phi(a))^{-1} = \phi(a^{-1})$, and it follows that $\phi(a^n) = (\phi(a))^n$ for all $n \in \mathbf{Z}$. We can
summarize by saying that any group isomorphism preserves general products, the
identity element, and inverses of elements. From now on we will generally use e
to denote the identity element of a group, and we will not distinguish between the
identity elements of various groups unless confusion would otherwise result.

We have emphasized that we really do not distinguish (algebraically) between
groups that are isomorphic. Thus isomorphism should be almost like equality. In
fact, it satisfies properties similar to those of an equivalence relation. There are
several things to prove, since our definition of isomorphism of groups involves a
function, say $\phi : G_1 \to G_2$, and this determines a direction to the isomorphism.

The reflexive property holds, since for any group G the identity mapping is
an isomorphism, and so $G \cong G$. If $G_1 \cong G_2$, then there is an isomorphism
$\phi : G_1 \to G_2$, and since ϕ is one-to-one and onto, there exists an inverse function
ϕ^{-1}, which by part (a) of the next proposition is an isomorphism. Thus $G_2 \cong G_1$,
and the symmetric property holds. Finally, the transitive property holds since if
$G_1 \cong G_2$ and $G_2 \cong G_3$, then by part (b) of the next proposition the composite of
the two given isomorphisms is an isomorphism, showing that $G_1 \cong G_3$.

3.4.2 Proposition.

(a) *The inverse of a group isomorphism is a group isomorphism.*

(b) *The composite of two group isomorphisms is a group isomorphism.*

Proof. (a) Let $\phi : G_1 \to G_2$ be a group isomorphism. Since ϕ is one-to-one and onto, by Proposition 2.1.7 there is an inverse function $\theta : G_2 \to G_1$. Recall that θ is defined as follows: for each element $g_2 \in G_2$ there exists a unique element $g_1 \in G_1$ such that $\phi(g_1) = g_2$, and then $\theta(g_2) = g_1$. As a direct consequence of the definition of θ, we have $\theta\phi$ equal to the identity on G_1 and $\phi\theta$ equal to the identity on G_2. The definition also implies that θ is one-to-one and onto. All that remains is to show that θ preserves products. Let $a_2, b_2 \in G_2$, and let $\theta(a_2) = a_1$ and $\theta(b_2) = b_1$. Then $\phi(a_1) = a_2$ and $\phi(b_1) = b_2$, so

$$\phi(a_1 b_1) = \phi(a_1)\phi(b_1) = a_2 b_2 ,$$

which shows, by the definition of θ, that

$$\theta(a_2 b_2) = a_1 b_1 = \theta(a_2)\theta(b_2) .$$

(b) Let $\phi : G_1 \to G_2$ and $\theta : G_2 \to G_3$ be group isomorphisms. By Proposition 2.1.5, the composite of two one-to-one and onto functions is again one-to-one and onto, so $\theta\phi$ is one-to-one and onto. If $a, b \in G_1$, then

$$\theta\phi(ab) = \theta(\phi(ab)) = \theta(\phi(a)\phi(b)) = \theta(\phi(a))\theta(\phi(b)) = \theta\phi(a)\theta\phi(b) ,$$

showing that $\theta\phi$ preserves products. $\square$

Example 3.4.1 ($\langle i \rangle \cong \mathbf{Z}_4$).

Consider the subgroup $\langle i \rangle = \{\pm 1, \pm i\}$ of $\mathbf{C}^\times$. Table 3.4.4 gives a multiplication table for this subgroup.

Table 3.4.4: Multiplication in $\langle i \rangle$

	1	-1	i	$-i$
1	1	-1	i	$-i$
-1	-1	1	$-i$	i
i	i	$-i$	-1	1
$-i$	$-i$	i	1	-1

Although Table 3.4.4 presents the group in what may be the most natural order, it is useful rearrange the table. It is interesting to arrange the table in the order

Table 3.4.5: Multiplication in $\langle i \rangle$

	i^0	i^1	i^2	i^3
i^0	i^0	i^1	i^2	i^3
i^1	i^1	i^2	i^3	i^0
i^2	i^2	i^3	i^0	i^1
i^3	i^3	i^0	i^1	i^2

$1, i, -1, -i$, as shown in Table 3.4.5. Actually, since $1 = i^0, i = i^1, -1 = i^2$, and $i^3 = -i$, we have used this representation in the table.

For the sake of comparison, in Table 3.4.6 we give the addition table for $\mathbf{Z}_4$. The elements of $\mathbf{Z}_4$ appear in precisely the same positions as the exponents of i did in the previous table. This illustrates in a concrete way the intuitive notion that multiplication of powers of i should correspond to addition of the exponents.

Table 3.4.6: Addition in $\mathbf{Z}_4$

	[0]	[1]	[2]	[3]
[0]	[0]	[1]	[2]	[3]
[1]	[1]	[2]	[3]	[0]
[2]	[2]	[3]	[0]	[1]
[3]	[3]	[0]	[1]	[2]

To make the isomorphism precise, define a function $\phi : \mathbf{Z}_4 \to \langle i \rangle$ by $\phi([n]) = i^n$. This formula depends on choosing a representative n of its equivalence class, so to show that the function is well-defined we must show that if $n \equiv m \pmod 4$, then $i^n = i^m$. This follows immediately from Proposition 3.2.8, since i has order 4. The function defines a one-to-one correspondence, so all that remains is to show that ϕ preserves the respective operations. We must first add elements $[n]$ and $[m]$ of $\mathbf{Z}_4$ (using the operation in $\mathbf{Z}_4$) and then apply ϕ, and compare this with the element we obtain by first applying ϕ and then multiplying (using the operation in $\langle i \rangle$):

$$\phi([n] + [m]) = \phi([n + m]) = i^{n+m} = i^n i^m = \phi([n])\phi([m]) \ .$$

We conclude that ϕ is a group isomorphism. $\square$

Example 3.4.2 (Exponential and logarithmic functions).

The groups $\mathbf{R}$ (under addition) and $\mathbf{R}^+$ (under multiplication) are isomorphic. Define $\phi : \mathbf{R} \to \mathbf{R}^+$ by $\phi(x) = e^x$. Then ϕ preserves the respective operations since $\phi(x + y) = e^{x+y} = e^x e^y = \phi(x)\phi(y)$. To show that ϕ is one-to-one, suppose that $\phi(x) = \phi(y)$. Then taking the natural logarithm of each side of the equation $e^x = e^y$ gives $x = y$. Finally, ϕ is an onto mapping since for any $y \in \mathbf{R}^+$ we have $y = e^{\ln(y)} = \phi(\ln(y))$. □

To show that two groups G_1 and G_2 are isomorphic, it is often necessary to actually define the function that gives the isomorphism. In practice, there is usually some natural correspondence between elements which suggests how to define the necessary function.

On the other hand, to show that the groups are not isomorphic, it is not practical to try to check all one-to-one correspondences between the groups to see that none of them preserve products. (Of course, if there is no one-to-one correspondence between the groups, they are not isomorphic. For example, no group with four elements can be isomorphic to a group with five elements.) What we need to do is to find a property of the first group which (i) the second group does not have and (ii) would be preserved by any isomorphism. The next proposition identifies several structural properties that are preserved by group isomorphisms.

3.4.3 Proposition. *Let $\phi : G_1 \to G_2$ be an isomorphism of groups.*
   (a) *If a has order n in G_1, then $\phi(a)$ has order n in G_2.*
   (b) *If G_1 is abelian, then so is G_2.*
   (c) *If G_1 is cyclic, then so is G_2.*

Proof. (a) Suppose that $a \in G_1$ with $a^n = e$. Then we must have $(\phi(a))^n = \phi(a^n) = \phi(e) = e$. This shows that the order of $\phi(a)$ is a divisor of the order of a. Since ϕ is an isomorphism, then there exists an inverse isomorphism that maps $\phi(a)$ to a, and a similar argument shows that the order of a is a divisor of the order of $\phi(a)$. It follows that a and $\phi(a)$ must have the same order.

(b) Assume that G_1 is abelian, and let $a_2, b_2 \in G_2$. Since ϕ is an onto mapping, there exist $a_1, b_1 \in G_1$ with $\phi(a_1) = a_2$ and $\phi(b_1) = b_2$. Then

$$a_2 b_2 = \phi(a_1)\phi(b_1) = \phi(a_1 b_1) = \phi(b_1 a_1) = \phi(b_1)\phi(a_1) = b_2 a_2 \ ,$$

showing that G_2 is abelian.

(c) Suppose that G_1 is cyclic, with $G_1 = \langle a \rangle$. For any element $y \in G_2$ we have $y = \phi(x)$ for some $x \in G_1$, since ϕ is onto. Using the assumption that G_1 is cyclic, we can write $x = a^n$ for some $n \in \mathbf{Z}$. Then $y = \phi(a^n) = (\phi(a))^n$, which shows that each element of G_2 can be expressed as a power of $\phi(a)$. Thus G_2 is cyclic, generated by $\phi(a)$. □

Example 3.4.3 (R $\not\cong$ R$^\times$).

The additive group **R** of real numbers is not isomorphic to the multiplicative group **R**$^\times$ of nonzero real numbers. One way to see this is to observe that **R**$^\times$ has an element of order 2, namely, -1. On the other hand, an element of order 2 in **R** must satisfy the equation $2x = 0$ (in additive notation). The only solution is $x = 0$, and so this shows that **R** has no element of order 2. Thus there cannot be an isomorphism between the two groups, since by Proposition 3.4.3 it would preserve the orders of all elements. $\square$

Example 3.4.4 (Z$_4$ $\not\cong$ Z$_2$ $\times$ Z$_2$).

The cyclic group $\mathbf{Z}_4$ and the Klein four-group $\mathbf{Z}_2 \times \mathbf{Z}_2$ are not isomorphic. In $\mathbf{Z}_4$ there is an element of order 4, namely, [1]. On the other hand, the order of an element in a direct product is the least common multiple of the orders of its components, and so any element of $\mathbf{Z}_2 \times \mathbf{Z}_2$ not equal to the identity must have order 2. $\square$

To motivate some further work with isomorphisms, let us ask the following question. Which of the groups S_3, $\mathrm{GL}_2(\mathbf{Z}_2)$, $\mathbf{Z}_6$, and $\mathbf{Z}_2 \times \mathbf{Z}_3$ are isomorphic? The first two groups we know to be nonabelian. On the other hand, any cyclic group is abelian, since powers of a fixed element will commute with each other. The element ([1], [1]) of $\mathbf{Z}_2 \times \mathbf{Z}_3$ has order 6 (the least common multiple of 2 and 3), and so $\mathbf{Z}_2 \times \mathbf{Z}_3$ is cyclic, as well as $\mathbf{Z}_6$. Thus the four groups represent at least two different isomorphism classes. Proposition 3.4.5 will show that $\mathbf{Z}_2 \times \mathbf{Z}_3$ is isomorphic to $\mathbf{Z}_6$, and the next example shows that the two nonabelian groups we are considering are also isomorphic.

Example 3.4.5 (GL$_2$(Z$_2$) $\cong$ S_3).

Refer to Table 3.3.5 for a multiplication table for $\mathrm{GL}_2(\mathbf{Z}_2)$. To establish the connection between S_3 and $\mathrm{GL}_2(\mathbf{Z}_2)$, let

$$e = \begin{bmatrix} 1 & 0 \\ 0 & 1 \end{bmatrix}, \qquad a = \begin{bmatrix} 1 & 1 \\ 1 & 0 \end{bmatrix}, \qquad \text{and} \qquad b = \begin{bmatrix} 0 & 1 \\ 1 & 0 \end{bmatrix}.$$

Then direct computations show that $a^3 = e$, $b^2 = e$, and $ba = a^2b$. Furthermore, each element of $\mathrm{GL}_2(\mathbf{Z}_2)$ can be expressed uniquely in one of the following forms: e, a, a^2, b, ab, a^2b. If we make these substitutions in the multiplication table for $\mathrm{GL}_2(\mathbf{Z}_2)$, we obtain Table 3.4.7.

In Section 3.3 we described S_3 by letting $a = (1, 2, 3)$ and $b = (1, 2)$, which allowed us to write

$$S_3 = \{e, a, a^2, b, ab, a^2b\}, \qquad \text{where } a^3 = e, \ b^2 = e, \ ba = a^2b,$$

Table 3.4.7: Multiplication in $GL_2(\mathbf{Z}_2)$

	e	a	a^2	b	ab	a^2b
e	e	a	a^2	b	ab	a^2b
a	a	a^2	e	ab	a^2b	b
a^2	a^2	e	a	a^2b	b	ab
b	b	a^2b	ab	e	a^2	a
ab	ab	b	a^2b	a	e	a^2
a^2b	a^2b	ab	b	a^2	a	e

without using permutations. This indicates how to define an isomorphism from S_3 to $GL_2(\mathbf{Z}_2)$. Let

$$\phi((1,2,3)) = \begin{bmatrix} 1 & 1 \\ 1 & 0 \end{bmatrix} \qquad \text{and} \qquad \phi((1,2)) = \begin{bmatrix} 0 & 1 \\ 1 & 0 \end{bmatrix}$$

and then extend this to all elements by letting

$$\phi((1,2,3)^i(1,2)^j) = \begin{bmatrix} 1 & 1 \\ 1 & 0 \end{bmatrix}^i \begin{bmatrix} 0 & 1 \\ 1 & 0 \end{bmatrix}^j$$

for $i = 0, 1, 2$ and $j = 0, 1$. Our remarks about the unique forms of the respective elements show that ϕ is a one-to-one correspondence. The fact that the multiplication tables are identical shows that ϕ respects the two operations. This verifies that ϕ is an isomorphism. □

Example 3.4.6 ($\mathbf{Z}_6 \cong \mathbf{Z}_2 \times \mathbf{Z}_3$).

Using the idea of the previous example, to show that $\mathbf{Z}_6$ and $\mathbf{Z}_2 \times \mathbf{Z}_3$ are isomorphic, we can look for elements that can be used to describe each group. Since we have already observed in Examples 3.2.8 and 3.3.5 that both groups are cyclic, we can let a be a generator for $\mathbf{Z}_6$ and b be a generator for $\mathbf{Z}_2 \times \mathbf{Z}_3$. Then the function $\phi(na) = nb$ can be shown to define an isomorphism. (Remember that we are using additive notation.) □

The next proposition gives an easier way to check that a function which preserves products is one-to-one. In additive notation, it depends on the fact that for any mapping which preserves sums we have $\phi(x_1) = \phi(x_2)$ if and only if $\phi(x_1 - x_2) = 0$. Any vector space is an abelian group under vector addition, and any linear transformation preserves sums. Thus the result that a linear transformation is one-to-one if and only if its null space is trivial is a special case of our next proposition.

3.4.4 Proposition. *Let G_1 and G_2 be groups, and let $\phi : G_1 \rightarrow G_2$ be a function such that $\phi(ab) = \phi(a)\phi(b)$ for all $a, b \in G_1$. Then ϕ is one-to-one if and only if $\phi(x) = e$ implies $x = e$, for all $x \in G_1$.*

Proof. Let $\phi : G_1 \rightarrow G_2$ satisfy the hypothesis of the proposition. If ϕ is one-to-one, then the only element that can map to the identity of G_2 is the identity of G_1. On the other hand, suppose that $\phi(x) = e$ implies $x = e$, for all $x \in G_1$. If $\phi(x_1) = \phi(x_2)$ for some $x_1, x_2 \in G_1$, then $\phi(x_1 x_2^{-1})\phi(x_2) = \phi(x_1 x_2^{-1} x_2) = \phi(x_1) = \phi(x_2) = e\phi(x_2)$, so we can cancel $\phi(x_2)$ to get $\phi(x_1 x_2^{-1}) = e$, which shows by assumption that $x_1 x_2^{-1} = e$, and thus $x_1 = x_2$. This shows that ϕ is one-to-one. $\square$

3.4.5 Proposition. *If m, n are positive integers such that $\gcd(m, n) = 1$, then $\mathbf{Z}_{mn}$ is isomorphic to $\mathbf{Z}_m \times \mathbf{Z}_n$.*

Proof. Define $\phi : \mathbf{Z}_{mn} \rightarrow \mathbf{Z}_m \times \mathbf{Z}_n$ by $\phi([x]_{mn}) = ([x]_m, [x]_n)$. If $a \equiv b \pmod{mn}$, then $a \equiv b \pmod{m}$ and $a \equiv b \pmod{n}$, and so ϕ is well-defined. It is easy to check that ϕ preserves sums. To show that ϕ is one-to-one we can use the previous proposition. If $\phi([x]_{mn}) = ([0]_m, [0]_n)$, then both m and n must be divisors of x. Since $\gcd(m, n) = 1$, it follows that mn must be a divisor of x, which shows that $[x]_{mn} = [0]_{mn}$. Since the two groups have the same number of elements, any one-to-one mapping must be onto, and thus ϕ is an isomorphism. $\square$

In most of our exercises, to show that two groups G_1 and G_2 are isomorphic, you must define a one-to-one function from one group to the other. Sometimes it is easier to define the function in one direction than in the other, so when you are working on a problem, it may be worth checking both ways. The inverse of the function we used in proving Proposition 3.4.5 is also interesting. Given $\phi : \mathbf{Z}_{mn} \rightarrow \mathbf{Z}_m \times \mathbf{Z}_n$ defined by $\phi([x]_{mn}) = ([x]_m, [x]_n)$, the inverse must assign to each $([a]_m, [b]_n) \in \mathbf{Z}_m \times \mathbf{Z}_n$ an element $[x]_{mn} \in \mathbf{Z}_{mn}$ such that $x \equiv a \pmod{m}$ and $x \equiv b \pmod{n}$. We could have applied the Chinese remainder theorem (Theorem 1.3.6) to define our function in this direction, but the other way seemed more natural.

EXERCISES: SECTION 3.4

1.† Show that the multiplicative group $\mathbf{Z}_{10}^{\times}$ is isomorphic to the additive group $\mathbf{Z}_4$.

    Hint: Find a generator $[a]_{10}$ of $\mathbf{Z}_{10}^{\times}$ and define $\phi : \mathbf{Z}_4 \rightarrow \mathbf{Z}_{10}^{\times}$ by $\phi([n]_4) = [a]_{10}^n$.

2. Show that the multiplicative group $\mathbf{Z}_7^{\times}$ is isomorphic to the additive group $\mathbf{Z}_6$.

3.† Show that the multiplicative group $\mathbf{Z}_8^{\times}$ is isomorphic to the group $\mathbf{Z}_2 \times \mathbf{Z}_2$.

4. Show that $\mathbf{Z}_5^\times$ is not isomorphic to $\mathbf{Z}_8^\times$ by showing that the first group has an element of order 4 but the second group does not.

5.† Is the additive group $\mathbf{C}$ of complex numbers isomorphic to the multiplicative group $\mathbf{C}^\times$ of nonzero complex numbers?

6. Let G_1 and G_2 be groups. Show that $G_2 \times G_1$ is isomorphic to $G_1 \times G_2$.

7. Let G be a group. Show that the group $(G, *)$ defined in Exercise 3 of Section 3.1 is isomorphic to G.

8. Prove that any group with three elements must be isomorphic to $\mathbf{Z}_3$.

9.† Find two abelian groups of order 8 that are not isomorphic.

10. Show that the group $\{f_{m,b} : \mathbf{R} \to \mathbf{R} \mid f(x) = mx + b, \ m \neq 0\}$ of affine functions from $\mathbf{R}$ to $\mathbf{R}$ (under composition of functions) is isomorphic to the group of all 2×2 matrices over $\mathbf{R}$ of the form $\begin{bmatrix} m & b \\ 0 & 1 \end{bmatrix}$ with $m \neq 0$ (under matrix multiplication). (See Exercises 10 and 11 of Section 3.1.)

11. Let G be the set of all matrices in $\mathrm{GL}_2(\mathbf{Z}_3)$ of the form $\begin{bmatrix} m & b \\ 0 & 1 \end{bmatrix}$. That is, $m, b \in \mathbf{Z}_3$ and $m \neq [0]_3$. Show that G is a subgroup of $\mathrm{GL}_2(\mathbf{Z}_3)$ that is isomorphic to S_3.

12. Let G be the following set of matrices over $\mathbf{R}$:

$$\begin{bmatrix} 1 & 0 \\ 0 & 1 \end{bmatrix}, \quad \begin{bmatrix} -1 & 0 \\ 0 & 1 \end{bmatrix}, \quad \begin{bmatrix} 1 & 0 \\ 0 & -1 \end{bmatrix}, \quad \begin{bmatrix} -1 & 0 \\ 0 & -1 \end{bmatrix}.$$

Show that G is isomorphic to $\mathbf{Z}_2 \times \mathbf{Z}_2$. (See Example 3.2.4.)

13. Let C_2 be the subgroup $\{\pm 1\}$ of the multiplicative group $\mathbf{R}^\times$. Show that $\mathbf{R}^\times$ is isomorphic to $\mathbf{R}^+ \times C_2$.

14. Let $G = \{x \in \mathbf{R} \mid x > 0 \text{ and } x \neq 1\}$, and define $*$ on G by $a * b = a^{\ln b}$. Show that G is isomorphic to the multiplicative group $\mathbf{R}^\times$. (See Exercise 9 of Section 3.1.)

15. Let G be any group, and let a be a fixed element of G. Define a function $\phi_a : G \to G$ by $\phi_a(x) = axa^{-1}$, for all $x \in G$. Show that ϕ_a is an isomorphism.

16. Let G be any group. Define $\phi : G \to G$ by $\phi(x) = x^{-1}$, for all $x \in G$.

    (a) Prove that ϕ is one-to-one and onto.

    (b) Prove that ϕ is an isomorphism if and only if G is abelian.

17. Let $\phi : G_1 \to G_2$ be a group isomorphism. Prove that if H is a subgroup of G_1, then $\phi(H) = \{y \in G_2 \mid y = \phi(h) \text{ for some } h \in H\}$ is a subgroup of G_2.

18. Define $\phi : \mathbf{C}^\times \to \mathbf{C}^\times$ by $\phi(a + bi) = a - bi$, for all nonzero complex numbers $a + bi$. Show that ϕ is an isomorphism.

19. Show that $\mathbf{C}^\times$ is isomorphic to the subgroup of $\mathrm{GL}_2(\mathbf{R})$ consisting of all matrices of the form $\begin{bmatrix} a & b \\ -b & a \end{bmatrix}$ such that $a^2 + b^2 \neq 0$.

20. Let G_1 and G_2 be groups. Show that G_1 is isomorphic to the subgroup of the direct product $G_1 \times G_2$ defined by $\{(x_1, x_2) \mid x_2 = e\}$.

21. Prove that if m, n are positive integers such that $\gcd(m, n) = 1$, then $\mathbf{Z}_{mn}^\times$ is isomorphic to $\mathbf{Z}_m^\times \times \mathbf{Z}_n^\times$.

22. Let a, b be positive integers, and let $d = \gcd(a, b)$ and $m = \mathrm{lcm}[a, b]$. Write $d = sa + tb$, $a = a'd$, and $b = b'd$. Prove that the function $f : \mathbf{Z}_m \times \mathbf{Z}_d \to \mathbf{Z}_a \times \mathbf{Z}_b$ defined by $f(([x]_m, [y]_d)) = ([x + ysa']_a, [x - ytb']_b)$ is an isomorphism.

    Note: This generalizes Proposition 3.4.5.

23. For each positive integer $n \geq 2$, let $C_n = \{(a, b) \in \mathbf{Z} \times \mathbf{Z} \mid a \equiv b \pmod{n}\}$. (See Exercise 9 of Section 3.3.) Show that C_n is isomorphic to $\mathbf{Z} \times \mathbf{Z}$.

24. Let $G = \mathbf{R} - \{-1\}$. Define $*$ on G by $a * b = a + b + ab$. Show that G is isomorphic to the multiplicative group $\mathbf{R}^\times$. (See Exercise 13 of Section 3.1.)

    Hint: Remember that an isomorphism maps identity to identity. Use this fact to help find the necessary mapping.

25. Let G be a group, and let S be any set for which there exists a one-to-one and onto function $\phi : G \to S$. Define an operation $*$ on S by setting $x_1 * x_2 = \phi(\phi^{-1}(x_1)\phi^{-1}(x_2))$, for all $x_1, x_2 \in S$.

    (a) Show that S is a group under this operation.

    (b) Show that $\phi : G \to S$ is a group isomorphism.

26. Let G_1 and G_2 be groups. A function from G_1 into G_2 that preserves products but is not necessarily a one-to-one correspondence will be called a **group homomorphism**, from the Greek word *homos* meaning same. Show that $\phi : \mathrm{GL}_2(\mathbf{R}) \to \mathbf{R}^\times$ defined by $\phi(A) = \det(A)$ for all matrices $A \in \mathrm{GL}_2(\mathbf{R})$ is a group homomorphism.

27. Using the definition of a group homomorphism given in Exercise 26, let $\phi : G_1 \to G_2$ be a group homomorphism. We define the **kernel** of ϕ to be

$$\ker(\phi) = \{x \in G_1 \mid \phi(x) = e\} .$$

    Prove that $\ker(\phi)$ is a subgroup of G_1.

28. Let $\phi : G_1 \to G_2$ be a group homomorphism. Prove that ϕ is a group isomorphism if and only if $\ker(\phi) = \{e\}$ and $\phi(G_1) = G_2$. (See Exercises 26, 27, and 17.)

3.5 Cyclic Groups

The class of cyclic groups will turn out to play a crucial role in studying the solution of equations by radicals. Yet this class can be characterized very simply, since we will show that a cyclic group must be isomorphic either to $\mathbf{Z}$ or $\mathbf{Z}_n$ for some n. This allows us to apply some elementary number theory to describe all subgroups of a cyclic group and to find all possible generators.

3.5.1 Theorem. *Every subgroup of a cyclic group is cyclic.*

Proof. Let G be a cyclic group with generator a, so that $G = \langle a \rangle$, and let H be any subgroup of G. If H is the trivial subgroup consisting only of e, then we are done since $H = \langle e \rangle$. If H is nontrivial, then it contains some element different from the identity, which can then be written in the form a^n for some integer $n \neq 0$. Since $a^{-n} = (a^n)^{-1}$ must also belong to H, we can assume that H contains some power a^k with $k > 0$.

Let m be the smallest positive integer such that $a^m \in H$. We claim that $H = \langle a^m \rangle$. Since $a^m \in H$, we have $\langle a^m \rangle \subseteq H$, and so the main point is to show that each element of H can be expressed as some power of a^m. Let $x \in H$. Then since $G = \langle a \rangle$, we have $x = a^k$ for some $k \in \mathbf{Z}$. By the division algorithm, $k = mq + r$ for $q, r \in \mathbf{Z}$ with $0 \leq r < m$. Then $x = a^k = a^{mq+r} = (a^m)^q a^r$. This shows that $a^r = (a^m)^{-q} x$ belongs to H (since a^m and x belong to H). This contradicts the definition of a^m as the smallest positive power of a in H unless $r = 0$. Therefore $k = mq$ and $x = (a^m)^q \in \langle a^m \rangle$. We conclude that $H = \langle a^m \rangle$ and so H is cyclic. $\square$

In our current terminology, Theorem 1.1.4 showed that every subgroup of $\mathbf{Z}$ is cyclic. This result can actually be used to give a very short proof of Theorem 3.5.1. Let $G = \langle a \rangle$ and let H be a subgroup of G. Let $I = \{n \in \mathbf{Z} \mid a^n \in H\}$. It follows from the rules for exponents that I is closed under addition and subtraction, so Theorem 1.1.4 implies that $I = m\mathbf{Z}$ for some integer m. We conclude that $H = \langle a^m \rangle$, and so H is cyclic.

The next theorem shows that any cyclic group is isomorphic either to $\mathbf{Z}$ or to $\mathbf{Z}_n$. Thus any two infinite cyclic groups are isomorphic to each other. Furthermore, two finite cyclic groups are isomorphic if and only if they have the same order.

3.5.2 Theorem. *Let G be a cyclic group.*

    (a) *If G is infinite, then $G \cong \mathbf{Z}$.*

    (b) *If $|G| = n$, then $G \cong \mathbf{Z}_n$.*

Proof. (a) Let $G = \langle a \rangle$ be an infinite cyclic group. Define $\phi : \mathbf{Z} \to G$ by $\phi(m) = a^m$, for all $m \in \mathbf{Z}$. The mapping ϕ is onto since $G = \langle a \rangle$, and Proposition 3.2.8 (a) shows that $\phi(m) \neq \phi(k)$ for $m \neq k$, so ϕ is also one-to-one. Finally, ϕ preserves the respective operations since

$$\phi(m + k) = a^{m+k} = a^m a^k = \phi(m)\phi(k) .$$

This shows that ϕ is an isomorphism.

(b) Let $G = \langle a \rangle$ be a finite cyclic group with n elements. Define $\phi : \mathbf{Z}_n \to G$ by $\phi([m]) = a^m$, for all $[m] \in \mathbf{Z}_n$. In order to show that ϕ is a function, we must check that the formula we have given is well-defined. That is, we must show that if $k \equiv m \pmod{n}$, then $a^k = a^m$. This follows from Proposition 3.2.8 (c). Furthermore, if $\phi([k]) = \phi([m])$, then the same proposition shows that $[k] = [m]$, and so ϕ is one-to-one. It is clear that ϕ is onto, since $G = \langle a \rangle$. Finally, ϕ preserves the respective operations since

$$\phi([m] + [k]) = a^{m+k} = a^m a^k = \phi([m])\phi([k]) .$$

This shows that ϕ is an isomorphism. $\square$

The subgroups of $\mathbf{Z}$ have the form $m\mathbf{Z}$, for $m \in \mathbf{Z}$. In addition, $m\mathbf{Z} \subseteq n\mathbf{Z}$ if and only if $n \mid m$. Thus $m\mathbf{Z} = n\mathbf{Z}$ if and only if $m = \pm n$.

The subgroups of $\mathbf{Z}_n$ take more work to describe. Given $m \in \mathbf{Z}$, we wish to find the multiples of $[m]$ in $\mathbf{Z}_n$. That is, we need to determine the integers b such that $[b] = k[m]$ for some $k \in \mathbf{Z}$. Equivalently, we need to know when $mx \equiv b \pmod{n}$ has a solution. By Theorem 1.3.5, the values of b are precisely the multiples of $\gcd(m, n)$.

In the next proposition, we have chosen to describe the subgroups of a cyclic group with n elements using multiplicative notation. If $G = \langle a \rangle$ is a finite cyclic group, and m is a positive divisor of the order of G, then $\langle a^m \rangle$ is a subgroup of G. The next proposition and corollary show that every subgroup of G can be written uniquely in this form.

3.5.3 Proposition. *Let $G = \langle a \rangle$ be a finite cyclic group of order n. If $m \in \mathbf{Z}$, then $\langle a^m \rangle = \langle a^d \rangle$, where $d = \gcd(m, n)$, and a^m has order n/d.*

Proof. Since $d \mid m$, we have $a^m \in \langle a^d \rangle$, and so $\langle a^m \rangle \subseteq \langle a^d \rangle$. On the other hand, there exist integers s, t such that $d = sm + tn$, and so

$$a^d = a^{sm+tn} = (a^m)^s (a^n)^t = (a^m)^s$$

since $a^n = e$. Thus $a^d \in \langle a^m \rangle$, and so $\langle a^d \rangle \subseteq \langle a^m \rangle$. The order of a^d is n/d, and so a^m has order n/d. $\square$

3.5.4 Corollary. *Let $G = \langle a \rangle$ be a finite cyclic group of order n.*

(a) *The element a^k generates G if and only if $\gcd(k, n) = 1$.*

(b) *If H is any subgroup of G, then $H = \langle a^k \rangle$ for some divisor k of n.*

(c) *If m and k are divisors of n, then $\langle a^m \rangle \subseteq \langle a^k \rangle$ if and only if $k \mid m$.*

Proof. The statements in parts (a) and (b) follow immediately from Proposition 3.5.3.

To prove part (c), first suppose that $k \mid m$. If $m = kq$, then $a^m = (a^k)^q \in \langle a^k \rangle$, and therefore $\langle a^m \rangle \subseteq \langle a^k \rangle$. Conversely, if $\langle a^m \rangle \subseteq \langle a^k \rangle$, then $a^m \in \langle a^k \rangle$, and so $m \equiv kt \pmod{n}$ for $t \in \mathbf{Z}$. It follows that $m = kt + nq$ for some $q \in \mathbf{Z}$, and so $k \mid m$ because by assumption k is a divisor of n. $\square$

We will use the notation $m\mathbf{Z}_n$ for the subgroup $\langle [m] \rangle$ consisting of all multiples of $[m]$. If m and k are divisors of n, then we have $m\mathbf{Z}_n \subseteq k\mathbf{Z}_n$ if and only if $k \mid m$. For small values of n, we can easily give a diagram showing all subgroups of $\mathbf{Z}_n$ and the inclusion relations between them. This is called a **subgroup diagram**. Since subgroups correspond to divisors of n and inclusions are the opposite of divisibility relations, we can find the diagram of divisors of n and simply turn it upside down.

Example 3.5.1.

In Example 1.2.2 we gave the diagram of all divisors of 12. This leads to the subgroup diagram given in Figure 3.5.1. $\square$

Figure 3.5.1: Subgroups of $\mathbf{Z}_{12}$

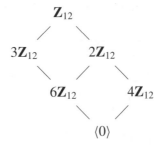

Example 3.5.2.

If n is a prime power, then the subgroup diagram of $\mathbf{Z}_n$ is particularly simple, since for any two subgroups, one is contained in the other. (Why?) We give the subgroup diagram for $\mathbf{Z}_{125}$ in Figure 3.5.2. $\square$

Figure 3.5.2: Subgroups of $\mathbf{Z}_{125}$

In Definition 3.3.3 we introduced the direct product of two groups. This definition can easily be extended to the direct product $G_1 \times \cdots \times G_n$ of n groups $G_1, \ldots, G_n$ by considering n-tuples in which the ith entry is an element of G_i, with componentwise multiplication. As with the direct product of two groups, the order of an element is the least common multiple of the orders of each component.

The following proposition implies that every finite cyclic group is isomorphic to a direct product of cyclic groups of prime power order. We could call this a structure theorem for finite cyclic groups, in the sense that we can show how they are built up from combinations of simpler cyclic groups. This is a special case of the general structure theorem for finite abelian groups, proved in Section 7.5, which states that any finite abelian group is isomorphic to a direct product of cyclic groups of prime power order.

3.5.5 Theorem. *Let n be a positive integer which has the prime decomposition* $n = p_1^{\alpha_1} p_2^{\alpha_2} \cdots p_m^{\alpha_m}$, *where* $p_1 < p_2 < \ldots < p_m$. *Then*

$$\mathbf{Z}_n \cong \mathbf{Z}_{p_1^{\alpha_1}} \times \mathbf{Z}_{p_2^{\alpha_2}} \times \cdots \times \mathbf{Z}_{p_m^{\alpha_m}} \ .$$

Proof. In the direct product of the given groups, the element with [1] in each component has order n, since the least common multiple of the given prime powers is n. Thus the direct product is cyclic of order n, so by Theorem 3.5.2, it must be isomorphic to $\mathbf{Z}_n$. □

For a positive integer n, the Euler φ-function $\varphi(n)$ is defined to be the number of positive integers less than or equal to n and relatively prime to n. Thus $\varphi(n)$ gives the number of elements of $\mathbf{Z}_n$ that are generators of $\mathbf{Z}_n$. In Section 1.4 we gave a formula for $\varphi(n)$, without proof. The proof is an easy consequence of our description of $\mathbf{Z}_n$.

3.5.6 Corollary. *Let n be a positive integer which has the prime decomposition* $n = p_1^{\alpha_1} p_2^{\alpha_2} \cdots p_m^{\alpha_m}$, *where* $p_1 < p_2 < \ldots < p_m$. *Then*

$$\varphi(n) = n \left(1 - \frac{1}{p_1} \right) \left(1 - \frac{1}{p_2} \right) \cdots \left(1 - \frac{1}{p_m} \right).$$

Proof. To count the generators of $\mathbf{Z}_n$, it is easier to use the isomorphic direct product

$$\mathbf{Z}_{p_1^{\alpha_1}} \times \mathbf{Z}_{p_2^{\alpha_2}} \times \cdots \times \mathbf{Z}_{p_m^{\alpha_m}} \cong \mathbf{Z}_n$$

obtained in Theorem 3.5.5 than it is to use $\mathbf{Z}_n$, since an isomorphism preserves generators. An element of this direct product is a generator if and only if it has order n, and so this means that the least common multiple of the orders of its components in their respective groups must be n. If $b = ([b_1]_{p_1^{\alpha_1}}, [b_2]_{p_2^{\alpha_2}}, \ldots, [b_m]_{p_m^{\alpha_m}})$ is an element of the direct product, then the order of each $[b_i]_{p_i^{\alpha_i}}$ must be a divisor of $p_i^{\alpha_i}$, say p_i^{β}, with $\beta_i \leq \alpha_i$. Then b has order $p_1^{\beta_1} p_2^{\beta_2} \cdots p_m^{\beta_m}$, and for this order to be equal to n we must have $\beta_i = \alpha_i$ for each i. It follows that an element of the direct product is a generator if and only if each of its components is a generator in its respective cyclic group of prime power order. Thus the total number of possible generators is equal to the product of the number of generators in each component.

We have reduced the problem to counting the number of generators in $\mathbf{Z}_{p^{\alpha}}$, for any prime p. The elements that are *not* generators are the multiples of p, and among the p^{α} elements of $\mathbf{Z}_{p^{\alpha}}$ there are $p^{\alpha-1}$ such multiples. Thus

$$\varphi(p^{\alpha}) = p^{\alpha} - p^{\alpha-1} = p^{\alpha} \left(1 - \frac{1}{p} \right).$$

Taking the product of these values for each of the primes in the decomposition of n gives the formula we want. $\square$

If G is a finite group, then as we noted following the definition of order, each element of G must have finite order. Thus for each $a \in G$, we have $a^{o(a)} = e$. If N is the least common multiple of the integers $o(a)$, for all $a \in G$, then $a^N = e$ for all $a \in G$. Since $o(a)$ is a divisor of $|G|$ for any $a \in G$, it follows that N is a divisor of $|G|$. Using this concept of the "exponent" of a group, we are able to characterize cyclic groups among all finite abelian groups. Note that some authors refer to every multiple of N as an exponent of G.

3.5.7 Definition. *Let G be a group. If there exists a positive integer N such that* $a^N = e$ *for all* $a \in G$, *then the smallest such positive integer is called the* **exponent** *of G.*

Example 3.5.3.

> The exponent of any finite group is the least common multiple of the orders of
> its elements. Thus the exponent of S_3 is 6, since S_3 has elements of order 1,
> 2, and 3. The exponent of $\mathbf{Z}_2 \times \mathbf{Z}_2$ is 2. Since we are using additive notation,
> remember to use multiples instead of powers. The exponent of $\mathbf{Z}_2 \times \mathbf{Z}_3$ is 6,
> since there are elements of order 1, 2, 3, and 6.   □

3.5.8 Lemma. *Let G be a group, and let $a, b \in G$ be elements such that $ab = ba$.
If the orders of a and b are relatively prime, then $o(ab) = o(a)o(b)$.*

Proof. Let $o(a) = n$ and $o(b) = m$. Then since $ab = ba$, we must have $(ab)^{mn} = a^{mn}b^{mn} = (a^n)^m(b^m)^n = e$, which shows that ab has finite order, say $o(ab) = k$.
Furthermore, $(ab)^{mn} = e$ implies that $k \mid mn$. On the other hand, $(ab)^k = e$, which
shows that $a^k = b^{-k}$. Therefore $a^{km} = (a^k)^m = (b^{-k})^m = (b^m)^{-k} = e$, showing
that $n \mid km$. Since $(n, m) = 1$, we must have $n \mid k$. A similar argument shows that
$m \mid k$, and then $mn \mid k$ since $(n, m) = 1$. Since m, n, k are positive integers with
$mn \mid k$ and $k \mid mn$, we have $k = mn$.   □

3.5.9 Proposition. *Let G be a finite abelian group.*

    (a) *The exponent of G is equal to the order of any element of G of largest order.*

    (b) *The group G is cyclic if and only if its exponent is equal to its order.*

Proof. (a) Choose an element $a \in G$ whose order is as large as possible. Let $b \in G$
and suppose that $o(b)$ is not a divisor of $o(a)$. Then in the prime factorizations of
$o(a)$ and $o(b)$, there exists a prime p with $o(a) = p^\alpha n$ and $o(b) = p^\beta m$, where
p is relatively prime to both n and m, and $\beta > \alpha \geq 0$. Then $o(a^{p^\alpha}) = n$ and
$o(b^m) = p^\beta$, so these orders are relatively prime. It follows from Lemma 3.5.8 that
$o(a^{p^\alpha} b^m) = np^\beta$, and this is greater than $o(a)$, a contradiction. Thus $o(b) \mid o(a)$ for
all $b \in G$, and $o(a)$ is therefore the exponent of G.

    (b) Part (b) follows immediately from part (a), since G is cyclic if and only if
there exists an element of order $|G|$.   □

EXERCISES: SECTION 3.5

1.†Let G be a group and let $a \in G$ be an element of order 12. What is the order of a^j
for $j = 2, \ldots, 11$?

2. Let G be a group and let $a \in G$ be an element of order 30. List the powers of a that
have order 2, order 3 or order 5.

3. Give the subgroup diagrams of the following groups.

   (a) $\mathbf{Z}_{24}$

   (b) $\mathbf{Z}_{36}$

4. Give the subgroup diagram of $\mathbf{Z}_{60}$.

5.†Find the cyclic subgroup of $\mathbf{C}^\times$ generated by $(\sqrt{2} + \sqrt{2}i)/2$.

6. Find the order of the cyclic subgroup of $\mathbf{C}^\times$ generated by $1 + i$.

7.†Which of the multiplicative groups $\mathbf{Z}_{15}^\times$, $\mathbf{Z}_{18}^\times$, $\mathbf{Z}_{20}^\times$, $\mathbf{Z}_{27}^\times$ are cyclic?

8. Find $\langle \pi \rangle$ in $\mathbf{R}^\times$.

9.†Find all cyclic subgroups of $\mathbf{Z}_4 \times \mathbf{Z}_2$.

10. Find all cyclic subgroups of $\mathbf{Z}_6 \times \mathbf{Z}_3$.

11. Which of the multiplicative groups $\mathbf{Z}_7^\times$, $\mathbf{Z}_{10}^\times$, $\mathbf{Z}_{12}^\times$, $\mathbf{Z}_{14}^\times$ are isomorphic?

12. Let a, b be positive integers, and let $d = \gcd(a, b)$ and $m = \text{lcm}[a, b]$. Use Proposition 3.5.5 to prove that $\mathbf{Z}_a \times \mathbf{Z}_b \cong \mathbf{Z}_d \times \mathbf{Z}_m$.

13. Show that in a finite cyclic group of order n, the equation $x^m = e$ has exactly m solutions, for each positive integer m that is a divisor of n.

14. Prove that any cyclic group with more than two elements has at least two different generators.

15. Prove that any finite cyclic group with more than two elements has an even number of distinct generators.

16. Let G be any group with no proper, nontrivial subgroups, and assume that $|G| > 1$. Prove that G must be isomorphic to $\mathbf{Z}_p$ for some prime p.

17. Let G be the set of all 3×3 matrices of the form $\begin{bmatrix} 1 & a & b \\ 0 & 1 & c \\ 0 & 0 & 1 \end{bmatrix}$.

   (a) Show that if $a, b, c \in \mathbf{Z}_3$, then G is a group with exponent 3.

   (b) Show that if $a, b, c \in \mathbf{Z}_2$, then G is a group with exponent 4.

18. Prove that $\sum_{d|n} \varphi(d) = n$ for any positive integer n.

   Hint: Interpret the equation in the cyclic group $\mathbf{Z}_n$, by considering all of its subgroups.

19. Let $n = 2^k$ for $k > 2$. Prove that $\mathbf{Z}_n^\times$ is not cyclic.

   Hint: Show that ± 1 and $(n/2) \pm 1$ satisfy the equation $x^2 = 1$, and that this is impossible in any cyclic group.

20. Let G be a group with p^k elements, where p is a prime number and $k \geq 1$. Prove that G has a subgroup of order p.

3.6 Permutation Groups

When groups were first studied, they were thought of as sets of permutations closed under products and including the identity, together with inverses of all elements. The abstract definition that we now use was not given until later. The content of Cayley's theorem, which we are about to prove, is the surprising result that this abstract definition is not any more general than the original concrete definition.

In the case of a finite group, a little thought about the group multiplication table may convince the reader that the theorem is not so surprising after all. As we have observed, each row in the multiplication table represents a permutation of the group elements. Furthermore, each row corresponds to multiplication by a given element, and so there is a natural way to assign a permutation to each element of the group.

3.6.1 Definition. *Any subgroup of the symmetric group* $\mathrm{Sym}(S)$ *on a set S is called a permutation group.*

In the following proof of Cayley's theorem, we must show that any group G is isomorphic to a subgroup of $\mathrm{Sym}(S)$ for some set S, so the first problem is to find an appropriate set S. Our choice is to let S be G itself. Next we must assign to each element a of G some permutation of G. The natural one is the function $\lambda_a : G \to G$ defined by $\lambda_a(x) = ax$ for all $x \in G$. (We use the notation λ_a to indicate multiplication on the left by a.) The values $\lambda_a(x)$ are the entries in the group table that occur in the row corresponding to multiplication by a, and this makes λ_a a permutation of G. Finally we must show that assigning λ_a to a respects the two operations and gives a one-to-one correspondence.

3.6.2 Theorem (Cayley). *Every group is isomorphic to a permutation group.*

Proof. Let G be any group. Given $a \in G$, define $\lambda_a : G \to G$ by $\lambda_a(x) = ax$, for all $x \in G$. Then λ_a is onto since the equation $ax = b$ has a solution for each $b \in G$, and it is one-to-one since the solution is unique, so we conclude that λ_a is a permutation of G. This shows that the function $\phi : G \to \mathrm{Sym}(G)$ defined by $\phi(a) = \lambda_a$ is well-defined.

We next want to show that $G_\lambda = \phi(G)$ is a subgroup of $\mathrm{Sym}(G)$, and to do so we need several facts. The formula $\lambda_a\lambda_b = \lambda_{ab}$ holds since for all $x \in G$ we have $\lambda_a(\lambda_b(x)) = a(bx) = (ab)x = \lambda_{ab}(x)$. Because λ_e is the identity function, this formula also implies that $(\lambda_a)^{-1} = \lambda_{a^{-1}}$. This shows that G_λ is closed, contains the identity, and contains inverses for its elements, so it is a subgroup.

To show that ϕ preserves products, we must show that $\phi(ab) = \phi(a)\phi(b)$. This follows from the formula $\lambda_{ab} = \lambda_a\lambda_b$. To complete the proof that $\phi : G \to G_\lambda$ is an isomorphism, it is only necessary to show that ϕ is one-to-one, since it is onto by the definition of G_λ. If $\phi(a) = \phi(b)$ for $a, b \in G$, then we have $\lambda_a(x) = \lambda_b(x)$ for all $x \in G$. In particular, $ae = \lambda_a(e) = \lambda_b(e) = be$, and so $a = b$.

In summary, we have found a subgroup G_λ of $\text{Sym}(G)$ and an isomorphism $\phi : G \to G_\lambda$ defined by assigning to each $a \in G$ the permutation λ_a. $\square$

In this section we will assume as a matter of course that all permutations in S_n are expressed in the natural way as a product of disjoint cycles. A formal proof that this can be done is given in Section 2.3. Recall that if $\sigma \in S_n$, and σ is written as a product of disjoint cycles, then the order of σ is the least common multiple of the lengths of its cycles.

Example 3.6.1 (Rigid motions of a square).

Groups of symmetries are very useful in geometry. We now look at the group of rigid motions of a square. Imagine a square of cardboard, placed in a box just large enough to contain it. Picking up the square and then replacing it in the box, in what may be a new position, gives what is called a **rigid motion** of the square. Each of the rigid motions determines a permutation of the vertices of the square, and the permutation notation gives a convenient way to describe these motions. To count the number of rigid motions, fix a vertex and label it A. Label one of the adjacent vertices B. We have a total of eight rigid motions, since we have four choices of a position in which to place vertex A, and then two choices for vertex B because it must be adjacent to A.

Figure 3.6.1: Rigid Motions of a Square

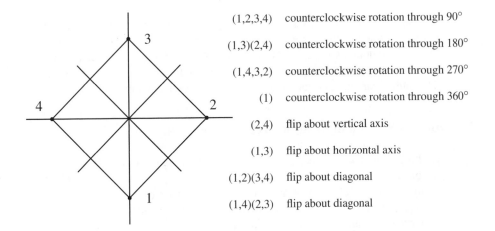

(1,2,3,4)	counterclockwise rotation through 90°
(1,3)(2,4)	counterclockwise rotation through 180°
(1,4,3,2)	counterclockwise rotation through 270°
(1)	counterclockwise rotation through 360°
(2,4)	flip about vertical axis
(1,3)	flip about horizontal axis
(1,2)(3,4)	flip about diagonal
(1,4)(2,3)	flip about diagonal

We have used $(1, 2, 3, 4)$ to describe the rigid motion in which the corner of the square currently occupying position 1 is placed in position 2, while the corner currently in position 2 is moved to position 3, the one in position 3 is moved to position 4, and the one in position 4 is moved to position 1. In the rigid motion (2, 4) the square is replaced so that the corners originally in positions 2 and 4 are interchanged, while the corners originally in positions 1 and 3 remain in the same positions. Note that we do not obtain all elements of S_4 as rigid motions, since, for example, (1, 2) would represent an impossible configuration.

It is possible to introduce an operation on the rigid motions, by simply saying that the "product" of two rigid motions will be given by first performing one and then the other. This defines a group, since following one rigid motion by another gives a third rigid motion, the identity is a rigid motion, and any rigid motion can be reversed, providing inverses. In terms of permutations of the vertices of the square, this operation just corresponds to ordinary multiplication of permutations. We give the multiplication table for this subgroup of S_4 in Table 3.6.1. In order to make the table smaller, we have found it necessary to use a more compact notation that omits commas.

In our notation, the motion σ carries the vertex currently in position i to position $\sigma(i)$. Thus we may think of a motion as a function from the set of position numbers into itself. As dictated by our convention for functions, the motion $\sigma\tau$ is the motion obtained by first performing τ and then σ. The reader should be warned that our convention of labeling positions is not followed by all authors; some prefer to follow the convention of labeling vertices. □

Example 3.6.2 (Rigid motions of an equilateral triangle).

The rigid motions of an equilateral triangle yield the group S_3. With the vertices labeled as in Figure 3.6.2, the counterclockwise rotations are given by the permutations (1, 2, 3), (1, 3, 2), and (1). Flipping the triangle about one of the angle bisectors gives one of the permutations (1, 2), (1, 3) or (2, 3). The multiplication table for S_3 has already been given in Table 3.1.1. □

Example 3.6.3 (Rigid motions of a regular polygon).

In this example we will determine the group of all rigid motions of a regular n-gon. In Section 3.3 we have seen that S_3, the group of rigid motions of an equilateral triangle, can be described using elements a (of order 3) and b (of order 2) which satisfy the equation $ba = a^2b$. The elements of S_3 can then be written (uniquely) as e, a, a^2, b, ab, and a^2b.

In Example 3.6.1, letting $a = (1, 2, 3, 4)$ and $b = (2, 4)$, we have elements of order 4 and 2, respectively, which can be shown to satisfy the equation $ba = a^3b$. Furthermore, using these elements the group can then be described as the set $\{e, a, a^2, a^3, b, ab, a^2b, a^3b\}$. The equation $ba = a^3b$ shows us

Table 3.6.1: Rigid Motions of a Square

	(1)	(1234)	(13)(24)	(1432)	(24)	(12)(34)	(13)	(14)(23)
(1)	(1)	(1234)	(13)(24)	(1432)	(24)	(12)(34)	(13)	(14)(23)
(1234)	(1234)	(13)(24)	(1432)	(1)	(12)(34)	(13)	(14)(23)	(24)
(13)(24)	(13)(24)	(1432)	(1)	(1234)	(13)	(14)(23)	(24)	(12)(34)
(1432)	(1432)	(1)	(1234)	(13)(24)	(14)(23)	(24)	(12)(34)	(13)
(24)	(24)	(14)(23)	(13)	(12)(34)	(1)	(1432)	(13)(24)	(1234)
(12)(34)	(12)(34)	(24)	(14)(23)	(13)	(1234)	(1)	(1432)	(13)(24)
(13)	(13)	(12)(34)	(24)	(14)(23)	(13)(24)	(1234)	(1)	(1432)
(14)(23)	(14)(23)	(13)	(12)(34)	(24)	(1432)	(13)(24)	(1234)	(1)

Figure 3.6.2: Rigid Motions of an Equilateral Triangle

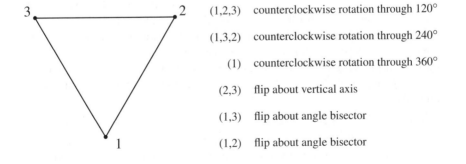

(1,2,3)	counterclockwise rotation through 120°
(1,3,2)	counterclockwise rotation through 240°
(1)	counterclockwise rotation through 360°
(2,3)	flip about vertical axis
(1,3)	flip about angle bisector
(1,2)	flip about angle bisector

how to multiply two elements in this form and then bring them back to the "standard form."

Now let us consider the general case of the rigid motions of a regular n-gon. Since a rigid motion followed by another rigid motion is again a rigid motion, and since any rigid motion can be reversed, the set of all rigid motions of a regular n-gon forms a group. To see that there are $2n$ rigid motions, fix two adjacent vertices. There are n places to send the first of these vertices, and then there are two choices for the adjacent vertex, giving a total of $2n$ motions. (The n-gon has been flipped over if the vertices appear in clockwise order.)

Figure 3.6.3:

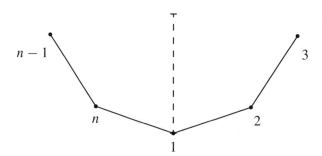

Figure 3.6.3 represents part of a regular n-gon. Let a be a counterclockwise rotation about the center, through $360/n$ degrees. Thus a is the cycle $(1, 2, 3, \ldots, n)$ of length n and has order n. Let b be a flip about the line of symmetry through position number 1. Thus b has order 2 and is given by the product of transpositions $(2, n)(3, n-1) \cdots$.

Consider the set $S = \{a^k, a^k b \mid 0 \le k < n\}$ of rigid motions. It is easy to see that the elements a^k for $0 \le k < n$ are all distinct, and that the elements $a^k b$ for $0 \le k < n$ are also distinct. Since the rigid motion represented by a^k does not flip the n-gon, while the motion represented by $a^j b$ does, it is never the case that $a^k = a^j b$. Thus $|S| = 2n$, and so $G = S$. Since we have listed (uniquely) all the elements of G, it only remains to show how they can be multiplied.

Clearly, $a^n = e$ and $b^2 = e$, and thus $a^{-1} = a^{n-1}$ and $b^{-1} = b$. After multiplying two elements, to bring the product into one of the standard forms listed above, we only need to know how to move b past a. That is, we must compute ba, and to do so it turns out to be easiest to compute bab. From Figure 3.6.4, it is easy to see that we obtain $bab = a^{-1}$, and then multiplying on the right by $b^{-1} = b$ we obtain the formula $ba = a^{-1}b$ (or $ba = a^{n-1}b$).

Figure 3.6.4:

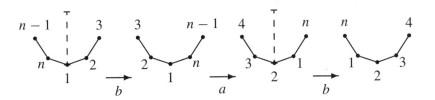

Thus, for example, if we want to multiply ab by a^2, we use the formula $ba = a^{-1}b$ as follows:

$$ab \cdot a^2 = a(ba)a = a(a^{-1}b)a = ba = a^{-1}b = a^{n-1}b \ .$$

We have obtained a complete description of the group of rigid motions of a regular n-gon in terms of elements a and b and the equations $a^n = e, b^2 = e$, and $ba = a^{-1}b$ that they satisfy. $\square$

3.6.3 Definition. *Let $n \geq 3$ be an integer. The group of rigid motions of a regular n-gon is called the nth **dihedral group**, denoted by D_n.*

Example 3.6.4 (Subgroups of S_3).

In Figure 3.6.5 we give the subgroup diagram of S_3, using the notation of Example 3.6.3. By Lagrange's theorem, the only possible orders of proper subgroups are 1, 2, or 3. Since subgroups of order 2 or 3 must be cyclic, it is relatively simple to find all subgroups. $\square$

Example 3.6.5 (Subgroups of D_4).

In Figure 3.6.6 we give the subgroup diagram of D_4, again using the notation of Example 3.6.3. The possible orders of proper subgroups are 1, 2, or 4. We first find all cyclic subgroups: a has order 4, while each of the elements a^2, b, ab, a^2b, a^3b has order 2. Any subgroup of order 4 that is not cyclic must be isomorphic to the Klein four-group, so it must contain two elements of order 2 and their product. By considering all possible pairs of elements of order 2 it is possible to find the remaining two subgroups of order 4. Just as a cyclic subgroup is the smallest subgroup containing the generator, these subgroups are the smallest ones containing the two elements used to construct it. In general, to find all subgroups, one would need to consider all possible combinations of elements, a difficult task in a large group. $\square$

Figure 3.6.5: Subgroups of S_3

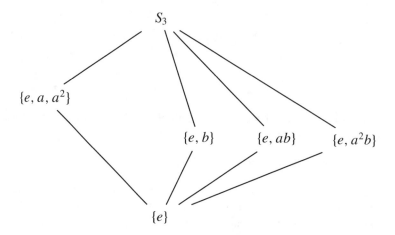

Figure 3.6.6: Subgroups of D_4

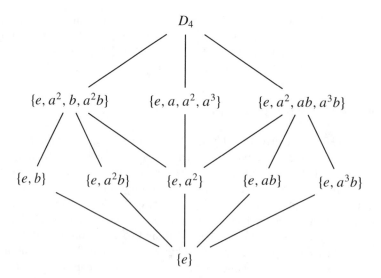

In Section 2.3 we proved that any permutation in S_n can be written as a product of transpositions (cycles of length two) and then proved that the number of transpositions in such a decomposition of a given permutation must either be always even or always odd. Thus we can call a permutation **even** if it can be expressed as an even number of transpositions, and **odd** otherwise.

3.6.4 Proposition. *The set of all even permutations of S_n is a subgroup of S_n.*

Proof. If σ and τ are even permutations, then each can be expressed as a product of an even number of transpositions. It follows that $\tau\sigma$ can be expressed as a product of an even number of transpositions, and so the set of all even permutations of S_n is closed under multiplication of permutations. Furthermore, the identity permutation is even. Since S_n is a finite set, this is enough to imply that we have a subgroup. $\square$

3.6.5 Definition. *The set of all even permutations of S_n is called the **alternating group** on n elements, and will be denoted by A_n.*

When we considered even and odd permutations in Chapter 2, our proof of Theorem 2.3.11 (justifying the definition of even and odd permutations) was different from the one usually given. We now give the standard approach to parity of elements of S_n.

Let Δ_n be the polynomial in n variables $x_1, x_2, \ldots, x_n$ defined by

$$\Delta_n = \prod_{1 \le i < j \le n} (x_i - x_j) .$$

Any permutation $\sigma \in S_n$ acts on Δ_n by permuting the subscripts, and we write

$$\sigma(\Delta_n) = \prod_{1 \le i < j \le n} (x_{\sigma(i)} - x_{\sigma(j)}) .$$

If $i < j$ and $\sigma(i) < \sigma(j)$, then the factors $x_i - x_j$ and $x_{\sigma(i)} - x_{\sigma(j)}$ of Δ_n have the same sign, but if $\sigma(i) > \sigma(j)$ then $x_{\sigma(i)} - x_{\sigma(j)} = -(x_{\sigma(j)} - x_{\sigma(i)})$. Because of such sign changes, we either have $\sigma(\Delta_n) = \Delta_n$ or $\sigma(\Delta_n) = -\Delta_n$.

For example, we have $\Delta_3 = (x_1 - x_2)(x_1 - x_3)(x_2 - x_3)$. Letting the permutation $(1, 2, 3)$ act on Δ_3 gives the new polynomial $(x_2 - x_3)(x_2 - x_1)(x_3 - x_1)$, in which the signs of two factors have been changed. On the other hand, the transposition $(1, 2)$ applied to Δ_3 gives the new polynomial $(x_2 - x_1)(x_2 - x_3)(x_1 - x_3)$, in which the sign of only one factor has changed.

3.6.6 Theorem. *A permutation in S_n is even if and only if it leaves the sign of Δ_n unchanged.*

Proof. Set $X = \{\Delta_n, -\Delta_n\}$. For $\sigma \in S_n$ we define $\widehat{\sigma} : X \to X$ by $\widehat{\sigma}(\Delta_n) = \prod_{1 \le i < j \le n} (x_{\sigma(i)} - x_{\sigma(j)})$ and $\widehat{\sigma}(-\Delta_n) = -\prod_{1 \le i < j \le n} (x_{\sigma(i)} - x_{\sigma(j)})$. If $\rho, \sigma \in S_n$, then it is routine to check that $\widehat{\sigma\rho}(\Delta_n) = \widehat{\sigma}(\widehat{\rho}(\Delta_n))$ by considering each case of whether or not σ or ρ changes the sign of Δ_n.

We next show that for any transposition $\tau = (r, s)$ we have $\widehat{\tau}(\Delta_n) = -\Delta_n$. Assume that $r < s$. We first observe that $\widehat{\tau}(\Delta_n) = \prod_{1 \le i < j \le n}(x_{\tau(i)} - x_{\tau(j)})$. Next we see that $x_{\tau(r)} - x_{\tau(s)} = x_s - x_r = -(x_r - x_s)$, and that if neither i nor j is r or s we have $x_{\tau(i)} - x_{\tau(j)} = x_i - x_j$. We analyze the remaining cases by considering pairs of factors of Δ_n: (1) if $i > s$, then $(x_{\tau(r)} - x_i)(x_{\tau(s)} - x_i) = (x_s - x_i)(x_r - x_i) = (x_r - x_i)(x_s - x_i)$; (2) if $r < i < s$, then $(x_{\tau(r)} - x_i)(x_i - x_{\tau(s)}) = (x_s - x_i)(x_i - x_r) = (x_r - x_i)(x_i - x_s)$; and (3) if $i < r$, then $(x_i - x_{\tau(r)})(x_i - x_{\tau(s)}) = (x_i - x_s)(x_i - x_r) = (x_i - x_r)(x_i - x_s)$. Thus $\widehat{\tau}(\Delta_n) = -\Delta_n$.

Given any $\sigma \in S_n$, we can write $\sigma = \tau_1\tau_2 \cdots \tau_k$, where each τ_i is a transposition. Then $\widehat{\sigma}(\Delta_n) = (-1)^k\Delta_n$. This completes the proof that σ is even if and only if σ leaves the sign of Δ_n unchanged. $\square$

EXERCISES: SECTION 3.6

1. Find the orders of each of these permutations.

   †(a) $(1, 2)(2, 3)(3, 4)$

   (b) $(1, 2, 5)(2, 3, 4)(5, 6)$

   †(c) $(1, 3,)(2, 6)(1, 4, 5)$

   (d) $(1, 2, 3)(2, 4, 3, 5)(1, 3, 2)$

2. Write out the addition tables for $\mathbf{Z}_4$ and for $\mathbf{Z}_2 \times \mathbf{Z}_2$. Use cycle notation to write out the permutation determined by each row of each of the addition tables, as in the discussion preceding Cayley's theorem.

3. Write out the addition table for $\mathbf{Z}_4 \times \mathbf{Z}_2$. Use cycle notation to write out the permutation determined by each row of the addition table, as in the discussion preceding Cayley's theorem.

4. Find the permutations that correspond to the rigid motions of a rectangle that is not a square. Do the same for the rigid motions of a rhombus (diamond) that is not a square.

5. Show that no proper subgroup of S_4 contains both $(1, 2, 3, 4)$ and $(1, 2)$.

6. Let the dihedral group D_n be given by elements a of order n and b of order 2, where $ba = a^{-1}b$.

   (a) Show that $a^{-m} = a^{n-m}$, for all $m \in \mathbf{Z}$.

   (b) Show that $ba^m = a^{-m}b$, for all $m \in \mathbf{Z}$.

   (c) Show that $ba^mb = a^{-m}$, for all $m \in \mathbf{Z}$.

7. Find the order of each element of D_6.

8.†Find the order of the group of rigid motions of a cube. (Imagine a wooden cube placed in a box just large enough to hold it. A rigid motion of the cube consists of picking up the cube and replacing it in the box, possibly in a different position.)

9. A rigid motion of a cube can be thought of either as a permutation of its eight vertices or as a permutation of its six sides. Find a rigid motion of the cube that has order 3, and express the permutation that represents it in both ways, as a permutation on eight elements and as a permutation on six elements.

10. Show that the following matrices form a subgroup of $GL_2(\mathbf{C})$ isomorphic to D_4:

$$\pm \begin{bmatrix} 1 & 0 \\ 0 & 1 \end{bmatrix}, \quad \pm \begin{bmatrix} i & 0 \\ 0 & -i \end{bmatrix}, \quad \pm \begin{bmatrix} 0 & 1 \\ 1 & 0 \end{bmatrix}, \quad \pm \begin{bmatrix} 0 & i \\ -i & 0 \end{bmatrix}.$$

11. Show that D_n is isomorphic to a subgroup of S_n, for $n \geq 3$.

12.†Find the largest possible order of an element in S_4. Answer the same question for S_5, S_6, S_7, S_8, and S_9.

13. List the elements of A_4.

14.†Without writing down all 60 elements of A_5, describe the possible shapes of the permutations (the number and lengths of their disjoint cycles) and how many of each type there are.

15. (a) Show that $A_4 = \{\sigma \in S_4 \mid \sigma = \tau^2 \text{ for some } \tau \in S_4\}$.

 (b) Show that $A_5 = \{\sigma \in S_5 \mid \sigma = \tau^2 \text{ for some } \tau \in S_5\}$.

 (c) Show that $A_6 \supset \{\sigma \in S_6 \mid \sigma = \tau^2 \text{ for some } \tau \in S_6\}$.

 (d) What can you say about A_n if $n > 6$?

16. Show that if G is any group of permutations, then the set of all even permutations in G forms a subgroup of G.

17. For any elements $\sigma, \tau \in S_n$, show that $\sigma \tau \sigma^{-1} \tau^{-1} \in A_n$.

18. Let S be an infinite set. Let H be the set of all elements $\sigma \in \text{Sym}(S)$ such that $\sigma(x) = x$ for all but finitely many $x \in S$. Prove that H is a subgroup of $\text{Sym}(S)$.

19. The center of a group is the set of all elements that commute with every other element of the group. That is, $Z(G) = \{x \in G \mid xg = gx \text{ for all } g \in G\}$. Show that if $n \geq 3$, then the center of S_n is trivial.

20. Let the dihedral group D_n be given by elements a of order n and b of order 2, where $ba = a^{-1}b$. Find the smallest subgroup of D_n that contains a^2 and b.

 Hint: Consider two cases, depending on whether n is odd or even.

21. Find the center of the dihedral group D_n.

 Hint: Consider two cases, depending on whether n is odd or even.

22. Show that in S_n the only elements which commute with the cycle $(1, 2, \ldots, n)$ are its powers.

23. Let $\tau = (a, b, c)$ and let σ be any permutation. Show that

$$\sigma \tau \sigma^{-1} = (\sigma(a), \sigma(b), \sigma(c)) \ .$$

24. Show that the product of two transpositions is one of (i) the identity; (ii) a 3-cycle; (iii) a product of two (nondisjoint) 3-cycles. Deduce that every element of A_n can be written as a product of 3-cycles.

25. Show that S_n is isomorphic to a subgroup of A_{n+2}.

26. Prove that every group of order n is isomorphic to a subgroup of $\mathrm{GL}_n(\mathbf{R})$.

27. Let permutations in S_4 act on polynomials in four variables by permuting the subscripts, as in Theorem 3.6.6.

    (a) Which permutations in S_4 leave the polynomial $(x_1 - x_2)(x_3 - x_4)$ unchanged?
    (b) Which permutations in S_4 leave the polynomial $\prod_{1 \le i < j \le 4}(x_i + x_j)$ unchanged?
    (c) Which permutations in S_4 leave the polynomial $\prod_{1 \le i < j \le 4}(x_i - x_j)$ unchanged?

3.7   Homomorphisms

In Section 3.4 we studied the notion of a group isomorphism. If G_1 and G_2 are groups, then a function $\phi : G_1 \to G_2$ is a group isomorphism if (i) ϕ is one-to-one and onto and (ii) ϕ respects the group operations, that is, if $\phi(ab) = \phi(a)\phi(b)$ for all $a, b \in G_1$. In this section we will relax these requirements and study functions that satisfy the second condition but not necessarily the first. In Definition 3.7.1 (which follows shortly), such functions are called *group homomorphisms*.

A group homomorphism carries algebraic information from one group to another. Since a homomorphism is not required to be one-to-one and onto, the second group can have a different structure from the first. We will often study a group G_1 by passing (via a homomorphism) from G_1 to a simpler group G_2. In this section we will determine properties that are carried from the first group to the second, and we will begin to consider the more difficult issues revolving around pulling information about the second group back to the first. Our first example of a homomorphism that is not an isomorphism comes from linear algebra.

Example 3.7.1 (Determinant of an invertible matrix).

Let G_1 be the group $\mathrm{GL}_n(\mathbf{R})$ of all invertible $n \times n$ matrices over the real numbers, and let G_2 be the multiplicative group $\mathbf{R}^\times$ of all nonzero real numbers. The formula $\det(AB) = \det(A)\det(B)$ for elements of $\mathrm{GL}_n(\mathbf{R})$ shows that the function $\phi : G_1 \to G_2$ defined by $\phi(A) = \det(A)$ is a group homomorphism.

To illustrate the information carried by the determinant function, consider the special case $n = 3$. For the associated linear transformation $L : \mathbf{R}^3 \to \mathbf{R}^3$ defined by $L(\mathbf{v}) = A\mathbf{v}$, for all $\mathbf{v} \in \mathbf{R}^3$, the following facts are usually mentioned in a linear algebra course. If S is a region in $\mathbf{R}^3$ with volume V, then the image $L(S)$ of S under the action of L is a region with volume $|\det(A)| \cdot V$. Furthermore, the sign of $\det(A)$ tells whether or not L preserves the orientation of the axes. The homomorphism property $\det(AB) = \det(A)\det(B)$ says that volume and orientation behave as expected when working with the composition of two linear transformations. $\square$

If the operations in both G_1 and G_2 are denoted additively, then the formula defining a homomorphism becomes $\phi(a + b) = \phi(a) + \phi(b)$. A familiar operation in calculus can be put into this context: the derivative of a sum is the sum of the derivatives. The next example also involves additive notation.

Example 3.7.2 (Parity of an integer).

The mapping $\phi : \mathbf{Z} \to \mathbf{Z}_2$ given by $\phi(n) = [n]_2$ enjoys the property that $\phi(n + m) = [n + m]_2 = [n]_2 + [m]_2 = \phi(n) + \phi(m)$ for all $n, m \in \mathbf{Z}$, but it is not one-to-one, and so ϕ is a homomorphism but not an isomorphism.

The information carried by ϕ involves the parity of an integer, since $n \in \mathbf{Z}$ is even if and only if $\phi(n) = [0]_2$, and odd if and only if $\phi(n) = [1]_2$. The homomorphism property describes how parity behaves under addition. For example, suppose that $n, m \in \mathbf{Z}$ are odd. Then $\phi(n) = [1]_2$ and $\phi(m) = [1]_2$, so $n + m$ is even since $\phi(n + m) = \phi(n) + \phi(m) = [1]_2 + [1]_2 = [0]_2$. $\square$

One of the most important examples of a group homomorphism is provided by the rule for exponents: $a^{n+m} = a^n a^m$. The next example considers the appropriate function that relates integers to powers of a group element a. This is an occasion when we will be comparing a group whose operation is denoted additively with one whose operation is denoted multiplicatively.

Example 3.7.3 (Exponential functions for groups).

Let G be a group, and let a be any element of G. Define $\phi : \mathbf{Z} \to G$ by $\phi(n) = a^n$, for all $n \in \mathbf{Z}$. The rules we have developed for exponents show that for all $n, m \in \mathbf{Z}$,

$$\phi(n + m) = a^{n+m} = a^n a^m = \phi(n) \cdot \phi(m) .$$

Thus ϕ is consistent with the operations in the respective groups.

If G is abelian, with its operation denoted additively, then we define $\phi : \mathbf{Z} \to G$ by $\phi(n) = na$. The fact that ϕ is a homomorphism is expressed by the formula $(n + m)a = na + ma$, which holds for all $n, m \in \mathbf{Z}$. After we have studied homomorphisms in more detail we will return to these examples to show how the ideas we have developed can be applied to help understand the order of an element and the cyclic subgroup generated by an element.   □

Example 3.7.4 (Linear transformations).

Many concepts from linear algebra provide examples of the general group theoretic concepts we are studying. Let V and W be vector spaces. Recall that a function $L : V \to W$ is called a **linear transformation** if $L(\mathbf{v}_1 + \mathbf{v}_2) = L(\mathbf{v}_1) + L(\mathbf{v}_2)$ and $L(a\mathbf{v}_1) = aL(\mathbf{v}_1)$ for all vectors $\mathbf{v}_1, \mathbf{v}_2 \in V$ and all scalars a. Since any vector space is an abelian group under vector addition, any linear transformation between vector spaces is actually a homomorphism of the underlying abelian groups. (The condition involving scalar multiplication is not involved in the group theory setting.) Linear differential equations fit into this context and thus provide examples of homomorphisms of abelian groups.

Note that the determinant function does not define a homomorphism on the abelian group of all $n \times n$ matrices under addition since it is possible to find matrices A, B for which $\det(A + B) \neq \det(A) + \det(B)$. On the other hand, the trace function $\mathrm{Tr} : M_n(\mathbf{R}) \to \mathbf{R}$ *is* a homomorphism, where $\mathrm{Tr}(A)$ is the sum of the diagonal elements of the matrix A.   □

Example 3.7.5 (Linear functions on $\mathbf{Z}_n$).

For a fixed integer m, define $\phi : \mathbf{Z}_n \to \mathbf{Z}_n$ by $\phi([x]) = [mx]$, for all $[x] \in \mathbf{Z}_n$. This is a function since if $a \equiv b \pmod{n}$, then $ma \equiv mb \pmod{n}$. It is a homomorphism since $\phi([a] + [b]) = \phi([a + b]) = [m(a + b)] = [ma] + [mb] = \phi([a]) + \phi([b])$.   □

We now formally record our definition of a homomorphism between groups. It follows immediately from the definition that an isomorphism is simply a homomorphism that is one-to-one and onto.

3.7.1 Definition. *Let G_1 and G_2 be groups, and let $\phi : G_1 \to G_2$ be a function. Then ϕ is said to be a **group homomorphism** if*

$$\phi(ab) = \phi(a)\phi(b)$$

for all $a, b \in G_1$.

3.7.2 Proposition. *If $\phi : G_1 \to G_2$ is a group homomorphism, then*

(a) $\phi(e) = e$;

(b) $(\phi(a))^{-1} = \phi(a^{-1})$ *for all $a \in G_1$;*

(c) *for any integer n and any $a \in G_1$, $\phi(a^n) = (\phi(a))^n$;*

(d) *if $a \in G_1$ and a has order n, then the order of $\phi(a)$ in G_2 is a divisor of n.*

Proof. (a) Since $\phi(e)\phi(e) = \phi(e^2) = \phi(e)$, cancellation gives $\phi(e) = e$.

(b) This follows since $\phi(a)\phi(a^{-1}) = \phi(aa^{-1}) = e$.

(c) This can be proved using a simple induction argument.

(d) Let $a \in G_1$ with $n = o(a)$. Since $\phi : G_1 \to G_2$ is a homomorphism, we must have $(\phi(a))^n = \phi(a^n) = \phi(e) = e$. Thus $o(\phi(a)) \mid n$. $\square$

Example 3.7.6 (Homomorphisms defined on cyclic groups).

In this example we will completely describe all homomorphisms defined on any cyclic group. Let C be a cyclic group, denoted multiplicatively, with generator a. If $\phi : C \to G$ is any group homomorphism, and $\phi(a) = g$, then the formula $\phi(a^m) = g^m$ must hold. Since every element of C is of the form a^m for some $m \in \mathbf{Z}$, this means that ϕ is completely determined by its value on a. Note that if a has finite order, then by the previous proposition the order of g must be a divisor of the order of a.

We next consider how to define homomorphisms on C. If C is infinite, then for an element g of any group G, the formula $\phi(a^m) = g^m$ defines a homomorphism since

$$\phi(a^m a^k) = \phi(a^{m+k}) = g^{m+k} = g^m g^k = \phi(a^m)\phi(a^k) .$$

If $|C| = n$ and g is any element of G whose order is a divisor of n, then the formula $\phi(a^m) = g^m$ defines a homomorphism. We must first show that the formula defines a function, since the formula depends on the choice of an exponent in writing an element $x \in C$ as a power of the generator a. If $x = a^m$ and $x = a^k$, then $m \equiv k \pmod{n}$, since a has order n. Thus we can write $m = k + qn$ for some integer q, and then

$$g^m = g^{k+qn} = g^k(g^n)^q = g^k$$

since $g^n = e$. This depends on the crucial assumption that the order of g is a divisor of n. Now the previous argument can be used to show that ϕ is a homomorphism. $\square$

Example 3.7.7 (Homomorphisms from $\mathbf{Z}_n$ to $\mathbf{Z}_k$).

As a particular case of the previous example, we now give explicit formulas for all homomorphisms $\phi : \mathbf{Z}_n \to \mathbf{Z}_k$. Any such homomorphism is completely determined by $\phi([1]_n)$, and this must be an element $[m]_k$ of $\mathbf{Z}_k$ whose order is a divisor of n. In an abelian group, with the operation denoted additively, Proposition 3.2.8 (b) states that if a has finite order, then $o(a) \mid n$ if and only if $n \cdot a = 0$. Applying this result to $[m]_k$ in $\mathbf{Z}_k$, we have $o([m]_k) \mid n$ if and only if $n \cdot [m]_k = [0]_k$, which happens if and only if $k \mid nm$. (Compare Exercise 11 of Section 2.1.)

Thus the formula $\phi([x]_n) = [mx]_k$, for all $[x]_n \in \mathbf{Z}_n$, defines a homomorphism if and only if $k \mid mn$. Furthermore, every homomorphism from $\mathbf{Z}_n$ into $\mathbf{Z}_k$ must be of this form. Note that $\phi(\mathbf{Z}_n)$ is the cyclic subgroup generated by $[m]_k$, and so ϕ will map $\mathbf{Z}_n$ onto $\mathbf{Z}_k$ if and only if $[m]_k$ is a generator of $\mathbf{Z}_k$. $\square$

Example 3.7.8 (Parity of a permutation).

We return once more to the theorem which states that the parity of a permutation in S_n is well-defined. (See Theorem 2.3.11 and Theorem 3.6.6.) We will give a proof of the theorem that uses the notion of a group homomorphism.

Let $\Delta_n = \prod_{1 \leq i < j \leq n}(x_i - x_j)$, and let G be the subgroup $\{\pm 1\}$ of $\mathbf{Q}^{\times}$. We define $\phi : S_n \to G$ by $\phi(\sigma) = 1$ if $\sigma(\Delta_n) = \Delta_n$ and $\phi(\sigma) = -1$ if $\sigma(\Delta_n) = -\Delta_n$, for each $\sigma \in S_n$. Then ϕ is a group homomorphism, since if $\rho, \sigma \in S_n$, then we showed in the proof of Theorem 3.6.6 that $\sigma\rho(\Delta_n) = \sigma(\rho(\Delta_n))$.

Let $\tau = (1, 2)$. Then $\tau(\Delta_n) = -\Delta_n$, and hence $\phi(\tau) = -1$. Let $\rho = (r, s)$ be any transposition. By Exercise 13 (b) of Section 2.3, there exists a permutation σ such that $\rho = \sigma\tau\sigma^{-1}$. Then $\phi(\rho) = \phi(\sigma)\phi(\tau)\phi(\sigma^{-1}) = -\phi(\sigma)\phi(\sigma^{-1}) = -1$. Hence $\phi(\rho) = -1$ for all transpositions $\rho \in S_n$.

Suppose that $\sigma = \tau_1\tau_2 \cdots \tau_m = \rho_1\rho_2 \cdots \rho_k$, where each ρ_i is a transposition. Then $\phi(\sigma) = \phi(\tau_1\tau_2 \cdots \tau_m) = (-1)^m$ and $\phi(\sigma) = \phi(\rho_1\rho_2 \cdots \rho_k) = (-1)^k$. Hence $(-1)^m = (-1)^k$, and so $m \equiv k \pmod 2$.

We conclude that if a permutation is written as a product of transpositions in two ways, then the number of transpositions is either even in both cases or odd in both cases. $\square$

Let $\phi : G_1 \to G_2$ be a group homomorphism. Recall the statement of Proposition 3.4.4: ϕ is one-to-one if and only if $\phi(x) = e$ implies $x = e$. The set $\{x \in G_1 \mid \phi(x) = e\}$ plays an important role in studying group homomorphisms. It should already be familiar to the student in the setting of linear algebra, where the kernel (or null space) of a linear transformation is studied.

3.7.3 Definition. *Let $\phi : G_1 \to G_2$ be a group homomorphism. Then*

$$\{x \in G_1 \mid \phi(x) = e\}$$

*is called the **kernel** of ϕ, and is denoted by $\ker(\phi)$.*

3.7.4 Proposition. *Let $\phi : G_1 \to G_2$ be a group homomorphism, with $K = \ker(\phi)$.*
(a) *K is a subgroup of G_1 such that $gkg^{-1} \in K$ for all $k \in K$ and $g \in G_1$.*
(b) *The homomorphism ϕ is one-to-one if and only if $K = \{e\}$.*

Proof. (a) The kernel of ϕ is nonempty since it contains e. If $a, b \in K$, then

$$\phi(ab^{-1}) = \phi(a)(\phi(b))^{-1} = e \cdot e = e$$

and this implies that K is a subgroup of G_1. Furthermore, if $k \in K$ and $g \in G_1$, then

$$\phi(gkg^{-1}) = \phi(g)\phi(k)(\phi(g))^{-1} = \phi(g)e(\phi(g))^{-1} = e.$$

Thus $gkg^{-1} \in K$.
(b) If ϕ is one-to-one, then the only element that can map to the identity of G_2 is the identity of G_1. On the other hand, suppose that $K = \{e\}$ and $\phi(a) = \phi(b)$ for some $a, b \in G_1$. Multiplying both sides of this equation by $(\phi(b))^{-1}$ gives us $\phi(ab^{-1}) = \phi(a)(\phi(b))^{-1} = e$, which shows that $ab^{-1} \in \ker(\phi)$. But then by assumption, $ab^{-1} = e$, and thus $a = b$. This shows that ϕ is one-to-one. □

The previous proposition shows that the kernel of a group homomorphism is a special type of subgroup, which we define below. We will study these subgroups in much greater detail in Section 3.8, so in this section our interest is only in their behavior with respect to group homomorphisms.

3.7.5 Definition. *A subgroup H of the group G is called a **normal** subgroup if $ghg^{-1} \in H$ for all $h \in H$ and $g \in G$.*

It is obvious from the definition that if $H = G$ or $H = \{e\}$, then H is normal. It is also clear that any subgroup of an abelian group is normal. As one of the exercises at the end of the section asks you to show, the only proper nontrivial normal subgroup of S_3 is its three element subgroup. The next proposition investigates how subgroups are related via a homomorphism.

3.7.6 Proposition. *Let $\phi : G_1 \to G_2$ be a group homomorphism.*
(a) *If H_1 is a subgroup of G_1, then $\phi(H_1)$ is a subgroup of G_2. If ϕ is onto and H_1 is normal in G_1, then $\phi(H_1)$ is normal in G_2.*
(b) *If H_2 is a subgroup of G_2, then $\phi^{-1}(H_2) = \{x \in G_1 \mid \phi(x) \in H_2\}$ is a subgroup of G_1. If H_2 is a normal in G_2, then $\phi^{-1}(H_2)$ is normal in G_1.*

Proof. (a) Let H_1 be a subgroup of G_1, and let $y, z \in \phi(H_1)$. Then there exist $a, b \in H_1$ with $\phi(a) = y$ and $\phi(b) = z$, and

$$yz^{-1} = \phi(a)(\phi(b))^{-1} = \phi(a)\phi(b^{-1}) = \phi(ab^{-1}) \in \phi(H_1).$$

Since $e \in \phi(H_1)$, this shows that $\phi(H_1)$ is a subgroup of G_2.

If ϕ is onto and H_1 is normal in G_1, let $y \in G_2$ and $z \in \phi(H_1)$. There exist $a \in G_1$ and $b \in H_1$ with $\phi(a) = y$ and $\phi(b) = z$. Then

$$yzy^{-1} = \phi(a)\phi(b)\phi(a^{-1}) = \phi(aba^{-1}) \in \phi(H_1)$$

since H_1 is normal and therefore $aba^{-1} \in H_1$.

(b) Let H_2 be a subgroup of G_2, and let

$$H_1 = \phi^{-1}(H_2) = \{x \in G_1 \mid \phi(x) \in H_2\} \ .$$

Then $e \in H_1$ since $\phi(e) = e \in H_2$. If $a, b \in H_1$, then $ab^{-1} \in H_1$ since $\phi(ab^{-1}) = \phi(a)(\phi(b))^{-1} \in H_2$ because H_2 is a subgroup. Thus H_1 is a subgroup.

If H_2 is a normal subgroup, then to show that H_1 is also normal, let g be any element of G_1, and let $h \in H_1$. Then $ghg^{-1} \in H_1$ because

$$\phi(ghg^{-1}) = \phi(g)\phi(h)(\phi(g))^{-1} \in H_2$$

since H_2 is normal. Thus H_1 is a normal subgroup. $\square$

If $\phi : G_1 \to G_2$ is a group homomorphism, then there is a natural equivalence relation on G_1 associated with the function ϕ given by defining $a \sim_\phi b$ if $\phi(a) = \phi(b)$, where $a, b \in G_1$. For arbitrary functions, this equivalence relation is studied in detail in Section 2.2, where the notation G_1/ϕ is used for the set of equivalence classes of the relation. We will use $[a]_\phi$ to denote the equivalence class of $a \in G_1$.

It may be useful to review the proof that we have in fact defined an equivalence relation. We have $a \sim_\phi a$ since $\phi(a) = \phi(a)$. If $a \sim_\phi b$, then $\phi(a) = \phi(b)$ implies $\phi(b) = \phi(a)$, which shows that $b \sim_\phi a$. Finally, if $a \sim_\phi b$ and $b \sim_\phi c$, then $\phi(a) = \phi(b)$ and $\phi(b) = \phi(c)$ implies $\phi(a) = \phi(c)$, so $a \sim_\phi c$.

The formula $[a]_n[b]_n = [ab]_n$ for multiplication of congruence classes in $\mathbf{Z}_n$ suggests that we might try a similar formula in G_1/ϕ, since we have a multiplication defined in G_1. Part of the next proposition shows that this natural multiplication is in fact well-defined. We will use the notation $[x]_\phi$ for the elements of the factor set G_1/ϕ, and then the natural projection $\pi : G_1 \to G_1/\phi$ (recall Definition 2.2.6) is defined by $\pi(x) = [x]_\phi$, for all $x \in G_1$.

3.7.7 Proposition. *Let $\phi : G_1 \to G_2$ be a group homomorphism. Then multiplication of equivalence classes in the factor set G_1/ϕ is well-defined, and G_1/ϕ is a group under this multiplication. The natural projection $\pi : G_1 \to G_1/\phi$ defined by $\pi(x) = [x]_\phi$ is a group homomorphism.*

Proof. To show that multiplication is well-defined, we must show that if $a \sim_\phi b$ and $c \sim_\phi d$, then $ac \sim_\phi bd$. If $\phi(a) = \phi(b)$ and $\phi(c) = \phi(d)$, then

$$\phi(ac) = \phi(a)\phi(c) = \phi(b)\phi(d) = \phi(bd) .$$

The associative law for G_1/ϕ follows from that of G_1, since

$$[a]_\phi([b]_\phi[c]_\phi) = [a]_\phi[bc]_\phi = [a(bc)]_\phi = [(ab)c]_\phi = [ab]_\phi[c]_\phi = ([a]_\phi[b]_\phi)[c]_\phi$$

for all $a, b, c \in G_1$. The class $[e]_\phi$ is an identity element since

$$[e]_\phi[a]_\phi = [ea]_\phi = [a]_\phi \quad \text{and} \quad [a]_\phi[e]_\phi = [ae]_\phi = [a]_\phi$$

for all $a \in G_1$. Finally, for any equivalence class $[a]_\phi$, there exists an inverse $[a^{-1}]_\phi$ since $[a^{-1}]_\phi[a]_\phi = [a^{-1}a]_\phi = [e]_\phi$ and $[a]_\phi[a^{-1}]_\phi = [e]_\phi$. Thus $([a]_\phi)^{-1} = [a^{-1}]_\phi$.

Since multiplication is well-defined, we have

$$\pi(ab) = [ab]_\phi = [a]_\phi[b]_\phi = \pi(a)\pi(b)$$

for all $a, b \in G_1$, and so π is a homomorphism. $\square$

The following theorem is extremely important, and we will return to it in the next section, where we give another proof. Theorem 2.2.7 shows that if $f : S \to T$ is any function, then there is a one-to-one correspondence between the elements of $f(S)$ and the elements of the factor set S/f determined by the equivalence relation $\sim_f$. Thus the basic one-to-one correspondence that we will give in Theorem 3.7.8 comes from set theory, and not from the algebraic structure of either group or the fact that ϕ is a homomorphism. (We choose to reprove this fact in Theorem 3.7.8 to make its proof self-contained.)

Now suppose that $\phi : G_1 \to G_2$ is a homomorphism. As in Figure 2.2.1 in Section 2.2, we can write ϕ as a composite function $\iota\overline{\phi}\pi$, where π is the function of Proposition 3.7.7 and ι is the inclusion mapping. This is illustrated in Figure 3.7.1, in which both π and ι are homomorphisms. We now show that the function $\overline{\phi}$ is an isomorphism of groups.

Figure 3.7.1:

$$G_1 \xrightarrow{\ \pi\ } G_1/\phi \xrightarrow{\ \overline{\phi}\ } \phi(G_1) \xrightarrow{\ \iota\ } G_2$$

3.7.8 Theorem. *Let $\phi : G_1 \to G_2$ be a group homomorphism. Then there exists a group isomorphism*

$$\overline{\phi} : G_1/\phi \to \phi(G_1) \, ,$$

where $\overline{\phi}$ is defined by $\overline{\phi}([a]_\phi) = \phi(a)$, for all $[a]_\phi \in G_1/\phi$.

Proof. For each equivalence class $[a]_\phi \in G_1/\phi$ we define $\overline{\phi}([a]_\phi) = \phi(a)$. This is a well-defined function since if $[a]_\phi = [b]_\phi$, then by definition $\phi(a) = \phi(b)$, and so $\overline{\phi}([a]_\phi) = \overline{\phi}([b]_\phi)$. If $\overline{\phi}([a]_\phi) = \overline{\phi}([b]_\phi)$, then $\phi(a) = \phi(b)$, and so $[a]_\phi = [b]_\phi$, which shows that $\overline{\phi}$ is one-to-one. The image of G_1/ϕ is

$$\{\overline{\phi}([a]_\phi) \mid a \in G_1\} = \{\phi(a) \mid a \in G_1\} = \phi(G_1)$$

so $\overline{\phi}$ maps G_1/ϕ onto $\phi(G_1)$. Finally, $\overline{\phi}$ is a homomorphism since

$$\overline{\phi}([a]_\phi)\overline{\phi}([b]_\phi) = \phi(a)\phi(b) = \phi(ab) = \overline{\phi}([ab]_\phi) = \overline{\phi}([a]_\phi[b]_\phi)$$

for all equivalence classes $[a]_\phi, [b]_\phi \in G_1/\phi$. □

Example 3.7.9 (Characterization of cyclic groups).

The power of Theorem 3.7.8 can be illustrated by giving another proof that every cyclic group is isomorphic to either $\mathbf{Z}$ or $\mathbf{Z}_n$, for some n. Given $G = \langle a \rangle$, define $\phi : \mathbf{Z} \to G$ by $\phi(m) = a^m$, as in Example 3.7.3. If a has infinite order, then ϕ is one-to-one, so in this case, $\mathbf{Z}$ is isomorphic to $\phi(\mathbf{Z}) = G$. If a has order n, then $a^m = a^k$ if and only if $m \equiv k \pmod{n}$. Thus $\phi(m) = \phi(k)$ if and only if $m \equiv k \pmod{n}$, which shows that $\mathbf{Z}/\phi$ is the additive group of congruence classes modulo n. Thus if a has order n, then $G \cong \mathbf{Z}_n$. □

Example 3.7.10 (Cayley's theorem).

Theorem 3.7.8 is also useful in giving a more concise proof of Cayley's theorem. Given any group G, define $\phi : G \to \mathrm{Sym}(G)$ by $\phi(a) = \lambda_a$, for any $a \in G$, where λ_a is the function defined by $\lambda_a(x) = ax$ for all $x \in G$. (It is necessary to check that λ_a is one-to-one and onto.) Then ϕ is a homomorphism since $\lambda_a \lambda_b = \lambda_{ab}$ for all $a, b \in G$. Because λ_a is the identity permutation only if $a = e$, we have $\ker(\phi) = \{e\}$. Since ϕ is one-to-one, the equivalence classes of the factor set G/ϕ are just the subsets of G consisting of single elements, and thus G itself is isomorphic to $\phi(G)$, which is a permutation group. □

Example 3.7.11.

Define $\phi : \mathbf{R} \to \mathbf{C}^{\times}$ by $\phi(\theta) = \cos\theta + i\sin\theta$, for all $\theta \in \mathbf{R}$. The trigonometric formulas for the cosine and sine of the sum of two angles can be
used (see Section A.5 of the appendix, on complex numbers) to show that
$\phi(\alpha + \beta) = \phi(\alpha) \cdot \phi(\beta)$, and so ϕ is a group homomorphism. Geometrically,
the function ϕ can be visualized as wrapping the real line around the unit
circle. In this process, numbers that differ by a multiple of 2π are identified.
It follows that $[\theta]_{\phi} = \{x \in \mathbf{R} \mid x = \theta + 2k\pi, \ k \in \mathbf{Z}\}$. $\quad\square$

We conclude the section with a proposition that gives a more complete description of the equivalence classes of the equivalence relation defined by a homomorphism. It shows that the equivalence relation defined by ϕ is the same as the one
defined in Lemma 3.2.9, with $H = \ker(\phi)$. Thus in Section 3.8 we will switch from
the notation G/ϕ to the more standard notation $G/\ker(\phi)$.

3.7.9 Proposition. *Let* $\phi : G_1 \to G_2$ *be a group homomorphism, and* $a, b \in G_1$.
The following conditions are equivalent:

    (1) $\phi(a) = \phi(b)$;

    (2) $ab^{-1} \in \ker(\phi)$;

    (3) $a = kb$ *for some* $k \in \ker(\phi)$;

    (4) $b^{-1}a \in \ker(\phi)$;

    (5) $a = bk$ *for some* $k \in \ker(\phi)$.

Proof. (1) implies (2): If $\phi(a) = \phi(b)$, then multiplying both sides of the equation
by $(\phi(b))^{-1}$ we have

$$e = \phi(a)(\phi(b))^{-1} = \phi(a)\phi(b^{-1}) = \phi(ab^{-1}) \, .$$

(2) implies (3): If $ab^{-1} = k \in \ker(\phi)$, then $a = kb$.

(3) implies (1): If $a = kb$ for some $k \in \ker(\phi)$, then

$$\phi(a) = \phi(kb) = \phi(k)\phi(b) = e\phi(b) = \phi(b) \, .$$

Similarly it can be shown that (1) implies (4) implies (5) implies (1). $\quad\square$

If $\phi : G_1 \to G_2$ is a homomorphism of abelian groups, with operations denoted additively, then Proposition 3.7.9 has the following form: For $a, b \in G_1$, the
following conditions are equivalent: (1) $\phi(a) = \phi(b)$; (2) $a - b \in \ker(\phi)$; and (3)
$a = b + k$ for some $k \in \ker(\phi)$. The following examples use additive notation.

Example 3.7.12 (Solution of nonhomogeneous linear systems).

Let A be an $m \times n$ matrix, and consider the nonhomogeneous equation $Ax = b$. We may view the matrix A as defining a homomorphism $\phi : \mathbf{R}^n \to \mathbf{R}^m$, where $\phi(\mathbf{x}) = A\mathbf{x}$. By Proposition 3.7.9, if we find a particular solution $\mathbf{x}_0$ with $A\mathbf{x}_0 = \mathbf{b}$, then the set of solutions to the nonhomogeneous equation $A\mathbf{x} = \mathbf{b}$ consists of all vectors of the form $\mathbf{x}_0 + \mathbf{k}$, where $\mathbf{k}$ is any solution of the homogeneous equation $A\mathbf{x} = \mathbf{0}$, since $\ker(\phi)$ is the solution space of the homogeneous equation $A\mathbf{x} = \mathbf{0}$. $\square$

Example 3.7.13 (Linear differential equations).

An analysis similar to the previous example shows that the standard theorem stating that the general solution to a nonhomogeneous linear differential equation is obtained by finding any particular solution and adding to it all solutions of the associated homogeneous equation is really just a consequence of the fact that linear differential operators preserve sums of functions. $\square$

EXERCISES: SECTION 3.7

1.†(a) Write down the formulas for all homomorphisms from $\mathbf{Z}_6$ into $\mathbf{Z}_9$.

   (b) Do the same for all homomorphisms from $\mathbf{Z}_{24}$ into $\mathbf{Z}_{18}$.

2. Write down the formulas for all homomorphisms from $\mathbf{Z}$ onto $\mathbf{Z}_{12}$.

3. Show that the following functions are homomorphisms. (Recall that $\mathbf{R}^+$ is the group of positive real numbers under multiplication.)

   (a) $\phi : \mathbf{R}^\times \to \mathbf{R}^+$ defined by $\phi(x) = |x|$

   (b) $\phi : \mathbf{R}^\times \to \mathbf{R}^\times$ defined by $\phi(x) = \dfrac{x}{|x|}$

   (c) $\phi : \mathbf{R}^\times \to \{\pm 1\}$ defined by $\phi(x) = \dfrac{x}{|x|}$

4. Let G be an abelian group, and let n be any positive integer. Show that the function $\phi : G \to G$ defined by $\phi(x) = x^n$ is a homomorphism.

5.†Let G be the multiplicative group $\mathbf{Z}_{15}^\times = \{1, 2, 4, 7, 8, 11, 13, 14\}$, and let $n = 2$. Compute the values of the function defined in Exercise 4, and find its kernel and the image of G.

6. Define $\phi : \mathbf{C}^\times \to \mathbf{R}^\times$ by $\phi(a + bi) = a^2 + b^2$, for all $a + bi \in \mathbf{C}^\times$. Show that ϕ is a homomorphism.

7. Which of the following functions are homomorphisms?

†(a) $\phi : \mathbf{R}^\times \to GL_2(\mathbf{R})$ defined by $\phi(a) = \begin{bmatrix} a & 0 \\ 0 & 1 \end{bmatrix}$

(b) $\phi : \mathbf{R} \to GL_2(\mathbf{R})$ defined by $\phi(a) = \begin{bmatrix} 1 & a \\ 0 & 1 \end{bmatrix}$

†(c) $\phi : M_2(\mathbf{R}) \to \mathbf{R}$ defined by $\phi\left(\begin{bmatrix} a & b \\ c & d \end{bmatrix}\right) = a$

(d) $\phi : GL_2(\mathbf{R}) \to \mathbf{R}^\times$ defined by $\phi\left(\begin{bmatrix} a & b \\ c & d \end{bmatrix}\right) = ab$

†(e) $\phi : GL_2(\mathbf{R}) \to \mathbf{R}$ defined by $\phi\left(\begin{bmatrix} a & b \\ c & d \end{bmatrix}\right) = a + d$

(f) $\phi : GL_2(\mathbf{R}) \to \mathbf{R}^\times$ defined by $\phi\left(\begin{bmatrix} a & b \\ c & d \end{bmatrix}\right) = ad - bc$

8. Let $\phi : G_1 \to G_2$ and $\theta : G_2 \to G_3$ be group homomorphisms. Prove that $\theta\phi : G_1 \to G_3$ is a homomorphism. Prove that $\ker(\phi) \subseteq \ker(\theta\phi)$.

9. Let ϕ be a group homomorphism of G_1 onto G_2. Prove that if G_1 is abelian then so is G_2; prove that if G_1 is cyclic then so is G_2. In each case, give a counterexample to the converse of the statement.

10. Let G be the group of affine functions from $\mathbf{R}$ into $\mathbf{R}$, as defined in Exercise 10 of Section 3.1. Define $\phi : G \to \mathbf{R}^\times$ as follows: for any function $f_{m,b} \in G$, let $\phi(f_{m,b}) = m$. Prove that ϕ is a group homomorphism, and find its kernel and image.

11. Let G be a group, and let H be a normal subgroup of G. Show that for each $g \in G$ and $h \in H$ there exist h_1 and h_2 in H with $gh = h_1 g$ and $hg = gh_2$.

12. Show that the only proper nontrivial normal subgroup of S_3 is the subgroup with three elements.

13. Let H be a subgroup of the group G. Prove that H is a normal subgroup of G if and only if for each $a \in G$ and each $h \in H$ there exists $h' \in H$ with $ah = h'a$.

14. Recall that the center of a group G is $\{x \in G \mid xg = gx \text{ for all } g \in G\}$. Prove that the center of any group is a normal subgroup.

15. Prove that the intersection of two normal subgroups is a normal subgroup.

16. Let G be a finite group of even order, with n elements, and let H be a subgroup with $n/2$ elements. Prove that H must be normal.

    Hint: Define $\phi : G \to \mathbf{R}^\times$ by $\phi(x) = 1$ if $x \in H$ and $\phi(x) = -1$ if $x \notin H$ and show that ϕ is a homomorphism with kernel H. To show that ϕ preserves products, show that if $g \notin H$ then $\{x \mid gx \in H\} = G - H$.

17.†Determine which subgroups of D_4 are normal.

18. Let the dihedral group D_n be given by elements a of order n and b of order 2, where $ba = a^{-1}b$. Show that any subgroup of $\langle a \rangle$ is normal in D_n.

19. Give an example to show that the assumption that ϕ is onto is needed in part (a) of Proposition 3.7.6.

20. Let G_1 and G_2 be groups.

    (a) Define $\pi_1 : G_1 \times G_2 \to G_1$ by $\pi_1((a_1, a_2)) = a_1$, for all $(a_1, a_2) \in G_1 \times G_2$ and define $\pi_2 : G_1 \times G_2 \to G_2$ by $\pi_2((a_1, a_2)) = a_2$, for all $(a_1, a_2) \in G_1 \times G_2$. Show that π_1 and π_2 are group homomorphisms.

    (b) Let G be any group, and let $\phi : G \to G_1 \times G_2$ be a function. Show that ϕ is a group homomorphism if and only if $\pi_1\phi$ and $\pi_2\phi$ are both group homomorphisms.

3.8   Cosets, Normal Subgroups, and Factor Groups

In this section we introduce the important notion of a coset of a subgroup, motivated by the results in the previous section on the equivalence relation defined by a homomorphism. We also show that the set of all cosets of any normal subgroup can be given a natural group structure, just as we did in the previous section for the equivalence classes determined by a homomorphism.

The congruence classes of the integers modulo 2 are the sets of even and odd integers. We have denoted the set of even integers by $2\mathbf{Z}$, and so we could denote the set of odd integers by $1 + 2\mathbf{Z}$, to show that each odd integer can be expressed as 1 plus an even integer. Of course, we could use any odd integer in place of 1, say $3 + 2\mathbf{Z}$ or $5 + 2\mathbf{Z}$, still giving the set of all odd integers. More generally, for any integer k we can express its congruence class in $\mathbf{Z}_n$ in the form $[k]_n = k + n\mathbf{Z}$.

Let $\phi : G_1 \to G_2$ be a group homomorphism, and let a be a fixed element in G_1. By Proposition 3.7.9, for $b \in G_1$ we have $\phi(b) = \phi(a)$ if and only if b can be written in the form $b = ak$ for some element $k \in \ker(\phi)$. We can express this by writing $b \in aK$, where $K = \ker(\phi)$, and aK consists of all elements that can be written in the form ak for some $k \in K$. (This is the product of the sets $\{a\}$ and K, as defined in Section 3.3.) If $\sim_\phi$ is the equivalence relation defined by letting $a \sim_\phi b$ if $\phi(a) = \phi(b)$, then the equivalence class $[a]_\phi$ defined by $\sim_\phi$ is precisely the set aK. Proposition 3.7.9 also shows that $aK = Ka$.

In the proof of Lagrange's theorem (Theorem 3.2.10), for a subgroup H of the finite group G, we introduced the equivalence relation $a \sim b$ on G by letting $a \sim b$ if $ab^{-1} \in H$. In the course of the proof we showed that the elements of the congruence class $[a]$ are precisely the elements of the form ha for $h \in H$. Thus we can write $[a] = Ha$.

We will now develop this idea further, for sets of the form aH. Let G be a group and let H be a subgroup of G. To review the work in Section 3.2 without retracing precisely the same steps, we will consider the relation $a \sim b$ if $a^{-1}b \in H$, for $a, b \in G$. This defines an equivalence relation: $a \sim a$ since $a^{-1}a \in H$; if $a \sim b$, then $a^{-1}b$ is in H, so its inverse $b^{-1}a$ is in H, and thus $b \sim a$; if $a \sim b$ and $b \sim c$, then $a^{-1}b$ and $b^{-1}c$ are in H, so their product is in H, and thus $a \sim c$. Our first proposition identifies the equivalence classes of this equivalence relation as sets of the form aH, for $a \in G$.

3.8.1 Proposition. *Let H be a subgroup of the group G, and let $a, b \in G$. Then the following conditions are equivalent:*
   (1) $bH = aH$;
   (2) $bH \subseteq aH$;
   (3) $b \in aH$;
   (4) $a^{-1}b \in H$.

Proof. It is obvious that (1) implies (2). Furthermore, (2) implies (3) since $b = be \in bH$.

(3) implies (4): If $b = ah$ for some $h \in H$, then $a^{-1}b = h \in H$.

(4) implies (1): Suppose that $a^{-1}b = h$ for some $h \in H$, so that $b = ah$ and $a = bh^{-1}$. To show that $bH \subseteq aH$, let $x \in bH$. Then $x = bh'$ for some $h' \in H$, and substituting for b gives $x = ahh'$, which shows that $x \in aH$. On the other hand, to show that $aH \subseteq bH$, let $x \in aH$. Then $x = ah''$ for some $h'' \in H$, and so $x = bh^{-1}h'' \in bH$. Thus we have shown that $bH = aH$. $\square$

3.8.2 Corollary. *Let H be a subgroup of the group G. The relation $\sim$ defined on G by setting $a \sim b$ if $aH = bH$, for all $a, b \in G$, is an equivalence relation on G.*

Proof. By Proposition 3.8.1 we have $aH = bH$ if and only if $a^{-1}b \in H$. Thus the relation defined in the corollary is the same as the one defined in the remarks preceding Proposition 3.8.1, and it was shown there that the relation is an equivalence relation. $\square$

In Proposition 3.8.1, the symmetry in the condition $bH = aH$ shows that the roles of a and b can be reversed. Thus $a^{-1}b \in H$ if and only if $b^{-1}a \in H$. Since we are working with an equivalence relation, the equivalence classes must partition G, and so we know that if $aH \cap bH \neq \emptyset$, then $aH = bH$.

3.8.3 Definition. *Let H be a subgroup of the group G, and let $a \in G$. The set*

$$aH = \{x \in G \mid x = ah \text{ for some } h \in H\}$$

*is called the **left coset** of H in G determined by a. Similarly, the **right coset** of H in G determined by a is the set*

$$Ha = \{x \in G \mid x = ha \text{ for some } h \in H\}.$$

*The number of left cosets of H in G is called the **index** of H in G, and is denoted by $[G : H]$.*

A result similar to Proposition 3.8.1 holds for right cosets. Let H be a subgroup of the group G, and let $a, b \in G$. Then the following conditions are equivalent: (1) $Ha = Hb$; (2) $Ha \subseteq Hb$; (3) $a \in Hb$; (4) $ab^{-1} \in H$; (5) $ba^{-1} \in H$; (6) $b \in Ha$; (7) $Hb \subseteq Ha$. The index of H in G could also be defined as the number of right cosets of H in G, since there is a one-to-one correspondence between left cosets and right cosets. (See Exercise 5.)

We next show that all left cosets of H have the same number of elements. Given any left coset aH of H, define the function $f : H \rightarrow aH$ by $f(h) = ah$, for all $h \in H$. Then f is one-to-one since if $f(h_1) = f(h_2)$, then $ah_1 = ah_2$ and so $h_1 = h_2$ by the cancellation law. It is obvious that f is onto, and so the one-to-one correspondence $f : H \rightarrow aH$ shows that aH has the same number of elements as H. If G is a finite group, this observation is at the heart of the proof of Lagrange's theorem, and shows that (in the notation introduced above) we always have $[G : H] = |G|/|H|$.

In the next example we list the left cosets of a given subgroup H of a finite group. For any $a \in H$ we have $aH = H$, so we begin by choosing any element a not in H. Then aH is found by listing all products of the form ah for $h \in H$. Now any element in aH determines the same coset, so for the next coset we choose any element not in H or aH (if possible). Continuing in this way provides a method for listing all cosets.

Example 3.8.1.

Let G be the multiplicative group $\mathbf{Z}_{11}^{\times}$ of nonzero elements of $\mathbf{Z}_{11}$. Let H be the subgroup $\{[1], [10]\}$ generated by $[10]$. The first coset we can identify is H itself. Choosing an element not in H, say $[2]$, we form the products $[2][1]$ and $[2][10] = [9]$, to obtain the coset $[2]H = \{[2], [9]\}$. Next we choose any element not in the first two cosets, say $[3]$, which gives us $[3]H = \{[3], [8]\}$, since $[3][1] = [3]$ and $[3][10] = [8]$. Continuing in this fashion, we obtain $[4]H = \{[4], [7]\}$ and $[5]H = \{[5], [6]\}$. Thus the cosets of H are the following sets:

$$H = \{[1], [10]\} , \qquad [2]H = \{[2], [9]\} , \qquad [3]H = \{[3], [8]\} ,$$

$$[4]H = \{[4], [7]\} , \qquad [5]H = \{[5], [6]\} .$$

As another example, let $K = \{[1], [3], [9], [5], [4]\}$ be the subgroup generated by $[3]$. Since the left cosets all have the same number of elements and we already have a coset with half of the total number of elements, there must be only one other coset, containing the rest of the elements. Thus the left cosets of K are the following sets:

$$K = \{[1], [3], [9], [5], [4]\} , \qquad [2]K = \{[2], [6], [7], [10], [8]\} . \quad \square$$

Example 3.8.2.

Let $G = S_3$, the group of all permutations on a set with three elements, and let $G = \{e, a, a^2, b, ab, a^2b\}$, where $a^3 = e$, $b^2 = e$, and $ba = a^2b$.

First, let H be the subgroup $\{e, b\}$. We must be careful since this is the first example in a non-abelian group, so we must distinguish between left and right cosets.

The left cosets of H are easily computed to be

$$H = \{e, b\}, \qquad aH = \{a, ab\}, \qquad a^2H = \{a^2, a^2b\}.$$

Since $ba = a^2b$ and $ba^2 = ab$, the right cosets of H are

$$H = \{e, b\}, \qquad Ha = \{a, a^2b\}, \qquad Ha^2 = \{a^2, ab\}.$$

Next, let N be the subgroup $\{e, a, a^2\}$. The left cosets of N are

$$N = \{e, a, a^2\}, \qquad bN = \{b, ba, ba^2\} = \{b, a^2b, ab\},$$

since $ba = a^2b$ and $ba^2 = ab$. The right cosets of N are

$$N = \{e, a, a^2\}, \qquad Nb = \{b, ab, a^2b\}.$$

Note that for the subgroup N the left and right cosets are the same. $\quad\square$

If G is an abelian group with the operation denoted by $+$, then the cosets of a subgroup H have the form

$$a + H = \{x \in G \mid x = a + h \text{ for some } h \in H\}.$$

Proposition 3.8.1 shows that in this case, $a + H = b + H$ if and only if $a - b \in H$.

Example 3.8.3.

Let $G = \mathbf{Z}_{12}$, and let H be the subgroup $4\mathbf{Z}_{12} = \{[0], [4], [8]\}$. To find all cosets of H, we begin by noting that $[4] + H = [8] + H = H$, so we start with an element not in H, say $[1]$. Then

$$[1] + H = \{[1] + [0], [1] + [4], [1] + [8]\} = \{[1], [5], [9]\}.$$

Next we choose an element not in H or $[1] + H$, say $[2]$. Then $[2] + H = \{[2], [6], [10]\}$, and we have only one coset remaining, $[3] + H = \{[3], [7], [11]\}$. Since we have used all elements of the group, we have found the following cosets:

$$H = \{[0], [4], [8]\}, \qquad [1] + H = \{[1], [5], [9]\},$$

$$[2] + H = \{[2], [6], [10]\}, \qquad [3] + H = \{[3], [7], [11]\}.$$

Since G is abelian, the right cosets are precisely the same as the left cosets. $\quad\square$

Example 3.8.4.

Let m be a real number, and consider the line $L = \{(x, y) \mid y = mx\}$ in the plane $\mathbf{R}^2$. Vector addition gives $\mathbf{R}^2$ a group structure, and L is then a subgroup of $\mathbf{R}^2$. The coset determined by a vector $(0, b)$ consists of all vectors of the form $(0, b) + (x, mx)$, or just $(x, mx + b)$. This is a line parallel to L, since it is the set $\{(x, y) \mid y = mx + b\}$. In fact, any line parallel to $y = mx$ is a coset of L. It is true in general that the cosets of a line through the origin are the lines parallel to the given line. □

For a homomorphism $\phi : G_1 \rightarrow G_2$, we saw in Section 3.7 that the set G_1/ϕ of equivalence classes defined by ϕ forms a group. Since $\phi(a) = \phi(b)$ if and only if $\phi(ab^{-1}) = e$, we have $a \sim_\phi b$ if and only if $ab^{-1} \in \ker(\phi)$. The equivalence classes of this equivalence relation are the right cosets of $\ker(\phi)$. Proposition 3.7.9 shows that the equivalence classes are also the left cosets of $\ker(\phi)$, since $a \sim b$ if and only if $a^{-1}b \in \ker(\phi)$.

Recall that a subgroup H of G is said to be normal if $aha^{-1} \in H$ for all $a \in G$ and $h \in H$. (See Definition 3.7.5.) Since $(a^{-1})^{-1} = a$, we can interchange the roles of a and a^{-1}, showing that H is normal if and only if $a^{-1}ha \in H$ for all $a \in G$ and $h \in H$. We will show that a subgroup is normal if and only if its left and right cosets coincide, that is, if and only if $aH = Ha$ for all $a \in G$. This provides an additional way to determine whether a subgroup is normal. When a subgroup is normal, we will see that multiplication of cosets is compatible with the structure of G, and that the set of cosets forms a group.

Since $\mathbf{Z}$ is abelian, the left and right cosets of any subgroup coincide. The equivalence relation defined by the subgroup $n\mathbf{Z}$, when stated in additive notation, just says that $a - b \in n\mathbf{Z}$, and so we have the well-known equivalence relation of congruence modulo n. We are already familiar with the resulting group $\mathbf{Z}_n$ defined by the corresponding equivalence classes.

Let N be a normal subgroup of the group G. We now want to introduce a multiplication for the cosets of N in G. We will do this by analogy with the multiplication in $\mathbf{Z}_n$ and in G/ϕ, where $\phi : G_1 \rightarrow G_2$ is a homomorphism. From a given coset of N we may choose any element a to use as a representative, so that we can write aN for the coset. The formula $aNbN = abN$ can then be interpreted in the following way: to multiply two cosets, choose representatives of the cosets, multiply them, and define the product to be the new coset in which the product of the representatives lies. There is a potential problem with the definition. We must make sure that it is independent of the choice of the representatives by which we have named the cosets. The following proposition takes care of the difficulty.

3.8.4 Proposition. *Let N be a normal subgroup of G, and let $a, b, c, d \in G$. If $aN = cN$ and $bN = dN$, then $abN = cdN$.*

Proof. If $aN = cN$ and $bN = dN$, then by Proposition 3.8.1 we have $a^{-1}c \in N$ and $b^{-1}d \in N$. Since N is normal, $d^{-1}(a^{-1}c)d \in N$. But then since $b^{-1}d \in N$, we have $(ab)^{-1}cd = (b^{-1}d)(d^{-1}a^{-1}cd) \in N$, and so $abN = cdN$. $\square$

3.8.5 Theorem. *If N is a normal subgroup of G, then the set of left cosets of N forms a group under the coset multiplication given by*

$$aNbN = abN$$

for all $a, b \in G$.

Proof. Proposition 3.8.4 shows that the given multiplication of left cosets is in fact well-defined. The subgroup N itself serves as an identity element, since $N = eN$ and therefore $eNaN = aN$ and $aNeN = aN$ for all $a \in G$. Furthermore, the inverse of aN is $a^{-1}N$ because $aNa^{-1}N = eN$ and $a^{-1}NaN = eN$. Finally, to show the associative law, let $a, b, c \in G$. Then

$$(aNbN)cN = abNcN = (ab)cN = a(bc)N = aNbcN = aN(bNcN) .$$

This completes the proof. $\square$

3.8.6 Definition. *If N is a normal subgroup of G, then the group of left cosets of N in G is called the **factor group** of G determined by N. It will be denoted by G/N.*

Example 3.8.5 (Order of an element in G/N).

> Let N be a normal subgroup of G. It is interesting to compute the order of an element of G/N. If $a \in G$, then the order of aN is the smallest positive integer n such that $(aN)^n = a^nN = N$. That is, the order of aN is the smallest positive integer n such that $a^n \in N$. $\square$

Let N be a normal subgroup of G. The mapping $\pi : G \to G/N$ defined by $\pi(x) = xN$, for all $x \in G$, is called the *natural projection* of G onto G/N. (This is consistent with Definition 2.2.6, since the elements of G/N are equivalence classes.) We showed in Proposition 3.7.4 that the kernel of any group homomorphism is a normal subgroup. The first part of the next proposition shows that the converse is true: any normal subgroup is the kernel of some group homomorphism.

3.8.7 Proposition. *Let N be a normal subgroup of G.*

(a) *The natural projection* $\pi : G \to G/N$ *defined by* $\pi(x) = xN$, *for all* $x \in G$, *is a group homomorphism, and* $\ker(\pi) = N$.

(b) *There is a one-to-one correspondence between subgroups of* G/N *and subgroups H of G with* $H \supseteq N$. *Specifically, if K is a subgroup of* G/N, *then* $\pi^{-1}(K)$ *is the corresponding subgroup of G; if H is a subgroup of G with* $H \supseteq N$, *then* $\pi(H)$ *is the corresponding subgroup of* G/N.

Under this correspondence, normal subgroups correspond to normal subgroups.

Proof. (a) Let $a, b \in G$. Then $\pi(ab) = abN = aNbN = \pi(a)\pi(b)$, showing that π is a homomorphism. Furthermore, we have $a \in \ker(\pi)$ if and only if $aN = N$, and this is equivalent to the statement that $a \in N$.

(b) Since π is a homomorphism, we can apply Proposition 3.7.6. If K is a subgroup of G/N, then $\pi^{-1}(K)$ is a subgroup of G that contains N, and if K is normal, then so is $\pi^{-1}(K)$. Since π is onto, it is clear that assigning to each subgroup of G/N its inverse image in G is a one-to-one mapping. To show that this mapping is onto, let H be a subgroup of G with $H \supseteq N$. We claim that $H = \pi^{-1}(\pi(H))$. By definition,

$$\pi^{-1}(\pi(H)) = \{x \in G \mid \pi(x) \in \pi(H)\} ,$$

and so it is clear that $H \subseteq \pi^{-1}(\pi(H))$. To show the reverse inclusion, let $a \in \pi^{-1}(\pi(H))$. Then $aN = hN$ for some $h \in H$, so we have $h^{-1}a \in N$. But since $N \subseteq H$, this implies $h^{-1}a \in H$, and so $a = h(h^{-1}a) \in H$. Finally, it follows directly from Proposition 3.7.6 that if H is normal, then so is its image H/N under the natural projection. $\square$

Example 3.8.6.

Let $G = \mathbf{Z}_{12}$ and let $N = \{[0], [3], [6], [9]\}$, the cyclic subgroup generated by the congruence class of 3. Then there are three elements of G/N, found by adding [1] and [2] to each element of N:

$$\{[0], [3], [6], [9]\} , \qquad \{[1], [4], [7], [10]\} , \qquad \{[2], [5], [8], [11]\} .$$

Since we only have three elements, the factor group G/N must be isomorphic to $\mathbf{Z}_3$. This can also be seen by considering the order of the equivalence class of [1]. Its order is the smallest positive multiple that gives the identity element of G/N, and so that is the smallest positive multiple of [1] that belongs to N. Thus [1] has order 3. $\square$

Let N be a normal subgroup of G. When we introduced the multiplication of cosets we did so by choosing representatives and multiplying them. It is very useful to have another way of viewing the product of cosets as products of sets.

For subsets S_1, S_2 of G, in Definition 3.3.1 we used $S_1 S_2$ to denote all elements $g \in G$ that have the form $g = s_1 s_2$ for some $s_1 \in S_1$ and $s_2 \in S_2$. Using our notation for multiplicative cosets, the elements of G/N have the form aN for elements $a \in G$. We can then consider the product $(aN)(bN)$, for $a, b \in G$, as the product of two subsets of G. To show that this product is equivalent to the one used in Theorem 3.8.5, we need to show that $(aN)(bN) = abN$.

3.8.8 Proposition. *Let H be a subgroup of the group G. The following conditions are equivalent:*

(1) *H is a normal subgroup of G;*

(2) *$aH = Ha$ for all $a \in G$;*

(3) *for all $a, b \in G$, abH is the set theoretic product $(aH)(bH)$;*

(4) *for all $a, b \in G$, $ab^{-1} \in H$ if and only if $a^{-1}b \in H$.*

Proof. (1) implies (2): Let $a \in G$. To show that $aH \subseteq Ha$, let $h \in H$. Then $aha^{-1} \in H$ since H is normal, and therefore $aha^{-1} = h'$ for some $h' \in H$. Thus $ah = h'a \in Ha$. The proof of the reverse inclusion is similar, so $aH = Ha$.

(2) implies (3): Assume that $Hb = bH$ for all $b \in G$. It is always true that $abH \subseteq (aH)(bH)$, since any element of the form abh, with $h \in H$, can be rewritten as $abh = (ae)(bh)$, and the latter form shows it to be in $(aH)(bH)$. To show the reverse inclusion, let $(ah_1)(bh_2) \in (aH)(bH)$, for $h_1, h_2 \in H$. Then $h_1 b \in Hb$ and $Hb = bH$, so $h_1 b = bh_3$ for some $h_3 \in H$. Thus $(ah_1)(bh_2) = ab(h_3 h_2) \in abH$.

(3) implies (1): If $(aH)(bH) = abH$ for all $a, b \in G$, then in particular we have $(aH)(a^{-1}H) = H$. Thus for any element $h \in H$, we have $aha^{-1} = aha^{-1}e \in H$, showing that H is normal.

(2) if and only if (4): Condition (2) holds if and only if the left and right cosets of H coincide. The left cosets of H are the equivalence classes of the equivalence relation determined by setting $a \sim b$ if $a^{-1}b \in H$, for all $a, b \in G$. The right cosets are the equivalence classes determined by the symmetric condition $ab^{-1} \in H$. Since the two equivalence relations coincide if and only if their equivalence classes are identical, condition (4) holds if and only if condition (2) holds. □

Example 3.8.7 (Normal subgroups of S_3).

As in Example 3.8.2, we view S_3 as the set $\{e, a, a^2, b, ab, a^2 b\}$, where $a^3 = e$, $b^2 = e$, and $ba = a^2 b$. The trivial subgroup $\{e\}$ and the improper subgroup G are normal. The proper nontrivial subgroups of S_3 are $H = \{e, b\}$, $K = \{e, ab\}$, $L = \{e, a^2 b\}$, and $N = \{e, a, a^2\}$ (see Figure 3.6.5).

In Example 3.8.2 we computed the left and right cosets of H. Since they do not coincide, it follows from Proposition 3.8.8 (b) that H is not a normal subgroup of G. Similarly, $aK = \{a, a^2 b\}$, while $Ka = \{a, aba\} = \{a, b\}$, and therefore

K is not normal in G. Furthermore, $aL = \{a, b\}$, but $La = \{a, ab\}$, and so L is not normal.

We showed in Example 3.8.2 that the left cosets of N are N and $bN = \{b, a^2b, ab\}$, while the right cosets are N and $Nb = \{b, ab, a^2b\}$. Since the left and right cosets coincide, we conclude that N is the only proper nontrivial normal subgroup of S_3. $\square$

Example 3.8.8 (Subgroups of index 2 are normal).

Let H be a subgroup of G, and assume that H has only two left cosets. Then these must be H and $G - H$, and since these must also be the right cosets, it follows from Proposition 3.8.8 that H is normal. This gives a much simpler proof than the one outlined in Exercise 16 of Section 3.7. $\square$

Example 3.8.9 (Normal subgroups of D_4).

Let G be the dihedral group D_4, given by elements a of order 4 and b of order 2, where $ba = a^{-1}b$. We refer to Figure 3.6.6 for the diagram of subgroups of D_4. The subgroups G and $\{e\}$ are normal, as always. The subgroups $\{e, a^2, b, a^2b\}$, $\{e, a, a^2, a^3\}$, and $\{e, a^2, ab, a^3b\}$ each have index 2, so they are normal by Example 3.8.8.

Let N be the subgroup $\{e, a^2\}$. The computation $b \cdot a^2 = a^3ba = a^3a^3b = a^2b$ shows that a^2 commutes with b, and since a^2 commutes with powers of a, it must commute with every element of G. This shows that N is contained in the center of G, and implies, in particular, that N is normal since its left and right cosets coincide.

To show that none of the subgroups $H = \{e, b\}$, $K = \{e, a^2b\}$, $L = \{e, ab\}$, and $M = \{e, a^3b\}$ is normal, we do not even need to find all of their left and right cosets. Short computations using the fact that $ba = a^3b$ will show that $aH \neq Ha, aK \neq Ka, aL \neq La$, and $aM \neq Ma$. $\square$

In Section 3.7, for any group homomorphism $\phi : G_1 \to G_2$ we constructed a group, which we denoted by G_1/ϕ. It consisted of the congruence classes of the equivalence relation $\sim_\phi$ defined by $a \sim_\phi b$ if $\phi(a) = \phi(b)$. By Proposition 3.7.9 this is precisely the relation of congruence modulo $\ker(\phi)$. Thus if we let $K = \ker(\phi)$, we have $[a]_\phi = aK$ for all $a \in G_1$. We now have in hand the notation in which the fundamental homomorphism theorem is usually stated. We include an outline of the proof using coset notation.

3.8.9 Theorem (Fundamental Homomorphism Theorem). *Let G_1, G_2 be groups. If $\phi : G_1 \to G_2$ is a homomorphism with $K = \ker(\phi)$, then $G_1/K \cong \phi(G_1)$.*

Proof. Define $\overline{\phi} : G_1/K \rightarrow \phi(G_1)$ by setting $\overline{\phi}(aK) = \phi(a)$, for each element aK of G_1/K. The function $\overline{\phi}$ is well-defined since if $aK = bK$ for $a, b \in G_1$, then $a = bk$ for some $k \in \ker(\phi)$, and so

$$\phi(a) = \phi(bk) = \phi(b)\phi(k) = \phi(b) \ .$$

It is a homomorphism since

$$\overline{\phi}(aKbK) = \overline{\phi}(abK) = \phi(ab) = \phi(a)\phi(b) = \overline{\phi}(aK)\overline{\phi}(bK)$$

for all $a, b \in G_1$. If $\phi(a) = \phi(b)$ for $a, b \in G_1$, then we have

$$\phi(ab^{-1}) = \phi(a)(\phi(b))^{-1} = e$$

and so $ab^{-1} \in K$, showing that $aK = bK$. Thus $\overline{\phi}$ is one-to-one, and it clearly maps G_1/K onto $\phi(G_1)$. $\square$

Let $\phi : G_1 \rightarrow G_2$ be a group homomorphism. Then we know that ϕ is one-to-one if and only if its kernel is the trivial subgroup of G_1, and in this case G_1 is isomorphic to $\phi(G_1)$. At the other extreme, if $\ker(\phi) = G_1$, then ϕ is the trivial mapping, which sends every element to the identity. Thus if G_1 has no proper nontrivial normal subgroups and $\phi : G_1 \rightarrow G_2$ is a group homomorphism, then ϕ is either one-to-one or trivial.

3.8.10 Definition. *The nontrivial group G is called a **simple** group if it has no proper nontrivial normal subgroups.*

For any prime p, the cyclic group $\mathbf{Z}_p$ is simple, since it has no proper nontrivial subgroups of any kind (every nonzero element is a generator). We will study simple groups in Chapter 7 because they play an important role in determining the structure of groups. We will show that the alternating group A_n is simple, if $n \geq 5$, and this fact will play an important role in the proof (in Chapter 8) that the general equation of degree 5 is not solvable by radicals.

Example 3.8.10 ($\mathbf{Z}_n/m\mathbf{Z}_n \cong \mathbf{Z}_m$ if $m \mid n$).

In Example 3.8.6 we computed a particular factor group of $\mathbf{Z}_{12}$. It is not difficult to do this computation for any factor group of $\mathbf{Z}_n$. Any homomorphic image of a cyclic group is again cyclic, and so all factor groups of $\mathbf{Z}_n$ must be cyclic, and hence isomorphic to $\mathbf{Z}_m$ for some m. However, we prefer to give a direct proof of the isomorphism. Let n be any positive integer. Then the subgroups of $\mathbf{Z}_n$ correspond to divisors of n, and so to describe all factor groups of $\mathbf{Z}_n$ we only need to describe $\mathbf{Z}_n/m\mathbf{Z}_n$ for all positive divisors m of n.

Define $\phi : \mathbf{Z}_n \to \mathbf{Z}_m$ by $\phi([x]_n) = [x]_m$. The mapping is well-defined since $m \mid n$. It is a homomorphism by Example 3.7.6, which describes homomorphisms of cyclic groups, and it is clearly onto. Finally,

$$\ker(\phi) = \{[x]_n \mid [x]_m = [0]_m\} = \{[x]_n \mid x \text{ is a multiple of } m\}$$

and so $\ker(\phi)$ is equal to $m\mathbf{Z}_n$. It follows from the fundamental homomorphism theorem that $\mathbf{Z}_n/m\mathbf{Z}_n \cong \mathbf{Z}_m$. □

Example 3.8.11 $(D_4/Z(D_4) \cong \mathbf{Z}_2 \times \mathbf{Z}_2)$**.**

In Example 3.8.9, we showed that $N = \{e, a^2\}$ is a normal subgroup of D_4, since it lies in the center $Z(D_4)$. The factor group G/N consists of the four cosets N, $aN = \{a, a^3\}$, $bN = \{b, a^2b\}$, and $abN = \{ab, a^3b\}$. We have $aNaN = N$, $bNbN = N$ and $abNabN = N$, which shows that every nontrivial element of G/N has order 2. □

Example 3.8.12.

Let $G = \mathbf{Z}_4 \times \mathbf{Z}_4$ and let $N = \{(0, 0), (2, 0), (0, 2), (2, 2)\}$. (We have omitted the brackets denoting congruence classes because that makes the notation too cumbersome.) There are four cosets of this subgroup, which we can denote as follows:

$$N, \quad (1, 0) + N, \quad (0, 1) + N, \quad (1, 1) + N.$$

The representatives of the cosets have been carefully chosen to show that each nontrivial element of the factor group has order 2, making the factor group G/N isomorphic to $\mathbf{Z}_2 \times \mathbf{Z}_2$.

Let K be the subgroup $\{(0, 0), (1, 0), (2, 0), (3, 0)\}$. It can be checked that the cosets of K have the following form:

$$K, \quad (0, 1) + K, \quad (0, 2) + K, \quad (0, 3) + K.$$

Having chosen these representatives, it is clear that the factor group must be cyclic, generated by $(0, 1) + K$, and so G/K is isomorphic to $\mathbf{Z}_4$. □

The previous example provides the motivation for a general comment concerning factor groups of direct products. One way to define a subgroup of a direct product $G_1 \times G_2$ is to use normal subgroups $N_1 \subseteq G_1$ and $N_2 \subseteq G_2$ to construct the following subgroup:

$$N_1 \times N_2 = \{(x_1, x_2) \mid x_1 \in N_1, x_2 \in N_2\} \subseteq G_1 \times G_2 .$$

In the previous example we computed the factor groups for the subgroups $2\mathbf{Z}_4 \times 2\mathbf{Z}_4$ and $\mathbf{Z}_4 \times \langle 0 \rangle$, respectively. The factor group is easy to describe in this case. (See Example 3.8.13 for a subgroup of a direct product that cannot be described in this manner.)

Let N_1 and N_2 be normal subgroups of the groups G_1 and G_2, respectively. Then $N_1 \times N_2$ is a normal subgroup of the direct product $G_1 \times G_2$ and

$$(G_1 \times G_2)/(N_1 \times N_2) \cong (G_1/N_1) \times (G_2/N_2) .$$

To prove this statement, define $\phi : G_1 \times G_2 \to G_1/N_1 \times G_2/N_2$ by $\phi(x_1, x_2) = (x_1 N_1, x_2 N_2)$ for all $x_1 \in G_1$, $x_2 \in G_2$. It is easy to verify the following: ϕ is a homomorphism, ϕ is onto, and $\ker(\phi) = N_1 \times N_2$. The desired conclusions follow from the fundamental homomorphism theorem.

Example 3.8.13.

Let $G = \mathbf{Z}_4 \times \mathbf{Z}_4$ and let N be the "diagonal" subgroup generated by $(1, 1)$. Then $N = \{(0, 0), (1, 1), (2, 2), (3, 3)\}$ and the factor group G/N will have four elements, so it must be isomorphic to either $\mathbf{Z}_4$ or $\mathbf{Z}_2 \times \mathbf{Z}_2$. The smallest positive multiple of $(1, 0)$ that belongs to N is $4 \cdot (1, 0) = (0, 0)$, showing that the coset $(1, 0) + N$ has order 4. Thus G/N is cyclic, and hence isomorphic to $\mathbf{Z}_4$. $\square$

Example 3.8.14 ($GL_n(\mathbf{R}) / SL_n(\mathbf{R}) \cong \mathbf{R}^\times$).

To give an example involving a group of matrices, consider $GL_n(\mathbf{R})$ and the subgroup $SL_n(\mathbf{R})$ of all $n \times n$ matrices with determinant 1. Define $\phi : GL_n(\mathbf{R}) \to \mathbf{R}^\times$ by letting $\phi(A) = \det(A)$, for any matrix $A \in GL_n(\mathbf{R})$. The formula $\det(AB) = \det(A) \det(B)$ shows that ϕ is a homomorphism. It is easy to see that ϕ is onto, and $\ker(\phi)$ is precisely $SL_n(\mathbf{R})$. Applying the fundamental homomorphism theorem, we see that $SL_n(\mathbf{R})$ is a normal subgroup, and we obtain $GL_n(\mathbf{R})/ SL_n(\mathbf{R}) \cong \mathbf{R}^\times$. $\square$

EXERCISES: SECTION 3.8

1. List all cosets in $\mathbf{Z}_{24}$ of each of the given subgroups.

   †(a) $\langle [3] \rangle$

   (b) $\langle [16] \rangle$

2. Let $G = \mathbf{Z}_3 \times \mathbf{Z}_6$, let $H = \langle (1, 2) \rangle$ and let $K = \langle (1, 3) \rangle$. List all cosets of H and K.

3.†For the subgroup $\{e, ab\}$ of S_3, list all left and right cosets.

4. For each of the subgroups $\{e, a^2\}$ and $\{e, b\}$ of D_4, list all left and right cosets.

5. Let G be a group with subgroup H. Prove that there is a one-to-one correspondence between the left and right cosets of H. (Your proof must include the case in which G is infinite.)

6. Prove that if N is a normal subgroup of G, and H is any subgroup of G, then $H \cap N$ is a normal subgroup of H.

7. Let H be a subgroup of G, and let $a \in G$. Show that aHa^{-1} is a subgroup of G that is isomorphic to H.

8. Let H be a subgroup of G. Show that H is normal in G if and only if $aHa^{-1} = H$ for all $a \in G$.

9. Let G be a finite group, and let n be a divisor of $|G|$. Show that if H is the only subgroup of G of order n, then H must be normal in G.

10. Let N be a normal subgroup of index m in G. Show that $a^m \in N$ for all $a \in G$.

11. Let N be a normal subgroup of G. Show that the order of any coset aN in G/N is a divisor of $o(a)$, when $o(a)$ is finite.

12. Let H and K be normal subgroups of G such that $H \cap K = \langle e \rangle$. Show that $hk = kh$ for all $h \in H$ and $k \in K$.

13. Let N be a normal subgroup of G. Prove that G/N is abelian if and only if N contains all elements of the form $aba^{-1}b^{-1}$ for $a, b \in G$.

14. Let N be a subgroup of the center of G. Show that if G/N is a cyclic group, then G must be abelian.

15.†Find all factor groups of the dihedral group D_4.

16. Let H and K be normal subgroups of G such that $H \cap K = \langle e \rangle$ and $HK = G$. Prove that $G \cong H \times K$.

17.†Compute the factor group $(\mathbf{Z}_6 \times \mathbf{Z}_4)/\langle (2, 2) \rangle$.

18. Compute the factor group $(\mathbf{Z}_6 \times \mathbf{Z}_4)/\langle (3, 2) \rangle$.

19. Show that $(\mathbf{Z} \times \mathbf{Z})/\langle (0, 1) \rangle$ is an infinite cyclic group.

20. Show that $(\mathbf{Z} \times \mathbf{Z})/\langle (1, 1) \rangle$ is an infinite cyclic group.

21. Show that $(\mathbf{Z} \times \mathbf{Z})/\langle (2, 2) \rangle$ is not a cyclic group.

22. Show that $\mathbf{R}^\times/\langle -1 \rangle$ is isomorphic to the group of positive real numbers under multiplication.

23. Let G be the set of all matrices in $\mathrm{GL}_2(\mathbf{Z}_5)$ of the form $\begin{bmatrix} m & b \\ 0 & 1 \end{bmatrix}$.

    (a) Show that G is a subgroup of $\mathrm{GL}_2(\mathbf{Z}_5)$.

    (b) Show that the subset N of all matrices in G of the form $\begin{bmatrix} 1 & c \\ 0 & 1 \end{bmatrix}$, with $c \in \mathbf{Z}_5$, is a normal subgroup of G.

    (c) Show that the factor group G/N is cyclic of order 4.

24. Let S be an infinite set. Let H be the set of all elements $\sigma \in \mathrm{Sym}(S)$ such that $\sigma(x) = x$ for all but finitely many $x \in S$. Prove that the subgroup H is normal in $\mathrm{Sym}(S)$. (See Exercise 18 of Section 3.6.)

25. Give an example of a finite group G with two normal subgroups H and K such that $G/H \cong G/K$ but $H \not\cong K$.

26. Let H and K be subgroups of the group G, and let $a, b \in G$. Show that either $aH \cap bK = \emptyset$ or else $aH \cap bK$ is a left coset of $H \cap K$.

27. Let H and N be subgroups of a group G, and assume that N is a normal subgroup of G. Prove the following statements.

    (a) N is a normal subgroup of HN.

    (b) Each element of HN/N has the form hN, for some $h \in H$.

    (c) $\phi : H \to HN/N$ defined by $\phi(h) = hN$, for all $h \in H$, is an onto homomorphism.

    (d) $HN/N \cong H/K$, where $K = H \cap N$.

28. Let H and N be normal subgroups of a group G, with $N \subseteq H$. Define $\phi : G/N \to G/H$ by $\phi(xN) = xH$, for all cosets $xN \in G/N$.

    (a) Show that ϕ is a well-defined onto homomorphism.

    (b) Show that $(G/N)/(H/N) \cong G/H$.

Notes

The publication in 1870 of *Traité des substitutions et des équations algébriques* by Camille Jordan (1838–1922) represented a fundamental change in the character of group theory. Prior to its publication, various individuals had studied groups of permutations and groups of geometric transformations, but after 1870 the abstract notion of a group was developed in several steps. The modern definition of a group, using an axiomatic approach, was given in the commutative case in 1870 by Leopold Kronecker (1823–1891), and in the general noncommutative case in 1893 by Heinrich Weber (1842–1913). Arthur Cayley (1821–1895) should be given credit for two papers that he published in 1854 on the theory of groups. He introduced the concept of a group table, which is sometimes still referred to as a "Cayley table."

We now return to our discussion (begun in the notes at the end of Chapter 2) of the use groups in the theory of equations. Certain "substitutions," or permutations, of the roots of a polynomial define its Galois group. For a polynomial with rational coefficients, every rational number must be left fixed by the substitution. Whether or not a particular permutation of the roots belongs to the Galois group depends on the coefficients of the equation, and is not generally easy to determine. We will see in Chapter 8 that the Galois group of a polynomial $p(x)$ determines whether or not $p(x) = 0$ is solvable by radicals.

To give an elementary illustration, we return to the discussion of the polynomial $x^3 - 3x + 1$ begun in the notes at the end of Chapter 2. From the equation

(1) $$x^3 - 3x + 1 = (x - r_1)(x - r_2)(x - r_3)$$

we obtain general identities for the roots r_1, r_2, r_3:

(2) $r_1 + r_2 + r_3 = 0$,   (3) $r_1 r_2 + r_2 r_3 + r_3 r_1 = -3$,   (4) $r_1 r_2 r_3 = -1$.

It is possible to find another identity that holds for the roots of this particular polynomial. To do so we will evaluate the discriminant

$$\Delta = (r_1 - r_2)^2 (r_1 - r_3)^2 (r_2 - r_3)^2$$

of $x^3 - 3x + 1$ (see the discussion of the discriminant in Appendix A.6). By differentiating equation (1) we obtain

$$3(x^2 - 1) = (x - r_2)(x - r_3) + (x - r_1)(x - r_3) + (x - r_1)(x - r_2).$$

Successive substitutions of $x = r_1, r_2, r_3$ yield $(r_1 - r_2)(r_1 - r_3) = 3(r_1^2 - 1)$, $(r_1 - r_2)(r_2 - r_3) = -3(r_2^2 - 1)$, and $(r_1 - r_3)(r_2 - r_3) = 3(r_3^2 - 1)$, so

$$\begin{aligned}
\Delta &= 3(r_1^2 - 1)(-3)(r_2^2 - 1)(3)(r_3^2 - 1) \\
&= (-27)\left[r_1^2 r_2^2 r_3^2 - (r_1^2 r_2^2 + r_1^2 r_3^2 + r_2^2 r_3^2) + (r_1^2 + r_2^2 + r_3^2) - 1 \right].
\end{aligned}$$

Squaring equation (4) gives us $r_1^2 r_2^2 r_3^2 = 1$. Squaring equation (3), we see that $(r_1 r_2 + r_2 r_3 + r_3 r_1)^2 = r_1^2 r_2^2 + r_1^2 r_3^2 + r_2^2 r_3^2 + 2 r_1 r_2 r_3 (r_1 + r_2 + r_3)$ and so it follows from (2) that $r_1^2 r_2^2 + r_1^2 r_3^2 + r_2^2 r_3^2 = 9$. Finally, by squaring equation (2) and using equation (3) we get $r_1^2 + r_2^2 + r_3^2 = 6$. Thus $\Delta = 81$, and so $(r_1 - r_2)(r_1 - r_3)(r_2 - r_3) = \sqrt{\Delta} \in \mathbf{Q}$. Any permutation of the roots that belongs to the Galois group must therefore leave $\sqrt{\Delta}$ fixed. The permutations (1), (1, 2, 3), and (1, 3, 2) applied to the subscripts leave $(r_1 - r_2)(r_1 - r_3)(r_2 - r_3)$ invariant, but the transpositions (1, 2), (1, 3), and (2, 3) each change the sign. As this argument suggests, we will be able to show in Example 8.6.5 that the Galois group of $x^3 - 3x + 1$ is the set of even permutations in S_3: $\{(1), (1, 2, 3), (1, 3, 2)\}$.

Chapter 4

POLYNOMIALS

In this chapter we will study polynomial equations from a concrete point of view. We will find it convenient, though, to introduce one abstract concept. We will study polynomials in which the coefficients can come from any set in which addition, subtraction, multiplication, and division are possible. In this setting we will prove numerous results for polynomials analogous to those for the set of integers: a prime factorization theorem, a division algorithm, and an analog of the Euclidean algorithm. Congruences for polynomials will be used to show that roots of polynomials can always be found, by enlarging (if necessary) the set over which the polynomials are considered.

4.1 Fields; Roots of Polynomials

We introduced the concept of a field in Section 3.3 in order to work with the group $GL_n(F)$ of invertible $n \times n$ matrices with entries in a field F. We now want to work with polynomials with coefficients in an arbitrary field, and so we need to develop this concept further. One very important tool is a division algorithm for polynomials (analogous to Theorem 1.1.3 for integers). To make sure that the division algorithm works, we need to be able to assume that we can add, subtract, multiply, and divide coefficients. Of course, all four operations are possible in the familiar fields $\mathbf{Q}$, $\mathbf{R}$, $\mathbf{C}$, and $\mathbf{Z}_p$, where p is prime. However, we also need to be able to consider other sets as coefficients.

For example, a typical procedure in our later work is the following. To find all solutions in $\mathbf{C}$ of a polynomial equation with coefficients in $\mathbf{Q}$, we proceed in stages. After we have found one solution, we let F be the smallest set of complex numbers that is a field and contains both $\mathbf{Q}$ and the solution. We then treat the original polynomial as if its coefficients come from F. Since the polynomial has a root in F, we will show that it can be factored into polynomials of lower degree (with coefficients in F). Then each of the factors can be investigated in a similar

179

manner, and the process is continued until a field E is obtained that is the smallest field inside $\mathbf{C}$ containing $\mathbf{Q}$ and all solutions of the original equation.

Example 4.1.1 ($\mathbf{Q}(\sqrt{2})$).

Given the equation $x^4 - 2x^3 - 3x^2 + 4x + 2 = 0$, we can factor to obtain $(x^2 - 2)(x^2 - 2x - 1) = 0$, and so $\sqrt{2}$ is a solution. We want to find the smallest set of complex numbers that contains all coefficients and roots of the equation and is closed under addition, subtraction, multiplication, and division. We will start by finding the smallest set of complex numbers that contains $\mathbf{Q}$ and $\sqrt{2}$ and forms a field (see Definition 3.3.5). We denote this set by $\mathbf{Q}(\sqrt{2})$. It is obvious that $\mathbf{Q}(\sqrt{2})$ must contain $\{a + b\sqrt{2} \mid a, b \in \mathbf{Q}\}$, since we must have closure under multiplication and addition.

We will show that $\mathbf{Q}(\sqrt{2}) = \{a + b\sqrt{2} \mid a, b \in \mathbf{Q}\}$. We first note that if $a_1 + b_1\sqrt{2} = a_2 + b_2\sqrt{2}$, then $a_1 - a_2 = (b_2 - b_1)\sqrt{2}$, so if $b_2 - b_1 \neq 0$ then we can divide by $b_2 - b_1$, which shows that $\sqrt{2}$ is rational, a contradiction. We conclude that $b_1 = b_2$, and $a_1 = a_2$, so there is only one way to represent an element of $\{a + b\sqrt{2} \mid a, b \in \mathbf{Q}\}$.

It is easy to see that $\{a + b\sqrt{2} \mid a, b \in \mathbf{Q}\}$ is closed under addition and subtraction. It is only slightly more difficult to check that the same is true for multiplication and division:

$$
\begin{aligned}
(a + b\sqrt{2})(c + d\sqrt{2}) &= ac + ad\sqrt{2} + bc\sqrt{2} + 2bd \\
&= (ac + 2bd) + (ad + bc)\sqrt{2},
\end{aligned}
$$

and, if c and d are not both zero, then

$$
\begin{aligned}
\frac{a + b\sqrt{2}}{c + d\sqrt{2}} &= \frac{(a + b\sqrt{2})(c - d\sqrt{2})}{(c + d\sqrt{2})(c - d\sqrt{2})} \\
&= \left(\frac{ac - 2bd}{c^2 - 2d^2}\right) + \left(\frac{bc - ad}{c^2 - 2d^2}\right)\sqrt{2}.
\end{aligned}
$$

In both cases the answer has the correct form: a rational number plus a rational number times the square root of 2. Note that $c^2 - 2d^2 \neq 0$ since $\sqrt{2}$ is not a rational number, which also shows that $c - d\sqrt{2} \neq 0$.

Thus $\mathbf{Q}(\sqrt{2})$ is the smallest field that contains $\mathbf{Q}$ and the one root $\sqrt{2}$. In $\mathbf{Q}(\sqrt{2})$, we can factor $x^2 - 2$ to get $x^2 - 2 = (x - \sqrt{2})(x + \sqrt{2})$. The quadratic formula can be used to show that the roots of $x^2 - 2x - 1$ are $1 + \sqrt{2}$, and $1 - \sqrt{2}$, which have the correct form to belong to $\mathbf{Q}(\sqrt{2})$. Thus by allowing coefficients from $\mathbf{Q}(\sqrt{2})$ we have the following factorization:

$$x^4 - 2x^3 - 3x^2 + 4x + 2 = (x - \sqrt{2})(x + \sqrt{2})(x - 1 - \sqrt{2})(x - 1 + \sqrt{2}). \quad \square$$

Example 4.1.2.

The polynomial $x^4 - x^2 - 2$ factors as $(x^2 - 2)(x^2 + 1)$. In this case the field $\mathbf{Q}(\sqrt{2})$ contains two roots $\pm\sqrt{2}$, but not the two remaining roots $\pm i$, and so we need a larger field to be certain of including all roots. Such a field will be constructed in Example 4.3.3. □

In Definition 3.3.5 we stated the definition of a field in group theoretic terms: the elements of a field form an abelian group under addition, the nonzero elements form an abelian group under multiplication, and addition and multiplication are connected by the distributive law. Roughly speaking, then, a field is a set in which the operations of addition, subtraction, multiplication, and division can be performed. We now repeat the definition of a field, with somewhat more detail, and without reference to groups.

4.1.1 Definition. *Let F be a set on which two binary operations are defined, called* **addition** *and* **multiplication**, *and denoted by $+$ and $\cdot$ respectively. That is, the following condition must be satisfied.*

(i) Closure: *For all $a, b \in F$ the sum $a + b$ and the product $a \cdot b$ are well-defined elements of F.*

Then F is called a **field** *with respect to these operations if the following properties hold.*

(ii) Associative laws: *For all $a, b, c \in F$,*

$$a + (b + c) = (a + b) + c \quad \text{and} \quad a \cdot (b \cdot c) = (a \cdot b) \cdot c .$$

(iii) Commutative laws: *For all $a, b \in F$,*

$$a + b = b + a \quad \text{and} \quad a \cdot b = b \cdot a .$$

(iv) Distributive laws: *For all $a, b, c \in F$,*

$$a \cdot (b + c) = (a \cdot b) + (a \cdot c) \quad \text{and} \quad (a + b) \cdot c = (a \cdot c) + (b \cdot c) .$$

(v) Identity elements: *The set F contains an element 0, called an* **additive identity element**, *such that for all $a \in F$,*

$$a + 0 = a \quad \text{and} \quad 0 + a = a .$$

The set F also contains an element 1 (required to be different from 0) called a **multiplicative identity element**, *such that for all $a \in F$,*

$$a \cdot 1 = a \quad \text{and} \quad 1 \cdot a = a .$$

(vi) _Inverse elements:_ _For each $a \in F$, the equations_

$$a + x = 0 \qquad and \qquad x + a = 0$$

have a solution $x \in F$, called an **additive inverse** of a, and denoted by $-a$.
For each nonzero element $a \in F$, the equations

$$a \cdot x = 1 \qquad and \qquad x \cdot a = 1$$

have a solution $x \in F$, called a **multiplicative inverse** of a, and denoted by a^{-1}.

We will shortly show that both additive inverses and multiplicative inverses are unique, and this will justify the notation in the preceding definition.

The basic examples you should keep in mind in this section are $\mathbf{Q}, \mathbf{R}, \mathbf{C}$, and $\mathbf{Z}_p$, where p is prime. In Section 4.3 we will show how to construct other important examples of fields, by generalizing the construction of $\mathbf{Q}(\sqrt{2})$ given in Example 4.1.1. This will also allow us to construct finite fields, which have many important applications. We now provide the first example of a finite field different from $\mathbf{Z}_p$, for p prime.

Example 4.1.3.

The following set of matrices, with entries from $\mathbf{Z}_2$, forms a field under the operations of matrix addition and multiplication:

$$F = \left\{ \begin{bmatrix} 0 & 0 \\ 0 & 0 \end{bmatrix}, \begin{bmatrix} 1 & 0 \\ 0 & 1 \end{bmatrix}, \begin{bmatrix} 1 & 1 \\ 1 & 0 \end{bmatrix}, \begin{bmatrix} 0 & 1 \\ 1 & 1 \end{bmatrix} \right\}.$$

Here we have omitted the brackets from the congruence classes [0] and [1], so that we simply have $1 + 1 = 0$, etc. You should check that F is closed under addition and multiplication. The associative and distributive laws hold for all matrices. You can check that these particular matrices commute under multiplication. The additive identity is $\begin{bmatrix} 0 & 0 \\ 0 & 0 \end{bmatrix}$, and the multiplicative identity is $\begin{bmatrix} 1 & 0 \\ 0 & 1 \end{bmatrix}$. Each element is its own additive inverse, and the multiplicative inverse of $\begin{bmatrix} 1 & 1 \\ 1 & 0 \end{bmatrix}$ is $\begin{bmatrix} 0 & 1 \\ 1 & 1 \end{bmatrix}$. $\square$

Each nonzero element of a field F is invertible; we will use $F^\times$ to denote the set of all nonzero elements of F. When there is no risk of confusion, we will often write ab instead of $a \cdot b$. If we assume that multiplication is done before addition, we can often eliminate parentheses. For example, the distributive laws can be written as $a(b + c) = ab + ac$ and $(a + b)c = ac + bc$. You should also note that the list of properties used to define a field is redundant in a number of places. For example, since multiplication is commutative, we really only need to state one of the distributive laws.

To simplify matters, we have dealt only with the two operations of addition and multiplication. To define subtraction we use the additive inverse of an element: $a - b = a + (-b)$. Similarly, we use the multiplicative inverse of an element to define division, as follows: if $b \neq 0$, then $a \div b = ab^{-1}$. We will sometimes write a/b in place of ab^{-1}. There is the possibility of a problem with these definitions, since $a - b$ and $a \div b$ should be unique, whereas we do not as yet even know that $-b$ and b^{-1} are unique. The next proposition takes care of this problem.

4.1.2 Proposition. *Let F be a field, with $a, b, c \in F$.*

(a) Cancellation laws: *If $a + c = b + c$, then $a = b$. If $c \neq 0$ and $a \cdot c = b \cdot c$, then $a = b$.*

(b) Uniqueness of identity elements: *If $a + b = a$, then $b = 0$. If $a \cdot c = a$ and $a \neq 0$, then $c = 1$.*

(c) Uniqueness of inverses: *If $a + b = 0$, then $b = -a$. If $a \neq 0$ and $ab = 1$, then $b = a^{-1}$.*

Proof. Proposition 3.1.2 shows that identity elements and inverses are unique for any binary operation. The axioms that hold for the field F show that F is an abelian group under addition, and so the additive cancellation law follows from Proposition 3.1.7 of Chapter 3.

Suppose that $a \cdot c = b \cdot c$, with $c \neq 0$. Since c is nonzero, an inverse c^{-1} exists, and we can rewrite the equation in the following way:

$$
\begin{aligned}
ac &= bc \\
(ac)c^{-1} &= (bc)c^{-1} \\
a(cc^{-1}) &= b(cc^{-1}) \\
a \cdot 1 &= b \cdot 1 \\
a &= b .
\end{aligned}
$$

This completes the proof. $\square$

4.1.3 Proposition. *Let F be a field.*

(a) *For all $a \in F$, $a \cdot 0 = 0$.*

(b) *If $a, b \in F$ with $a \neq 0$ and $b \neq 0$, then $ab \neq 0$.*

(c) *For all $a \in F$, $-(-a) = a$.*

(d) *For all $a, b \in F$, $(a)(-b) = (-a)(b) = -ab$.*

(e) *For all $a, b \in F$, $(-a)(-b) = ab$.*

Proof. (a) We will use the fact that $0 + 0 = 0$. By the distributive law,

$$a \cdot 0 + a \cdot 0 = a \cdot (0 + 0) = a \cdot 0 = a \cdot 0 + 0 ,$$

so the cancellation law for addition shows that $a \cdot 0 = 0$.

(b) If $a \neq 0$ and $a \cdot b = 0$, then $a \cdot b = a \cdot 0$ and the cancellation law for multiplication shows that $b = 0$.

(c) In words, the equation $-(-a) = a$ states that the additive inverse of $-a$ is a, and this follows from the equation $-a + a = 0$ which defines $-a$.

(d) Using the distributive law,

$$a \cdot b + a \cdot (-b) = a \cdot (b + (-b)) = a \cdot 0 = 0 ,$$

which shows that $(a)(-b)$ is the additive inverse of ab, and so $(a)(-b) = -(ab)$. Similarly, $(-a)(b) = -ab$.

(e) Now consider $(-a)(-b)$. By what we have just shown,

$$(-a)(-b) = -((-a)(b)) = -(-ab) = ab ,$$

and this completes the proof. $\square$

Having proved some elementary results on fields, we are now ready to discuss polynomials with coefficients in a field. In high school algebra we talk about x as an "unknown" quantity. Sometimes this is appropriate, but we need to think in a more general context, not limiting x to be an element of a specific field. We should usually think of x as a symbol on which various operations can be performed, and to encourage this we use the word "indeterminate" in place of "unknown".

4.1.4 Definition. *Let F be a field. If $a_m, a_{m-1}, \ldots, a_1, a_0 \in F$ (where m is a nonnegative integer), then any expression of the form*

$$a_m x^m + a_{m-1} x^{m-1} + \ldots + a_1 x + a_0$$

*is called a **polynomial over** F in the **indeterminate** x with **coefficients** $a_m, a_{m-1}, \ldots, a_0$. The subscript i of the coefficient a_i is called its **index**.*

*If n is the largest nonnegative integer such that $a_n \neq 0$, then we say that the polynomial $f(x) = a_n x^n + \ldots + a_0$ has **degree** n, written $\deg(f(x)) = n$, and a_n is called the **leading coefficient** of $f(x)$. If a_0 is the leading coefficient of $f(x)$, then $f(x)$ is called a **constant polynomial**.*

*If the leading coefficient of $f(x)$ is 1, then $f(x)$ is said to be a **monic polynomial**.*

The set of all polynomials with coefficients in F is denoted by $F[x]$.

According to this definition, the zero polynomial (each of whose coefficients is zero) has no degree. For convenience, it is often assigned $-\infty$ as a degree. A constant polynomial $f(x) = a_0$ has degree 0 if $a_0 \neq 0$. Thus a polynomial belongs to the coefficient field if and only if it has degree 0 or $-\infty$.

Two polynomials are equal by definition if they have the same degree and all corresponding coefficients are equal. It is important to distinguish between the polynomial $f(x)$ as an element of $F[x]$ and the corresponding **polynomial function** from F into F defined by substituting elements of F in place of x. If $f(x) = a_m x^m + \ldots + a_0$ and $c \in F$, then $f(c) = a_m c^m + \ldots + a_0$. In fact, if F is a finite field, it is possible to have two different polynomials that define the same polynomial function.

Example 4.1.4.

Let F be the field $\mathbf{Z}_5$ and consider the polynomials x^5 and x. For any $c \in \mathbf{Z}_5$, by Fermat's theorem (Theorem 1.4.12) we have $c^5 \equiv c \pmod 5$, and so the polynomial functions $f(x) = x^5$ and $g(x) = x$ are equal when considered as functions from $\mathbf{Z}_5$ into $\mathbf{Z}_5$.

As another example, consider the polynomials $x^5 - 2x + 1$ and $4x + 1$. Then for any $c \in \mathbf{Z}_5$ we have

$$c^5 - 2c + 1 \equiv -c + 1 \equiv 4c + 1 \pmod 5 ,$$

which shows that $x^5 - 2x + 1$ and $4x + 1$ define the same function.   □

For the polynomials

$$f(x) = a_m x^m + a_{m-1} x^{m-1} + \ldots + a_1 x + a_0$$

and

$$g(x) = b_n x^n + b_{n-1} x^{n-1} + \ldots + b_1 x + b_0 ,$$

the sum of $f(x)$ and $g(x)$ is defined by just adding corresponding coefficients. The product $f(x)g(x)$ is defined to be

$$a_m b_n x^{n+m} + \ldots + (a_2 b_0 + a_1 b_1 + a_0 b_2)x^2 + (a_1 b_0 + a_0 b_1)x + a_0 b_0 .$$

The coefficient c_k of x^k in $f(x)g(x)$ can be described by the formula

$$c_k = \sum_{i=0}^{k} a_i b_{k-i} = \sum_{i+j=k} a_i b_j .$$

This definition of the product is consistent with what we would expect to obtain using a naive approach: expand the product using the distributive law repeatedly (this amounts to multiplying each term by every other) and then collect similar terms.

With this addition and multiplication, $F[x]$ has properties similar to those of the integers. Checking that the following properties hold is tedious, though not difficult. The necessary proofs use the definitions, and depend on the properties of the coefficient field. We have omitted all details.

(i) Associative laws: For any polynomials $f(x), g(x), h(x)$ over F,

$$f(x) + (g(x) + h(x)) = (f(x) + g(x)) + h(x) ,$$
$$f(x) \cdot (g(x) \cdot h(x)) = (f(x) \cdot g(x)) \cdot h(x) .$$

(ii) Commutative laws: For any polynomials $f(x), g(x)$ over F,

$$f(x) + g(x) = g(x) + f(x) ,$$
$$f(x) \cdot g(x) = g(x) \cdot f(x) .$$

(iii) Distributive laws: For any polynomials $f(x), g(x), h(x)$ over F,

$$f(x) \cdot (g(x) + h(x)) = (f(x) \cdot g(x)) + (f(x) \cdot h(x)) ,$$
$$(f(x) + g(x)) \cdot h(x) = (f(x) \cdot h(x)) + (g(x) \cdot h(x)) .$$

(iv) Identity elements: The additive and multiplicative identities of F, considered as constant polynomials, serve as identity elements.

(v) Additive inverses: For each polynomial $f(x)$ over F, the polynomial $-f(x)$ serves as an additive inverse.

In the formula above, two polynomials are multiplied by multiplying each term of the first by each term of the second, and then collecting similar terms. In Figure 4.1.1 we show an efficient way to do this, by arranging the similar terms in columns.

4.1.5 Proposition. *If $f(x)$ and $g(x)$ are nonzero polynomials in $F[x]$, then their product $f(x)g(x)$ is nonzero and*

$$\deg(f(x)g(x)) = \deg(f(x)) + \deg(g(x)) .$$

Proof. Suppose that

$$f(x) = a_m x^m + \ldots + a_1 x + a_0$$

and

$$g(x) = b_n x^n + \ldots + b_1 x + b_0 ,$$

with $\deg(f(x)) = m$ and $\deg(g(x)) = n$. Thus $a_m \neq 0$ and $b_n \neq 0$. It follows from the general formula for multiplication of polynomials that the leading coefficient of $f(x)g(x)$ must be $a_m b_n$, which must be nonzero by Proposition 4.1.3 (b). Thus the degree of $f(x)g(x)$ is $m+n$, since in $f(x)g(x)$ the coefficient of x^{m+n} is $a_m b_n$. □

Figure 4.1.1:

$$
\begin{array}{rrrr}
2x^4 & -3x^2 & +5x & +1 \\
\times & x^2 & -4x & -2 \\
\hline
-4x^4 & +6x^2 & -10x & -2 \\
-8x^5 +12x^3 & -20x^2 & -4x & \\
2x^6 -3x^4 +5x^3 & +x^2 & & \\
\hline
2x^6 \quad -8x^5 \quad -7x^4 \quad +17x^3 & -13x^2 & -14x & -2 \\
\end{array}
$$

Note that we could extend the statement of Proposition 4.1.5 to include the zero polynomial, provided we would use the convention that assigns to the zero polynomial the degree $-\infty$.

4.1.6 Corollary. *If $f(x)$, $g(x)$, $h(x) \in F[x]$, and $f(x)$ is not the zero polynomial, then $f(x)g(x) = f(x)h(x)$ implies $g(x) = h(x)$.*

Proof. If $f(x)g(x) = f(x)h(x)$, then we can use the distributive law to rewrite the equation as $f(x)(g(x) - h(x)) = 0$, and since $f(x) \neq 0$, the previous proposition implies that $g(x) - h(x) = 0$, or simply $g(x) = h(x)$. $\square$

Having proved Proposition 4.1.5, we can make some further observations about $F[x]$. If $f(x)g(x) = 1$, then both $f(x)$ and $g(x)$ must be constant polynomials, since the sum of their degrees must be 0. This shows that the only polynomials that have multiplicative inverses are those of degree 0, which correspond to the nonzero elements of F. In this sense $F[x]$ is very far away from being a field itself, although all of the other properties of a field are satisfied.

4.1.7 Definition. *Let $f(x)$, $g(x) \in F[x]$. If $f(x) = q(x)g(x)$ for some $q(x) \in F[x]$, then we say that $g(x)$ is a **factor** or **divisor** of $f(x)$, and we write $g(x) \mid f(x)$.*

The set of all polynomials divisible by $g(x)$ will be denoted by $\langle g(x) \rangle$.

4.1.8 Lemma. *For any element $c \in F$, and any positive integer k,*

$$
(x - c) \mid (x^k - c^k) .
$$

Proof. A direct computation shows that we have the following factorization:

$$
x^k - c^k = (x - c)(x^{k-1} + cx^{k-2} + \ldots + c^{k-2}x + c^{k-1}) .
$$

Note that the quotient $x^{k-1} + cx^{k-2} + \ldots + c^{k-2}x + c^{k-1}$ has coefficients in F. $\square$

The next theorem shows that for any polynomial $f(x)$, the remainder when $f(x)$ is divided by $x - c$ is $f(c)$. That is, $f(x) = q(x)(x - c) + f(c)$. The **remainder** $f(c)$ and **quotient** $q(x)$ are unique.

4.1.9 Theorem (Remainder Theorem). *Let $f(x) \in F[x]$ be a nonzero polynomial, and let $c \in F$. Then there exists a polynomial $q(x) \in F[x]$ such that*

$$f(x) = q(x)(x - c) + f(c) .$$

Moreover, if $f(x) = q_1(x)(x - c) + k$, where $q_1(x) \in F[x]$ and $k \in F$, then $q_1(x) = q(x)$ and $k = f(c)$.

Proof. If $f(x) = a_m x^m + \ldots + a_0$, then

$$f(x) - f(c) = a_m(x^m - c^m) + \ldots + a_1(x - c) .$$

By Lemma 4.1.8, $x - c$ is a divisor of each term on the right-hand side of the equation, and so it must be a divisor of $f(x) - f(c)$. Thus

$$f(x) - f(c) = q(x)(x - c)$$

for some polynomial $q(x) \in F[x]$, or equivalently,

$$f(x) = q(x)(x - c) + f(c) .$$

If $f(x) = q_1(x)(x - c) + k$, then

$$(q(x) - q_1(x))(x - c) = k - f(c) .$$

If $q(x) - q_1(x) \neq 0$, then by Proposition 4.1.5 the left-hand side of the equation has degree ≥ 1, which contradicts the fact that the right-hand side of the equation is a constant. Thus $q(x) - q_1(x) = 0$, which also implies that $k - f(c) = 0$, and so the quotient and remainder are unique. $\square$

Example 4.1.5.

Let us work through the previous proof in the case $F = \mathbf{Q}$, for $f(x) = x^2 + 5x - 2$ and $c = 5$:

$$\begin{aligned}
f(x) - f(5) &= (x^2 + 5x - 2) - (5^2 + 5 \cdot 5 - 2) \\
&= (x^2 - 5^2) + (5x - 25) \\
&= (x + 5)(x - 5) + 5(x - 5) \\
&= (x + 10)(x - 5) .
\end{aligned}$$

Thus we obtain $x^2 + 5x - 2 = (x + 10)(x - 5) + 48$. $\square$

4.1.10 Definition. *Let $f(x) = a_m x^m + \ldots + a_0 \in F[x]$. An element $c \in F$ is called a **root** of the polynomial $f(x)$ if $f(c) = 0$.*

4.1.11 Corollary. *Let $f(x) \in F[x]$ be a nonzero polynomial, and let $c \in F$. Then c is a root of $f(x)$ if and only if $x - c$ is a factor of $f(x)$. That is, $f(c) = 0$ if and only if $(x - c) \mid f(x)$.*

4.1.12 Corollary. *A polynomial of degree n with coefficients in the field F has at most n distinct roots in F.*

Proof. The proof will use induction on the degree of the polynomial $f(x)$. The result is certainly true if $f(x)$ has degree 0, that is, if $f(x)$ is a nonzero constant. Now suppose that the result is true for all polynomials of degree $n-1$. If c is a root of $f(x)$, then by Corollary 4.1.11 we can write $f(x) = q(x)(x - c)$, for some polynomial $q(x)$. If a is any root of $f(x)$, then substituting shows that $q(a)(a - c) = 0$, which implies that either $q(a) = 0$ or $a = c$. By assumption, $q(x)$ has at most $n - 1$ distinct roots and so this shows that $f(x)$ can have at most $n - 1$ distinct roots which are different from c. $\square$

EXERCISES: SECTION 4.1

1. Let $f(x)$, $g(x)$, $h(x) \in F[x]$. Show that the following properties hold.
   (a) If $g(x) \mid f(x)$ and $h(x) \mid g(x)$, then $h(x) \mid f(x)$.
   (b) If $h(x) \mid f(x)$ and $h(x) \mid g(x)$, then $h(x) \mid (f(x) \pm g(x))$.
   (c) If $g(x) \mid f(x)$, then $g(x) \cdot h(x) \mid f(x) \cdot h(x)$.
   (d) If $g(x) \mid f(x)$ and $f(x) \mid g(x)$, then $f(x) = kg(x)$ for some $k \in F$.

2. Let p be a prime number, and let n be a positive integer. How many polynomials are there of degree n over $\mathbf{Z}_p$?

3.†For $f(x) = 2x^3 + x^2 - 2x + 1$, use the method of Theorem 4.1.9 to write $f(x) = q(x)(x - 1) + f(1)$.

4. For $f(x) = x^3 + 3x^2 - 10x + 5$, use the method of Theorem 4.1.9 to write $f(x) = q(x)(x - 2) + f(2)$.

5. Over the given field F, write $f(x) = q(x)(x - c) + f(c)$ for
   †(a) $f(x) = 2x^3 + x^2 - 4x + 3$; $\ c = 1$; $\ F = \mathbf{Q}$;
   (b) $f(x) = x^3 - 5x^2 + 6x + 5$; $\ c = 2$; $\ F = \mathbf{Q}$;
   †(c) $f(x) = x^3 + 1$; $\ c = 1$; $\ F = \mathbf{Z}_3$;
   (d) $f(x) = x^3 + 2x + 3$; $\ c = 2$; $\ F = \mathbf{Z}_5$.

$Z_p = \{0, 1, \ldots p-1\}$

6. Let p be a prime number. Find all roots of $x^{p-1} - 1$ in $\mathbf{Z}_p$.

7. Show that if c is any element of the field F and $k > 2$ is an odd integer, then $x + c$ is a factor of $x^k + c^k$.

8. Show that rational numbers correspond to decimals which are either repeating or terminating.

   Hint: If $q = m/n$, then when dividing m by n to put q into decimal form there are at most n different remainders. Conversely, if d is a repeating decimal, then find s, t such that $10^s d - 10^t d$ is an integer.

9. Let a be a nonzero element of a field F. Show that $(a^{-1})^{-1} = a$ and $(-a)^{-1} = -a^{-1}$.

10. Let a, b, c be elements of a field F. Prove that if $a \neq 0$, then the equation $ax + b = c$ has a unique solution.

11. Show that the set $\mathbf{Q}(\sqrt{3}) = \{a + b\sqrt{3} \mid a, b \in \mathbf{Q}\}$ is closed under addition, subtraction, multiplication, and division.

12. Let F be any field. An $n \times n$ matrix with entries in F is called a **scalar** matrix if it has the form aI, where I is the $n \times n$ identity matrix, and $a \in F$. Prove that the set of all $n \times n$ scalar matrices over F is a field under the operations of matrix addition and multiplication.

13. Show that the set of matrices of the form $\begin{bmatrix} a & b \\ -b & a \end{bmatrix}$, where $a, b \in \mathbf{R}$, is a field under the operations of matrix addition and multiplication.

14. Show that the set of matrices of the form $\begin{bmatrix} a & b \\ 2b & a \end{bmatrix}$, where $a, b \in \mathbf{Q}$, is a field under the operations of matrix addition and multiplication.

15. Complete the proof that the set of matrices in Example 4.1.3 is a field.

16. Show that the following set of matrices over $\mathbf{Z}_2$ is a field under the operations of matrix addition and multiplication:

$$\begin{bmatrix} 0 & 0 & 0 \\ 0 & 0 & 0 \\ 0 & 0 & 0 \end{bmatrix}, \quad \begin{bmatrix} 1 & 0 & 0 \\ 0 & 1 & 0 \\ 0 & 0 & 1 \end{bmatrix}, \quad \begin{bmatrix} 0 & 1 & 0 \\ 0 & 0 & 1 \\ 1 & 1 & 0 \end{bmatrix}, \quad \begin{bmatrix} 1 & 1 & 0 \\ 0 & 1 & 1 \\ 1 & 1 & 1 \end{bmatrix},$$

$$\begin{bmatrix} 0 & 0 & 1 \\ 1 & 1 & 0 \\ 0 & 1 & 1 \end{bmatrix}, \quad \begin{bmatrix} 1 & 0 & 1 \\ 1 & 0 & 0 \\ 0 & 1 & 0 \end{bmatrix}, \quad \begin{bmatrix} 0 & 1 & 1 \\ 1 & 1 & 1 \\ 1 & 0 & 1 \end{bmatrix}, \quad \begin{bmatrix} 1 & 1 & 1 \\ 1 & 0 & 1 \\ 1 & 0 & 0 \end{bmatrix}.$$

17. Let $(x_0, y_0), (x_1, y_1), (x_2, y_2)$ be points in the Euclidean plane $\mathbf{R}^2$ such that x_0, x_1, x_2 are distinct. Show that the formula

$$f(x) = \frac{y_0(x - x_1)(x - x_2)}{(x_0 - x_1)(x_0 - x_2)} + \frac{y_1(x - x_0)(x - x_2)}{(x_1 - x_0)(x_1 - x_2)} + \frac{y_2(x - x_0)(x - x_1)}{(x_2 - x_0)(x_2 - x_1)}$$

defines a polynomial $f(x)$ such that $f(x_0) = y_0$, $f(x_1) = y_1$, and $f(x_2) = y_2$. (This is a special case of Lagrange's interpolation formula.)

18. Use Lagrange's interpolation formula to find a polynomial $f(x)$ such that $f(1) = 0$, $f(2) = 1$, and $f(3) = 4$.

19.†Find a polynomial $f(x)$ such that $f(1) = -15$, $f(0) = 3$, $f(2) = -3$, and $f(4) = 15$.

20. Show that $f(x) = b_1 + \frac{1}{2}(b_2 - b_0)\left(\frac{x-a}{h}\right) + \frac{1}{2}(b_2 - 2b_1 + b_0)\left(\frac{x-a}{h}\right)^2$ has the property that $f(a - h) = b_0$, $f(a) = b_1$, and $f(a + h) = b_2$.

    Note: Exercises 20 and 21 provide the basis for Simpson's rule for numerical integration via parabolic approximations.

21.†For the polynomial $f(x)$ in Exercise 20, find $\int_{a-h}^{a+h} f(x)dx$.

22. Is it possible to define a quadratic polynomial whose graph contains the four points $(-1, -2)$, $(0, -2)$, $(1, 0)$, and $(2, 2)$?

23. (a) Extend the formula of Exercise 17 to the case of four points in the plane.

    (b) Extend the formula of Exercise 17 to the case of k points in the plane.

24. From Exercises 2 and 3 of Section A.4 of the appendix, it appears to be a reasonable conjecture that $\sum_{i=1}^{n} i^k$ is a polynomial of degree $k+1$ in n, e.g. $\sum_{i=1}^{n} i = (n^2+n)/2$ and $\sum_{i=1}^{n} i^2 = (2n^3+3n^2+n)/6$. We will assume, for the moment, that $\sum_{i=1}^{n} i^k = P_{k+1}(n)$ is a polynomial in n of degree $k + 1$, and then attempt to find $P_{k+1}(n)$ by using the formula from Exercise 23. For this purpose we need $k + 2$ points that the polynomial passes through. We get these by evaluating the sums $\sum_{i=1}^{n} i^k = P_{k+1}(n)$ for $n = 1, 2, \ldots, k+2$. (Any $k+2$ values of n will do, but these are easy to compute.)

    For example, to find $\sum_{i=1}^{n} i = P_2(n)$ we have $P_2(1) = \sum_{i=1}^{1} i = 1$, $P_2(2) = \sum_{i=1}^{2} i = 3$, $P_2(3) = \sum_{i=1}^{3} i = 6$. Hence $P_2(n)$ passes through $(1, 1)$, $(2, 3)$, and $(3, 6)$. Thus

    $$P_2(n) = \frac{1(n-2)(n-3)}{(1-2)(1-3)} + \frac{3(n-1)(n-3)}{(2-1)(2-3)} + \frac{6(n-1)(n-2)}{(3-1)(3-2)} = \frac{1}{2}n^2 + \frac{1}{2}n.$$

    Note that the derivation was based on the unproved assumption that $P_{k+1}(n)$ is a polynomial. Once we have $P_{k+1}(n)$, we can prove that $P_{k+1}(n) = \sum_{i=1}^{n} i^k$ by induction.

    Use the above method to find formulas for the following sums.

    (a) $\sum_{i=1}^{n} i^2$

    (b) $\sum_{i=1}^{n} i^3$.

4.2   Factors

Our aim in this section is to obtain, for polynomials, results analogous to some of
the theorems we proved in Chapter 1 for integers. In particular, we are looking for a
division algorithm, an analog of the Euclidean algorithm, and a prime factorization
theorem. Many of the arguments in Chapter 1 involved finding the smallest number
in some set of integers, and so the size, in terms of absolute value, was important.
Very similar arguments can be given for polynomials, with the notion of degree
replacing that of absolute value.

   Our first goal is to formulate and prove a division algorithm. The following
example is included just to remind you of the procedure that you probably learned
in high-school algebra.

Example 4.2.1.

   In dividing the polynomial $6x^4 - 2x^3 + x^2 + 5x - 18$ by $2x^2 - 3$, the first step
   is to divide $6x^4$ by $2x^2$, to get $3x^2$. The next step is to multiply $2x^2 - 3$ by
   $3x^2$ and subtract the result from $6x^4 - 2x^3 + x^2 + 5x - 18$. The algorithm for
   division of polynomials then proceeds much like the algorithm for division of
   integers, as shown in Figure 4.2.1.

Figure 4.2.1:

$$
\begin{array}{rr|rrrrr}
 & & & & 3x^2 & -x & +5 \\
\hline
2x^2 & -3 & 6x^4 & -2x^3 & +x^2 & +5x & -18 \\
 & & 6x^4 & & -9x^2 & & \\
\hline
 & & & -2x^3 & +10x^2 & +5x & \\
 & & & -2x^3 & & +3x & \\
\hline
 & & & & 10x^2 & +2x & -18 \\
 & & & & 10x^2 & & -15 \\
\hline
 & & & & & 2x & -3
\end{array}
$$

   Thus $6x^4 - 2x^3 + x^2 + 5x - 18 = (3x^2 - x + 5)(2x^2 - 3) + (2x - 3)$, where
   the last term is the remainder. □

   The proof of Theorem 4.2.1 is merely a formal verification, using induction,
that the procedure followed in Example 4.2.1 will always work. The polynomials

$q(x)$ and $r(x)$ given by the theorem, with $f(x) = q(x)g(x) + r(x)$, are called (as expected) the **quotient** and **remainder** when $f(x)$ is divided by $g(x)$. Notice that if we divide polynomials with coefficients in a given field, then the quotient and remainder must have coefficients from the same field.

The division algorithm for integers (Theorem 1.1.3) was stated for integers a and b, with $b > 0$. It is easily extended to a statement that is parallel to the next theorem for polynomials: For any $a, b \in \mathbf{Z}$ with $b \neq 0$, there exist unique integers q and r such that $a = bq + r$, with $0 \leq r < |b|$. The role played by the absolute value of an integer is now played by the degree of a polynomial. Note that assigning the degree $-\infty$ to the zero polynomial would simplify the statement of the division algorithm, requiring simply that the degree of the remainder be less than the degree of the divisor.

4.2.1 Theorem (Division Algorithm). *For any polynomials $f(x)$ and $g(x)$ in $F[x]$, with $g(x) \neq 0$, there exist unique polynomials $q(x)$, $r(x) \in F[x]$ such that*

$$f(x) = q(x)g(x) + r(x) ,$$

where either $\deg(r(x)) < \deg(g(x))$ *or* $r(x) = 0$.

Proof. Let

$$f(x) = a_m x^m + \ldots + a_1 x + a_0$$

and

$$g(x) = b_n x^n + \ldots + b_0 ,$$

where $a_m \neq 0$ and $b_n \neq 0$. In case $f(x)$ has lower degree than $g(x)$, we can simply take $q(x) = 0$ and $r(x) = f(x)$. The proof of the other case will use induction on the degree of $f(x)$.

If $f(x)$ has degree zero, it is easy to see that the theorem holds. Now assume that the theorem is true for all polynomials $f(x)$ of degree less than m. (We are assuming that $m \geq n$.) The reduction to a polynomial of lower degree is achieved by using the procedure outlined in Example 4.2.1. We divide $a_m x^m$ by $b_n x^n$ to get $a_m b_n^{-1} x^{m-n}$, then multiply by $g(x)$ and subtract from $f(x)$. This gives

$$f_1(x) = f(x) - a_m b_n^{-1} x^{m-n} g(x) ,$$

where $f_1(x)$ has degree less than m since the leading term of $f(x)$ has been cancelled by $a_m b_n^{-1} x^{m-n} b_n x^n$. Now by the induction hypothesis we can write

$$f_1(x) = q_1(x)g(x) + r(x) ,$$

where the degree of $r(x)$ is less than n, unless $r(x) = 0$. Since

$$f(x) = f_1(x) + a_m b_n^{-1} x^{m-n} g(x) ,$$

substitution gives the desired result:

$$f(x) = \left(q_1(x) + a_m b_n^{-1} x^{m-n}\right) g(x) + r(x) .$$

The quotient $q(x) = q_1(x) + a_m b_n^{-1} x^{m-n}$ has coefficients in F, since $a_m, b_n \in F$, and $q_1(x)$ has coefficients in F by the induction hypothesis. Finally, the remainder $r(x)$ has coefficients in F by the induction hypothesis.

To show that the quotient $q(x)$ and remainder $r(x)$ are unique, suppose that

$$f(x) = q_1(x)g(x) + r_1(x)$$

and

$$f(x) = q_2(x)g(x) + r_2(x) .$$

Thus

$$(q_1(x) - q_2(x))g(x) = r_2(x) - r_1(x) ,$$

and if $q_2(x) - q_1(x) \neq 0$, then the degree of $(q_2(x) - q_1(x))g(x)$ is greater than or equal to the degree of $g(x)$ (by Proposition 4.1.5), whereas the degree of $r_2(x) - r_1(x)$ is less than the degree of $g(x)$. This is a contradiction, so we can conclude that $q_2(x) = q_1(x)$, and this forces

$$r_2(x) - r_1(x) = (q_1(x) - q_2(x))g(x) = 0 ,$$

completing the proof. $\square$

Theorem 4.1.9 is a particular case of the general division algorithm. If $g(x)$ is the linear polynomial $x - c$, then the remainder must be a constant when $f(x)$ is divided by $x - c$. Substituting c into the equation $f(x) = q(x)(x - c) + r(x)$ shows that $r(c) = f(c)$, so the remainder on division by $x - c$ is the same element of F as $f(x)$ evaluated at $x = c$.

Example 4.2.2.

In this example we illustrate the division algorithm for the polynomials in Example 4.2.1, over the finite field $\mathbf{Z}_7$. Here, as on previous occasions, we simplify the notation by omitting the notation for congruence classes modulo 7. Our first step is to reduce the coefficients of $f(x) = 6x^4 - 2x^3 + x^2 + 5x - 18$ and $g(x) = 2x^2 - 3$ modulo 7, to obtain $f(x) = 6x^4 + 5x^3 + x^2 + 5x + 3$ and $g(x) = 2x^2 + 4$. Since it is much easier to divide by a monic polynomial, we multiply $g(x)$ by 4 (the inverse of 2 modulo 7) and work with $x^2 + 2$. Proceeding as in Example 4.2.1, we obtain

$$6x^4 + 5x^3 + x^2 + 5x + 3 = (6x^2 + 5x + 3)(x^2 + 2) + (2x + 4) ,$$

where the last term is the remainder. The work is shown in Figure 4.2.2.

Figure 4.2.2:

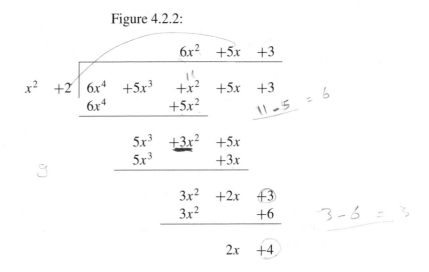

To take care of the fact that we divided by $g(x)/2$, we need to multiply the divisor $x^2 + 2$ by 2 and divide the quotient $6x^2 + 5x + 3$ by 2. This finally gives us

$$6x^4 + 5x^3 + x^2 + 5x + 3 = (3x^2 + 6x + 5)(2x^2 + 4) + (2x + 4) ,$$

with the remainder being unchanged.   □

The next result is parallel to Theorem 1.1.4, which shows that every subgroup of **Z** is cyclic. It will play an important role in Chapters 5 and 6.

4.2.2 Theorem. *Let I be a subset of $F[x]$ that satisfies the following conditions:*
   (i) *I contains a nonzero polynomial;*
   (ii) *if $f(x), g(x) \in I$, then $f(x) + g(x) \in I$;*
   (iii) *if $f(x) \in I$ and $q(x) \in F[x]$, then $q(x)f(x) \in I$.*
If $d(x)$ is any nonzero polynomial in I of minimal degree, then

$$I = \{f(x) \in F[x] \mid f(x) = q(x)d(x) \text{ for some } q(x) \in F[x]\} .$$

Proof. If I contains a nonzero polynomial, then the set of all natural numbers n such that I contains a polynomial of degree n is a nonempty set, so by the well-ordering principle it must contain a smallest element, say m. Thus we can find a nonzero polynomial of minimal degree m in I, say $d(x)$.

Every multiple $q(x)d(x)$ of $d(x)$ must be in I by condition (iii). Next we need to show that $d(x)$ is a divisor of any other polynomial $h(x) \in I$. One way to proceed is to simply divide $h(x)$ by $d(x)$, using the division algorithm, and then show that the remainder must be zero.

We can write

$$h(x) = q(x)d(x) + r(x) ,$$

where $r(x)$ is either zero or has lower degree than $d(x)$. Solving for $r(x)$, we have

$$r(x) = h(x) + (-q(x))d(x) .$$

This shows that $r(x) \in I$, since $h(x) \in I$ and I is closed under addition and under multiplication by any polynomial. But then $r(x)$ must be zero, since I cannot contain a nonzero polynomial of lower degree than the degree of $d(x)$. This shows that $h(x)$ is a multiple of $d(x)$. □

We should note that in any set of polynomials of the form

$$\{f(x) \in F[x] \mid f(x) = q(x)d(x) \text{ for some } q(x) \in F[x]\} ,$$

the degree of $d(x)$ must be minimal (among degrees of nonzero elements). Multiplying by the inverse of the leading coefficient of $d(x)$ gives a monic polynomial of the same degree that is still in the set.

4.2.3 Definition. *A monic polynomial $d(x) \in F[x]$ is called the **greatest common divisor** of $f(x)$, $g(x) \in F[x]$ if*

(i) *$d(x)$ is a divisor of both $f(x)$ and $g(x)$, and*

(ii) *any divisor of both $f(x)$ and $g(x)$ is also a divisor of $d(x)$.*

The greatest common divisor of $f(x)$ and $g(x)$ is denoted by $\gcd(f(x), g(x))$.

If $\gcd(f(x), g(x)) = 1$, then the polynomials $f(x)$ and $g(x)$ are said to be relatively prime.

Note that if both $f(x)$ and $g(x)$ are the zero polynomial, then by our definition there is no greatest common divisor, since the zero polynomial is not monic. We can show the uniqueness of the greatest common divisor as follows. Suppose that $c(x)$ and $d(x)$ are both greatest common divisors of $f(x)$ and $g(x)$. Then $c(x) \mid d(x)$ and $d(x) \mid c(x)$, say $d(x) = a(x)c(x)$ and $c(x) = b(x)d(x)$, and so we have $d(x) = a(x)b(x)d(x)$. Therefore $a(x)b(x) = 1$, and so Proposition 4.1.5 shows that $a(x)$ and $b(x)$ are both of degree zero. Thus $c(x)$ is a constant multiple of $d(x)$, and since both are monic, the constant must be 1, which shows that $c(x) = d(x)$.

4.2.4 Theorem. *For any nonzero polynomials $f(x)$, $g(x) \in F[x]$, the greatest common divisor $\gcd(f(x), g(x))$ exists and can be expressed as a linear combination of $f(x)$ and $g(x)$, in the form*

$$\gcd(f(x), g(x)) = a(x)f(x) + b(x)g(x)$$

for some $a(x)$, $b(x) \in F[x]$.

Proof. It is easy to check that

$$I = \{a(x)f(x) + b(x)g(x) \mid a(x), \ b(x) \in F[x]\}$$

satisfies the conditions of Theorem 4.2.2, and so I consists of all polynomial multiples of any monic polynomial in I of minimal degree, say $d(x)$.

Since $f(x)$, $g(x) \in I$, we have $d(x) \mid f(x)$ and $d(x) \mid g(x)$. On the other hand, since $d(x)$ is a linear combination of $f(x)$ and $g(x)$, it follows that if $h(x) \mid f(x)$ and $h(x) \mid g(x)$, then $h(x) \mid d(x)$. Thus $d(x) = \gcd(f(x), g(x))$. $\square$

Example 4.2.3 (Euclidean algorithm for polynomials).

Let $f(x)$, $g(x) \in F[x]$ be nonzero polynomials. We can use the division algorithm to write $f(x) = q(x)g(x) + r(x)$, with $\deg(r(x)) < \deg(g(x))$ or $r(x) = 0$. If $r(x) = 0$, then $g(x)$ is a divisor of $f(x)$, and so $\gcd(f(x), g(x)) = cg(x)$, for some $c \in F$. If $r(x) \neq 0$, then it is easy to check that the common divisors of $f(x)$ and $g(x)$ are the same as the common divisors of $g(x)$ and $r(x)$, so $\gcd(f(x), g(x)) = \gcd(g(x), r(x))$. This step reduces the degrees of the polynomials involved, and so repeating the procedure leads to the greatest common divisor of the two polynomials in a finite number of steps.

The Euclidean algorithm for polynomials is similar to the Euclidean algorithm for finding the greatest common divisor of nonzero integers. The polynomials $a(x)$ and $b(x)$ for which $\gcd(f(x), g(x)) = a(x)f(x) + b(x)g(x)$ can be found just as for integers (see Example 1.1.2). $\square$

Example 4.2.4.

To find $\gcd(2x^4 + x^3 - 6x^2 + 7x - 2, \ 2x^3 - 7x^2 + 8x - 4)$ over **Q**, divide the polynomial of higher degree by the one of lower degree, to get the quotient $x + 4$ and remainder $14x^2 - 21x + 14$. The answer will be unchanged by dividing through by a nonzero constant, so we can use the polynomial $2x^2 - 3x + 2$.

As for integers, we now have

$$\gcd(2x^4 + x^3 - 6x^2 + 7x - 2, \ 2x^3 - 7x^2 + 8x - 4)$$

$$= \gcd(2x^3 - 7x^2 + 8x - 4, \ 2x^2 - 3x + 2) \ .$$

Dividing as before gives the quotient $x - 2$, with remainder zero. This shows that the greatest common divisor that we are looking for is $x^2 - (3/2)x + 1$ (we divided through by 2 to obtain a monic polynomial).   □

4.2.5 Proposition. *Let $p(x), f(x), g(x) \in F[x]$. If $\gcd(p(x), f(x)) = 1$ and $p(x) \mid f(x)g(x)$, then $p(x) \mid g(x)$.*

Proof. If $\gcd(p(x), f(x)) = 1$, then

$$1 = a(x)p(x) + b(x)f(x)$$

for some $a(x), \ b(x) \in F[x]$. Thus

$$g(x) = a(x)g(x)p(x) + b(x)f(x)g(x) \ ,$$

which shows that if $p(x) \mid f(x)g(x)$, then $p(x) \mid g(x)$.   □

4.2.6 Definition. *A nonconstant polynomial is said to be **irreducible over the field** F if it cannot be factored in $F[x]$ into a product of polynomials of lower degree. It is said to be **reducible** over F if such a factorization exists.*

All polynomials of degree 1 are irreducible. On the other hand, any polynomial of greater degree that has a root in F is reducible over F, since by the remainder theorem it can be factored into polynomials of lower degree. To check that a polynomial is irreducible over a field F, in general it is not sufficient to merely check that it has no roots in F. For example, $x^4 + 4x^2 + 4 = (x^2 + 2)^2$ is reducible over $\mathbf{Q}$, but it certainly has no rational roots. However, a polynomial of degree 2 or 3 can be factored into a product of polynomials of lower degree if and only if one of the factors is linear, which then gives a root. This remark proves the next proposition.

4.2.7 Proposition. *A polynomial of degree 2 or 3 is irreducible over the field F if and only if it has no roots in F.*

The field F is crucial in determining irreducibility. The polynomial $x^2 + 1$ is irreducible over $\mathbf{R}$, since it has no real roots, but considered as a polynomial over $\mathbf{C}$, it factors as $x^2 + 1 = (x + i)(x - i)$. Over the field $\mathbf{Z}_2$, the polynomial $x^2 + x + 1$ is irreducible, since it has no roots in $\mathbf{Z}_2$. But on the other hand, it is reducible over the field $\mathbf{Z}_3$, since $x^2 + x + 1 = (x + 2)^2$ when the coefficients are viewed as representing congruence classes in $\mathbf{Z}_3$.

4.2.8 Lemma. *The nonconstant polynomial $p(x) \in F[x]$ is irreducible over F if and only if for all $f(x)$, $g(x) \in F[x]$, $p(x) \mid (f(x)g(x))$ implies $p(x) \mid f(x)$ or $p(x) \mid g(x)$.*

Proof. First assume that $p(x) \mid f(x)g(x)$. If $p(x)$ is irreducible and $p(x) \nmid f(x)$, then $\gcd(p(x), f(x)) = 1$, and so $p(x) \mid g(x)$ by Proposition 4.2.5.

Conversely, if the given condition holds, then $p(x) \neq f(x)g(x)$ for polynomials of lower degree, since $p(x) \nmid f(x)$ and $p(x) \nmid g(x)$. $\square$

Because of the similarity between Lemma 4.2.8 and Lemma 1.2.5, it is evident that irreducible polynomials should play a role analogous to that of prime numbers, and one of the results we should look for is a unique factorization theorem. The proof of Theorem 1.2.7 can be carried over to polynomials by using irreducible polynomials in place of prime numbers and by using the degree of a polynomial in place of the absolute value of a number. For this reason we have chosen to omit the proof of the next theorem, even though it is extremely important.

4.2.9 Theorem (Unique Factorization). *Any nonconstant polynomial with coefficients in the field F can be expressed as an element of F times a product of monic polynomials, each of which is irreducible over the field F. This expression is unique except for the order in which the factors occur.*

We will show in Section 4.4 that the irreducible polynomials in $\mathbf{C}[x]$ are precisely the linear polynomials, and that any irreducible polynomial in $\mathbf{R}[x]$ has degree one or two. These facts are a consequence of the fundamental theorem of algebra, which states that any polynomial over $\mathbf{C}$ of positive degree has a root in $\mathbf{C}$. (A proof is given in Theorem 8.3.10.)

Polynomials cannot be factored as easily over the field of rational numbers as over the field of real numbers, so the theory of irreducible polynomials over the field of rational numbers is much richer than the corresponding theory over the real numbers. For example, $x^2 - 2$ and $x^4 + x^3 + x^2 + x + 1$ are irreducible over the field of rational numbers, the first since $\sqrt{2}$ is irrational and the second as a consequence of a criterion we will develop in the next section.

In studying roots and factors of polynomials, it is often of interest to know whether there are any repeated roots or factors. The derivative $p'(x)$ of the polynomial $p(x)$ can be used to check for repeated roots and factors. It is possible to formally define the derivative of a polynomial over any field, and we will do so in Section 8.2. For the moment we will restrict ourselves to the case of polynomials with real coefficients, so that we can feel free to use any formulas we might need from calculus.

4.2.10 Definition. *Let* $f(x) \in F[x]$. *An element* $c \in F$ *is said to be a **root of multiplicity*** $m \geq 1$ *of* $f(x)$ *if*

$$(x - c)^m \mid f(x) \quad but \quad (x - c)^{m+1} \nmid f(x) .$$

4.2.11 Proposition. *A nonconstant polynomial* $f(x)$ *over the field* **R** *of real numbers has no repeated factors if and only if* $\gcd(f(x), f'(x)) = 1$.

Proof. We will prove an equivalent statement: a nonconstant polynomial $f(x)$ over **R** has a repeated factor if and only if $\gcd(f(x), f'(x)) \neq 1$.

Suppose that $\gcd(f(x), f'(x)) = d(x) \neq 1$ and that $p(x)$ is an irreducible factor of $d(x)$. Then $f(x) = a(x)p(x)$ and $f'(x) = b(x)p(x)$ for some $a(x), b(x) \in F[x]$. Using the product rule to differentiate gives

$$f'(x) = a'(x)p(x) + a(x)p'(x) = b(x)p(x) .$$

This shows that $p(x) \mid a(x)p'(x)$, since

$$a(x)p'(x) = b(x)p(x) - a'(x)p(x) ,$$

and thus $p(x) \mid a(x)$ because $p(x)$ is irreducible and $p(x) \nmid p'(x)$. Therefore $f(x) = c(x)p(x)^2$ for some $c(x) \in F[x]$, and so $f(x)$ has a repeated factor.

Conversely, if $f(x)$ has a repeated factor, say $f(x) = g(x)^n q(x)$, with $n > 1$, then

$$f'(x) = n \cdot g(x)^{n-1} g'(x)q(x) + g(x)^n q'(x)$$

and $g(x)$ is a common divisor of $f(x)$ and $f'(x)$. $\square$

EXERCISES: SECTION 4.2

1. Use the division algorithm to find the quotient and remainder when $f(x)$ is divided by $g(x)$ over the field of rational numbers **Q**.

   †(a) $f(x) = 2x^4 + 5x^3 - 7x^2 + 4x + 8$ $\qquad\qquad$ $g(x) = 2x - 1$

   (b) $f(x) = 2x^7 - 5x^6 + 5x^5 - x^3 - x^2 + 4x - 5$ $\quad$ $g(x) = x^2 - x + 1$

   †(c) $f(x) = x^5 + 1$ $\qquad\qquad\qquad$ $g(x) = x + 1$

   (d) $f(x) = 2x^4 + x^3 - 6x^2 - x + 2$ $\qquad\qquad$ $g(x) = 2x^2 - 5$

$2x$ ⟌ $2x^4 + x^3 + x^2 + 6x + 2$

2. Use the division algorithm to find the quotient and remainder when $f(x)$ is divided by $g(x)$, over the indicated field.

(a) $f(x) = x^4 + 1$ $g(x) = x + 1$ over $\mathbf{Z}_2$

(b) $f(x) = x^5 + 4x^4 + 2x^3 + 3x^2$ $g(x) = x^2 + 3$ over $\mathbf{Z}_5$

(c) $f(x) = x^5 + 2x^3 + 3x^2 + x - 1$ $g(x) = x^2 + 5$ over $\mathbf{Z}_7$

(d) $f(x) = 2x^4 + x^3 + x^2 + 6x + 2$ $g(x) = 2x^2 + 2$ over $\mathbf{Z}_7$

3. Find the greatest common divisor of $f(x)$ and $f'(x)$, over $\mathbf{Q}$.

†(a) $f(x) = x^4 - x^3 - x + 1$

(b) $f(x) = x^3 - 3x - 2$

†(c) $f(x) = x^3 + 2x^2 - x - 2$

(d) $f(x) = x^4 + 2x^3 + 3x^2 + 2x + 1$

4. Find the greatest common divisor of the given polynomials, over $\mathbf{Q}$.

(a) $2x^3 + 2x^2 - x - 1$ and $2x^4 - x^2$

(b) $4x^3 - 2x^2 - 3x + 1$ and $2x^2 - x - 2$

(c) $x^{10} - x^7 - x^5 + x^3 + x^2 - 1$ and $x^8 - x^5 - x^3 + 1$

(d) $x^5 + x^4 + 2x^2 - x - 1$ and $x^3 + x^2 - x$

5. Find the greatest common divisor of the given polynomials, over the given field.

†(a) $x^4 + x^3 + x + 1$ and $x^3 + x^2 + x + 1$ over $\mathbf{Z}_2$

(b) $x^3 - 2x^2 + 3x + 1$ and $x^3 + 2x + 1$ over $\mathbf{Z}_5$

†(c) $x^5 + 4x^4 + 6x^3 + 6x^2 + 5x + 2$ and $x^4 + 3x^2 + 3x + 6$ over $\mathbf{Z}_7$

(d) $x^5 + x^4 + 2x^2 + 4x + 4$ and $x^3 + x^2 + 4x$ over $\mathbf{Z}_5$

6. In each part of Exercise 4, write the greatest common divisor as a linear combination of the given polynomials. That is, given $f(x)$ and $g(x)$, find $a(x)$ and $b(x)$ such that $d(x) = a(x)f(x) + b(x)g(x)$, where $d(x)$ is the greatest common divisor of $f(x)$ and $g(x)$.

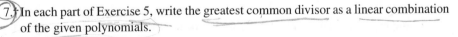

7. In each part of Exercise 5, write the greatest common divisor as a linear combination of the given polynomials.

8. Let F be a field, let $f(x)$, $g_1(x)$, $g_2(x)$ be nonzero polynomials in $F[x]$. Let $q_1(x)$ and $r_1(x)$ be the quotient and remainder when $f(x)$ is divided by $g_1(x)$, and let $q_2(x)$ and $r_2(x)$ be the quotient and remainder when $q_1(x)$ is divided by $g_2(x)$. Show that the quotient when $f(x)$ is divided by the product $g_1(x)g_2(x)$ is $q_2(x)$. What is the remainder?

9. Let $a \in \mathbf{R}$, and let $f(x) \in \mathbf{R}[x]$, with derivative $f'(x)$. Show that the remainder when $f(x)$ is divided by $(x - a)^2$ is $f'(a)(x - a) + f(a)$.

10. Let $p(x) = a_n x^n + a_{n-1} x^{n-1} + \ldots + a_1 x + a_0$ be a polynomial with rational coefficients such that a_n and a_0 are nonzero. Show that $p(x)$ is irreducible over the field of rational numbers if and only if $q(x) = a_0 x^n + a_1 x^{n-1} + \ldots + a_{n-1} x + a_n$ is irreducible over the field of rational numbers.

11. Find the irreducible factors of $x^6 - 1$ over $\mathbf{R}$. $(x-1)(x+1)(x^2-x+1)(x^2+x+1)$

12.† Find all monic irreducible polynomials of degree ≤ 5 over $\mathbf{Z}_2$. Show that the product of all such polynomials of degree ≤ 2 is $x^4 - x$.

   Hint: First develop a criterion that allows you to tell at a glance whether or not a polynomial has no roots. Among the polynomials with no roots, use irreducible factors of degree $\leq n$ to find reducible polynomials of degree $n + 1$.

13. Find all monic irreducible polynomials of degree ≤ 3 over $\mathbf{Z}_3$.

   Using your list, write each of the following polynomials as a product of irreducible polynomials.

   (a) $x^2 - 2x + 1$

   (b) $x^4 + 2x^2 + 2x + 2$

   (c) $2x^3 - 2x + 1$

   (d) $x^4 + 1$

   (e) $x^9 - x$

14. Show that there are exactly $(p^2 - p)/2$ monic irreducible polynomials of degree 2 over $\mathbf{Z}_p$ (where p is any prime number).

15. Show that for any real number $a \neq 0$, the polynomial $x^n - a$ has no multiple roots in $\mathbf{R}$.

Use the following definition in Exercises 16–20.

Definition. *Let $p(x)$ be a nonzero polynomial in $F[x]$. For any $f(x)$, $g(x) \in F[x]$, we write $f(x) \equiv g(x) \pmod{p(x)}$ if $f(x)$ and $g(x)$ have the same remainder when divided by $p(x)$. That is, $f(x) \equiv g(x) \pmod{p(x)}$ if and only if $p(x) \mid (f(x) - g(x))$.*

16. Let $p(x)$ be a nonzero polynomial. Show that for each polynomial $f(x)$ there is a unique polynomial $r(x)$ with $r(x) = 0$ or $\deg(r(x)) < \deg(p(x))$ such that $r(x) \equiv f(x) \pmod{p(x)}$.

17. Let $p(x)$ be a nonzero polynomial. Show that if $f(x) \equiv c(x) \pmod{p(x)}$ and $g(x) \equiv d(x) \pmod{p(x)}$, then

$$f(x) + g(x) \equiv c(x) + d(x) \pmod{p(x)}$$

and

$$f(x)g(x) \equiv c(x)d(x) \pmod{p(x)} .$$

18. Compute the following products. (Your answer should have degree 1.)

    †(a) $(a + bx)(c + dx) \equiv$??? (mod $x^2 + 1$) over **Q**

    (b) $(a + bx)(c + dx) \equiv$??? (mod $x^2 - 2$) over **Q**

    †(c) $(a + bx)(c + dx) \equiv$??? (mod $x^2 + x + 1$) over $\mathbf{Z}_2$

    (d) $(a + bx)(c + dx) \equiv$??? (mod $x^2 + 1$) over $\mathbf{Z}_3$

19. Let $f(x)$ be a nonzero polynomial. Show that there exists a polynomial $g(x)$ with $f(x)g(x) \equiv 1$ (mod $p(x)$) if and only if $\gcd(f(x), p(x)) = 1$.

20. Find a polynomial $q(x)$ such that

    †(a) $(a + bx)q(x) \equiv 1$ (mod $x^2 + 1$) over **Q**

    (b) $(a + bx)q(x) \equiv 1$ (mod $x^2 - 2$) over **Q**

    †(c) $(a + bx)q(x) \equiv 1$ (mod $x^2 + x + 1$) over $\mathbf{Z}_2$

    (d) $(x^2 + 2x + 1)q(x) \equiv 1$ (mod $x^3 + x^2 + 1$) over $\mathbf{Z}_3$

4.3 Existence of Roots

The polynomial $x^2 + 1$ has no roots in the field **R** of real numbers. However, we can obtain a root by constructing an element i for which $i^2 = -1$ and adding it (in some way) to the field **R**. This leads to the field **C**, which contains elements of the form $a + bi$, for $a, b \in \mathbf{R}$. The only problem is to find a way of constructing the root i.

In this section we will show that for any polynomial, over any field, it is possible to construct a larger field in which the polynomial has a root. To do this we will use congruence classes of polynomials. The construction is similar in many ways to the construction of the field $\mathbf{Z}_p$ as a set of congruence classes of **Z**. By iterating the process, it is possible to find a field that contains all of the roots of the polynomial, so that over this field the polynomial factors into a product of linear polynomials.

4.3.1 Definition. *Let E and F be fields. If F is a subset of E and has the operations of addition and multiplication induced by E, then F is called a **subfield** of E, and E is called an **extension field** of F.*

4.3.2 Definition. *Let F be a field, and let $p(x)$ be a fixed polynomial over F. If $a(x), b(x) \in F[x]$, then we say that $a(x)$ and $b(x)$ are **congruent modulo** $p(x)$, written $a(x) \equiv b(x)$ (mod $p(x)$), if $p(x) \mid (a(x) - b(x))$.*

*The set $\{b(x) \in F[x] \mid a(x) \equiv b(x)$ (mod $p(x)$)$\}$ is called the **congruence class** of $a(x)$, and will be denoted by $[a(x)]$.*

The set of all congruence classes modulo $p(x)$ will be denoted by $F[x]/\langle p(x)\rangle$.

The reason for the notation $F[x]/\langle p(x)\rangle$ will become clear in Chapter 5.

We first note that congruence of polynomials defines an equivalence relation. Then since $a(x) \equiv b(x)$ (mod $p(x)$) if and only if $a(x) - b(x) = q(x)p(x)$ for some $q(x) \in F[x]$, the polynomials in the congruence class of $a(x)$ modulo $p(x)$ must be precisely the polynomials of the form $b(x) = a(x) + q(x)p(x)$, for some $q(x)$. We gave a similar description for the congruence classes of $\mathbf{Z}_n$.

When working with congruence classes modulo n, we have often chosen to work with the smallest nonnegative number in the class. Similarly, when working with congruence classes of polynomials, the polynomial of lowest degree in the congruence class is a natural representative. The next proposition guarantees that this representative is unique.

4.3.3 Proposition. *Let F be a field, let $p(x)$ be a nonzero polynomial in $F[x]$, and let $a(x)$ be any polynomial in $F[x]$. If $p(x)$ is not a factor of $a(x)$, then the congruence class $[a(x)]$ modulo $p(x)$ contains exactly one polynomial $r(x)$ with $\deg(r(x)) < \deg(p(x))$.*

Proof. Given $a(x) \in F[x]$, we can use the division algorithm to write

$$a(x) = q(x)p(x) + r(x) ,$$

with $\deg(r(x)) < \deg(p(x))$ or $r(x) = 0$. The assumption that $p(x)$ is not a divisor of $a(x)$ eliminates the case in which $r(x) = 0$.

Solving for $r(x)$ in the above equation shows it to be in the congruence class $[a(x)]$. The polynomial $r(x)$ is the only representative with this property, since if

$$b(x) \equiv a(x) \text{ (mod } p(x))$$

and $\deg(b(x)) < \deg(p(x))$, then

$$b(x) \equiv r(x) \text{ (mod } p(x))$$

and so $p(x) \mid (b(x) - r(x))$. This is a contradiction unless $b(x) = r(x)$, since either $\deg(b(x) - r(x)) < \deg(p(x))$ or $b(x) - r(x) = 0$. $\square$

4.3.4 Proposition. *Let F be a field, and let $p(x)$ be a nonzero polynomial in $F[x]$. For any polynomials $a(x)$, $b(x)$, $c(x)$, and $d(x)$ in $F[x]$, the following conditions hold.*

(a) *If $a(x) \equiv c(x)$ (mod $p(x)$) and $b(x) \equiv d(x)$ (mod $p(x)$), then*

$$a(x)+b(x) \equiv c(x)+d(x) \text{ (mod } p(x)) \quad and \quad a(x)b(x) \equiv c(x)d(x) \text{ (mod } p(x)) .$$

(b) *If $a(x)b(x) \equiv a(x)c(x)$ (mod $p(x)$) and $\gcd(a(x), p(x)) = 1$, then*

$$b(x) \equiv c(x) \text{ (mod } p(x)) .$$

Proof. The proof is similar to that of Theorem 1.3.3 and will be omitted. □

Proposition 4.3.4 allows us to define addition and multiplication in $F[x]/\langle p(x)\rangle$, for any nonzero polynomial $p(x) \in F[x]$. We make the following definitions, analogous to those in Proposition 1.4.2:

$$[a(x)] + [b(x)] = [a(x) + b(x)] \quad \text{and} \quad [a(x)] \cdot [b(x)] = [a(x)b(x)] .$$

Example 4.3.1 ($\mathbf{R}[x]/\langle x^2 + 1\rangle$).

Let $F = \mathbf{R}$, the field of real numbers, and let $p(x) = x^2 + 1$. Then every congruence class in $\mathbf{R}[x]/\langle x^2 + 1\rangle$ can be represented by a linear polynomial of the form $a + bx$, by Proposition 4.3.3. If we multiply the congruence classes represented by $a + bx$ and $c + dx$, we have

$$ac + (bc + ad)x + bdx^2 .$$

Dividing by $x^2 + 1$ gives the remainder

$$(ac - bd) + (bc + ad)x ,$$

which is a representative of the product of the two congruence classes. An easier way to make this computation is to note that

$$x^2 + 1 \equiv 0 \pmod{x^2 + 1} ,$$

and so

$$x^2 \equiv -1 \pmod{x^2 + 1} ,$$

which means that we can replace x^2 by -1 in the product

$$ac + (bc + ad)x + bdx^2 .$$

This multiplication is the same as the multiplication of complex numbers, and gives another way to define $\mathbf{C}$. Note that the congruence class $[x]$ has the property that its square is the congruence class $[-1]$, and so if we identify the set of real numbers with the set of congruence classes of the form $[a]$, where $a \in \mathbf{R}$, then the class $[x]$ would be identified with i. We can formalize this identification after we define the concept of an isomorphism of fields. □

4.3.5 Proposition. *Let F be a field, and let $p(x)$ be a nonzero polynomial in $F[x]$. For any $a(x) \in F[x]$, the congruence class $[a(x)]$ has a multiplicative inverse in $F[x]/\langle p(x)\rangle$ if and only if $\gcd(a(x), p(x)) = 1$.*

Proof. To find a multiplicative inverse for $[a(x)]$ we must find a congruence class $[b(x)]$ with $[a(x)][b(x)] = [1]$. Since

$$a(x)b(x) \equiv 1 \pmod{p(x)}$$

if and only if there exists $t(x) \in F[x]$ with

$$a(x)b(x) = 1 + t(x)p(x) \, ,$$

this shows that $[a(x)]$ has a multiplicative inverse if and only if 1 can be written as a linear combination of $a(x)$ and $p(x)$, which occurs if and only if $\gcd(a(x), p(x)) = 1$. The inverse $[b(x)] = [a(x)]^{-1}$ can be found by using the Euclidean algorithm. $\square$

4.3.6 Theorem. *Let F be a field, and let $p(x)$ be a nonconstant polynomial over F. Then $F[x]/\langle p(x)\rangle$ is a field if and only if $p(x)$ is irreducible over F.*

Proof.   Proposition 4.3.4 shows that addition and multiplication of congruence classes are well-defined. The associative, commutative, and distributive laws follow easily from the corresponding laws for addition and multiplication of polynomials. For example,

$$[a(x)][b(x)] = [a(x)b(x)] = [b(x)a(x)] = [b(x)][a(x)]$$

for all $a(x), b(x) \in F[x]$. The additive identity is $[0]$ and the multiplicative identity is $[1]$, while the additive inverse of $[a(x)]$ is $[-a(x)]$. All that remains to show that $F[x]/\langle p(x)\rangle$ is a field is to show that each nonzero congruence class has a multiplicative inverse. Since by Proposition 4.3.3 we can work with representatives of lower degree than $\deg(p(x))$, by Proposition 4.3.5 each nonzero congruence class $[a(x)]$ has a multiplicative inverse if and only if $\gcd(a(x), p(x)) = 1$ for all nonzero polynomials $a(x)$ with $\deg(a(x)) < \deg(p(x))$. This occurs if and only if $p(x)$ is irreducible, completing the proof. $\square$

We note that whenever $p(x)$ is irreducible, the congruence class $[a(x)]$ is invertible if $a(x) \neq 0$ and $\deg(a(x)) < \deg(p(x))$. Conversely, if $[a(x)]$ is invertible, then $[a(x)] = [r(x)]$ where $r(x) \neq 0$ and $\deg(r(x)) < \deg(p(x))$.

Our final remarks about Theorem 4.3.6 tie these ideas back to ideas we have already met in Chapter 3. The discussion following Definition 4.1.4 shows that the set $F[x]$ of polynomials forms an abelian group under addition. Furthermore, since the set $\langle p(x)\rangle$ of all polynomials divisible by $p(x)$ is closed under addition and subtraction, it is a subgroup of $F[x]$. Since every subgroup of an abelian group is normal, it follows that addition of congruence classes modulo $p(x)$ is just the operation defined in the corresponding factor group $F[x]/\langle p(x)\rangle$. Thus Theorem 3.8.5 shows that $F[x]/\langle p(x)\rangle$ is a group under addition, and it is clear that this factor group is abelian. Note that we could have shortened the proof of Theorem 4.3.6 by using these results. We cannot apply results on groups to the multiplication of cosets of $\langle p(x) \rangle$, since the set of nonzero polynomials is not a group under multiplication. In Section 5.3 we will handle multiplication of cosets in the more general setting of commutative rings.

4.3.7 Definition. *Let F_1 and F_2 be fields. A function $\phi : F_1 \to F_2$ is called an* ***isomorphism of fields*** *if it is one-to-one and onto,*

$$\phi(a + b) = \phi(a) + \phi(b) , \quad and \quad \phi(ab) = \phi(a)\phi(b)$$

for all $a, b \in F_1$.

Our first use of the notion of an isomorphism of fields is in comparing two different constructions of the set of complex numbers.

Example 4.3.2 (Construction of the complex numbers).

> We can now give the full story of Example 4.3.1. In Section A.5 of the appendix, we define **C** to be the set of all expressions of the form $a + bi$, where $a, b \in \mathbf{R}$ and $i^2 = -1$. Since $x^2 + 1$ is irreducible over **R**, it follows from Theorem 4.3.6 that $\mathbf{R}[x]/\langle x^2 + 1 \rangle$ is a field. By Proposition 4.3.3 its elements are in one-to-one correspondence with polynomials of the form $a + bx$. Furthermore, the mapping $\phi : \mathbf{R}[x]/\langle x^2 + 1 \rangle \to \mathbf{C}$ defined by $\phi([a + bx]) = a + bi$ can be shown to be an isomorphism of fields (see Exercise 7). Since $x^2 \equiv -1 \pmod{x^2 + 1}$, the congruence class $[x]$ of the polynomial x satisfies the condition $[x]^2 = -1$. Thus the construction of $\mathbf{R}[x]/\langle x^2 + 1 \rangle$ using congruence classes allows us to provide a concrete model for the construction usually given in high school, in which we merely conjure up a symbol i for which $i^2 = -1$. □

When we think of the set of complex numbers as $\mathbf{C} = \{a + bi \mid a, b \in \mathbf{R}\}$, we think of a as the "real" part of $a + bi$, so that we have $\mathbf{R} \subseteq \mathbf{C}$. In a similar way, in $\mathbf{R}[x]/\langle x^2 + 1 \rangle$ we can think of the cosets $[a]$ coming the from constant polynomials as coming from **R**. To make this precise, we can define an isomorphism of fields between **R** and $\{[a] \in \mathbf{R}[x]/\langle x^2 + 1 \rangle \mid a \in \mathbf{R}\}$, by letting $\phi(a) = [a]$, for all $a \in \mathbf{R}$. We often say that we can "identify" these two fields, in order to be able to think of **R** as a subfield of $\mathbf{R}[x]/\langle x^2 + 1 \rangle$.

In the next theorem we will make this sort of identification. If $p(x)$ is an irreducible polynomial in $F[x]$, then in $F[x]/\langle p(x) \rangle$ the subset $\{[a] \mid a \in F\}$ is a subfield isomorphic to F (see Exercise 6).

4.3.8 Theorem (Kronecker). *Let F be a field, and let $f(x)$ be any nonconstant polynomial in $F[x]$. Then there exists an extension field E of F and an element $u \in E$ such that $f(u) = 0$.*

Proof. The polynomial $f(x)$ can be written as a product of irreducible polynomials, and so we let $p(x)$ be one of the irreducible factors of $f(x)$. It is sufficient to find an extension field E containing an element u such that $p(u) = 0$.

By Theorem 4.3.6, $F[x]/\langle p(x)\rangle$ is a field, which we will denote by E. The field F is easily seen to be isomorphic to the subfield of E consisting of all congruence classes of the form $[a]$, where $a \in F$. We make this identification of F with the corresponding subfield of E so that we can consider E to be an extension of F. Let u be the congruence class $[x]$. If $p(x) = a_n x^n + \ldots + a_0$, where $a_i \in F$, then we must compute $p(u)$. We obtain

$$p(u) = a_n([x])^n + \ldots + a_1([x]) + a_0 = [a_n x^n + \ldots + a_1 x + a_0] = [0]$$

since $p(x) \equiv 0 \pmod{p(x)}$. Thus $p(u) = 0$ and the proof is complete. □

4.3.9 Corollary. *Let F be a field, and let $f(x)$ be any nonconstant polynomial in $F[x]$. Then there exists an extension field E over which $f(x)$ can be factored into a product of linear factors.*

Proof. Factor out all linear factors of $f(x)$ and let $f_1(x)$ be the remaining factor. We can find an extension E_1 in which $f_1(x)$ has a root, say u_1. Then we can write $f_1(x) = (x - u_1) f_2(x)$, and by considering $f_2(x)$ as an element of $E_1[x]$, we can continue the same procedure for $f_2(x)$. We will finally arrive at an extension E that contains enough of the roots of $f(x)$ so that over this extension $f(x)$ can be factored into a product of linear factors. □

$(x^2 - 2)(x^2 + 1)$

Example 4.3.3.

Consider the polynomial $x^4 - x^2 - 2$ of Example 4.1.2, with coefficients in $F = \mathbf{Q}$. It factors as $(x^2 - 2)(x^2 + 1)$, and as our first step we let $E_1 = \mathbf{Q}[x]/\langle x^2 - 2\rangle$, which is isomorphic to $\mathbf{Q}(\sqrt{2})$ (see Exercise 12). Although E_1 contains the roots $\pm\sqrt{2}$ of the factor $x^2 - 2$, it does not contain the roots $\pm i$ of the factor $x^2 + 1$, and so we must obtain a further extension $E_2 = E_1[x]/\langle x^2 + 1\rangle$. In Chapter 6 we will see that E_2 is isomorphic to the smallest subfield of $\mathbf{C}$ that contains $\sqrt{2}$ and i, which is denoted by $\mathbf{Q}(\sqrt{2}, i)$. □

Example 4.3.4.

Let $F = \mathbf{Z}_2$ and let $p(x) = x^2 + x + 1$. Then $p(x)$ is irreducible over $\mathbf{Z}_2$ since it has no roots in $\mathbf{Z}_2$, and so $\mathbf{Z}_2[x]/\langle x^2 + x + 1\rangle$ is a field by Theorem 4.3.6. It follows from Proposition 4.4.4 that the congruence classes modulo $x^2 + x + 1$ can be represented by $[0]$, $[1]$, $[x]$ and $[1 + x]$, since these are the only polynomials of degree less than 2 over $\mathbf{Z}_2$. Addition and multiplication are given in Tables 4.3.1 and 4.3.2. To simplify these tables, all brackets have been omitted in listing the congruence classes. □

Table 4.3.1: Addition in $\mathbf{Z}_2[x]/\langle x^2 + x + 1\rangle$

+	0	1	x	$1+x$
0	0	1	x	$1+x$
1	1	0	$1+x$	x
x	x	$1+x$	0	1
$1+x$	$1+x$	x	1	0

Table 4.3.2: Multiplication in $\mathbf{Z}_2[x]/\langle x^2 + x + 1\rangle$

×	0	1	x	$1+x$
0	0	0	0	0
1	0	1	x	$1+x$
x	0	x	$1+x$	1
$1+x$	0	$1+x$	1	x

If $q(x)$ is irreducible over $\mathbf{Z}_p$, then $\mathbf{Z}_p[x]/\langle q(x)\rangle$ has p^n elements if $\deg(q(x)) = n$, since there are exactly $p^n - 1$ polynomials over $\mathbf{Z}_p$ of degree less than n (including 0 gives p^n elements). It is possible to show that there exist polynomials of degree n irreducible over $\mathbf{Z}_p$ for each integer $n > 0$. This guarantees the existence of a finite field having p^n elements, for each prime number p and each positive integer n. Finite fields will be investigated in greater detail in Section 6.5.

EXERCISES: SECTION 4.3

1. Let F be a field. Given $p(x) \in F[x]$, prove that congruence modulo $p(x)$ defines an equivalence relation on $F[x]$.

2. Prove Proposition 4.3.4.

3. Let E be a field, and let F be a subfield of E. Prove that the multiplicative identity of F must be the same as that of E.

4. Let E be a field, and let F be a subset of E that contains a nonzero element. Prove that F is a subfield of E if and only if F is closed under the addition, subtraction, multiplication, and division of E.

5. Let $\phi : F_1 \to F_2$ be an isomorphism of fields. Prove that $\phi(1) = 1$ (that is, prove that ϕ must map the multiplicative identity of F_1 to the multiplicative identity of F_2).

6. Let F be a field, let $p(x)$ be an irreducible polynomial in $F[x]$, and let

$$E = \{[a] \in F[x]/\langle p(x)\rangle \mid a \in F\} .$$

   Show that E is a subfield of $F[x]/\langle p(x)\rangle$. Then show that $\phi : F \to E$ defined by $\phi(a) = [a]$, for all $a \in F$, is an isomorphism of fields.

7.† Verify that the function $\phi : \mathbf{C} \to \mathbf{R}[x]/\langle x^2 + 1\rangle$ defined by $\phi(a + bi) = [a + bx]$ is an isomorphism of fields.

8. Prove that $\mathbf{R}[x]/\langle x^2 + 2\rangle$ is isomorphic to $\mathbf{C}$.

9. Prove that $\mathbf{R}[x]/\langle x^2 + x + 1\rangle$ is isomorphic to $\mathbf{C}$.

10. Is $\mathbf{Q}[x]/\langle x^2 + 2\rangle$ isomorphic to $\mathbf{Q}[x]/\langle x^2 + 1\rangle$?

11. Let F be any field. Prove that the field of $n \times n$ scalar matrices over F (defined in Exercise 12 of Section 4.1) is isomorphic to F.

12. Prove that $\mathbf{Q}[x]/\langle x^2 - 2\rangle$ is isomorphic to $\mathbf{Q}(\sqrt{2}) = \{a + b\sqrt{2} \mid a, b \in \mathbf{Q}\}$, which was shown to be a field in Example 4.1.1.

13. Prove that $\mathbf{Q}[x]/\langle x^2 - 3\rangle$ is isomorphic to $\mathbf{Q}(\sqrt{3}) = \{a + b\sqrt{3} \mid a, b \in \mathbf{Q}\}$, which was shown to be a field in Exercise 11 of Section 4.1.

14. Show that the polynomial $x^2 - 3$ has a root in $\mathbf{Q}(\sqrt{3})$ but not in $\mathbf{Q}(\sqrt{2})$. Explain why this implies that $\mathbf{Q}(\sqrt{3})$ is not isomorphic to $\mathbf{Q}(\sqrt{2})$.

15. Prove that the field of all matrices over $\mathbf{Q}$ of the form $\begin{bmatrix} a & b \\ 2b & a \end{bmatrix}$ (as defined in Exercise 14 of Section 4.1) is isomorphic to $\mathbf{Q}(\sqrt{2})$.

16.† Prove that the field given in Example 4.3.4 is isomorphic to the following field of the four matrices given in Example 4.1.3:

$$\left\{ \begin{bmatrix} 0 & 0 \\ 0 & 0 \end{bmatrix}, \begin{bmatrix} 1 & 0 \\ 0 & 1 \end{bmatrix}, \begin{bmatrix} 1 & 1 \\ 1 & 0 \end{bmatrix}, \begin{bmatrix} 0 & 1 \\ 1 & 1 \end{bmatrix} \right\} .$$

17. Find an irreducible polynomial $p(x)$ of degree 3 over $\mathbf{Z}_2$, and list all elements of $\mathbf{Z}_2[x]/\langle p(x)\rangle$. Give the identities necessary to multiply elements.

18. Give addition and multiplication tables for the field $\mathbf{Z}_3[x]/\langle x^2 + x + 2\rangle$.

19.† Find a polynomial of degree 3 irreducible over $\mathbf{Z}_3$, and use it to construct a field with 27 elements. List the elements of the field; give the identities necessary to multiply elements.

20. As in Exercise 19, construct a field having 125 elements.

21. Find multiplicative inverses of the given elements in the given fields.

    †(a) $[a + bx]$ in $\mathbf{R}[x]/\langle x^2 + 1\rangle$

    (b) $[a + bx]$ in $\mathbf{Q}[x]/\langle x^2 - 2\rangle$

    †(c) $[x^2 - 2x + 1]$ in $\mathbf{Q}[x]/\langle x^3 - 2\rangle$

    (d) $[x^2 - 2x + 1]$ in $\mathbf{Z}_3[x]/\langle x^3 + x^2 + 2x + 1\rangle$

    †(e) $[x]$ in $\mathbf{Z}_5[x]/\langle x^2 + x + 1\rangle$

    (f) $[x + 4]$ in $\mathbf{Z}_5[x]/\langle x^3 + x + 1\rangle$

22. For which values of $a = 1, 2, 3, 4$ is $\mathbf{Z}_5[x]/\langle x^2 + a\rangle$ a field? Show your work.

23.† For which values of $k = 2, 3, 5, 7, 11$ is $\mathbf{Z}_k[x]/\langle x^2 + 1\rangle$ a field? Show your work.

24. Let F be a finite field. Show that $F[x]$ has irreducible polynomials of arbitrarily high degree.

    Hint: Imitate Euclid's proof that there exist infinitely many prime numbers.

4.4 Polynomials over Z, Q, R, and C

In this section we will give several criteria for determining when polynomials with integer coefficients have rational roots or are irreducible over the field of rational numbers. We will use the notation $\mathbf{Z}[x]$ for the set of all polynomials with integer coefficients. We will also investigate polynomials with complex coefficients, in which case we can show that allowing complex numbers as coefficients makes it possible to factor a polynomial completely (as a product of linear factors).

4.4.1 Proposition. *Let* $f(x) = a_n x^n + a_{n-1} x^{n-1} + \ldots + a_1 x + a_0$ *be a polynomial with integer coefficients. If* r/s *is a rational root of* $f(x)$, *with* $(r, s) = 1$, *then* $r \mid a_0$ *and* $s \mid a_n$.

Proof. If $f(r/s) = 0$, then multiplying $f(r/s)$ by s^n gives the equation

$$a_n r^n + a_{n-1} r^{n-1} s + \cdots + a_1 r s^{n-1} + a_0 s^n = 0 .$$

It follows that $r \mid a_0 s^n$ and $s \mid a_n r^n$, so $r \mid a_0$ and $s \mid a_n$ since $(r, s) = 1$. □

Example 4.4.1.

Suppose that we wish to find all integral roots of

$$f(x) = x^3 - 3x^2 + 2x - 6 .$$

Using Proposition 4.4.1, all rational roots of $f(x)$ can be found by testing only a finite number of values. By considering the signs we can see that $f(x)$ cannot have any negative roots, so we only need to check the positive factors of 6. Substituting, we obtain $f(1) = -6$, $f(2) = -6$, and $f(3) = 0$. Thus 3 is a root of $f(x)$, and so we can use the division algorithm to show that

$$x^3 - 3x^2 + 2x - 6 = (x^2 + 2)(x - 3) .$$

It is now clear that 6 is not a root, and we are done.   □

Example 4.4.2.

Let $f(x) \in \mathbf{Z}[x]$. If c is an integral root of $f(x)$, then $f(x) = q(x)(x - c)$ for some polynomial $q(x)$. The proof of the remainder theorem (Theorem 4.1.9) uses the fact that

$$x^n - c^n = (x - c)(x^{n-1} + cx^{n-2} + \ldots + c^{n-2}x + c^{n-1}) .$$

Since c is an integer, a further analysis of the proof shows that $q(x) \in \mathbf{Z}[x]$. For any integer n, we must have $f(n) = q(n)(n - c)$, and since $f(n)$, $q(n) \in \mathbf{Z}$, this shows that $(c - n) \mid f(n)$.

This observation can be combined with Proposition 4.4.1 to find the integer (and thus rational) roots of monic equations such as

$$x^3 + 15x^2 - 3x - 6 = 0 .$$

By Proposition 4.4.1, the possible rational roots are ± 1, ± 2, ± 3, and ± 6. Letting $f(x) = x^3 + 15x^2 - 3x - 6$, we find that $f(1) = 7$, so for any root c, we have $(c - 1)\mid 7$. This eliminates all of the possible values except $c = 2$ and $c = -6$. We find that $f(2) = 56$, so 2 is not a root. This shows, in addition, that $(c - 2)\mid 56$ for any root c, but -6 still passes this test. Finally, $f(-6) = 336$, and so this eliminates -6, and $f(x)$ has no rational roots. We have also shown, by Proposition 4.2.7, that the polynomial

$$f(x) = x^3 + 15x^2 - 3x - 6$$

is irreducible in $\mathbf{Q}[x]$.   □

4.4.2 Definition. *A polynomial with integer coefficients is called **primitive** if 1 and -1 are the only common divisors of its coefficients.*

At this point it is useful to extend the definition of the greatest common divisor to the case of more than two integers. For $a_1, \ldots, a_n \in \mathbf{Z}$, the greatest common divisor

$\gcd(a_1, \ldots, a_n)$ is defined to be a positive integer d such that $d \mid a_i$, for $1 \leq i \leq n$, and if $c \in \mathbf{Z}$ with $c \mid a_i$, for $1 \leq i \leq n$, then $c \mid d$. Note that $\gcd(a_1, a_2, a_3) = \gcd(\gcd(a_1, a_2), a_3)$, and so we could have given an inductive definition. It is easy to check that the integer d in this definition is unique, so we are justified in referring to *the* greatest common divisor of a finite set of integers.

For any polynomial $p(x)$ with integer coefficients, the greatest common divisor of its coefficients will be called the **content** of $p(x)$. We can always factor out the content of $p(x)$, leaving a primitive polynomial.

For example, the polynomial $12x^2 - 18x + 30$ has content 6, and so we can write

$$12x^2 - 18x + 30 = 6(2x^2 - 3x + 5),$$

where $2x^2 - 3x + 5$ is a primitive polynomial.

Our immediate goal is to prove Gauss's lemma, which states that the product of two primitive polynomials is again primitive. In the proof, given two primitive polynomials we will show that no prime number can be a divisor of all of the coefficients of the product. To do this we will use a lemma illustrated by the following example. Recall that in the polynomial $a_m x^m + \ldots + a_1 x + a_0$ the subscript attached to a coefficient is called its index (see Definition 4.1.4).

If $g(x) = x^2 - 2x + 6$ and $h(x) = x^3 - 5x^2 + 3x + 12$, let $f(x) = g(x)h(x) = x^5 - 7x^4 + 19x^3 - 24x^2 - 6x + 72$. In the product $f(x)$, the coefficients 72, -6, and -24 are each divisible by 2. The coefficient of x^3 is 19, and this is the coefficient of least index that is not divisible by 2. In $g(x)$, the coefficient of least index that is not divisible by 2 is the coefficient 1 of x^2. In $h(x)$, the coefficient of least index that is not divisible by 2 is the coefficient 3 of x. The following lemma shows that it is no coincidence that 19 is the coefficient in $g(x)h(x)$ of $x^3 = x^2 \cdot x$.

As another example, for the above polynomials $f(x)$, $g(x)$, $h(x)$ consider the prime 5. The constant terms of $g(x)$ and $h(x)$ are the coefficients of least index not divisible by 5, and so the constant term of $f(x)$ is the coefficient of least index not divisible by 5.

4.4.3 Lemma. *Let p be a prime number, and let $f(x) = g(x)h(x)$, where $f(x) = a_m x^m + \ldots + a_1 x + a_0$, $g(x) = b_n x^n + \ldots + b_1 x + b_0$, and $h(x) = c_k x^k + \ldots + c_1 x + c_0$. If b_s and c_t are the coefficients of $g(x)$ and $h(x)$ of least index not divisible by p, then a_{s+t} is the coefficient of $f(x)$ of least index not divisible by p.*

Proof. For the coefficient a_{s+t} of $f(x)$, we have

$$a_{s+t} = b_0 c_{s+t} + b_1 c_{s+t-1} + \ldots + b_{s-1} c_{t+1} + b_s c_t + b_{s+1} c_{t-1} + \ldots + b_{s+t} c_0.$$

By assumption, each of the coefficients $b_0, b_1, \ldots, b_{s-1}$ and $c_{t-1}, \ldots, c_0$ is divisible by p. Thus, with the exception of $b_s c_t$, each term in the above sum is divisible by p. This implies that a_{s+t} is not divisible by p.

In any coefficient of $f(x)$ of smaller index, each term in the sum $a_k = \sum_{i=0}^{k} b_i c_{k-i}$ is divisible by p, and thus a_{s+t} is the coefficient of least index not divisible by p. $\square$

4.4.4 Theorem (Gauss's Lemma). *The product of two primitive polynomials is itself primitive.*

Proof. Let p be any prime number, and let $f(x) = g(x)h(x)$ be a product of primitive polynomials. Since $g(x)$ and $h(x)$ are primitive, each one has a coefficient not divisible by p, and then by Lemma 4.4.3 it follows that $f(x)$ has at least one coefficient not divisible by p. Since this is true for every prime, we conclude that $f(x)$ is primitive.   □

4.4.5 Theorem. *A polynomial with integer coefficients that can be factored into polynomials with rational coefficients can also be factored into polynomials of the same degree with integer coefficients.*

Proof. Let $f(x) \in \mathbf{Z}[x]$, and assume that $f(x) = g(x)h(x)$ in $\mathbf{Q}[x]$. By factoring out the appropriate least common multiples of denominators and greatest common divisors of numerators, we can assume that $f(x) = (m/n)g^*(x)h^*(x)$, where $(m, n) = 1$ and $g^*(x)$, $h^*(x)$ are primitive, with the same degrees as $g(x)$ and $h(x)$, respectively. If d_i is any coefficient of $g^*(x)h^*(x)$, then $n|md_i$ since $f(x)$ has integer coefficients, so $n|d_i$ since $(n, m) = 1$. By Gauss's lemma, $g^*(x)h^*(x)$ is primitive, so we have $n = 1$, and thus $f(x)$ has a factorization $f(x) = (mg^*(x))(h^*(x))$ into a product of polynomials in $\mathbf{Z}[x]$. The general result, for any number of factors, can be proved by using an induction argument.   □

4.4.6 Theorem (Eisenstein's Irreducibility Criterion). *Let*

$$f(x) = a_n x^n + a_{n-1} x^{n-1} + \ldots + a_0$$

be a polynomial with integer coefficients. If there exists a prime number p such that

$$a_{n-1} \equiv a_{n-2} \equiv \ldots \equiv a_0 \equiv 0 \pmod{p}$$

but $a_n \not\equiv 0 \pmod{p}$ and $a_0 \not\equiv 0 \pmod{p^2}$, then $f(x)$ is irreducible over the field of rational numbers.

Proof. Suppose that $f(x)$ can be factored as $f(x) = g(x)h(x)$, where $g(x) = b_m x^m + \ldots + b_0$ and $h(x) = c_k x^k + \ldots + c_0$. By Theorem 4.4.5 we can assume that both factors have integer coefficients. Furthermore, we can assume that either b_0 or c_0 is not divisible by p, since $b_0 c_0 = a_0$ is not divisible by p^2. Let us assume that $p \nmid b_0$. If c_t is the coefficient of $h(x)$ of least index that is not divisible by p, then by Lemma 4.4.3 it follows that $a_t = a_{0+t}$ is the coefficient of $f(x)$ of least index that is not divisible by p. By assumption a_i is divisible by p for $i < n$, so $t = n$, showing that $h(x)$ and $f(x)$ have the same degree. Thus $f(x)$ is irreducible because it cannot be factored into a product of polynomials of lower degree.   □

In Theorem 4.4.6, the condition $a_{n-1} \equiv \cdots \equiv a_0 \equiv 0 \pmod{p}$ can be summed up as saying that p is a divisor of the greatest common divisor of these coefficients. The theorem cannot always be applied. For example, if $f(x) = x^3 - 5x^2 - 3x + 6$, then $\gcd(5, 3, 6) = 1$ and no prime can be found for which the necessary conditions are satisfied. Yet $f(x)$ is irreducible, since Propositions 4.4.1 can be used to show that $f(x)$ has no rational roots.

To show that $p(x)$ is irreducible, it is sufficient to show that $p(x+c)$ is irreducible for some integer c, since if $p(x) = f(x)g(x)$, then $p(x + c) = f(x + c)g(x + c)$. For example, Eisenstein's criterion cannot be applied to $x^2 + 1$, but substituting $x + 1$ for x gives another proof that $x^2 + 1$ is irreducible over the field of rational numbers.

4.4.7 Corollary. *If p is prime, then the polynomial*

$$\Phi_p(x) = x^{p-1} + x^{p-2} + \ldots + x + 1$$

is irreducible over the field of rational numbers.

Proof. Note that

$$\Phi_p(x) = \frac{x^p - 1}{x - 1}$$

and consider

$$\Phi_p(x + 1) = \frac{(x + 1)^p - 1}{x} = x^{p-1} + \binom{p}{1}x^{p-2} + \binom{p}{2}x^{p-3} + \ldots + p \ .$$

For $1 \le i \le p - 1$, the prime p is a factor of the binomial coefficient $\binom{p}{i}$, and so Eisenstein's criterion can now be applied to $\Phi_p(x + 1)$, proving that $\Phi_p(x)$ is irreducible over **Q**. $\square$

If p is prime, Corollary 4.4.7 shows that

$$x^p - 1 = (x - 1)(x^{p-1} + \ldots + 1)$$

gives the factorization over **Q** of $x^p - 1$ into two irreducible factors. This is not the case when the degree is composite. For example,

$$x^4 - 1 = (x - 1)(x + 1)(x^2 + 1)$$

and

$$x^{15} - 1 = (x - 1)(x^2 + x + 1)(x^4 + x^3 + x^2 + x + 1)(x^8 - x^7 + x^5 - x^4 + x^3 - x + 1) \ .$$

4.4.8 Definition. *The roots in $\mathbf{C}$ of the polynomial $x^n - 1$ are called the* **complex** **nth roots of unity**.

A complex nth root of unity is said to be **primitive** *if it is a root of the polynomial $x^n - 1$, but is not a root of $x^m - 1$ for any positive integer $m < n$.*

For example, the 4th roots of unity are $\pm 1, \pm i$. Of these, i and $-i$ are primitive 4th roots of unity, and they are roots of the factor $x^2 + 1$ of $x^4 - 1$. The reader is referred to Section 8.5 for a full discussion of the factors of $x^n - 1$, and their relationship to primitive roots of unity. Note that if p is a prime number, then every pth root of unity except 1 is primitive.

In order to characterize the irreducible polynomials over the complex numbers and over the real numbers, we need to use the theorem that is usually referred to as the "fundamental theorem of algebra." It was discovered by D'Alembert in 1746, although he gave an incorrect proof. The first acceptable proof was given by Gauss in 1799. We will postpone our proof until Section 8.3, at which point we will have developed enough theory to give an algebraic proof using Galois theory.

Using analytic techniques, a short proof is usually given in a beginning course in complex variables. This proof is an application of a theorem of Liouville which follows from an estimate (due to Cauchy) of the derivative of an analytic function. A standard reference for this material is the book *Functions of One Complex Variable*, by Conway. There is a topological proof based on the notion of a winding number. A nice discussion of this proof appears in the book *Algebraic Topology: A First Course*, by Fulton.

4.4.9 Theorem (Fundamental Theorem of Algebra). *Any polynomial of positive degree with complex coefficients has a complex root.*

Proof. See Theorem 8.3.10. □

4.4.10 Corollary. *Any polynomial $f(z)$ of degree $n > 0$ with complex coefficients can be expressed as a product of linear factors, in the form*

$$f(z) = c(z - z_1)(z - z_2) \cdots (z - z_n) .$$

Proof. We need to use Corollary 4.1.1, which shows that roots of $f(z)$ correspond to linear factors. A detailed proof would use Theorem 4.4.9 and induction on the degree of $f(z)$. □

Corollary 4.4.10 can be used to give formulas relating the roots and coefficients of a polynomial. For example, if $f(z) = z^2 + a_1 z + a_0$ has roots z_1, z_2, then

$$z^2 + a_1 z + a_0 = (z - z_1)(z - z_2) = z^2 + (-z_1 - z_2)z + z_1 z_2 .$$

Thus,

$$z_1 + z_2 = -a_1 \qquad \text{and} \qquad z_1 z_2 = a_0 \ .$$

Similarly, if $f(z) = z^3 + a_2 z^2 + a_1 z + a_0$ has roots z_1, z_2, z_3, then

$$z_1 + z_2 + z_3 = -a_2 \ , \qquad z_1 z_2 + z_1 z_3 + z_2 z_3 = a_1 \ , \qquad z_1 z_2 z_3 = -a_0 \ .$$

This pattern can be extended easily to the general case.

If $z = a + bi$ is a complex number, then its **complex conjugate**, denoted by $\overline{z}$, is $\overline{z} = a - bi$. Note that $z\overline{z} = a^2 + b^2$ and $z + \overline{z} = 2a$ are real numbers, whereas $z - \overline{z} = (2b)i$ is a purely imaginary number. Furthermore, $z = \overline{z}$ if and only if z is a real number (i.e., $b = 0$). Since $a^2 + b^2 \geq 0$, and we define $|a + bi| = \sqrt{a^2 + b^2}$, we have $|z| = \sqrt{z\overline{z}}$. It can be checked that $\overline{z + w} = \overline{z} + \overline{w}$ and $\overline{zw} = \overline{z}\ \overline{w}$.

4.4.11 Proposition. *Let $f(x)$ be a polynomial with real coefficients. Then a complex number z is a root of $f(x)$ if and only if its complex conjugate $\overline{z}$ is a root of $f(x)$.*

Proof. If $f(x) = a_n x^n + \ldots + a_0$, then $a_n z^n + \ldots + a_0 = 0$ for any root z of $f(x)$. Taking the complex conjugate of both sides shows that

$$\overline{a}_n (\overline{z})^n + \ldots + \overline{a}_1 \overline{z} + \overline{a}_0 = a_n (\overline{z})^n + \ldots + a_1 \overline{z} + a_0 = 0$$

and thus $\overline{z}$ is a root of $f(x)$.

Conversely, if $\overline{z}$ is a root of $f(x)$, then so is $z = \overline{\overline{z}}$. $\square$

4.4.12 Theorem. *Any polynomial of positive degree with real coefficients can be factored into a product of linear and quadratic terms with real coefficients.*

Proof. Let $f(x)$ be a polynomial with real coefficients, of degree $n > 0$. By Corollary 4.4.10 we can write $f(x) = c(x - z_1)(x - z_2) \cdots (x - z_n)$, where $c \in \mathbf{R}$. If z_i is not a real root, then by Proposition 4.4.11, $\overline{z}_i$ is also a root, and so $x - \overline{z}_i$ occurs as one of the factors. But then

$$(x - z_i)(x - \overline{z}_i) = x^2 - (z_i + \overline{z}_i)x + z_i \overline{z}_i$$

has real coefficients. Thus if we pair each nonreal root with its conjugate, the remaining roots will be real, and so $f(x)$ can be written as a product of linear and quadratic polynomials each having real coefficients. $\square$

As an immediate consequence of the above result, note that any polynomial of odd degree that has real coefficients must have at least one real root. This follows from the fact that such a polynomial must have a linear factor with real coefficients, and this factor yields a real root.

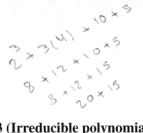

Example 4.4.3 (Irreducible polynomials in R[x]).

In $\mathbf{R}[x]$, a polynomial $ax^2 + bx + c$ with $a \neq 0$ has roots

$$x = \frac{-b \pm \sqrt{b^2 - 4ac}}{2a} \, ,$$

and these are real numbers if and only if $b^2 - 4ac \geq 0$. Since any factors of $ax^2 + bx + c$ must be linear and hence correspond to roots, we can see that the polynomial is reducible over $\mathbf{R}$ if and only if $b^2 - 4ac \geq 0$. For example, $x^2 + 1$ is irreducible over $\mathbf{R}$.

In summary, irreducible polynomials in $\mathbf{R}[x]$ must have one of the forms $ax + b$, with $a \neq 0$ or $ax^2 + bx + c$, with $a \neq 0$ and $b^2 - 4ac < 0$.   □

EXERCISES: SECTION 4.4

1. If $f(x)$ has integer coefficients and m is an integer solution of the polynomial equation $f(x) = 0$, then for any n we can reduce both m and the coefficients of $f(x)$ modulo n, and the equation becomes a congruence that still holds. For the following equations, verify that the given roots modulo 3 and 5 are in fact all such roots. Use this information to eliminate some of the integer roots, and then find all integer roots.

   (a) $x^3 - 7x^2 + 4x - 28 = 0$ (roots are 1 (mod 3) and ± 1, 2 (mod 5)).

   (b) $x^3 - 9x^2 + 10x - 16 = 0$ (roots are 2 (mod 3) and 3 (mod 5)).

2. Find all integer roots of the following equations (use any method).

   (a) $x^3 + 8x^2 + 13x + 6 = 0$

   (b) $x^3 - 5x^2 - 2x + 24 = 0$

   (c) $x^3 - 10x^2 + 27x - 18 = 0$

   (d) $x^4 + 4x^3 + 8x + 32 = 0$

   (e) $x^7 + 2x^5 + 4x^4 - 8x^2 - 32 = 0$

3. Find all integer roots of the following equations (use any method).

   †(a) $x^4 - 2x^3 - 21x^2 + 22x + 40 = 0$

   (b) $y^3 - 9y^2 - 24y + 216 = 0$

   †(c) $x^5 + 47x^4 + 423x^3 + 140x^2 + 1213x - 420 = 0$

   (d) $x^5 - 34x^3 + 29x^2 + 212x - 300 = 0$

   †(e) $x^4 - 23x^3 + 187x^2 - 653x + 936 = 0$

4. Use Eisenstein's criterion to show that each of these polynomials is irreducible over the field of rational numbers. (You may need to make a substitution.)

(a) $x^4 - 12x^2 + 18x - 24$ $\;= 3$

(b) $4x^3 - 15x^2 + 60x + 180 \; = 5$

(c) $2x^{10} - 25x^3 + 10x^2 - 30$

(d) $x^2 + 2x - 5$ (substitute $x - 1$ or $x + 1$)

5. Use Eisenstein's criterion to show that each of these polynomials is irreducible over the field of rational numbers. (You may need to make a substitution.)

†(a) $x^4 + 1$ (substitute $x + 1$)

(b) $x^6 + x^3 + 1$ (substitute $x + 1$)

†(c) $x^3 + 3x^2 + 5x + 5$

(d) $x^3 - 3x^2 + 9x - 10$

6. Show that if the positive integer n is not a square, then for some prime p and some integer k, the polynomial $x^2 - \dfrac{n}{p^{2k}}$ satisfies Eisenstein's criterion. Conclude that $\sqrt{n}$ is not a rational number.

7. Let $f(x) = x^2 + 100x + n$.

(a) Give an infinite set of integers n such that $f(x)$ is reducible over **Q**.

(b) Give an infinite set of integers n such that $f(x)$ is irreducible over **Q**.

8. Find the irreducible factors of $x^4 - 5x^2 + 6$ over **Q**, over **Q**($\sqrt{2}$), and over **R**.

9. Let $f(x) = a_n x^n + a_{n-1}x^{n-1} + \ldots + a_1 x + a_0$ be a polynomial with rational coefficients. Show that if $c \neq 0$ and c is a root of $f(x)$, then $1/c$ is a root of $g(x) = a_0 x^n + a_1 x^{n-1} + \ldots + a_{n-1}x + a_n$.

10. Let $f(x) = a_n x^n + a_{n-1}x^{n-1} + \ldots + a_1 x + a_0$ be a polynomial with rational coefficients. Show that if c is a root of $f(x)$, and k is a nonzero constant, then kc is a root of $g(x) = b_n x^n + b_{n-1}x^{n-1} + \ldots + b_1 x + b_0$, where $b_i = k^{n-i}a_i$.

11. Verify each of the following, for complex numbers z and w.

(a) $\overline{zw} = \overline{z}\,\overline{w}$

(b) $|zw| = |z||w|$

12. Let a and b be integers, each of which can be written as the sum of two squares. Show that ab has the same property.

Hint: Use part (b) of Exercise 11.

13. For $f(x) = a_n x^n + \ldots + a_0 \in$ **C**$[x]$, let $\overline{f}(x) = \overline{a_n}x^n + \ldots + \overline{a_0}$ be the polynomial obtained by taking the complex conjugate of each coefficient. Show that the product $f(x)\overline{f}(x)$ has coefficients in **R**.

14. Let m and n be positive integers. Prove that $x^m - 1$ is a factor of $x^n - 1$ in **Q**$[x]$ if and only if $m|n$.

15.†Find the irreducible factors of $x^8 - 1$ over $\mathbf{Q}$.

16. Find the irreducible factors of $x^9 - 1$ over $\mathbf{Q}$.

17. One way to substitute $x + c$ for x in a polynomial $f(x)$ is to use Taylor's formula. Recall that for a polynomial $f(x)$ of degree n,

$$f(x) = \frac{f^{(n)}(c)}{n!}(x - c)^n + \frac{f^{(n-1)}(c)}{(n-1)!}(x - c)^{n-1} + \ldots + f'(c)(x - c) + f(c),$$

where $f^{(k)}(x)$ denotes the kth derivative of $f(x)$. Thus

$$f(x + c) = \frac{f^{(n)}(c)}{n!}x^n + \ldots + f'(c)x + f(c).$$

This can sometimes be an efficient method to check Eisenstein's criterion for the coefficients that occur in several different substitutions.

(a) For the polynomial $x^6 + x^3 + 1$ in Exercise 5 (b), use Taylor's formula to determine which of the substitutions $x + 1$, $x - 1$, $x + 2$, $x - 2$ lead to coefficients that satisfy Eisenstein's criterion for some prime p.

(b) Repeat part (a) for the polynomial $x^3 - 3x^2 + 9x - 10$ in Exercise 5 (d).

18. Verify that $x^{10} - 1 = (x - 1)(x + 1)(x^4 + x^3 + x^2 + x + 1)(x^4 - x^3 + x^2 - x + 1)$. Use the method of Exercise 17 to show that $x^4 - x^3 + x^2 - x + 1$ is irreducible over $\mathbf{Q}$.

19. Verify that $x^{12} - 1 = (x - 1)(x + 1)(x^2 + 1)(x^2 + x + 1)(x^2 - x + 1)(x^4 - x^2 + 1)$. Use the method of Exercise 17 to show that for the polynomial $x^4 - x^2 + 1$, no substitution of the form $x + n$ will lead to a polynomial that satisfies Eisenstein's criterion. Find another way to show that $x^4 - x^2 + 1$ is irreducible over $\mathbf{Q}$.

20. Let $f(x)/g(x)$ be a rational function, where $f(x)$, $g(x) \in \mathbf{R}[x]$.

(a) Show that if $g(x) = h(x)k(x)$, with $\gcd(h(x), k(x)) = 1$, then there exist polynomials $s(x), t(x)$ such that $\dfrac{f(x)}{g(x)} = \dfrac{s(x)}{h(x)} + \dfrac{t(x)}{k(x)}$.

Hint: If $p(x)$ and $q(x)$ are relatively prime, then there exist polynomials $a(x)$ and $b(x)$ with $a(x)p(x) + b(x)q(x) = 1$. Therefore $\dfrac{1}{p(x)q(x)} = \dfrac{a(x)}{q(x)} + \dfrac{b(x)}{p(x)}$.

(b) Show that if $h(x) = p(x)^m$, where $p(x)$ is irreducible, then there exist polynomials $q(x), r_0(x), r_1(x), \ldots, r_{m-1}(x)$ such that for each i either $r_i(x) = 0$ or $\deg(r_i(x)) < \deg(p(x))$ and $\dfrac{s(x)}{h(x)} = q(x) + \dfrac{r_{m-1}(x)}{p(x)} + \dfrac{r_{m-2}(x)}{p(x)^2} + \ldots + \dfrac{r_0(x)}{p(x)^m}$.

(c) Show that $f(x)/g(x)$ can be expressed as a polynomial plus a sum of **partial fractions** of the form $\dfrac{c}{(x + a)^m}$ or $\dfrac{cx + d}{(x^2 + ax + b)^m}$.

Note: This result is used in calculus to prove that the indefinite integral of any rational function can be expressed in terms of algebraic, trigonometric, or exponential functions and their inverses.

Notes

In Section A.6 of the appendix we give solutions by radicals to the general cubic and quartic equations. We now give a brief historical note on the developments leading up to the formulation of these solutions.

Leonardo of Pisa (1170–1250) published *Liber abbaci* in 1202. (He was called Fibonacci, as a member of the Bonacci family.) His book contained, among other topics, an introduction to Hindu-Arabic numerals, a variety of problems of interest in trade, a chapter on calculations with square roots and cube roots, and a chapter containing a systematic treatment of linear and quadratic equations. Leonardo's book built on work written in Arabic and set the stage for further development.

Mathematicians in the Arab world, including the Persian Omar Khayyam (about 1100), had studied cubic equations. Their solutions were found as intersections of conic sections. The Italians were also interested in solving cubic equations. For example, in a book published in the middle of the fourteenth century there are problems involving interest rates which lead to cubic equations. At that date, solutions to certain very special cases of cubic equations were already known.

The general cubic equation

$$ax^3 + bx^2 + cx + d = 0$$

can be reduced to the form

$$x^3 + px + q = 0$$

by dividing through by a and then introducing a new variable $y = x + b/3a$. If only positive coefficients and positive values of x are allowed, then there are three cases:

$$
\begin{aligned}
x^3 + px &= q \\
x^3 &= px + q \\
x^3 + q &= px .
\end{aligned}
$$

The first case was solved by Scipione del Ferro, a professor at the University of Bologna, who died in 1526. His solution was never published, although he told it to a few friends.

In 1535 there was a mathematical contest between one of del Ferro's students and a Venetian mathematics teacher named Niccolò Fontana (1499–1557). (He is usually known as Tartaglia, which means the "stammerer.") Tartaglia found the general solution to the equations (of the first type above) posed by del Ferro's student and won the contest. Then four years later Tartaglia was approached by Cardano, who lived in Milan, with a request to share his method of solution. Tartaglia finally did so, but only after obtaining a sworn oath from Cardano that he would never publish the solution.

Shortly afterward, Cardano succeeded in extending the method of solution of cubics of the first type to the remaining two types. In doing so, he was probably the first person to use complex numbers in the form $a + \sqrt{-b}$. Cardano's younger friend and secretary, Lodovico Ferrari (1522–1565), discovered that the general fourth degree equation can be reduced to a cubic equation, and hence could be solved. This put Cardano and Ferrari in a difficult position, being in possession of important new results which they could not publish because of the oath Cardano had sworn. In 1543 they were able to examine del Ferro's papers in Bologna, and found out that he had indeed discovered a solution to the one case of the cubic.

Cardano decided to publish the solution of the cubic and fourth-degree equation in a book entitled *Ars Magna* (1545), in which he stated that del Ferro deserved credit for the solution of cubic equations of the first type, although it was from Tartaglia, who had rediscovered the solution, that he had learned it. He stated that he had extended the solution to the two remaining cases, and that Ferrari had given him the solution to the fourth-degree equation. This marked the beginning of a dispute between Cardano and Tartaglia, and a year later Tartaglia published the story of the oath, including its full text.

In Section A.6 of the appendix we also include the solution of the general fourth-degree equation that was given by René Descartes (1596–1650). Much of our modern notation is due to Descartes, who used the last letters of the alphabet to denote unknown quantities, and the first letters to denote known quantities. He used the term "imaginary numbers" for expressions of the form $a \pm \sqrt{-b}$, and was the first to systematically write powers with exponents in the modern form. Analytic geometry, whose primary aim is to solve geometric problems using algebraic methods, was invented independently (and very nearly simultaneously) by Descartes and Fermat.

Chapter 5

COMMUTATIVE RINGS

Many of the algebraic properties of the set of integers are also valid for the set of all polynomials with coefficients in any field. By working with these properties in an abstract setting, it is sometimes possible to prove one theorem that can be applied to both situations, rather than proving two separate theorems, one for each case. As an additional bonus, the abstract setting for the theorem can then be applied to new situations.

For example, we have shown that any integer greater than 1 can be written as a product of prime numbers. It is also true that any nonconstant polynomial with real coefficients can be written as a product of irreducible polynomials. Since the same basic principles are used in each proof, it should be possible to give one proof that would cover both cases. A mathematician should try to recognize such similarities and take advantage of them, and should also ask whether the techniques can be applied to other questions.

The interplay between specific examples and abstract theories is critical to the development of useful mathematics. As more and more is learned about various specific examples, it becomes necessary to synthesize the knowledge, so that it is easier to grasp the essential character of each example and to relate the examples to each other. By generalizing and abstracting from concrete examples, it may be possible to present a unified theory that can be more easily understood than seemingly unrelated pieces of information from a variety of situations.

In a general theory of abstract algebraic objects of a particular type, the main problem is that of classifying and describing the objects. This often includes determining the simplest sort of building blocks and describing all of the ways in which they can be put together. In this chapter we give a very simple example of a theorem of this sort, when we show that, as a ring, $\mathbf{Z}_n$ is isomorphic to a direct sum of similar rings of prime power order. These rings have a particularly simple structure, and cannot themselves be expressed as a direct sum of smaller rings.

The concept of an abstract commutative ring will be introduced in Section 5.1 to provide a common framework for studying a variety of questions. The set of integers modulo n will be shown to form a commutative ring. This is an example of an important procedure by which new rings can be constructed—the use of a congruence relation on a given ring. The notion of a factor group will be extended to that of a factor ring. In the construction of factor rings, the role of normal subgroups in forming factor groups will be played by subsets called "ideals." Examples are $n\mathbf{Z}$ in $\mathbf{Z}$ and $\langle f(x) \rangle$ in $F[x]$. Then the notions of prime number and irreducible polynomial motivate the definition of a "prime ideal," which allows us to tie together a number of facts about integers and polynomials. In Section 5.4 we construct quotient fields for integral domains, and thus characterize all subrings of fields.

5.1   Commutative Rings; Integral Domains

In Chapter 1, we began our study of abstract algebra by concentrating on one of the most familiar algebraic structures, the set of integers. In both $\mathbf{Z}$ and $\mathbf{Z}_n$ we have two basic operations—addition and multiplication. Subtraction and division (when possible) are defined in terms of these two operations. After studying groups in Chapter 3, where we have only one operation to deal with, we returned to systems with two operations when we worked with fields and polynomials in Chapter 4.

We will now undertake a systematic study of systems in which there are two operations that generalize the familiar operations of addition and multiplication. The examples you should have in mind are these: the set of integers $\mathbf{Z}$; the set $\mathbf{Z}_n$ of integers modulo n; any field F (in particular the set $\mathbf{Q}$ of rational numbers and the set $\mathbf{R}$ of real numbers); the set $F[x]$ of all polynomials with coefficients in a field F. The axioms we will use are the same as those for a field, with one crucial exception. We have dropped the requirement that each nonzero element has a multiplicative inverse (see Definition 4.1.1), in order to include integers and polynomials in the class of objects we want to study. Because we are now considering *two* operations at the same time, rather than studying one operation at a time, we need to have a connection between the operations; the distributive laws accomplish this.

5.1.1 Definition. *Let R be a set on which two binary operations are defined, called addition and multiplication, and denoted by $+$ and $\cdot$. Then we say that the **distributive laws** hold for addition and multiplication if*

$$a \cdot (b + c) = a \cdot b + a \cdot c \quad and \quad (a + b) \cdot c = a \cdot c + b \cdot c$$

for all $a, b, c \in R$.

The distributive laws should be familiar, since we have already used them in the set of integers (see Appendix A.1) and in the definition of a field (see Definition 4.1.1). You should have met them in other examples too, since they hold for the usual addition and multiplication of polynomials, as in Chapter 4, and also for addition and multiplication of matrices.

5.1.2 Definition. *Let R be a set on which two binary operations are defined, called* ***addition*** *and* ***multiplication*** *and denoted by* $+$ *and* $\cdot$ *respectively. Then R is called a* ***commutative ring*** *with respect to these operations if the following properties hold:*

> **(i)** *R is an abelian group under addition;*
>
> **(ii)** *multiplication is associative and commutative;*
>
> **(iii)** *R has a multiplicative identity element;*
>
> **(iv)** *the distributive laws hold.*

Since any commutative ring R determines an abelian group by just considering the set R together with the single operation of addition, we call this group the **underlying additive group** of R. Although we require that multiplication in R is commutative, the set of nonzero elements certainly need not define an abelian group under multiplication.

As you are learning the definition of a commutative ring, it may help to refer to the expanded version of the definition, given below, in which all of the conditions are written out explicitly. If you need to determine whether or not a set is a commutative ring under two given operations, this expanded version gives you a "check list" of conditions that you need to go through.

5.1.2′ (Expanded version of Definition 5.1.2) *Let R be a set on which two binary operations are defined, denoted by* $+$ *and* $\cdot$ *respectively. That is, the following condition must be satisfied:*

> **(i)** ***Closure:*** *If* $a, b \in R$, *then the sum* $a + b$ *and the product* $a \cdot b$ *are well-defined elements of R.*

Then R is called a ***commutative ring*** *with respect to these operations if the following properties hold.*

> **(ii)** ***Associative laws:*** *For all* $a, b, c \in R$,
>
> $$a + (b + c) = (a + b) + c \quad \text{and} \quad a \cdot (b \cdot c) = (a \cdot b) \cdot c .$$
>
> **(iii)** ***Commutative laws:*** *For all* $a, b \in R$,
>
> $$a + b = b + a \quad \text{and} \quad a \cdot b = b \cdot a .$$
>
> **(iv)** ***Distributive laws:*** *For all* $a, b, c \in R$,
>
> $$a \cdot (b + c) = a \cdot b + a \cdot c \quad \text{and} \quad (a + b) \cdot c = a \cdot c + b \cdot c .$$

(v) Identity elements: The set R contains an element 0, called an **additive identity element**, such that for all $a \in R$,

$$a + 0 = a \quad \text{and} \quad 0 + a = a .$$

The set R contains an element 1, called a **multiplicative identity element**, such that for all $a \in R$,

$$a \cdot 1 = a \quad \text{and} \quad 1 \cdot a = a .$$

(vi) Additive inverses: For each $a \in R$, the equations

$$a + x = 0 \quad \text{and} \quad x + a = 0$$

have a solution x in R, called the **additive inverse** of a, and denoted by $-a$.

We usually refer to the element 1 simply as the *identity* of the ring R. To avoid any possible confusion with the additive identity 0, we will refer to 0 as the **zero element** of R. Since we do not require that $1 \neq 0$, we could have $R = \{0\}$, with $0 + 0 = 0$ and $0 \cdot 0 = 0$. We will refer to this ring as the **zero ring**.

A set with two binary operations that satisfy conditions (i)–(vi) of 5.1.2′, with the exception of the commutative law for multiplication, is called a **ring**. Although we will not discuss them here, there are many interesting examples of noncommutative rings. From your work in linear algebra, you should already be familiar with one such example, the set of all 2×2 matrices over **R**. The standard rules for matrix arithmetic provide all of the axioms for a commutative ring, with the exception of the commutative law for multiplication. Although this is certainly an important example worthy of study, we have chosen to work only with commutative rings, with emphasis on integral domains, fields, and polynomial rings over them.

We should note that in a commutative ring, either one of the distributive laws implies the other. If you are checking the axioms for a ring, if you first prove that multiplication is commutative, then you only need to check one of the distributive laws. The definition requires you to check that there is an identity element. We should point out that many textbooks do not include the existence of an identity element in the definition of a commutative ring.

Before giving some further examples of commutative rings, it is helpful to have some additional information about them. Our observation that any commutative ring is an abelian group under addition implies that the cancellation law holds for addition. This proves part (a) of the next statement. Just as in the case of a field, various uniqueness statements follow from Proposition 3.1.2.

Let R be a commutative ring, with elements $a, b, c \in R$.
(a) If $a + c = b + c$, then $a = b$.
(b) If $a + b = 0$, then $b = -a$.

(c) If $a + b = a$ for some $a \in R$, then $b = 0$.

In Proposition 4.1.3 the following properties were shown to hold for any field. The proof remains valid for any commutative ring. Note that (d) and (f) involve connections between addition and multiplication. Their proofs make use of the distributive law, since at the beginning it provides the only link between the two operations.

Let R be a commutative ring, with elements $a, b \in R$.

(d) For all $a \in R$, $a \cdot 0 = 0$.

(e) For all $a \in R$, $-(-a) = a$.

(f) For all $a, b \in R$, $(-a) \cdot (-b) = a \cdot b$.

We will follow the usual convention of performing multiplications before additions unless parentheses intervene.

Example 5.1.1 (Z_n as a ring).

In Section 1.4 we listed the properties of addition and multiplication of congruence classes, which show that the set Z_n of integers modulo n is a commutative ring. From our study of groups we know that Z_n is a factor group of Z (under addition), and so it is an abelian group under addition. To verify that the necessary properties hold for multiplication, it is necessary to use the corresponding properties for Z. We checked the distributive law in Section 1.4. It is worth commenting on the proof of the associative law, to point out the crucial parts of the proof. To check that the associative law holds for all $[a], [b], [c] \in Z_n$, we have

$$[a]([b][c]) = [a][bc] = [a(bc)] \quad \text{and} \quad ([a][b])[c] = [ab][c] = [(ab)c] ,$$

and so these two expressions are equal because the associative law holds for multiplication in Z. In effect, as soon as we have established that multiplication is well-defined (Proposition 1.4.2), it is easy to show that the necessary properties are inherited by the set of congruence classes. Note that [1] is the identity of Z_n.

The rings Z_n form a class of commutative rings that is a good source of counterexamples. For instance, it provides an easy example showing that the cancellation law may fail for multiplication. In the commutative ring Z_6 we have $[2][3] = [4][3]$, but $[2] \neq [4]$. $\square$

Example 5.1.2 (Polynomial rings).

Let R be a commutative ring. We let T denote the set of infinite tuples

$$(a_0, a_1, a_2, \ldots)$$

such that $a_i \in R$ for all nonnegative integers i, and $a_i \neq 0$ for only finitely many terms a_i. We say that two infinite tuples are equal if and only if the corresponding entries are equal. We introduce addition and multiplication in T as follows:

$$(a_0, a_1, a_2, \ldots) + (b_0, b_1, b_2, \ldots) = (a_0 + b_0, a_1 + b_1, a_2 + b_2, \ldots)$$

$$(a_0, a_1, a_2, \ldots) \cdot (b_0, b_1, b_2, \ldots) = (c_0, c_1, c_2, \ldots), \text{ for } c_k = \sum_{i+j=k} a_i b_j .$$

Then $(1, 0, 0, \ldots)$ is the identity of T, where 1 is the identity of R, and it can be shown that T is a commutative ring under the above operations.

We will follow the usual conventions of writing a for the element $(a, 0, 0, \ldots)$, when $a \in R$, and x for the element $(0, 1, 0, \ldots)$. Then

$$
\begin{aligned}
x^2 &= (0, 1, 0, \ldots) \cdot (0, 1, 0, \ldots) \\
&= (0 \cdot 0, \ 0 \cdot 1 + 1 \cdot 0, \ 0 \cdot 0 + 1 \cdot 1 + 0 \cdot 0, \\
&\qquad 0 \cdot 0 + 1 \cdot 0 + 0 \cdot 1 + 0 \cdot 0, \ldots) \\
&= (0, 0, 1, 0, \ldots) .
\end{aligned}
$$

Similarly, $x^3 = (0, 0, 0, 1, 0, \ldots)$, and so on. We can then write

$$
\begin{aligned}
(a_0, a_1, \ldots, a_m, 0, 0, \ldots) &= a_0(1, 0, 0, \ldots) + a_1(0, 1, 0, \ldots) \\
&\quad + a_2(0, 0, 1, \ldots) + \ldots \\
&= a_0 + a_1 x + \ldots + a_{m-1} x^{m-1} + a_m x^m ,
\end{aligned}
$$

allowing us to use our previous notation $R[x]$ for the **ring of polynomials over R in the indeterminate** x. We say that R is the **coefficient ring**. As in Definition 4.1.4, if n is the largest integer such that $a_n \neq 0$, then we say that the polynomial has **degree** n, and a_n is called the **leading coefficient** of the polynomial. An element of the form $a = (a, 0, 0, \ldots)$ is called a **constant polynomial**. We can, of course, use any symbol to represent the tuple $(0, 1, 0, \ldots)$.

Once we know that $R[x]$ is a commutative ring, it is easy to work with polynomials in two indeterminates x and y. We can simply use $R[x]$ as the coefficient ring, and consider all polynomials over $R[x]$ in the indeterminate y. For example, by factoring out the appropriate terms we have

$$2x - 4xy + y^2 + xy^2 + x^2 y^2 - 3xy^3 + x^3 y^2 + 2x^2 y^3 =$$

$$2x + (-4x)y + (1 + x + x^2 + x^3)y^2 + (-3x + 2x^2)y^3 .$$

The ring of polynomials in two indeterminates with coefficients in R is usually denoted by $R[x, y]$, rather than by $(R[x])[y]$. $\square$

The next proposition will make it easier for us to give examples, by giving a simple criterion for testing subsets of known commutative rings to determine whether they are also commutative rings. It seems easiest to just use *ab* to denote the product $a \cdot b$, as we have already been doing. But you must remember that this can represent any operation that merely behaves in certain ways like ordinary multiplication.

5.1.3 Definition. *Let S be a commutative ring. A subset R of S is called a **subring** of S if it is a commutative ring under the addition and multiplication of S, and has the same identity element as S.*

Looking at the familiar sets $\mathbf{Z} \subseteq \mathbf{Q} \subseteq \mathbf{R} \subseteq \mathbf{C}$, it is easy to check that each one is a subring of the next larger set. If F is any field, then in the polynomial ring $F[x]$ we can identify the elements of F with the constant polynomials. This allows us to think of F as a subring of $F[x]$.

Let F and E be fields. If F is a subring of E, according to the above definition, then we usually say (more precisely) that F is a **subfield** of E (as in Definition 4.4.1). Of course, there may be other subrings of fields that are not necessarily subfields. Any subring is a subgroup of the underlying additive group of the larger ring, so the two commutative rings must have the same zero element.

If S is not the zero ring, then it contains the zero ring $R = \{0\}$ as a proper subset. We note that the zero ring is not a subring of S, since it does not contain the identity element 1 of S. This is in distinct contrast to the situation for groups.

5.1.4 Proposition. *Let S be a commutative ring, and let R be a subset of S. Then R is a subring of S if and only if*

   (i) *R is closed under addition and multiplication;*

   (ii) *if $a \in R$, then $-a \in R$;*

   (iii) *R contains the identity of S.*

Proof. If R is a subring, then the closure axioms must certainly hold. Suppose that z is the zero element of R. Then $z + z = z = z + 0$, where 0 is the zero element of S, so $z = 0$, since the cancellation law for addition holds in S. If $a \in R$ and b is the additive inverse of a in R, then $a + b = 0$, so $b = -a$ by Proposition 3.1.2, and this shows that $-a \in R$. Finally, the definition of a subring demands that the subring must have the same multiplicative identity, and therefore the identity of S must belong to R.

Conversely, suppose that the given conditions hold. The first condition shows that condition (i) of 5.1.2′ is satisfied. Conditions (ii)–(iv) of 5.1.2′ are inherited from S. The element 1 serves as an identity for R, and then $-1 \in R$ by assumption, so $0 = 1 + (-1) \in R$ since R is closed under addition. Thus conditions (v) and (vi) of 5.1.2′ are also satisfied. $\square$

Example 5.1.3 (Subrings of $\mathbf{Z}_n$).

If R is any subring of $\mathbf{Z}_n$, then according to our definition [1] must belong to R. Since R is a subgroup of the underlying additive group of $\mathbf{Z}_n$, and [1] is a generator of this group, it follows that $R = \mathbf{Z}_n$. Thus the only subring of $\mathbf{Z}_n$ is $\mathbf{Z}_n$ itself. $\square$

Example 5.1.4.

Let S be the commutative ring $\mathbf{Z}_6$ and let R be the subset $\{[0], [2], [4]\}$. Then R is closed under addition and multiplication and contains the additive inverse of each element in R. Since $[4][0] = [0]$, $[4][2] = [2]$, and $[4][4] = [4]$, the subset R also has an identity element, namely $[4]$. This shows that R can be considered to be a commutative ring under the operations on S, but we do not consider it to be a subring, since its identity element is not the same as the one in S. $\square$

Example 5.1.5 (Gaussian integers).

Let $\mathbf{Z}[i]$ be the set of complex numbers of the form $m + ni$, where $m, n \in \mathbf{Z}$. Since

$$(m + ni) + (r + si) = (m + r) + (n + s)i$$

$$\text{and} \quad (m + ni)(r + si) = (mr - ns) + (nr + ms)i \, ,$$

for all $m, n, r, s \in \mathbf{Z}$, the usual sum and product of numbers in $\mathbf{Z}[i]$ have the correct form to belong to $\mathbf{Z}[i]$. This shows that $\mathbf{Z}[i]$ is closed under addition and multiplication of complex numbers. The negative of any element in $\mathbf{Z}[i]$ again has the correct form, as does $1 = 1 + 0i$, so $\mathbf{Z}[i]$ is a commutative ring by Proposition 5.1.4. $\square$

Example 5.1.6 ($\mathbf{Z}[\sqrt{2}]$).

In Example 4.1.1 we verified that $\mathbf{Q}(\sqrt{2}) = \{a + b\sqrt{2} \mid a, b \in \mathbf{Q}\}$ is a field. It has an interesting subset

$$\mathbf{Z}[\sqrt{2}] = \{m + n\sqrt{2} \mid m, n \in \mathbf{Z}\}$$

which is obviously closed under addition. The product of two elements is given by

$$(m_1 + n_1\sqrt{2})(m_2 + n_2\sqrt{2}) = (m_1 m_2 + 2n_1 n_2) + (m_1 n_2 + m_2 n_1)\sqrt{2}$$

and so the set is also closed under multiplication. Proposition 5.1.4 can be applied to show that $\mathbf{Z}[\sqrt{2}]$ is a subring of $\mathbf{Q}(\sqrt{2})$, since $1 \in \mathbf{Z}[\sqrt{2}]$.

Since $\mathbf{Q}(\sqrt{2})$ is a field, it contains $1/(m + n\sqrt{2})$ whenever $m + n\sqrt{2} \neq 0$, but $1/(m + n\sqrt{2}) \in \mathbf{Z}[\sqrt{2}]$ if and only if $m/(m^2 - 2n^2)$ and $n/(m^2 - 2n^2)$ are integers. It can be shown that this occurs if and only if $m^2 - 2n^2 = \pm 1$. (See Exercise 4.) $\square$

5.1.5 Definition. *Let R be a commutative ring. An element $a \in R$ is said to be* **invertible** *if there exists an element $b \in R$ such that $ab = 1$.*

In this case, the element a is also called a **unit** *of R, and the element b is called a* **multiplicative inverse** *of a, usually denoted by a^{-1}.*

It follows from Proposition 3.1.2 that if $a \in R$ is invertible, then the multiplicative inverse of a is unique. Since $0 \cdot b = 0$ for all $b \in R$, it is impossible for 0 to be invertible (except in the zero ring). Furthermore, if $a \in R$ and $ab = 0$ for some nonzero $b \in R$, then a cannot be a unit since multiplying both sides of the equation by the inverse of a (if it existed) would show that $b = 0$.

An element a such that $ab = 0$ for some $b \neq 0$ is called a **divisor of zero**.

Example 5.1.7.

Let R be the set of all functions from the set of real numbers into the set of real numbers, with ordinary addition and multiplication of functions (not composition of functions). It is not hard to show that R is a commutative ring, since addition and multiplication are defined pointwise, and the addition and multiplication of real numbers satisfy all of the field axioms. It is easy to find divisors of zero in this ring: let $f(x) = 0$ for $x < 0$ and $f(x) = 1$ for $x \geq 0$, and let $g(x) = 0$ for $x \geq 0$ and $g(x) = 1$ for $x < 0$. Then $f(x)g(x) = 0$ for all x, which shows that $f(x)g(x)$ is the zero function.

The identity element of R is the function $f(x) = 1$ (for all x). Then a function $g(x)$ has a multiplicative inverse if and only if $g(x) \neq 0$ for all x. Thus, for example, $g(x) = 2 + \sin(x)$ has a multiplicative inverse, but $h(x) = \sin(x)$ does not.   □

When thinking of the units of a commutative ring, here are some good examples to keep in mind. The only units of $\mathbf{Z}$ are 1 and -1. We showed in Proposition 1.4.5 that the set of units of $\mathbf{Z}_n$ consists of the congruence classes $[a]$ for which $(a, n) = 1$. We showed in Example 3.1.4 that $\mathbf{Z}_n^{\times}$ is a group under multiplication of congruence classes.

We will use the notation $R^{\times}$ for the set of units of any commutative ring R.

5.1.6 Proposition. *Let R be a commutative ring. Then the set $R^{\times}$ of units of R is an abelian group under the multiplication of R.*

Proof. As usual, let 1 denote the identity of R. If $a, b \in R^{\times}$, then a^{-1} and b^{-1} exist in R, and so $ab \in R^{\times}$ since $(ab)(b^{-1}a^{-1}) = 1$. We certainly have $1 \in R^{\times}$, and $a^{-1} \in R^{\times}$ since $(a^{-1})^{-1} = a$. Finally, the associative and commutative laws hold in $R^{\times}$ since they hold for all elements of R.   □

In the context of commutative rings we can give the following definition. A field is a commutative ring in which $1 \neq 0$ and every nonzero element is invertible. We can say, loosely, that a field is a set on which the operations of addition, subtraction, multiplication, and division can be defined. For example, the real point of Corollary 1.4.6 (c) is that $\mathbf{Z}_n$ is a field if and only if n is a prime number.

We have already observed that the cancellation law for addition follows from the existence of additive inverses. A similar result holds for multiplication. If $ab = ac$ and a is a unit, then multiplying by a^{-1} gives $a^{-1}(ab) = a^{-1}(ac)$, and then by using the associative law for multiplication, the fact that $a^{-1}a = 1$, and the fact that 1 is an identity element, we see that $b = c$.

If the cancellation law for multiplication holds in a commutative ring R, then for any elements $a, b \in R$, $ab = 0$ implies that $a = 0$ or $b = 0$. Conversely, if this condition holds and $ab = ac$, then $a(b - c) = 0$, so if $a \neq 0$ then $b - c = 0$ and $b = c$. Thus the cancellation law for multiplication holds in R if and only if R has no nonzero divisors of zero.

5.1.7 Definition. *A commutative ring R is called an **integral domain** if $1 \neq 0$ and for all $a, b \in R$,*

$$ab = 0 \text{ implies } a = 0 \text{ or } b = 0.$$

The ring of integers $\mathbf{Z}$ is the most fundamental example of an integral domain. The ring of all polynomials with real coefficients is also an integral domain, since the product of any two nonzero polynomials is again nonzero. As shown in Example 5.1.7, the ring of all real valued functions is not an integral domain.

Our definition of a commutative ring allows the identity element to be equal to the zero element. Of course, if $1 = 0$, then every element of the ring is equal to zero. In this respect the definition of an integral domain parallels the definition of a field, as given in Definition 3.5.6.

Example 5.1.8 ($D[x]$ is an integral domain if D is an integral domain).

Let D be any integral domain. The ring $D[x]$ of all polynomials with coefficients in D is also an integral domain. To show this we note that if $f(x)$ and $g(x)$ are nonzero polynomials with leading coefficients a_m and b_n, respectively, then since D is an integral domain, the product $a_m b_n$ is nonzero. This shows that the leading coefficient of the product $f(x)g(x)$ is nonzero, and so $f(x)g(x) \neq 0$. Just as in Proposition 4.1.5, we have $f(x)g(x) \neq 0$ because the degree of $f(x)g(x)$ is equal to $\deg(f(x)) + \deg(g(x))$. Since the constant polynomial 1 is the identity of $D[x]$, we certainly have $1 \neq 0$. □

The next theorem gives a condition that is very useful in studying integral domains. It shows immediately, for example, that $\mathbf{Z}[i]$ and $\mathbf{Z}[\sqrt{2}]$ are integral domains.

A converse to Theorem 5.1.8 will be given in Section 5.4, showing that all integral domains can essentially be viewed as being subrings of fields. Proving this converse involves constructing a field of fractions in much the same way that the field of rational numbers can be constructed from the integers.

5.1.8 Theorem. *Any subring of a field is an integral domain.*

Proof. If R is a subring of the field F, then it inherits the condition $1 \neq 0$. If $a, b \in R$ with $ab = 0$ (in R), then of course the same equation holds in F. Either $a = 0$ or $a \neq 0$, and in the latter case a has a multiplicative inverse a^{-1} in F, even though the inverse may not be in R. Multiplying both sides of the equation $ab = 0$ (in F) by a^{-1} gives $b = 0$, and this equation is the same in R as in F. $\square$

Corollary 1.4.8 (b) shows that $\mathbf{Z}_n$ is an integral domain if and only if n is a prime number. It may be useful to go over the proof again. If we use the condition that $ab \equiv 0 \pmod{n}$ implies that $a \equiv 0 \pmod{n}$ or $b \equiv 0 \pmod{n}$, or equivalently, the condition that $n \mid ab$ implies $n \mid a$ or $n \mid b$, then we can see why n must be prime if and only if $\mathbf{Z}_n$ is an integral domain. Why should the notions of field and integral domain be the same for the rings $\mathbf{Z}_n$? The next theorem gives an answer, at least from one point of view.

5.1.9 Theorem. *Any finite integral domain must be a field.*

Proof. Let D be a finite integral domain, and let D^* be the set of nonzero elements of D. If $d \in D$ and $d \neq 0$, then multiplication by d defines a function from D^* into D^*, since $ad \neq 0$ if $a \neq 0$. Let $f : D^* \to D^*$ be defined by $f(x) = xd$, for all $x \in D^*$. Then f is a one-to-one function, since $f(x) = f(y)$ implies $xd = yd$, and so $x = y$ since the cancellation law holds in an integral domain. But then f must map D^* onto D^*, since by Proposition 2.1.8 any one-to-one function from a finite set into itself must be onto, and so $1 = f(a)$ for some $a \in D^*$. That is, $ad = 1$ for some $a \in D$, and so d is invertible. Since we have shown that each nonzero element of D is invertible, it follows that D is a field. $\square$

EXERCISES: SECTION 5.1

1. Which of the following sets are subrings of the field $\mathbf{Q}$ of rational numbers? Assume that m, n are integers with $n \neq 0$ and $(m, n) = 1$.

†(a) $\{\frac{m}{n} \mid n \text{ is odd}\}$

(b) $\{\frac{m}{n} \mid n \text{ is even}\}$

‡(c) $\{\frac{m}{n} \mid 4 \nmid n\}$

(d) $\{\frac{m}{n} \mid (n, k) = 1\}$ where k is a fixed positive integer

2. Which of the following sets are subrings of the field $\mathbf{R}$ of real numbers?

   (a) $A = \{m + n\sqrt{2} \mid m, n \in \mathbf{Z}$ and n is even$\}$

   (b) $B = \{m + n\sqrt{2} \mid m, n \in \mathbf{Z}$ and m is odd$\}$

   (c) $C = \{a + b\sqrt[3]{2} \mid a, b \in \mathbf{Q}\}$

   (d) $D = \{a + b\sqrt[3]{3} + c\sqrt[3]{9} \mid a, b, c \in \mathbf{Q}\}$

   (e) $E = \{m + nu \mid m, n \in \mathbf{Z}\}$, where $u = (1 + \sqrt{3})/2$

   (f) $F = \{m + nv \mid m, n \in \mathbf{Z}\}$, where $v = (1 + \sqrt{5})/2$

3. Consider the following conditions on the set of all 2×2 matrices $\begin{bmatrix} a & b \\ c & d \end{bmatrix}$ with rational entries. Which conditions below define a commutative ring? If the set is a ring, find all units.

   Hint: From your previous work in linear algebra, you may assume that the set of 2×2 matrices over $\mathbf{Q}$ satisfies all of the properties of Definition 5.1.2 except the commutative law for multiplication. Thus it is sufficient to check the commutative law and the conditions of Proposition 5.1.4.

   †(a) all matrices with $d = a, c = 0$

   (b) all matrices with $d = a, c = b$

   †(c) all matrices with $d = a, c = -2b$

   (d) all matrices with $d = a, c = -b$

   †(e) all matrices with $c = 0$

   (f) all matrices with $a = 0$ and $d = 0$

4. Let $R = \{m + n\sqrt{2} \mid m, n \in \mathbf{Z}\}$.

   (a) Show that $m + n\sqrt{2}$ is a unit in R if and only if $m^2 - 2n^2 = \pm 1$.

   Hint: Show that if $(m + n\sqrt{2})(x + y\sqrt{2}) = 1$, then $(m - n\sqrt{2})(x - y\sqrt{2}) = 1$ and multiply the two equations.

   (b) Show that $1 + 2\sqrt{2}$ has infinite order in $R^\times$.

   (c) Show that 1 and -1 are the only units that have finite order in $R^\times$.

5. Let R be a subset of an integral domain D. Prove that if R is a ring under the operations of D, then R is a subring of D.

6. Let D be a finite integral domain. Give another proof of Theorem 5.1.9 by showing that if d is a nonzero element of D, then $d^{-1} = d^k$, for some positive integer k.

7. An element a of a commutative ring R is called **nilpotent** if $a^n = 0$ for some positive integer n. Prove that if u is a unit in R and a is nilpotent, then $u - a$ is a unit in R.

   Hint: First try the case when $u = 1$.

8. Let R be a commutative ring such that $a^2 = a$ for all $a \in R$. Show that $a + a = 0$ for all $a \in R$.

9. Let I be any set and let R be the collection of all subsets of I. Define addition and multiplication of subsets $A, B \subseteq I$ as follows:

$$A + B = (A \cup B) \cap \overline{A \cap B} \quad \text{and} \quad A \cdot B = A \cap B.$$

Show that R is a commutative ring under this addition and multiplication.

10. For the ring R defined in Exercise 9, write out addition and multiplication tables for the following cases:

 †(a) I has two elements;

 (b) I has three elements.

11. A commutative ring R is called a **Boolean ring** if $a^2 = a$ for all $a \in R$. Show that in a Boolean ring the commutative law follows from the other axioms.

12. Let I be any set and let R be the collection of all subsets of I. Define addition and multiplication of subsets $A, B \subseteq I$ as follows:

$$A + B = A \cup B \quad \text{and} \quad A \cdot B = A \cap B.$$

Is R a commutative ring under this addition and multiplication?

13. Let R be the set of all continuous functions from the set of real numbers into itself.

 (a) Show that R is a commutative ring if the formulas $(f + g)(x) = f(x) + g(x)$ and $(f \cdot g)(x) = f(x)g(x)$ for all $x \in \mathbf{R}$, are used to define addition and multiplication of functions.

 (b) Which properties in the definition of a commutative ring fail if the product of two functions is defined to be $(fg)(x) = f(g(x))$, for all x?

14. Define new operations on $\mathbf{Q}$ by letting $a \oplus b = a + b$ and $a \odot b = 2ab$, for all $a, b \in \mathbf{Q}$. Show that $\mathbf{Q}$ is a commutative ring under these operations.

15. Define new operations on $\mathbf{Z}$ by letting $m \oplus n = m + n - 1$ and $m \odot n = m + n - mn$, for all $m, n \in \mathbf{Z}$. Is $\mathbf{Z}$ a commutative ring under these operations?

16. Let R and S be commutative rings. Prove that the set of all ordered pairs (r, s) such that $r \in R$ and $s \in S$ can be given a ring structure by defining

$$(r_1, s_1) + (r_2, s_2) = (r_1 + r_2, s_1 + s_2) \quad \text{and} \quad (r_1, s_1) \cdot (r_2, s_2) = (r_1 r_2, s_1 s_2).$$

This is called the **direct sum** of R and S, denoted by $R \oplus S$.

17. Give addition and multiplication tables for $\mathbf{Z}_2 \oplus \mathbf{Z}_2$.

18. Generalizing to allow the direct sum of three commutative rings, give addition and multiplication tables for $\mathbf{Z}_2 \oplus \mathbf{Z}_2 \oplus \mathbf{Z}_2$.

19. Find all units of the following rings.

 (a) $\mathbf{Z} \oplus \mathbf{Z}$

 †(b) $\mathbf{Z}_4 \oplus \mathbf{Z}_9$

20. An element e of a ring R is said to be **idempotent** if $e^2 = e$. Find all idempotent elements of the following rings.

    (a) $\mathbf{Z}_8$ and $\mathbf{Z}_9$

    (b) $\mathbf{Z}_{10}$ and $\mathbf{Z}_{12}$

    (c) $\mathbf{Z} \oplus \mathbf{Z}$

    (d) $\mathbf{Z}_{10} \oplus \mathbf{Z}_{12}$

21. Let A be an abelian group, and let $R = \{(a, n) \mid a \in A \text{ and } n \in \mathbf{Z}\}$. Define binary operations $+$ and $\cdot$ on R by $(a, n) + (b, m) = (a + b, n + m)$ and $(a, n) \cdot (b, m) = (am + nb, nm)$, for all (a, n) and (b, m) in R. Show that R is a commutative ring.

22. Let R be a set that satisfies all of the axioms of a commutative ring, with the exception of the existence of a multiplicative identity element. Define binary operations $+$ and $\cdot$ on $R_1 = \{(r, n) \mid r \in R, n \in \mathbf{Z}\}$ by $(r, n) + (s, m) = (r + s, n + m)$ and $(r, n) \cdot (s, m) = (rs + ns + mr, nm)$, for all (r, n) and (s, m) in R_1. Show that R_1 is a commutative ring with identity $(0, 1)$ and that $\{(r, 0) \mid r \in R\}$ satisfies all of the conditions of a subring, with the exception that it does not have the multiplicative identity of R.

5.2 Ring Homomorphisms

In Chapter 3, we found that homomorphisms played an important role in the study of groups. Now in studying commutative rings we have two operations to consider. As with groups, we will be interested in functions which preserve the algebraic properties that we are studying. We begin the section with two examples, each of which involves an isomorphism.

Example 5.2.1.

The definition of the set of complex numbers usually involves the introduction of a symbol i that satisfies $i^2 = -1$, and then we let

$$\mathbf{C} = \{a + bi \mid a, b \in \mathbf{R}\} .$$

In Section 4.3 we gave another definition by using the field $\mathbf{R}[x]/\langle x^2 + 1\rangle$, in which the congruence class $[x]$ plays the role of i. To look for a more concrete description of $\mathbf{C}$, we can try to find such an element i with $i^2 = -1$ in some familiar setting. If we identify real numbers with scalar 2×2 matrices over $\mathbf{R}$, then the matrix

$$\begin{bmatrix} 0 & 1 \\ -1 & 0 \end{bmatrix}$$

has the property that its square is equal to the matrix corresponding to -1. This suggests that we should consider the set T of matrices of the form

$$a \begin{bmatrix} 1 & 0 \\ 0 & 1 \end{bmatrix} + b \begin{bmatrix} 0 & 1 \\ -1 & 0 \end{bmatrix} = \begin{bmatrix} a & b \\ -b & a \end{bmatrix} .$$

You should verify that T is a commutative ring. To show the connection with $\mathbf{C}$, we define $\phi : \mathbf{C} \to T$ by

$$\phi(a + bi) = \begin{bmatrix} a & b \\ -b & a \end{bmatrix} .$$

To add complex numbers we just add the corresponding real and imaginary parts, and since matrix addition is componentwise, it is easy to show that ϕ preserves sums. To show that it preserves products we give the following computations:

$$\phi((a+bi)(c+di)) = \phi((ac-bd)+(ad+bc)i) = \begin{bmatrix} ac - bd & ad + bc \\ -(ad + bc) & ac - bd \end{bmatrix} ,$$

$$\phi(a + bi)\phi(c + di) = \begin{bmatrix} a & b \\ -b & a \end{bmatrix} \begin{bmatrix} c & d \\ -d & c \end{bmatrix} = \begin{bmatrix} ac - bd & ad + bc \\ -(ad + bc) & ac - bd \end{bmatrix} .$$

Thus $\phi((a + bi)(c + di)) = \phi(a + bi)\phi(c + di)$, and since it is clear that ϕ is one-to-one and onto, we could compute sums and products of complex numbers by working with the corresponding matrices. This gives a concrete model of the complex numbers. Note that the function ϕ preserves both the zero element and the identity element, since

$$\phi(0) = \begin{bmatrix} 0 & 0 \\ 0 & 0 \end{bmatrix} \quad \text{and} \quad \phi(1) = \begin{bmatrix} 1 & 0 \\ 0 & 1 \end{bmatrix} .$$

The notion of a ring isomorphism that will be given in Definition 5.2.1 makes this correspondence precise, and justifies the remarks in Appendix A.5 regarding this matrix model of the field of complex numbers. □

Example 5.2.2.

In Section 4.4, in some cases in which Eisenstein's criterion for irreducibility did not apply to a polynomial $f(x)$ directly, we were able to apply the criterion after making a substitution of the form $x + c$. It is useful to examine this procedure more carefully. Let F be a field and let $c \in F$. We can define a function $\phi : F[x] \to F[x]$ by $\phi(f(x)) = f(x + c)$, for each polynomial $f(x) \in F[x]$. To show that ϕ is one-to-one and onto we only need to observe that it has an inverse function ϕ^{-1} defined by $\phi^{-1}(f(x)) = f(x - c)$, for each $f(x) \in F[x]$. It can be checked that ϕ preserves addition and multiplication, in the following sense: adding or multiplying two polynomials first and then substituting $x + c$ is the same as first substituting $x + c$ into each polynomial and then adding or multiplying. In symbols,

$$\phi(f(x) + g(x)) = \phi(f(x)) + \phi(g(x))$$

and

$$\phi(f(x)g(x)) = \phi(f(x))\phi(g(x))$$

for all $f(x)$, $g(x) \in F[x]$. Furthermore, the function ϕ leaves the constant polynomial 1 unchanged.

Using the above correspondence, we see that if $f(x)$ has a nontrivial factorization $f(x) = g(x)h(x)$, then $\phi(f(x))$ has the nontrivial factorization $\phi(f(x)) = \phi(g(x))\phi(h(x))$. A similar condition holds with ϕ^{-1} in place of ϕ, and so we have shown that $f(x)$ is irreducible if and only if $\phi(f(x)) = f(x + c)$ is irreducible. $\square$

5.2.1 Definition. *Let R and S be commutative rings. A function $\phi : R \rightarrow S$ is called a **ring homomorphism** if*

(i) $\phi(a + b) = \phi(a) + \phi(b)$, *for all $a, b \in R$;*
(ii) $\phi(ab) = \phi(a)\phi(b)$, *for all $a, b \in R$; and*
(iii) $\phi(1) = 1$.

*A ring homomorphism that is one-to-one and onto is called a **ring isomorphism**. If there is a ring isomorphism from R onto S, we say that R is **isomorphic** to S, and write $R \cong S$.*

*A ring isomorphism from the commutative ring R onto itself is called an **automorphism** of R.*

The condition that states that a ring homomorphism must preserve addition is equivalent to the statement that a ring homomorphism must be a group homomorphism of the underlying additive group of the ring. This means that we have at our disposal all of the results that we have obtained for group homomorphisms.

A word of warning similar to that given for group homomorphisms is probably in order. If $\phi : R \rightarrow S$ is a ring homomorphism, with $a, b \in R$, then we need to note that in the equation

$$\phi(a + b) = \phi(a) + \phi(b)$$

the sum $a+b$ occurs in R, using the addition of that ring, whereas the sum $\phi(a)+\phi(b)$ occurs in S, using the appropriate operation of S. A similar remark applies to the respective operations of multiplication in the equation

$$\phi(ab) = \phi(a)\phi(b) ,$$

where the product ab uses the operation in R and the product $\phi(a)\phi(b)$ uses the operation in S. We have chosen not to use different symbols for the operations in the two rings, since there is generally not much chance for confusion, and it is customary to make this simplification. Finally, in both rings we have used the symbol 1 to denote the identity element.

Note that Exercise 5 of Section 4.3 shows that our definition of a ring isomorphism is consistent with the definition of an isomorphism of fields given in Definition 4.3.7. Section 4.3 thus contains a number of examples of ring isomorphisms.

We begin with some basic results on ring isomorphisms. It follows from the next proposition that "is isomorphic to" is reflexive, symmetric, and transitive. Recall that a function is one-to-one and onto if and only if it has an inverse (see Proposition 2.1.7). Thus a ring isomorphism always has an inverse, but it is not evident that this inverse must preserve addition and multiplication.

5.2.2 Proposition.

(a) *The inverse of a ring isomorphism is a ring isomorphism.*

(b) *The composite of two ring isomorphisms is a ring isomorphism.*

Proof. (a) Let $\phi : R \to S$ be an isomorphism of commutative rings. We have shown in Proposition 3.4.2 that ϕ^{-1} is an isomorphism of the underlying additive groups. To show that ϕ^{-1} is a ring homomorphism, let $s_1, s_2 \in S$. Since ϕ is onto, there exist $r_1, r_2 \in R$ such that $\phi(r_1) = s_1$ and $\phi(r_2) = s_2$. Then $\phi^{-1}(s_1 s_2)$ must be the unique element $r \in R$ for which $\phi(r) = s_1 s_2$. Since ϕ preserves multiplication,

$$\phi^{-1}(s_1 s_2) = \phi^{-1}(\phi(r_1)\phi(r_2)) = \phi^{-1}(\phi(r_1 r_2)) = r_1 r_2 = \phi^{-1}(s_1)\phi^{-1}(s_2) .$$

Finally, $\phi^{-1}(1) = 1$ since $\phi(1) = 1$.

(b) If $\phi : R \to S$ and $\theta : S \to T$ are isomorphisms of commutative rings, then

$$\theta\phi(ab) = \theta(\phi(ab)) = \theta(\phi(a)\phi(b)) = \theta(\phi(a)) \cdot \theta(\phi(b)) = \theta\phi(a) \cdot \theta\phi(b) .$$

Furthermore, $\theta\phi(1) = \theta(\phi(1)) = \theta(1) = 1$. The remainder of the proof follows immediately from the corresponding result for group homomorphisms. $\square$

To show that commutative rings R and S are isomorphic, we usually construct the isomorphism. To show that they are not isomorphic, it is necessary to show that no isomorphism can possibly be constructed. Sometimes this can be done by just considering the sets involved. For example, if n and m are different positive integers, then $\mathbf{Z}_n$ is not isomorphic to $\mathbf{Z}_m$ since no one-to-one correspondence can be defined between $\mathbf{Z}_n$ and $\mathbf{Z}_m$.

One way to show that two commutative rings are not isomorphic is to find an algebraic property that is preserved by all isomorphisms and that is satisfied by one commutative ring but not the other. We now look at several very elementary examples of such properties.

If $\phi : R \to S$ is an isomorphism, then from our results on group isomorphisms we know that ϕ must map 0 to 0 and must preserve additive inverses. If $a \in R$ is a unit, then there exists an element $b \in R$ with $ab = 1$, so $\phi(a)\phi(b) = \phi(ab) = \phi(1) = 1$. This shows that $\phi(a)$ is a unit in S, and thus ϕ preserves units. In fact, by using the same argument for ϕ^{-1}, we see that a is a unit in R if and only if $\phi(a)$ is a unit in S. This implies that R is a field if and only if S is a field.

As an immediate consequence of the remarks in the previous paragraph, we can see that $\mathbf{Z}$ is not isomorphic to $\mathbf{Q}$, because $\mathbf{Q}$ is a field but $\mathbf{Z}$ is not. Another interesting problem is to show that $\mathbf{R}$ and $\mathbf{C}$ are not isomorphic. Since both are fields, we cannot use the previous argument. Let us suppose that we could define an isomorphism $\phi : \mathbf{C} \to \mathbf{R}$. Then we would have

$$\phi(i)^2 = \phi(i^2) = \phi(-1) = -\phi(1) = -1 ,$$

and so $\mathbf{R}$ would have a square root of -1, which we know to be impossible. Thus $\mathbf{R}$ and $\mathbf{C}$ cannot be isomorphic.

In many important cases we will be interested in functions that preserve addition and multiplication of commutative rings but are not necessarily one-to-one and onto.

Example 5.2.3 (Natural projection $\pi : \mathbf{Z} \to \mathbf{Z}_n$).

We already know by results on factor groups in Section 3.8 that the mapping $\pi : \mathbf{Z} \to \mathbf{Z}_n$ given by $\pi(x) = [x]_n$, for all $x \in \mathbf{Z}$, is a group homomorphism. The formula $[x]_n[y]_n = [xy]_n$, which defines multiplication of congruence classes, shows immediately that π also preserves multiplication. Since $\pi(1) = [1]_n$, it follows that π is a ring homomorphism. Note that π is onto, but not one-to-one, since $\pi(n) = [0]_n$.   □

Example 5.2.4 (Natural inclusion $\iota : R \to R[x]$).

Let R be any commutative ring, and define $\iota : R \to R[x]$ by $\iota(a) = a$, for all $a \in R$. That is, $\iota(a)$ is defined to be the constant polynomial a. It is easy to check that ι is a ring homomorphism that is one-to-one but not onto.   □

In high school the operation of "substituting numerical values for the unknown x" in polynomial functions (and more general functions) plays an important role. A precise understanding of "substitution" depends on the idea of a ring homomorphism, as the next two examples illustrate. Example 5.2.6 will be particularly important in later work.

Example 5.2.5 (Evaluation at $\sqrt{2}$).

Consider the mapping $\phi : \mathbf{Q}[x] \to \mathbf{R}$ defined by $\phi(f(x)) = f(\sqrt{2})$, for all polynomials $f(x) \in \mathbf{Q}[x]$. That is, the mapping ϕ is defined on a polynomial with rational coefficients by substituting $x = \sqrt{2}$. It is easy to check that adding (or multiplying) two polynomials first and then substituting in $x = \sqrt{2}$ is the same as substituting first in each polynomial and then adding (or multiplying). Thus ϕ preserves sums and products, and so it is a ring homomorphism since we also have $\phi(1) = 1$. Note that ϕ is not one-to-one, since $\phi(x^2 - 2) = 0$. For any polynomial $f(x)$, in computing $\phi(f(x)) = f(\sqrt{2})$ we can use the fact that $(\sqrt{2})^n = 2^{n/2}$ if n is an even positive integer, and $(\sqrt{2})^n = 2^{(n-1)/2}\sqrt{2}$ if n is an odd positive integer. It follows that the image of ϕ is $\mathbf{Q}(\sqrt{2}) = \{a + b\sqrt{2} \mid a, b \in \mathbf{Q}\}$. □

Example 5.2.6 (Evaluation mapping).

The previous example can be generalized, since there is nothing special about the particular fields we chose or the particular element we worked with. Let F and E be fields, with F a subfield of E. For any element $u \in E$ we can define a function $\phi_u : F[x] \to E$ by letting $\phi_u(f(x)) = f(u)$, for each $f(x) \in F[x]$. Then ϕ_u preserves sums and products since

$$\phi_u(f(x) + g(x)) = f(u) + g(u) = \phi_u(f(x)) + \phi_u(g(x))$$

and

$$\phi_u(f(x) \cdot g(x)) = f(u) \cdot g(u) = \phi_u(f(x)) \cdot \phi_u(g(x)) ,$$

for all $f(x), g(x) \in F[x]$. Furthermore, $\phi(1) = 1$, and thus ϕ_u is a ring homomorphism.

Since the polynomials in $F[x]$ are evaluated at u, the homomorphism ϕ_u is called an **evaluation mapping**. □

Let $\phi : R \to S$ be a ring homomorphism. By elementary results on group theory, we know that ϕ must map the zero element of R onto the zero element of S, that ϕ must preserve additive inverses, and that $\phi(R)$ must be an additive subgroup of S. It is convenient to list these results formally in the next proposition. To complete the proof of part (c), the fact that $\phi(R)$ is closed under multiplication follows since ϕ preserves multiplication, and $\phi(R)$ has the same identity as S since $\phi(1) = 1$.

5.2.3 Proposition. *Let $\phi : R \to S$ be a ring homomorphism. Then*

(a) $\phi(0) = 0$;

(b) $\phi(-a) = -\phi(a)$ *for all $a \in R$;*

(c) $\phi(R)$ *is a subring of S.*

Example 5.2.7.

Any isomorphism $\phi : R \to S$ of commutative rings induces a group isomorphism from $R^{\times}$ onto $S^{\times}$. For any $a \in R^{\times}$ we have

$$\phi(a)\phi(a^{-1}) = \phi(aa^{-1}) = \phi(1) = 1 ,$$

and so ϕ maps $R^{\times}$ into $S^{\times}$. Applying this argument to ϕ^{-1}, which is also a ring homomorphism, shows that it maps $S^{\times}$ into $R^{\times}$. We conclude that ϕ is one-to-one and onto when restricted to $R^{\times}$, and this shows that $R^{\times}$ and $S^{\times}$ are isomorphic (as groups).   □

Before giving some additional examples, we need to give a definition analogous to one for groups. The kernel of a ring homomorphism will be defined to be the kernel of the mapping when viewed as a group homomorphism of the underlying additive groups of the rings.

5.2.4 Definition. *Let $\phi : R \to S$ be a ring homomorphism. The set*

$$\{a \in R \mid \phi(a) = 0\}$$

*is called the **kernel** of ϕ, denoted by* $\ker(\phi)$.

For example, the kernel of the natural projection $\pi : \mathbf{Z} \to \mathbf{Z}_n$ discussed in Example 5.2.3 is $\{x \mid [x]_n = [0]_n\} = n\mathbf{Z}$. Note that $\ker(\pi)$ is not a subring of $\mathbf{Z}$. In fact, the kernel of a ring homomorphism is a subring only in the trivial case, when the codomain is the zero ring.

Example 5.2.8 (Kernel of the evaluation mapping).

Let F be a subfield of the field E. For an element $u \in E$, let $\phi_u : F[x] \to E$ be the evaluation mapping defined by setting $\phi_u(f(x)) = f(u)$, for each $f(x) \in F[x]$ (see Example 5.2.6). In this case

$$\ker(\phi_u) = \{f(x) \in F[x] \mid f(u) = 0\}$$

is the set of all polynomials $f(x) \in F[x]$ for which u is a root (when $f(x)$ is viewed as a polynomial in $E[x]$).   □

5.2.5 Proposition. *Let $\phi : R \to S$ be a ring homomorphism.*

(a) *If $a, b \in \ker(\phi)$ and $r \in R$, then $a + b$, $a - b$, and ra belong to $\ker(\phi)$.*

(b) *The homomorphism ϕ is an isomorphism if and only if $\ker(\phi) = \{0\}$ and $\phi(R) = S$.*

Proof. (a) If $a, b \in \ker(\phi)$, then

$$\phi(a \pm b) = \phi(a) \pm \phi(b) = 0 \pm 0 = 0 ,$$

and so $a \pm b \in \ker(\phi)$. If $r \in R$, then

$$\phi(ra) = \phi(r) \cdot \phi(a) = \phi(r) \cdot 0 = 0 ,$$

showing that $ra \in \ker(\phi)$.

(b) This part follows from the fact that ϕ is a group homomorphism, since ϕ is one-to-one if and only if $\ker(\phi) = 0$ and ϕ is onto if and only if $\phi(R) = S$. $\square$

Let $\phi : R \to S$ be a ring homomorphism. The fundamental homomorphism theorem for groups implies that the abelian group $R/\ker(\phi)$ is isomorphic to the abelian group $\phi(R)$, which is a subgroup of S. In order to obtain a homomorphism theorem for commutative rings we need to consider the cosets of $\ker(\phi)$. Intuitively, the situation may be easiest to understand if we consider the cosets to be defined by the equivalence relation $\sim_\phi$ given by $a \sim_\phi b$ if $\phi(a) = \phi(b)$, for all $a, b \in R$. The sum of equivalence classes $[a]$ and $[b]$ in $R/\ker(\phi)$ is well-defined, using the formula $[a] + [b] = [a + b]$, for all $a, b \in R$. The product of equivalence classes $[a]$ and $[b]$ in $R/\ker(\phi)$ is defined by the expected formula $[a] \cdot [b] = [ab]$, for all $a, b \in R$. To show that this multiplication is well-defined, we note that if $a \sim_\phi c$ and $b \sim_\phi d$, then $ab \sim_\phi cd$ since

$$\phi(ab) = \phi(a)\phi(b) = \phi(c)\phi(d) = \phi(cd) .$$

Since our earlier results on groups imply that $R/\ker(\phi)$ is an abelian group, to show that $R/\ker(\phi)$ is a commutative ring we only need to verify the distributive law, the associative and commutative laws for multiplication, and check that there is an identity element. We have

$$[a]([b]+[c]) = [a][b+c] = [a(b+c)] = [ab+ac] = [ab]+[ac] = [a][b]+[a][c] ,$$

showing that the distributive law follows directly from the definitions of addition and multiplication for equivalence classes, and the distributive law in R. The proofs that the associative and commutative laws hold are similar. It is easy to check that $[1]$ serves as an identity element.

The coset notation $a + \ker(\phi)$ is usually used for the equivalence class $[a]$. With this notation, addition and multiplication of cosets are expressed by the formulas

$$(a + \ker(\phi)) + (b + \ker(\phi)) = (a + b) + \ker(\phi)$$

and

$$(a + \ker(\phi)) \cdot (b + \ker(\phi)) = (ab) + \ker(\phi) .$$

It is important to remember that these are additive cosets, not multiplicative cosets.

5.2.6 Theorem (Fundamental Homomorphism Theorem for Rings). *Let* $\phi : R \to S$ *be a ring homomorphism. Then* $R/\ker(\phi) \cong \phi(R)$.

Proof. Define $\overline{\phi} : R/\ker(\phi) \to \phi(R)$ by setting $\overline{\phi}(a + \ker(\phi)) = \phi(a)$, for all $a \in R$. Since we have already shown that $R/\ker(\phi)$ is a commutative ring, we can apply the fundamental homomorphism theorem for groups to show that $\overline{\phi}$ is an isomorphism of the abelian groups $R/\ker(\phi)$ and $\phi(R)$. We only need to show that $\overline{\phi}$ preserves multiplication, and this follows from the computation

$$
\begin{aligned}
\overline{\phi}((a + \ker(\phi)) \cdot (b + \ker(\phi))) &= \overline{\phi}(ab + \ker(\phi)) \\
&= \phi(ab) = \phi(a)\phi(b) \\
&= \overline{\phi}(a + \ker(\phi)) \cdot \overline{\phi}(b + \ker(\phi)) \, .
\end{aligned}
$$

This completes the proof. $\square$

The evaluation mapping in Example 5.2.6 can perhaps be better understood in a more general context. In working with $F[x]$, we never used the fact that F was a field, so we can consider polynomials over any commutative ring, as in Example 5.1.2. In Example 5.2.6 we *did* use the fact that the inclusion mapping $F \to E$ is a ring homomorphism, so we will consider the situation in that generality.

5.2.7 Proposition. *Let R and S be commutative rings, let $\theta : R \to S$ be a ring homomorphism, and let s be any element of S. Then there exists a unique ring homomorphism $\widehat{\theta}_s : R[x] \to S$ such that $\widehat{\theta}_s(r) = \theta(r)$ for all $r \in R$, and $\widehat{\theta}_s(x) = s$.*

Proof. We will first show the uniqueness. If $\phi : R[x] \to S$ is any ring homomorphism with $\phi(r) = \theta(r)$ for all $r \in R$ and $\phi(x) = s$, then for any polynomial

$$ f(x) = a_0 + a_1 x + \ldots + a_m x^m $$

in $R[x]$ we must have

$$
\begin{aligned}
\phi(a_0 + a_1 x + \cdots + a_m x^m) &= \phi(a_0) + \phi(a_1 x) + \ldots + \phi(a_m x^m) \\
&= \phi(a_0) + \phi(a_1)\phi(x) + \ldots + \phi(a_m)(\phi(x))^m \\
&= \theta(a_0) + \theta(a_1)s + \ldots + \theta(a_m)s^m \, .
\end{aligned}
$$

This shows that the only possible way to define $\widehat{\theta}_s$ is the following:

$$ \widehat{\theta}_s(a_0 + a_1 x + \ldots + a_m x^m) = \theta(a_0) + \theta(a_1)s + \ldots + \theta(a_m)s^m \, . $$

Given this definition, we must show that $\widehat{\theta}_s$ is a ring homomorphism. Since addition of polynomials is defined componentwise, and θ preserves sums, it is easy to check that $\widehat{\theta}_s$ preserves sums of polynomials. If

$$ g(x) = b_0 + b_1 x + \ldots + b_n x^n \, , $$

then the coefficient c_k of the product $h(x) = f(x)g(x)$ is given by the formula

$$c_k = \sum_{i+j=k} a_i b_j .$$

Applying θ to both sides gives

$$\theta(c_k) = \theta\left(\sum_{i+j=k} a_i b_j\right) = \sum_{i+j=k} \theta(a_i)\theta(b_j)$$

since θ preserves both sums and products. This formula is precisely what we need to check that

$$\widehat{\theta}_s(f(x)g(x)) = \widehat{\theta}_s(h(x)) = \widehat{\theta}_s(f(x))\widehat{\theta}_s(g(x)) .$$

Since $\widehat{\theta}_s(1) = 1$, this finishes the proof that $\widehat{\theta}_s$ is a ring homomorphism. $\square$

Example 5.2.9 ($\mathbf{Z}_n \cong \mathbf{Z}/n\mathbf{Z}$).

The natural projection $\pi : \mathbf{Z} \to \mathbf{Z}_n$ defined by $\pi(x) = [x]_n$, for all $x \in \mathbf{Z}$, is onto with $\ker(\pi) = n\mathbf{Z}$. $\square$

Example 5.2.10 ($\mathbf{Q}(\sqrt{2}) \cong \mathbf{Q}[x]/I$ for $I = \{f(x) \in \mathbf{Q}[x] \mid f(\sqrt{2}) = 0\}$).

It was shown in Example 5.2.5 that evaluation at $\sqrt{2}$ defines a ring homomorphism $\phi : \mathbf{Q}[x] \to \mathbf{R}$ with $\phi(\mathbf{Q}[x]) = \mathbf{Q}(\sqrt{2})$, and in Example 5.2.8 that $\ker(\phi) = \{f(x) \in \mathbf{Q}[x] \mid f(\sqrt{2}) = 0\}$. By the fundamental homomorphism theorem we have $\mathbf{Q}(\sqrt{2}) \cong \mathbf{Q}[x]/\ker(\phi)$. $\square$

Example 5.2.11 (Roots of polynomials).

Let R be a subring of the ring S, and let $\theta : R \to S$ be the inclusion mapping. If $s \in S$, then the ring homomorphism $\widehat{\theta}_s : R[x] \to S$ defined in Proposition 5.2.7 should be thought of as an evaluation mapping, since $\widehat{\theta}_s(f(x)) = f(s)$, for any polynomial $f(x) \in R[x]$. If $f(s) = 0$, then we say that s is a **root** of the polynomial $f(x)$. (Compare Definition 4.1.10.)

We must be careful when considering roots of polynomials whose coefficients come from an arbitrary commutative ring. If $(x - 2)(x - 3) = 0$ and x is an integer, then we can conclude that either $x - 2 = 0$ or $x - 3 = 0$, and so either $x = 2$ or $x = 3$. But if x represents an element of $\mathbf{Z}_6$, then we might have $x - [2] = [3]$ and $x - [3] = [2]$, since this still gives $(x - [2])(x - [3]) = [3][2] = [0]$ in $\mathbf{Z}_6$. In addition to the obvious roots $[2]$ and $[3]$, it is easy to see that $[0]$ and $[5]$ are also roots, so the polynomial $x^2 - [5]x + [6]$ has four distinct roots over $\mathbf{Z}_6$. $\square$

Example 5.2.12 (Reduction modulo n).

Let $\pi : \mathbf{Z} \to \mathbf{Z}_n$ be the natural projection considered in Example 5.2.3. In Example 5.1.2 we discussed the ring of polynomials with coefficients in a commutative ring, and so in this context we can consider the polynomial rings $\mathbf{Z}[x]$ and $\mathbf{Z}_n[x]$. Define $\widehat{\pi} : \mathbf{Z}[x] \to \mathbf{Z}_n[x]$ as follows: For any polynomial

$$f(x) = a_0 + a_1 x + \ldots + a_m x^m$$

in $\mathbf{Z}[x]$, set

$$\widehat{\pi}(f(x)) = \pi(a_0) + \pi(a_1)x + \ldots + \pi(a_m)x^m \ .$$

That is, $\widehat{\pi}$ simply reduces all of the coefficients of $f(x)$ modulo n. This is actually a special case of the result obtained in Proposition 5.2.7. We can think of π as a homomorphism from $\mathbf{Z}$ into $\mathbf{Z}_n[x]$, and then we have extended π to $\mathbf{Z}[x]$ by mapping $x \in \mathbf{Z}[x]$ to $x \in \mathbf{Z}_n[x]$. It follows that $\widehat{\pi}$ is a homomorphism. Furthermore, it is easy to see that the kernel of $\widehat{\pi}$ is the set of all polynomials for which each coefficient is divisible by n.

To illustrate the power of homomorphisms, suppose that $f(x)$ has a nontrivial factorization $f(x) = g(x)h(x)$ in $\mathbf{Z}[x]$. Then

$$\widehat{\pi}(f(x)) = \widehat{\pi}(g(x)h(x)) = \widehat{\pi}(g(x))\widehat{\pi}(h(x)) \ .$$

If $\deg(\widehat{\pi}(f(x))) = \deg(f(x))$, then this gives a nontrivial factorization of $\widehat{\pi}(f(x))$ in $\mathbf{Z}_n[x]$.

This means that a polynomial $f(x)$ with integer coefficients can be shown to be irreducible over the field $\mathbf{Q}$ of rational numbers by finding a positive integer n such that when the coefficients of $f(x)$ are reduced modulo n, the new polynomial has the same degree and cannot be factored nontrivially in $\mathbf{Z}_n[x]$. Eisenstein's irreducibility criterion (Theorem 4.4.6) for the prime p can be interpreted as a condition which states that the polynomial cannot be factored when reduced modulo p^2. □

We next introduce a construction for commutative rings that is analogous to the direct product of groups.

5.2.8 Proposition. *Let $R_1, R_2, \ldots, R_n$ be commutative rings. The set of n-tuples $(a_1, a_2, \ldots, a_n)$ such that $a_i \in R_i$ for each i is a commutative ring under the following addition and multiplication:*

$$(a_1, a_2, \ldots, a_n) + (b_1, b_2, \ldots, b_n) = (a_1 + b_1, \ a_2 + b_2, \ldots, a_n + b_n)$$

$$(a_1, a_2, \ldots, a_n) \cdot (b_1, b_2, \ldots, b_n) = (a_1 b_1, \ a_2 b_2, \ldots, a_n b_n) \ ,$$

for n-tuples $(a_1, a_2, \ldots, a_n)$ and $(b_1, b_2, \ldots, b_n)$.

Proof. The proof that the given addition defines a group is an easy extension of the proof of Proposition 3.3.4. The remainder of the proof is left as an exercise. □

We say that the addition and multiplication defined in Proposition 5.2.8 are defined "componentwise", since there is no interaction between different components. (Compare this, for example, with the definition of multiplication of complex numbers thought of as ordered pairs: $(a_1, a_2) \cdot (b_1, b_2) = (a_1b_1 - a_2b_2, a_1b_2 + a_2b_1)$.)

5.2.9 Definition. *Let $R_1, R_2, \ldots, R_n$ be commutative rings. The set of n-tuples $(a_1, a_2, \ldots, a_n)$ such that $a_i \in R_i$ for each i, under the operations of componentwise addition and multiplication, is called the **direct sum** of the commutative rings $R_1, R_2, \ldots, R_n$, and is denoted by*

$$R_1 \oplus R_2 \oplus \cdots \oplus R_n .$$

Let $R_1, R_2, \ldots, R_n$ be commutative rings. Then $(1, 1, \ldots, 1)$ is the identity of the direct sum

$$R = R_1 \oplus R_2 \oplus \cdots \oplus R_n .$$

Furthermore, an element $(a_1, a_2, \ldots, a_n)$ in the direct sum is a unit if and only if each component a_i is a unit in R_i. This can be shown by observing that

$$(a_1, a_2, \ldots, a_n)(b_1, b_2, \ldots, b_n) = (1, 1, \ldots, 1)$$

if and only if $a_i b_i = 1$ for each i. It then follows easily that

$$R^\times \cong R_1^\times \times R_2^\times \times \cdots \times R_n^\times$$

as groups. (This is part of Exercise 21.)

Example 5.2.13.

Let n be a positive integer with prime decomposition $n = p_1^{\alpha_1} p_2^{\alpha_2} \cdots p_m^{\alpha_m}$. Then the isomorphism

$$\mathbf{Z}_n \cong \mathbf{Z}_{p_1^{\alpha_1}} \oplus \mathbf{Z}_{p_2^{\alpha_2}} \oplus \cdots \oplus \mathbf{Z}_{p_m^{\alpha_m}}$$

can be shown easily by referring to Theorem 3.5.5. The mapping ϕ defined by

$$\phi([x]_n) = ([x]_{p_1^{\alpha_1}} , [x]_{p_2^{\alpha_2}} , \ldots, [x]_{p_m^{\alpha_m}})$$

for all $[x]_n \in \mathbf{Z}_n$ is easily seen to preserve multiplication.

Recall that if n is a positive integer with prime decomposition

$$n = p_1^{\alpha_1} p_2^{\alpha_2} \cdots p_m^{\alpha_m} ,$$

then

$$\varphi(n) = n \left(1 - \frac{1}{p_1} \right) \left(1 - \frac{1}{p_2} \right) \cdots \left(1 - \frac{1}{p_m} \right) ,$$

where $\varphi(n)$ is the number of positive integers less than or equal to n and relatively prime to n. This was proved in Corollary 3.5.6 by counting the generators of the direct product of the component groups rather than counting the generators of $\mathbf{Z}_n$ directly. We are now in a position to give a proof based on results for commutative rings. From the general remarks preceding this example, we have

$$\mathbf{Z}_n^\times \cong \mathbf{Z}_{p_1^{\alpha_1}}^\times \times \mathbf{Z}_{p_2^{\alpha_2}}^\times \times \cdots \times \mathbf{Z}_{p_m^{\alpha_m}}^\times .$$

Then $\varphi(n)$, which is the order of $\mathbf{Z}_n^\times$, can be found by multiplying the orders of the given groups $\mathbf{Z}_{p_i^{\alpha_i}}^\times$. From here this argument proceeds just as the one in Corollary 3.5.6. □

As an application of the results in this section we study the characteristic of a commutative ring.

5.2.10 Definition. *Let R be a commutative ring. The smallest positive integer n such that $n \cdot 1 = 0$ is called the **characteristic** of R, denoted by* char(R).

*If no such positive integer exists, then R is said to have **characteristic zero**.*

The characteristic of a commutative ring R is closely connected to a definition from group theory: if char(R) is nonzero, then it is just the order of 1 in the underlying additive group of R. If char$(R) = n \neq 0$, then it follows from the distributive law that $n \cdot a = (n \cdot 1) \cdot a = 0 \cdot a = 0$ for all $a \in R$. This implies that n is the exponent of the underlying abelian group, giving another way to think of char(R).

A more sophisticated way to view the characteristic is to define a ring homomorphism $\phi : \mathbf{Z} \to R$ by $\phi(n) = n \cdot 1$. The rules we developed in Chapter 3 for considering multiples of an element in an abelian group show that ϕ is a homomorphism. The characteristic of R is just the (nonnegative) generator of ker(ϕ).

5.2.11 Proposition. *An integral domain has characteristic 0 or p, for some prime number p.*

Proof. Let D be an integral domain, and consider the mapping $\phi : \mathbf{Z} \to D$ defined by $\phi(n) = n \cdot 1$. Note that ϕ is a ring homomorphism, since for any $m, n \in \mathbf{Z}$ we have $\phi(m + n) = (m + n) \cdot 1 = m \cdot 1 + n \cdot 1 = \phi(m) + \phi(n)$, $\phi(mn) = (mn) \cdot 1 = (m \cdot 1)(n \cdot 1) = \phi(m)\phi(n)$, and $\phi(1) = 1 \cdot 1 = 1$.

The fundamental homomorphism theorem for rings shows that $\mathbf{Z}/\ker(\phi)$ is isomorphic to the subring $\phi(\mathbf{Z})$ of D. Since $\phi(\mathbf{Z})$ inherits the property that D has no nontrivial divisors of zero, this shows that $\mathbf{Z}/\ker(\phi)$ must be an integral domain. Thus either $\ker(\phi) = 0$, in which case $\text{char}(D) = 0$, or $\ker(\phi) = n\mathbf{Z}$ for some positive number n. Then $\mathbf{Z}/\ker(\phi) \cong \mathbf{Z}_n$, and Corollary 1.4.6 implies that n is prime, so in this case $\text{char}(D)$ is a prime number. $\square$

$$\phi(x+y) = \phi(x) + \phi(y)$$
$$\phi(x)(y) = \phi(x)\phi(y)$$
$$\phi(0) = 0$$

EXERCISES: SECTION 5.2

$$\phi(1) = 1$$

1. Let R be a commutative ring, and let D be an integral domain. Let $\phi : R \to D$ be a nonzero function such that $\phi(a+b) = \phi(a) + \phi(b)$ and $\phi(ab) = \phi(a)\phi(b)$, for all $a, b \in R$. Show that ϕ is a ring homomorphism.

2. Let F be a field and let $\phi : F \to R$ be a ring homomorphism. Show that ϕ is either zero or one-to-one.

3. Let F, E be fields, and let $\phi : F \to E$ be a ring homomorphism. Show that if ϕ is onto, then ϕ must be an isomorphism.

4. Show that taking complex conjugates defines an automorphism of $\mathbf{C}$. That is, for $z \in \mathbf{C}$, define $\phi(z) = \bar{z}$, and show that ϕ is an automorphism of $\mathbf{C}$.

5. Show that the identity mapping is the only ring homomorphism from $\mathbf{Z}$ into $\mathbf{Z}$.

6. Show that the set of all matrices over $\mathbf{Z}$ of the form $\begin{bmatrix} m & n \\ 2n & m \end{bmatrix}$ is a ring isomorphic to the ring $\mathbf{Z}[\sqrt{2}]$ defined in Example 5.1.6.

7. Define $\phi : \mathbf{Z}[\sqrt{2}] \to \mathbf{Z}[\sqrt{2}]$ by $\phi(m + n\sqrt{2}) = m - n\sqrt{2}$, for all $m, n \in \mathbf{Z}$. Show that ϕ is an automorphism of $\mathbf{Z}[\sqrt{2}]$.

8. Let F be a field, and let $a \in F$. Define $\phi : F[x] \to F[x]$ by $\phi(f(x)) = f(x+a)$, for all $f(x) \in F[x]$. Show that ϕ is an automorphism of $F[x]$.

9. Show that the composite of two ring homomorphisms is a ring homomorphism.

10. Let R and S be rings, and let $\phi, \theta : R \to S$ be ring homomorphisms. Show that $\{r \in R \mid \phi(r) = \theta(r)\}$ is a subring of R.

11. Show that the direct sum of two nonzero rings is never an integral domain.

12. Let R_1 and R_2 be commutative rings.

(a) Define $\pi_1 : R_1 \oplus R_2 \to R_1$ by $\pi_1((r_1, r_2)) = r_1$, for all $(r_1, r_2) \in R_1 \oplus R_2$ and define $\pi_2 : R_1 \oplus R_2 \to R_2$ by $\pi_2((r_1, r_2)) = r_2$, for all $(r_1, r_2) \in R_1 \oplus R_2$. Show that π_1 and π_2 are ring homomorphisms.

(b) Let R be any ring, and let $\phi : R \to R_1 \oplus R_2$ be a function. Show that ϕ is a ring homomorphism if and only if $\pi_1\phi$ and $\pi_2\phi$ are both ring homomorphisms.

13.†Find all ring homomorphisms from $\mathbf{Z} \oplus \mathbf{Z}$ into $\mathbf{Z}$. That is, find all possible formulas, and show why no others are possible.

14. Find all ring homomorphisms from $\mathbf{Z} \oplus \mathbf{Z}$ into $\mathbf{Z} \oplus \mathbf{Z}$.

15. For the rings $\mathbf{Z}_n$ and $\mathbf{Z}_k$, show that if $k \mid n$, then the function $\phi : \mathbf{Z}_n \rightarrow \mathbf{Z}_k$ defined by $\phi([x]_n) = [x]_k$, for all $[x]_n \in \mathbf{Z}_n$, is a ring homomorphism. Show that this is the only ring homomorphism from $\mathbf{Z}_n$ to $\mathbf{Z}_k$.

16. Are $\mathbf{Z}_9$ and $\mathbf{Z}_3 \oplus \mathbf{Z}_3$ isomorphic as rings?

17. Let S be the subset of $\mathbf{Z}_4 \oplus \mathbf{Z}_4$ given by $\{([m]_4, [n]_4) \mid m \equiv n \pmod 2)\}$.

    (a) Show that S is a subring of $\mathbf{Z}_4 \oplus \mathbf{Z}_4$.

    (b) Show that S is not isomorphic (as a ring) to any ring of the form $\mathbf{Z}_n$, nor to any direct sum of such rings.

18. Define $\phi : \mathbf{Z} \rightarrow \mathbf{Z}_m \oplus \mathbf{Z}_n$ by $\phi(x) = ([x]_m, [x]_n)$. Find the kernel and image of ϕ. Show that ϕ is onto if and only if $\gcd(m, n) = 1$.

19. Let R be the ring given by defining new operations on $\mathbf{Z}$ by letting $m \oplus n = m + n - 1$ and $m \odot n = m + n - mn$. Define $\phi : \mathbf{Z} \rightarrow R$ by $\phi(n) = 1 - n$. Show that ϕ is an isomorphism.

$m \oplus d$ $\phi(m+n) = 1 - (m+n)$

20. Let I be any set and let R be the collection of all subsets of I. Define addition and multiplication of subsets $A, B \subseteq I$ as follows:

$$A + B = (A \cup B) \cap \overline{A \cap B} \quad \text{and} \quad A \cdot B = A \cap B .$$

    †(a) Show that if I has two elements, then R is isomorphic to $\mathbf{Z}_2 \oplus \mathbf{Z}_2$.

    (b) Show that if I has three elements, then R is isomorphic to $\mathbf{Z}_2 \oplus \mathbf{Z}_2 \oplus \mathbf{Z}_2$.

21. Let $R_1, R_2, \ldots, R_n$ be commutative rings. Complete the proof of Proposition 5.2.8, to show that $R = R_1 \oplus R_2 \oplus \cdots \oplus R_n$ is a commutative ring. Then show that $R^\times \cong R_1^\times \times R_2^\times \times \cdots \times R_n^\times$.

22. Let R be an integral domain. Show that R contains a subring isomorphic to $\mathbf{Z}_p$ for some prime number p if and only if $\text{char}(R) = p$.

23. Show that if R is an integral domain with characteristic $p > 0$, then for all $a, b \in R$ we must have $(a + b)^p = a^p + b^p$. Show by induction that we must also have $(a + b)^{p^n} = a^{p^n} + b^{p^n}$ for all positive integers n.

5.3 Ideals and Factor Rings

We have shown that the kernel of any ring homomorphism is closed under sums, differences, and "scalar" multiples (see Proposition 5.2.5). In other words, the kernel of a ring homomorphism is an additive subgroup that is closed under multiplication by any element of the ring. For any integer n, the subset $n\mathbf{Z}$ of the ring of integers satisfies the same properties. The same is true for the subset $\langle f(x)\rangle$ of the ring $F[x]$ of polynomials over a field F, where $f(x)$ is any polynomial. In each of these examples we have been able to make the cosets of the given set into a commutative ring, and this provides the motivation for the next definition and several subsequent results.

5.3.1 Definition. *Let R be a commutative ring. A nonempty subset I of R is called an **ideal** of R if*

  (i) $a \pm b \in I$ *for all $a, b \in I$, and*

  (ii) $ra \in I$ *for all $a \in I$ and $r \in R$.*

For any commutative ring R, it is clear that the set $\{0\}$ is an ideal of R, which we will refer to as the **trivial** ideal. The set R is also always an ideal of R; we say that it is not a *proper* ideal since it is not a proper subset of R. Among commutative rings (with $1 \neq 0$), fields are characterized by the property that these two ideals are the only ideals of the ring.

5.3.2 Proposition. *Let R be a commutative ring with $1 \neq 0$. Then R is a field if and only if it has no proper nontrivial ideals.*

Proof. First assume that R is a field, and let I be any ideal of R. Either $I = \{0\}$, or else there exists $a \in I$ such that $a \neq 0$. In the second case, since R is a field, there exists an inverse a^{-1} for a, and then for any $r \in R$ we have $r = r \cdot 1 = r(a^{-1}a) = (ra^{-1})a$, so by the definition of an ideal we have $r \in I$. We have shown that either $I = \{0\}$ or $I = R$.

Conversely, assume that R has no proper nontrivial ideals, and let a be a nonzero element of R. We will show that the set

$$I = \{x \in R \mid x = ra \text{ for some } r \in R\}$$

is an ideal. First, I is nonempty since $a = 1 \cdot a \in I$. If $r_1 a, r_2 a \in I$, then we have $r_1 a \pm r_2 a = (r_1 \pm r_2)a$, showing that I is an additive subgroup of R. Finally, if $x = ra \in I$, then for any $s \in R$ we have $sx = (sr)a \in I$, and so I is an ideal. By assumption we must have $I = R$, since $I \neq \{0\}$, and since $1 \in R$, we have $1 = ra$ for some $r \in R$. This implies that a is invertible, and so we have shown that R is a field. $\square$

Let R be a commutative ring. For any $a \in R$, we use the notation

$$Ra = \{x \in R \mid x = ra \text{ for some } r \in R\} \,.$$

The proof of the previous theorem shows that Ra is an ideal of R that contains a. It is obvious from the definition of an ideal that any ideal that contains a must also contain Ra, and so we are justified in saying that Ra is the smallest ideal that contains a.

Note that $R \cdot 1$ consists of all of R, since every element $r \in R$ can be expressed in the form $r \cdot 1$. Thus R is the smallest ideal (in fact, the only ideal) that contains the identity of R.

5.3.3 Definition. *Let R be a commutative ring, and let $a \in R$. The ideal*

$$Ra = \{x \in R \mid x = ra \text{ for some } r \in R\}$$

*is called the **principal ideal** generated by a. The notation $\langle a \rangle$ will also be used.*

*An integral domain in which every ideal is a principal ideal is called a **principal ideal domain**.*

Example 5.3.1 (Z is a principal ideal domain).

In the terminology of the above definition, we see that Theorem 1.1.4 shows that the ring of integers $\mathbf{Z}$ is a principal ideal domain. Moreover, given any nonzero ideal I of $\mathbf{Z}$, the smallest positive integer in I is a generator for the ideal. □

Example 5.3.2 ($F[x]$ is a principal ideal domain).

We showed in Theorem 4.2.2 that if F is any field, then the ring $F[x]$ of polynomials over F is a principal ideal domain. If I is any nonzero ideal of $F[x]$, then $f(x)$ is a generator for I if and only if it has minimal degree among the nonzero elements of I. Since a generator of I is a divisor of every element of I, there is only one monic generator for I (see Exercise 6 (a)). □

In addition to studying the ring of polynomials $F[x]$ over a field, as in Example 5.3.2, we have also considered the ring of polynomials $R[x]$ over any commutative ring R. However, if the coefficients do not come from a field, then the proof of the division algorithm is no longer valid, and so we should not expect $R[x]$ to be a principal ideal domain. The ring $\mathbf{Z}[x]$ of polynomials with integer coefficients is an integral domain, but not every ideal is principal (see Exercise 24).

Let I be an ideal of the commutative ring R. Then I is a subgroup of the underlying additive group of R, and so by Theorem 3.8.5 the cosets of I in R determine a factor group R/I, which is again an abelian group. The cosets of I are usually denoted additively, in the form $a + I$, for all elements $a \in R$. We know from Proposition 3.8.1 that elements $a, b \in R$ determine the same coset of I if and only if $a - b \in I$. Since the cosets of I partition R, they determine an equivalence relation on R, by defining two elements of R to be equivalent if their difference is in I.

In Chapter 1 we used the notion of congruence modulo n to define the sets $\mathbf{Z}_n$, which we now know to be commutative rings. Since $a \equiv b \pmod{n}$ if and only if $a - b$ is a multiple of n, this is precisely the equivalence relation determined by the cosets of the ideal $n\mathbf{Z}$.

In Section 4.3, given an irreducible polynomial $p(x)$, we used the notion of congruence modulo $p(x)$ to construct a new field from the congruence classes, in which we could find a root of $p(x)$. Again in this case, the congruence classes are the cosets of the principal ideal $\langle p(x) \rangle$ of $F[x]$, consisting of all polynomial multiples of $p(x)$.

You should keep these examples in mind as you become familiar with the coset notation for elements of R/I, when I is an ideal of a commutative ring R.

5.3.4 Proposition. *Let I be an ideal of the commutative ring R. The operation defined on the abelian group R/I by setting*

$$(a + I) \cdot (b + I) = ab + I \, ,$$

for $a, b \in R$, is a binary operation.

Proof. To show that the given operation is well-defined, let $a, b \in R$. If $c \in a + I$ and $d \in b + I$, then by definition $a - c \in I$ and $b - d \in I$. Multiplying $a - c$ by b and $b - d$ by c gives us elements that still belong to I. Then using the distributive law and adding gives $(ab - cb) + (cb - cd) = ab - cd$, and this is an element of I, showing that $cd \in ab + I$. Thus the definition of the operation $\cdot$ is independent of the choice of representatives of the cosets, and so it is a well-defined operation on the factor group R/I. $\square$

Given any ideal, we can now construct a *factor ring* relative to the ideal. This parallels the construction (in Theorem 3.8.5) of a factor group relative to a normal subgroup.

5.3.5 Theorem. *If I is an ideal of the commutative ring R, then R/I is a commutative ring under the operations defined for $a, b \in R$ by*

$$(a + I) + (b + I) = (a + b) + I \quad \text{and} \quad (a + I) \cdot (b + I) = ab + I \, .$$

Proof. It follows from Theorem 3.8.5 that R/I is a group under the addition of congruence classes. Proposition 5.3.4 shows that multiplication of congruence classes is well-defined. To show that the distributive law holds, let $a, b, c \in R$. Then

$$
\begin{aligned}
(a + I) \cdot ((b + I) + (c + I)) &= (a + I) \cdot ((b + c) + I) \\
&= a(b + c) + I \\
&= (ab + ac) + I \\
&= (ab + I) + (ac + I) \\
&= (a + I) \cdot (b + I) + (a + I) \cdot (c + I) .
\end{aligned}
$$

Note that we have used the definitions of addition and multiplication of cosets, together with the fact that the distributive law holds in R.

Similar computations show that the associative and commutative laws hold for multiplication. The coset $1 + I$ is a multiplicative identity for R/I, and this observation completes the proof that R/I is a commutative ring. $\square$

5.3.6 Definition. *Let I be an ideal of the commutative ring R. The ring R/I is called the **factor ring** of R modulo I.*

The most familiar factor ring is $\mathbf{Z}/n\mathbf{Z}$, for which we will continue to use the notation $\mathbf{Z}_n$. In this ring, multiplication can be viewed as repeated addition, and it is easy to show that any subgroup is an ideal. Thus we already know the diagram of ideals of $\mathbf{Z}_n$, in which ideals correspond to the divisors of n. Recall that in $\mathbf{Z}$ we have $m \mid k$ if and only if $m\mathbf{Z} \supseteq k\mathbf{Z}$, and so the diagram of ideals of $\mathbf{Z}_n$ corresponds to the diagram of all ideals of $\mathbf{Z}$ that contain $n\mathbf{Z}$. As shown by the next proposition, the analogous result holds in any factor ring.

5.3.7 Proposition. *Let I be an ideal of the commutative ring R.*

(a) The natural projection $\pi : R \to R/I$ defined by $\pi(a) = a + I$ for all $a \in R$ is a ring homomorphism, and $\ker(\pi) = I$.

(b) There is a one-to-one correspondence between the ideals of R/I and ideals of R that contain I. The correspondence is defined as follows: to each ideal J of R/I we assign the ideal $\pi^{-1}(J)$ of R; to each ideal J of R that contains I we assign the ideal $\pi(J)$ of R/I.

Proof. The parts of the proposition that involve addition follow directly from Proposition 3.8.7. To prove part (a), the natural projection must be shown to preserve multiplication, and this follows directly from the definition of multiplication of congruence classes.

To prove part (b), we can use the one-to-one correspondence between subgroups that is given by Proposition 3.8.7. We must show that this correspondence preserves ideals. If J is an ideal of R that contains I, then it corresponds to the additive subgroup

$$\pi(J) = \{a + I \mid a \in J\} .$$

For any element $r+I \in R/I$ and any element $a+I \in \pi(J)$, we have $(r+I)(a+I) = ra + I$, and then $ra + I \in \pi(J)$ since $ra \in J$. On the other hand, if J is an ideal of R/I, then it corresponds to the subgroup

$$\pi^{-1}(J) = \{a \in R \mid a + I \in J\} .$$

If $r \in R$ and $a \in \pi^{-1}(J)$, then $ra \in \pi^{-1}(J)$ since $ra+I = (r+I)(a+I) \in J$. $\quad\square$

Example 5.3.3 ($\mathbf{Q}[x]/\langle x^2 - 2x + 1\rangle$).

Let $R = \mathbf{Q}[x]$ and let $I = \langle x^2 - 2x + 1\rangle$. Using the division algorithm, it is possible to show that the cosets of I correspond to the possible remainders upon division by $x^2 - 2x + 1$ (see Proposition 4.3.3). Thus we only need to consider cosets of the form $a+bx+I$, for all $a, b \in \mathbf{Q}$. Since $x^2-2x+1 \in I$, we can use the formula $x^2 + I = -1 + 2x + I$ to simplify products. This gives us the following formulas:

$$(a + bx + I) + (c + dx + I) = (a + c) + (b + d)x + I$$

and

$$(a + bx + I) \cdot (c + dx + I) = (ac - bd) + (bc + ad + 2bd)x + I .$$

By the previous proposition, the ideals of R/I correspond to the ideals of R that contain I. Since $\mathbf{Q}[x]$ is a principal ideal domain, these ideals are determined by the divisors of $x^2 - 2x + 1$, showing that there is only one proper nontrivial ideal in R/I, corresponding to the ideal generated by $x - 1$. $\quad\square$

Example 5.3.4 ($\mathbf{Q}[x, y]/\langle y\rangle \cong \mathbf{Q}[x]$).

Let $R = \mathbf{Q}[x, y]$, the ring of polynomials in two indeterminates with rational coefficients, and let $I = \langle y\rangle$. That is, I is the set of all polynomials that have y as a factor. In forming R/I we make the elements of I congruent to 0, and so in some sense we should be left with just polynomials in x. This is made precise in the following way: define $\phi : \mathbf{Q}[x, y] \to \mathbf{Q}[x]$ by $\phi(f(x, y)) = f(x, 0)$. It is necessary to check that ϕ is a ring homomorphism. Then it is clear that $\ker(\phi) = \langle y\rangle$, and so we can conclude from the fundamental homomorphism theorem for rings that $R/I \cong \mathbf{Q}[x]$. $\quad\square$

Example 5.3.5.

Let $\phi : R \to S$ be an isomorphism of commutative rings, let I be any ideal of R, and let $J = \phi(I)$. We will show that R/I is isomorphic to S/J. To do this, let π be the natural projection from S onto S/J, and consider $\overline{\phi} = \pi\phi$. Then $\overline{\phi}$ is onto since both π and ϕ are onto, and

$$\ker(\overline{\phi}) = \{r \in R \mid \phi(r) \in J\} = I \, ,$$

so it follows from the fundamental homomorphism theorem for rings that R/I is isomorphic to S/J. □

To motivate the next definition, consider the ring of integers $\mathbf{Z}$. We know that the proper nontrivial ideals of $\mathbf{Z}$ correspond to the positive integers, with $n\mathbf{Z} \subseteq m\mathbf{Z}$ if and only if $m \mid n$. Euclid's lemma (Lemma 1.2.5) states that an integer $p > 1$ is prime if and only if it satisfies the following property: if $p \mid ab$ for integers a, b, then either $p \mid a$ or $p \mid b$. In the language of ideals, this says that p is prime if and only if $ab \in p\mathbf{Z}$ implies $a \in p\mathbf{Z}$ or $b \in p\mathbf{Z}$, for all integers a, b.

Lemma 4.2.8 gives a similar characterization of irreducible polynomials, stating that a polynomial $p(x) \in F[x]$ is irreducible if and only if $p(x) \mid f(x)g(x)$ implies $p(x) \mid f(x)$ or $p(x) \mid g(x)$. When formulated in terms of the principal ideal $\langle p(x) \rangle$, this shows that $p(x)$ is irreducible over F if and only if $f(x)g(x) \in \langle p(x) \rangle$ implies $f(x) \in \langle p(x) \rangle$ or $g(x) \in \langle p(x) \rangle$ Thus irreducible polynomials play the same role in $F[x]$ as do prime numbers in $\mathbf{Z}$.

The prime numbers in $\mathbf{Z}$ can be characterized in another way. Since a prime p has no divisors except $\pm p$ and ± 1, there cannot be any ideals properly contained between $p\mathbf{Z}$ and $\mathbf{Z}$. We refer to ideals with this property as maximal ideals.

5.3.8 Definition. *Let I be a proper ideal of the commutative ring R. Then I is said to be a* **prime ideal** *of R if for all $a, b \in R$ it is true that $ab \in I$ implies $a \in I$ or $b \in I$.*

The ideal I is said to be a **maximal ideal** *of R if for all ideals J of R such that $I \subseteq J \subseteq R$, either $J = I$ or $J = R$.*

Note that if R is a commutative ring with $1 \neq 0$, then R is an integral domain if and only if the trivial ideal $\{0\}$ of R is a prime ideal. This observations shows that in $\mathbf{Z}$ the trivial ideal is a prime ideal that is not maximal. On the other hand, the characterization of prime and maximal ideals given in the next proposition shows that in a commutative ring any maximal ideal is also a prime ideal, since any field is an integral domain.

5.3.9 Proposition. *Let I be a proper ideal of the commutative ring R.*

(a) *The factor ring R/I is a field if and only if I is a maximal ideal of R.*

(b) *The factor ring R/I is an integral domain if and only if I is a prime ideal of R.*

(c) *If I is a maximal ideal, then it is a prime ideal.*

Proof. (a) Since I is assumed to be a proper ideal of R, it does not contain 1, and thus $1 + I \neq 0 + I$ in R/I. Therefore Proposition 5.3.2 implies that R/I is a field if and only if it has no proper nontrivial ideals. Using the one-to-one correspondence between ideals given by Proposition 5.3.7 (b), this occurs if and only if there are no ideals properly between I and R, which is precisely the statement that I is a maximal ideal.

(b) Assume that R/I is an integral domain, and let $a, b \in R$ with $ab \in I$. Remember that the zero element of R/I is the coset consisting of all elements of I. Thus in R/I we have a product $(a + I)(b + I)$ of cosets that is equal to the zero coset, and so by assumption this implies that either $a + I$ or $b + I$ is the zero coset. This implies that either $a \in I$ or $b \in I$, and so I must be a prime ideal.

Conversely, assume that I is a prime ideal. If $a, b \in R$ with $(a+I)(b+I) = 0+I$ in R/I, then we have $ab \in I$, and so by assumption either $a \in I$ or $b \in I$. This shows that either $a + I = 0 + I$ or $b + I = 0 + I$, and so R/I is an integral domain.

(c) This follows from (a) and (b), since every field is an integral domain. $\square$

Example 5.3.6 (Ring isomorphisms preserve prime (or maximal) ideals).

Let $\phi : R \to S$ be an isomorphism of commutative rings. The isomorphism gives a one-to-one correspondence between ideals of the respective rings, and it is not hard to show directly that prime (or maximal) ideals of R correspond to prime (or maximal) ideals of S. This can also be proved as an application of the previous proposition, since in Example 5.3.5 we observed that if I is any ideal of R, then $\phi(I)$ is an ideal of S with $R/I \cong S/\phi(I)$. It then follows immediately from Proposition 5.3.9 that I is prime (or maximal) in R if and only if $\phi(I)$ is prime (or maximal) in S. $\square$

We have already observed that in the ring $\mathbf{Z}$, prime numbers determine maximal ideals. One reason behind this fact is that any finite integral domain is a field, which we proved in Theorem 5.1.9. It is also true that if F is a field, then irreducible polynomials in $F[x]$ determine maximal ideals. The next proposition gives a general reason applicable in both of these special cases.

5.3.10 Theorem. *Every nonzero prime ideal of a principal ideal domain is maximal.*

Proof. Let P be a nonzero prime ideal of a principal ideal domain R, and let J be any ideal with $P \subseteq J \subseteq R$. Since R is a principal ideal domain, we can assume that $P = Ra$ and $J = Rb$ for some elements $a, b \in R$. Since $a \in P$, we have $a \in J$, and so there exists $r \in R$ such that $a = rb$. This implies that $rb \in P$, and so either $b \in P$ or $r \in P$. In the first case, $b \in P$ implies that $J = P$. In the second case, $r \in P$ implies that $r = sa$ for some $s \in R$, since P is generated by a. This gives $a = sab$, and using the assumption that R is an integral domain allows us to cancel a to get $1 = sb$. This shows that $1 \in J$, and so $J = R$. $\square$

Example 5.3.7 (Ideals of $F[x]$).

We can now summarize the information we have regarding polynomials over a field, using a ring theoretic point of view. Let F be any field. The nonzero ideals of $F[x]$ are all principal, of the form $\langle f(x) \rangle$, where $f(x)$ is any polynomial of minimal degree in the ideal. The ideal is prime (and hence maximal) if and only if $f(x)$ is irreducible. If $p(x)$ is irreducible, then the factor ring $F[x]/\langle p(x) \rangle$ is a field. $\square$

Example 5.3.8 ($F[x, y]$ is not a principal ideal domain).

To show that $F[x, y]$ is not a principal ideal domain we will show that the conclusion of Theorem 5.3.10 does not hold. The ideal $P = \{f(x, y) \in F[x, y] \mid f(0, y) = 0\}$ is a prime ideal, since $F[x, y]/P \cong F[y]$. To see this, define $\phi : F[x, y] \to F[y]$ by $\phi(f(x, y)) = f(0, y)$, for all $f(x, y)$ in $F[x, y]$. Then P is a nonzero prime ideal that is not maximal, since it is properly contained in the ideal $M = \{f(x, y) \in F[x, y] \mid f(0, 0) = 0\}$. Note that M is a maximal ideal since it is the kernel of the onto mapping $\theta : F[x, y] \to F$ defined by $\theta(f(x, y)) = f(0, 0)$, for all $f(x, y) \in F[x, y]$. $\square$

Example 5.3.9 (Kernel and image of the evaluation mapping).

Let F be a subfield of E, and for any element $u \in E$ define the evaluation mapping $\phi_u : F[x] \to E$ by $\phi_u(f(x)) = f(u)$, for all $f(x) \in F[x]$. We have already seen in Example 5.2.6 that ϕ_u defines a ring homomorphism. Since $\phi_u(F[x])$ is a subring of E, it follows from Theorem 5.1.8 that it must be an integral domain. By the fundamental homomorphism theorem for rings this image is isomorphic to $F[x]/\ker(\phi_u)$, and so by Proposition 5.3.9, the kernel of ϕ_u must be a prime ideal. If $\ker(\phi_u)$ is nonzero, then it follows from Theorem 5.3.10 that it is a maximal ideal. By Proposition 5.3.9, we know that $F[x]/\ker(\phi_u)$ is a field, so it follows from the fundamental homomorphism theorem for rings that the image of ϕ_u is in fact a subfield of E. This fact will be very important in our study of fields in Chapter 6, where we denote $\phi_u(F[x])$ by $F(u)$. $\square$

We motivated the definition of a prime ideal by looking at the principal ideal generated by a prime number in $\mathbf{Z}$ and at the principal ideal generated by an irreducible polynomial in $F[x]$. These two examples, by themselves, probably would not provide sufficient motivation for the abstract definition. In both $\mathbf{Z}$ and $F[x]$, where F is a field, we have shown that each element can be expressed as a product of primes or irreducibles, respectively. This is not true in general for all commutative rings. In fact, certain subrings of $\mathbf{C}$ which turn out to be important in number theory do not have this property. One of the original motivations for introducing the notion of an ideal (or "ideal number") was to be able to salvage at least the property that every ideal can be expressed as a product of prime ideals.

EXERCISES: SECTION 5.3

1.†Give a multiplication table for the ring $\mathbf{Z}_2[x]/\langle x^2 + 1\rangle$.

2. Give a multiplication table for the ring $\mathbf{Z}_2[x]/\langle x^3 + x^2 + x + 1\rangle$.

3.†Let R be the ring $\mathbf{Q}[x]/\langle x^3 + 2x^2 - x - 3\rangle$. Describe the elements of R and give the formulas necessary to describe the product of any two elements.

4. Give a multiplication table for the ring $\mathbf{Z}_3[x]/\langle x^2 - 1\rangle$.

5. Show that $\mathbf{Q}[x]/\langle x^2 - 2\rangle \cong \mathbf{Q}[x]/\langle x^2 + 4x + 2\rangle$.

   Hint: Use Example 5.3.5 and Exercise 8 of Section 5.2.

6. Let $R = F[x]$ and let I be any ideal of R.

   (a) Prove that there is a unique monic polynomial $f(x)$ with $I = \langle f(x)\rangle$.

   (b) Prove that if I is a maximal ideal of R, then $I = \langle p(x)\rangle$ for some monic irreducible polynomial $p(x)$.

7. Show that the intersection of two ideals of a commutative ring is again an ideal.

8. Show that if R is a finite ring, then every prime ideal of R is maximal.

9. Find a nonzero prime ideal of $\mathbf{Z} \oplus \mathbf{Z}$ that is not maximal.

10. Let P be a prime ideal of the commutative ring R. Prove that if I and J are ideals of R and $I \cap J \subseteq P$, then either $I \subseteq P$ or $J \subseteq P$.

11. Let R be a commutative ring, with $a \in R$. The **annihilator** of a is defined by

$$\mathrm{Ann}(a) = \{x \in R \mid xa = 0\}.$$

Prove that $\mathrm{Ann}(a)$ is an ideal of R.

12. Recall that an element of a commutative ring is said to be *nilpotent* if $a^n = 0$ for some positive integer n. (See Exercise 7 of Section 5.1.)

(a) Show that the set N of all nilpotent elements of a commutative ring forms an ideal of the ring.

(b) Show that R/N has no nonzero nilpotent elements.

(c) Show that $N \subseteq P$ for each prime ideal P of R.

13. Let R be a commutative ring with ideals I, J. Let

$$I + J = \{x \in R \mid x = a + b \text{ for some } a \in I, b \in J\}.$$

(a) Show that $I + J$ is an ideal.

†(b) Determine $n\mathbf{Z} + m\mathbf{Z}$ in the ring of integers.

14. Let R be a commutative ring with ideals I, J. Define the product of the two ideals by

$$IJ = \left\{\sum_{i=1}^{n} a_i b_i \mid a_i \in I, b_i \in J, n \in \mathbf{Z}^+\right\}.$$

(a) Show that IJ is an ideal that is contained in $I \cap J$.

(b) Determine $(n\mathbf{Z})(m\mathbf{Z})$ in the ring of integers.

15. Let $M = \{f(x, y) \in F[x, y] \mid f(0, 0) = 0\}$ be the maximal ideal of $F[x, y]$ defined in Example 5.3.8.

(a) Show that $M = \{s(x, y)x + t(x, y)y \mid s(x, y), t(x, y) \in F[x, y]\}$.

(b) Using the definition in Exercise 14, find M^2.

16. Let $R = \{m + n\sqrt{2} \mid m, n \in \mathbf{Z}\}$ and let $I = \{m + n\sqrt{2} \mid m, n \in \mathbf{Z} \text{ and } m \text{ is even }\}$.

(a) Show that I is an ideal of R.

†(b) Find the well-known commutative ring to which R/I is isomorphic.

Hint: How many congruence classes does I determine?

17. Let R be the set of all matrices $\begin{bmatrix} a & b \\ c & d \end{bmatrix}$ over $\mathbf{Q}$ such that $a = d$ and $c = 0$.

(a) Verify that R is a commutative ring.

(b) Let I be the set of all such matrices for which $a = d = 0$. Show that I is an ideal of R.

(c) Use the fundamental homomorphism theorem for rings to show that $R/I \cong \mathbf{Q}$.

18. Let R be a commutative ring with ideals I, J such that $I \subseteq J \subseteq R$.

(a) Show that J/I is an ideal of R/I.

(b) Show that the factor ring $(R/I)/(J/I)$ is isomorphic to R/J.

Hint: Define a ring homomorphism from R/I onto R/J and apply the fundamental homomorphism theorem for rings.

(c) Show that J/I is a prime (or maximal) ideal of R/I if and only if J is a prime (or maximal) ideal of R.

19. Use Exercise 18 together with Proposition 5.3.9 to determine all prime ideals and all maximal ideals of $\mathbf{Z}_n$.

20. In the ring $\mathbf{Z}[i]$ of Gaussian integers (see Example 5.1.5) let $\langle p \rangle$ be the ideal generated by a prime number. Show that $\mathbf{Z}[i]/\langle p \rangle$ has p^2 elements, and has characteristic p.

21. In the ring $\mathbf{Z}[i]$ of Gaussian integers find necessary and sufficient conditions on integers m and n for the element $m + ni$ to belong to the ideal $\langle 1 + 2i \rangle$. Use these conditions to determine the ideal $\langle 1 + 2i \rangle \cap \mathbf{Z}$ of $\mathbf{Z}$.

22. In the ring $\mathbf{Z}[i]$ of Gaussian integers show that the ideal $\langle 5 - i \rangle$ is not a prime ideal.
    Hint: Show that $\mathbf{Z}[i]/\langle 5 - i \rangle \cong \mathbf{Z}_{26}$ by defining an onto ring homomorphism $\phi :$ $\mathbf{Z} \to \mathbf{Z}[i]/\langle 5 - i \rangle$ by $\phi(n) = n + \langle 5 - i \rangle$.

23. Let R be the set of all continuous functions from the set of real numbers into itself. In Exercise 13 of Section 5.1, we have shown that R is a commutative ring if the following formulas

$$(f + g)(x) = f(x) + g(x) \quad \text{and} \quad (f \cdot g)(x) = f(x)g(x)$$

for all x, are used to define addition and multiplication of functions. Let a be a fixed real number, and let I be the set of all functions $f(x) \in R$ such that $f(a) = 0$. Show that I is a maximal ideal of R.

24. Let I be the smallest ideal of $\mathbf{Z}[x]$ that contains both 2 and x. Show that I is not a principal ideal.

25. Let R and S be commutative rings, let I be an ideal of R, and let J be an ideal of S.
    (a) Show that $I \oplus J$ is an ideal of $R \oplus S$.
    (b) Show that $(R \oplus S)/(I \oplus J) \cong (R/I) \oplus (S/J)$.
    (c) Show that $I \oplus J$ is a prime ideal of $R \oplus S$ if and only if either $I = R$ and J is a prime ideal of S, or else I is a prime ideal of R and $J = S$.
    (d) Show that if K is any ideal of $R \oplus S$, then there exists an ideal I of R and an ideal J of S such that $K = I \oplus J$.

26. Let R be the set of all rational numbers m/n such that n is odd.
    (a) Show that R is a subring of $\mathbf{Q}$.
    (b) Let $2^k R = \{m/n \in R \mid m$ is a multiple of 2^k and n is odd$\}$, for any positive integer k. Show that $2^k R$ is an ideal of R.
    (c) Show that each proper nonzero ideal of R has the form $2^k R$, for some positive integer k.
    (d) Show that $R/2^k R$ is isomorphic to $\mathbf{Z}_{2^k}$.
    (e) Show that $2R$ is the unique maximal ideal of R.

5.4   Quotient Fields

In Section 5.1 we showed that any subring of a field is an integral domain. In this section we will show that any integral domain is isomorphic to a subring of a field. This can be done in such a way that the subring and field are very closely connected; the field contains no more elements than one would necessarily have in order for each element of the domain to have an inverse.

The best example we can give is provided by considering the ring $\mathbf{Z}$ of integers as a subring of the field $\mathbf{Q}$ of rational numbers. Every rational number has the form m/n for integers m, n, with $n \neq 0$. Another way to say this is to observe that if q is any rational number, then there exists some nonzero integer n such that $nq \in \mathbf{Z}$, showing that $\mathbf{Z}$ and $\mathbf{Q}$ are closely related. This statement is false when $\mathbf{Z}$ is viewed as a subring of $\mathbf{R}$, for example.

We start with any integral domain D. (You may want to keep in mind the ring $\mathbf{R}[x]$ of polynomials with real coefficients.) We have information only about D and the operations defined on it, so we must be very careful not to make implicit assumptions that are not warranted. The situation is something like that in constructing the complex numbers from the real numbers. At a naive level we can simply introduce a symbol i that does what we want it to do, namely, provide a root of the equation $x^2 + 1 = 0$. However, to give a completely rigorous development using only facts about the real numbers, we found it necessary to use the notion of a factor ring and work in $\mathbf{R}[x]/\langle x^2 + 1 \rangle$.

To formally construct fractions with numerator and denominator in D we will consider ordered pairs (a, b), where a is to represent the numerator and b is to represent the denominator. Of course, we need to require that $b \neq 0$. In $\mathbf{Q}$, two fractions m/n and r/s may be equal even though their corresponding numerators and denominators are not equal. We can express the fact that $m/n \;=\; r/s$ by writing $ms = nr$. This shows that we must make certain identifications within the set of ordered pairs (a, b), and the appropriate way to do this is to introduce an equivalence relation.

5.4.1 Lemma. *Let D be an integral domain, and let*

$$W = \{(a, b) \mid a, b \in D \ \text{and} \ b \neq 0\} \ .$$

The relation $\sim$ defined on W by $(a, b) \sim (c, d)$ if $ad = bc$ is an equivalence relation.

Proof. Given $(a, b) \in W$ we have $(a, b) \sim (a, b)$ since $ab = ba$. The symmetric law holds since if $(c, d) \in W$ with $(a, b) \sim (c, d)$, then $ad = bc$, and so $cb = da$, showing that $(c, d) \sim (a, b)$. To show the transitive law, suppose that $(a, b) \sim (c, d)$ and $(c, d) \sim (u, v)$ for ordered pairs in W. Then we have $ad = bc$ and $cv = du$, and so multiplying the first equation by v gives $adv = bcv$, while multiplying the second equation by b gives $bcv = bdu$. Thus we have $adv = bdu$, and since d is a nonzero element of an integral domain, we can cancel it to obtain $av = bu$, which shows that $(a, b) \sim (u, v)$. $\square$

5.4.2 Definition. *Let D be an integral domain. The equivalence classes of the set*

$$\{(a, b) \mid a, b \in D \text{ and } b \neq 0\}$$

under the equivalence relation defined by $(a, b) \sim (c, d)$ if $ad = bc$ will be denoted by $[a, b]$.

The set of all such equivalence classes will be denoted by $Q(D)$.

Since our ultimate goal is to show that $Q(D)$ is a field that contains a copy of D as a subring, we must decide how to define addition and multiplication of equivalence classes. Returning to **Q**, we have

$$\frac{m}{n} + \frac{p}{q} = \frac{mq}{nq} + \frac{np}{nq} = \frac{mq + np}{nq}.$$

With this as motivation, we define

$$[a, b] + [c, d] = [ad + bc, bd],$$

since a and c represent the "numerators" of our equivalence classes, while b and d represent the "denominators," and by analogy with **Q** the "numerator" and "denominator" of the sum should be $ad + bc$ and bd, respectively. Similarly, since

$$\frac{m}{n} \cdot \frac{p}{q} = \frac{mp}{nq},$$

we define $[a, b] \cdot [c, d] = [ac, bd]$.

5.4.3 Lemma. *Let D be an integral domain. Define binary operations $+$ and $\cdot$ on $Q(D)$ by*

$$[a, b] + [c, d] = [ad + bc, bd] \quad \text{and} \quad [a, b] \cdot [c, d] = [ac, bd],$$

for $[a, b], [c, d] \in Q(D)$. Then $+$ and $\cdot$ are well-defined operations on $Q(D)$.

Proof. Let $[a, b] = [a', b']$ and $[c, d] = [c', d']$ be elements of $Q(D)$. To show that addition is well-defined, we must show that

$$[a, b] + [c, d] = [a', b'] + [c', d'],$$

or, equivalently, that

$$[ad + bc, bd] = [a'd' + b'c', b'd'].$$

In terms of the equivalence relation on $Q(D)$, this means that we must show that

$$(ad + bc)b'd' = bd(a'd' + b'c') .$$

We are given that $ab' = ba'$ and $cd' = dc'$. Multiplying the first of these two equations by dd' and the second by bb' and then adding terms gives us the desired equation.

In order to show that multiplication is well-defined, we must show that $[ac, bd] = [a'c', b'd']$. This is left to the reader, as Exercise 1. $\square$

5.4.4 Theorem. *Let D be an integral domain. Then $Q(D)$ is a field that contains a subring isomorphic to D.*

Proof. Showing that the commutative and associative laws hold for addition is left as Exercise 2. Let 1 be the identity element of D. For all $[a, b] \in Q(D)$, we have $[0, 1] + [a, b] = [a, b]$, and so $[0, 1]$ serves as an additive identity element. Furthermore,

$$[-a, b] + [a, b] = [-ab + ba, b^2] = [0, b^2] ,$$

which shows that $[-a, b]$ is the additive inverse of $[a, b]$ since $[0, b^2] = [0, 1]$.

Verifying the commutative and associative laws for multiplication is left as an exercise. The equivalence class $[1, 1]$ clearly acts as a multiplicative identity element. Note that we have $[1, 1] = [d, d]$ for any nonzero $d \in D$. If $[a, b]$ is a nonzero element of $Q(D)$, that is, if $[a, b] \neq [0, 1]$, then we must have $a \neq 0$. Therefore $[b, a]$ is an element of $Q(D)$, and since $[b, a] \cdot [a, b] = [ab, ab]$, we have $[b, a] = [a, b]^{-1}$. Thus every nonzero element of $Q(D)$ is invertible, and so to complete the proof that $Q(D)$ is a field we only need to check that the distributive law holds.

To show that the distributive law holds, let $[a, b]$, $[c, d]$, and $[u, v]$ be elements of $Q(D)$. We have

$$([a, b] + [c, d]) \cdot [u, v] = [(ad + bc)u, (bd)v]$$

and

$$[a, b] \cdot [u, v] + [c, d] \cdot [u, v] = [(au)(dv) + (bv)(cu), (bv)(dv)] .$$

We can factor $[v, v]$ out of the second expression, showing equality.

Finally, consider the mapping $\phi : D \to Q(D)$ defined by $\phi(d) = [d, 1]$, for all $d \in D$. It is easy to show that ϕ preserves sums and products, and since $\phi(1) = [1, 1]$ is the multiplicative identity of $Q(D)$, it follows that ϕ is a ring homomorphism. If $[d, 1] = [0, 1]$, then we must have $d = 0$, which shows that $\ker(\phi) = \{0\}$. By the fundamental homomorphism theorem for rings, $\phi(D)$ is a subring of $Q(D)$ that is isomorphic to D. $\square$

5.4.5 Definition. *Let D be an integral domain. The field $Q(D)$ defined in Definition 5.4.2 is called the **field of quotients** or **field of fractions** of D.*

5.4.6 Theorem. *Let D be an integral domain, and let $\phi : D \to Q(D)$ be the ring homomorphism defined by $\phi(d) = [d, 1]$, for all $d \in D$.*

If $\theta : D \to F$ is any one-to-one ring homomorphism from D into a field F, then there exists a unique ring homomorphism $\widehat{\theta} : Q(D) \to F$ that is one-to-one and satisfies $\widehat{\theta}\phi(d) = \theta(d)$, for all $d \in D$.

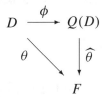

Proof. For $[a, b] \in Q(D)$, define $\widehat{\theta}([a, b]) = \theta(a)\theta(b)^{-1}$. Since $b \neq 0$ and θ is one-to-one, $\theta(b)^{-1}$ exists in F, and the definition makes sense. We must show that $\widehat{\theta}$ is well-defined. If $[a, b] = [a', b']$, then $ab' = ba'$, and applying θ to both sides of this equation gives $\theta(a)\theta(b') = \theta(b)\theta(a')$ since θ is a ring homomorphism. Both $\theta(b)^{-1}$ and $\theta(b')^{-1}$ exist, so we must have $\theta(a)\theta(b)^{-1} = \theta(a')\theta(b')^{-1}$.

The proof that $\widehat{\theta}$ is a one-to-one ring homomorphism is left as an exercise. For any $d \in D$, we have $\widehat{\theta}(\phi(d)) = \widehat{\theta}([d, 1]) = \theta(d)\theta(1)^{-1} = \theta(d)$, as required.

To prove the uniqueness of $\widehat{\theta}$, suppose that $\psi : Q(D) \to F$ with $\psi\phi(d) = \theta(d)$ for all $d \in D$. It follows from the definition of ϕ that $\psi([d, 1]) = \theta(d)$ for all $d \in D$. For any element $[a, b] \in Q(D)$ we have

$$\psi([a, 1]) = \psi([a, b][b, 1]) = \psi([a, b])\psi([b, 1]) ,$$

and so substituting $\theta(a)$ for $\psi([a, 1])$ and $\theta(b)$ for $\psi([b, 1])$ gives us

$$\theta(a) = \psi([a, b])\theta(b)$$

or

$$\psi([a, b]) = \theta(a)\theta(b)^{-1} ,$$

which completes the proof. □

We have continued to use the notation $[a, b]$ for elements of $Q(D)$ in order to emphasize that equivalence classes are involved. From this point on we will use the more familiar notation a/b in place of $[a, b]$. We identify an element $d \in D$ with the fraction $d/1$, and this allows us to assume that D is a subring of $Q(D)$. If $b \in D$ is nonzero, then $1/b \in Q(D)$, and $(1/b) \cdot (b/1) = b/b = 1$ shows that $1/b = b^{-1}$, (where we have identified b and $b/1$). Thus we can also write $a/b = (a/1) \cdot (1/b) = ab^{-1}$, for $a, b \in D$, with $b \neq 0$.

5.4.7 Corollary. *Let D be an integral domain that is a subring of a field F. If each element of F has the form a/b for some $a, b \in D$, then F is isomorphic to the field of quotients $Q(D)$ of D.*

Proof. By Theorem 5.4.6, the inclusion mapping $\theta : D \to F$ can be extended to a one-to-one ring homomorphism $\widehat{\theta} : Q(D) \to F$. The condition that each element of F has the form a/b for some $a, b \in D$ is precisely the one necessary to guarantee that $\widehat{\theta}$ is onto. □

Example 5.4.1.

Let D be the integral domain consisting of all fractions $m/n \in \mathbf{Q}$ such that n is odd. (See Exercise 26 in Section 5.3.) If a/b is any element of $\mathbf{Q}$ such that $\gcd(a, b) = 1$, then either b is odd, in which case $a/b \in D$, or b is even, in which case a is odd and $a/b = 1 \cdot (b/a)^{-1}$, with $b/a \in D$. Applying Corollary 5.4.7 shows that $\mathbf{Q} \cong Q(D)$. □

Example 5.4.2 (Field of rational functions).

If F is any field, then we know that the ring of polynomials $F[x]$ is an integral domain. Applying Theorem 5.4.4 shows that we can construct a field that contains $F[x]$ by considering all fractions of the form $f(x)/g(x)$, where $f(x), g(x)$ are polynomials with $g(x) \neq 0$.

This field is called the **field of rational functions** in x, over the field F, and is denoted by $F(x)$. Note that the elements of $F(x)$ are no more functions than are the elements of the polynomial ring $F[x]$. □

5.4.8 Corollary. *Any field contains a subfield isomorphic to $\mathbf{Q}$ or $\mathbf{Z}_p$, for some prime number p.*

Proof. Let F be any field, and, as in Proposition 5.2.11, let ϕ be the homomorphism from $\mathbf{Z}$ into F defined by $\phi(n) = n \cdot 1$. If $\ker(\phi) \neq \{0\}$, then as in Proposition 5.2.11, we have $\ker(\phi) = p\mathbf{Z}$ for some prime p, and so the image of ϕ is a subfield isomorphic to $\mathbf{Z}_p$. If ϕ is one-to-one, then by Theorem 5.4.6 it extends to a one-to-one ring homomorphism from $\mathbf{Q}$ (the field of quotients of $\mathbf{Z}$) into F. The image of this ring homomorphism is a subfield of F isomorphic to $\mathbf{Q}$. □

EXERCISES: SECTION 5.4

Throughout these exercises, D will denote an integral domain, and $Q(D)$ will denote its quotient field.

1. Complete the proof in Lemma 5.4.3, to show that multiplication of equivalence classes in $Q(D)$ is well-defined.

2. Show that the associative and commutative laws hold for addition in $Q(D)$.

3. Show that the associative and commutative laws hold for multiplication in $Q(D)$.

4. Let $\phi : D \to Q(D)$ be the mapping $\phi(d) = [d, 1]$ defined in Theorem 5.4.4. Show that ϕ is an isomorphism if and only if D is a field.

5. In Theorem 5.4.6, verify that $\widehat{\theta}$ is a one-to-one ring homomorphism.

6. Let D_1 and D_2 be integral domains, with quotient fields $Q(D_1)$ and $Q(D_2)$, respectively. Let $\theta : D_1 \to D_2$ be a ring homomorphism.

   (a) Prove that if θ is one-to-one, then there exists a ring homomorphism $\widehat{\theta} : Q(D_1) \to Q(D_2)$ such that $\widehat{\theta}([d, 1]) = [\theta(d), 1]$ for all $d \in D_1$.

   (b) Prove that if θ is *not* one-to-one, then it is impossible to find a ring homomorphism $\widehat{\theta} : Q(D_1) \to Q(D_2)$ that satisfies the conditions of part (a).

7.† Determine $Q(D)$ for $D = \{m + n\sqrt{2} \mid m, n \in \mathbf{Z}\}$. (See Example 5.1.6.)

8. Let p be a prime number, and let $D = \{m/n \mid m, n \in \mathbf{Z} \text{ and } p \nmid n\}$. Verify that D is an integral domain and find $Q(D)$.

9.† Determine $Q(D)$ for $D = \{m + ni \mid m, n \in \mathbf{Z}\} \subseteq \mathbf{C}$.

10. Considering $\mathbf{Z}[x]$ as a subring of $\mathbf{Q}[x]$, show that both rings have the same quotient field.

11. Show that if P is a prime ideal of D, then $D_P = \{a/b \in Q(D) \mid b \notin P\}$ is an integral domain with $D \subseteq D_P \subseteq Q(D)$.

12. In the ring D_P defined in Exercise 11, let $M = \{a/b \in D_P \mid a \in P\}$.

    (a) Show that M is an ideal of D_P.

    (b) Show that $D_P/M \cong Q(R/P)$, and conclude that M is a maximal ideal of D_P.

13. Let R be a commutative ring. A **derivation** on R is a function $\partial : R \to R$ such that (i) $\partial(x + y) = \partial(x) + \partial(y)$ and (ii) $\partial(xy) = \partial(x)y + x\partial(y)$. Show that if ∂ is a derivation on an integral domain D with quotient field $Q(D)$, then ∂ can be extended to a derivation $\overline{\partial}$ of $Q(D)$ by defining $\overline{\partial}(a/b) = (b\partial(a) - a\partial(b))/b^2$ for all $a, b \in D$ with $b \neq 0$.

14. Show that $\partial : \mathbf{Q}[x] \to \mathbf{Q}[x]$ defined by $\partial(f(x)) = f'(x)$ for all $f(x) \in \mathbf{Q}[x]$ is a derivation. Describe the derivation $\overline{\partial}$ defined on the quotient field of $\mathbf{Q}[x]$ (see Exercise 13).

Notes

The terminology for the structures we studied in this chapter apparently came originally from an 1897 paper by David Hilbert (1862–1943). He used the name "number ring" for sets of the form $\{m + n\alpha \mid m, n \in \mathbf{Z}\}$, where α is a root of a quadratic polynomial with integer coefficients. The more general definition of a ring evolved over the next quarter of a century.

Of course, earlier work had been done in specific contexts. In particular, there was a great deal of interest in determining whether or not a unique factorization theorem (similar to Theorem 1.2.7 and Theorem 4.2.9) holds in various subsets of the complex numbers. This question was answered in the negative as early as 1844. (See the introduction to Chapter 9 for further information.)

Hilbert used ideal theory to reformulate and solve some problems in the theory of invariants, which deals with finding polynomials that remain fixed under particular sets of transformations. One of the problems was to find a finite set of invariants that could be used in determining all the rest, and this is done in the "Hilbert basis theorem." (See Exercise 9 of Section 9.2 for a statement of the theorem.)

Hilbert also recognized that the theory of ideals was important in the study of algebraic curves. Even today, commutative ring theory has maintained very close connections with the field of algebraic geometry. As with much of abstract algebra, the general notions that we now use took their present form in the work of the great algebraist Emmy Noether (1882–1935).

Chapter 6

FIELDS

In Corollary 4.3.9 we showed that for any polynomial over a field K, a larger field F can be constructed, which contains enough of the roots of the given polynomial so that the polynomial then "splits" into a product of linear polynomials with co-efficients in F. Thus the roots of any polynomial in $K[x]$ can always be found in some extension field of K.

A polynomial in $\mathbf{Q}[x]$ is said to be *solvable by radicals* if its roots can be obtained from the coefficients of the polynomial by allowing field operations and the extraction of nth roots, for various n. In Chapter 8 we will answer the question of which polynomials in $\mathbf{Q}[x]$ are solvable by radicals. To do this we must study the interplay between the field of coefficients of the polynomial and the field of roots of the polynomial. This chapter therefore studies field extensions and splitting fields.

If F is an extension field of K, then F can be viewed as a vector space over K. We will exploit the concept of the dimension of a vector space to show that several geometric constructions are impossible. These constructions were already studied by the Greeks in the fifth century B.C. It is impossible to find a general method to trisect an angle; given a circle, it is impossible to construct a square of the same area; and given a cube, it is impossible to construct a cube with double the given volume. The method of construction in each case is limited to using a straightedge and compass. The points constructed at any stage of the procedure represent solutions of quadratic equations with coefficients from the smallest field containing the previously constructed numbers. We will show that any constructible number lies in an extension of $\mathbf{Q}$ whose dimension over $\mathbf{Q}$ is a power of 2, and thus to show that a particular number cannot be constructed, we only need to use a dimension argument.

As a further application we will be able to give a complete list of all finite fields. Such fields are used in algebraic coding theory, which provides an approach to a problem encountered in the transmission of encoded data. When data is transmitted over telephone lines or via satellite connections, there is a substantial chance that errors will occur. To make it possible to detect (and even correct) these errors, additional information can be sent along. In the most naive approach, the entire message could simply be repeated numerous times. The real task is to find efficient algorithms for encoding, and one successful algorithm involves the use of polynomials over finite fields.

6.1   Algebraic Elements

We recall from Chapter 5 that a commutative ring (with $1 \neq 0$) is called a *field* if every nonzero element is invertible. (See Definition 4.1.1 for a complete list of the properties of a field.) Thus the operations of addition, subtraction, multiplication, and division (by nonzero elements) are all possible within a field. We should also note that the elements of a field form an abelian group under addition, while the nonzero elements form an abelian group under multiplication. This observation allows us to make use of the results we have proved for groups in Chapter 3.

Our primary interest is to study roots of polynomials. This usually involves the interplay of two fields: one that contains the coefficients of the polynomial, and another that contains the roots of the polynomial. In many situations we will start with a known field K, and then construct a larger field F. At this point we need to recall Definition 4.3.1. The field F is said to be an *extension field* of the field K if K is a subset of F which is a field under the operations of F. This is equivalent to saying that K is a *subfield* of F. In this context K is often called the *base field.* Note that if K is a subfield of F, then the additive and multiplicative groups that determine K are subgroups of the corresponding additive and multiplicative groups of F.

If F is an extension field of the base field K, then those elements of F that are roots of nonzero polynomials in $K[x]$ will be called *algebraic over* K. If $u \in F$ is a root of some nonzero polynomial $f(x) \in K[x]$, then let $p(x)$ have minimal degree among all polynomials of which u is a root. Using the division algorithm, we can write $f(x) = q(x)p(x) + r(x)$, where either $r(x) = 0$ or $\deg(r(x)) < \deg(p(x))$. Solving for $r(x)$ and substituting u shows that u is a root of $r(x)$, which violates the definition of $p(x)$ unless $r(x) = 0$, and so we have shown that $p(x)$ is a divisor of $f(x)$.

With the notation above, we next observe that $p(x)$ must be an irreducible polynomial. To show this, suppose that $p(x) = g(x)h(x)$ for polynomials $g(x), h(x)$ in $K[x]$ with $\deg(g(x)) < \deg(p(x))$ and $\deg(h(x)) < \deg(p(x))$. Then substituting u gives $g(u)h(u) = p(u) = 0$, and so either $g(u) = 0$ or $h(u) = 0$ since these are elements of F, which is a field. This contradicts the definition of $p(x)$ as a polynomial of minimal degree that has u as a root.

These facts about the polynomials that have a given element as a root can be proved in another way by using the concept of an ideal from Chapter 5. We will show in the proof of Proposition 6.1.2 that if F is an extension field of K, and $u \in F$, then $\{f(x) \in F[x] \mid f(u) = 0\}$ is an ideal of the ring $F[x]$. The arguments in the preceding paragraphs can be expressed more neatly by using results from ring theory. This approach also lends itself to more powerful applications, and so we will adopt the ring-theoretic point of view.

Recall that the nonzero ideals of $F[x]$ are all principal, of the form $\langle f(x) \rangle = \{ q(x) f(x) \mid q(x) \in F[x] \}$, where $f(x)$ is any polynomial of minimal degree in the ideal (see Example 5.3.7). A nonzero ideal is prime (and hence maximal) if and only if its generator $f(x)$ is irreducible. Furthermore, if $p(x)$ is irreducible, then the factor ring $F[x]/\langle p(x) \rangle$ is a field.

6.1.1 Definition. *Let F be an extension field of K and let $u \in F$. If there exists a nonzero polynomial $f(x) \in K[x]$ such that $f(u) = 0$, then u is said to **satisfy the polynomial** $f(x)$ and to be **algebraic over** K.*

*If u does not satisfy any nonzero polynomial in $K[x]$, then u is said to be **transcendental over** K.*

The familiar constants e and π are transcendental over $\mathbf{Q}$. These are not easy results, and the analytic proofs lie beyond the scope of this book. (Proofs can be found in the book by I. Niven called *Irrational Numbers*.) That e is transcendental was proved by Charles Hermite (1822–1882) in 1873, and the corresponding result for π was proved by Ferdinand Lindemann (1852–1939) in 1882.

6.1.2 Proposition. *Let F be an extension field of K, and let $u \in F$ be algebraic over K. Then there exists a unique monic irreducible polynomial $p(x) \in K[x]$ such that $p(u) = 0$. It is characterized as the monic polynomial of minimal degree that has u as a root. Furthermore, if $f(x)$ is any polynomial in $K[x]$ with $f(u) = 0$, then $p(x) \mid f(x)$.*

Proof. Let I be the set of all polynomials $f(x) \in K[x]$ such that $f(u) = 0$. It is easy to see that I is closed under sums and differences, and if $f(x) \in I$, then $g(x) f(x) \in I$ for all $g(x) \in K[x]$. Thus I is an ideal of $K[x]$, and so $I = \langle p(x) \rangle$ for any nonzero polynomial $p(x) \in I$ that has minimal degree. If $f(x), g(x) \in K[x]$ with $f(x) g(x) \in I$, then we have $f(u) g(u) = 0$, which implies that either $f(u) = 0$ or $g(u) = 0$, and so we see that I is a prime ideal. This implies that the unique monic generator $p(x)$ of I must be irreducible. Finally, since $I = \langle p(x) \rangle$, we have $p(x) \mid f(x)$ for any $f(x) \in I$. $\square$

6.1.3 Definition. *Let F be an extension field of K, and let u be an algebraic element of F. The monic polynomial $p(x)$ of minimal degree in $K[x]$ such that $p(u) = 0$ is called the **minimal polynomial** of u over K. The degree of the minimal polynomial of u over K is called the **degree** of u over K.*

Example 6.1.1 ($\sqrt{2}$ has degree 2 over Q).

Considering the set of real numbers **R** as an extension field of the set of rational numbers **Q**, the number $\sqrt{2} \in \mathbf{R}$ has minimal polynomial $x^2 - 2$ over **Q**, and so it has degree 2 over **Q**. □

Example 6.1.2 ($\sqrt[4]{2}$ has degree 2 over Q($\sqrt{2}$)).

Recall that $\mathbf{Q}(\sqrt{2}) = \{a + b\sqrt{2} \mid a, b \in \mathbf{Q}\}$ was shown to be an extension field of **Q** in Example 4.1.1. Considering **R** as an extension field of $\mathbf{Q}(\sqrt{2})$, the number $\sqrt[4]{2} \in \mathbf{R}$ has minimal polynomial $x^2 - \sqrt{2}$ over $\mathbf{Q}(\sqrt{2})$, and so it has degree 2 over $\mathbf{Q}(\sqrt{2})$.

We note that $\sqrt[4]{2}$ has degree 4 over **Q**, since Eisenstein's criterion shows that $x^4 - 2$ is irreducible over **Q**, and so it is the minimal polynomial of $\sqrt[4]{2}$ over **Q**. The minimal polynomial over **Q** and the minimal polynomial over $\mathbf{Q}(\sqrt{2})$ are related since $x^2 - \sqrt{2}$ is a factor of $x^4 - 2$ in $\mathbf{Q}(\sqrt{2})[x]$. □

Example 6.1.3 ($\sqrt{2} + \sqrt{3}$ has degree 4 over Q).

In this example we will compute the minimal polynomial of $\sqrt{2} + \sqrt{3}$ over **Q**. If we let $x = \sqrt{2} + \sqrt{3}$, then we must find a nonzero polynomial with rational coefficients that has x as a root. We begin by rewriting our equation as $x - \sqrt{2} = \sqrt{3}$. Squaring both sides gives $x^2 - 2\sqrt{2}x + 2 = 3$. Since we still need to eliminate the square root to obtain coefficients over **Q**, we can again rewrite the equation to obtain $x^2 - 1 = 2\sqrt{2}x$. Then squaring both sides and rewriting the equation gives $x^4 - 10x^2 + 1 = 0$.

To show that $x^4 - 10x^2 + 1$ is the minimal polynomial of $\sqrt{2} + \sqrt{3}$ over **Q**, we must show that it is irreducible over **Q**. It is easy to check that there are no rational roots, so it could only be the product of two quadratic polynomials, which by Theorem 4.4.5 can be assumed to have integer coefficients. Unfortunately, Eisenstein's irreducibility criterion cannot be applied, and so we must try to verify directly that the polynomial is irreducible over **Q**. A factorization over **Z** of the form

$$x^4 - 10x^2 + 1 = (x^2 + ax + b)(x^2 + cx + d)$$

leads to the equations $a + c = 0$, $b + ac + d = -10$, $ad + bc = 0$, and $bd = 1$. Substituting for c in the second equation, we get $a^2 = b + d + 10$. Either $b = d = 1$, and $a^2 = 12$, or $b = d = -1$, and $a^2 = 8$, a contradiction in either case since $a \in \mathbf{Z}$. We conclude that $x^4 - 10x^2 + 1$ is irreducible over **Q**, and so it must be the minimal polynomial of $\sqrt{2} + \sqrt{3}$ over **Q**. □

In a field F, the intersection of any collection of subfields of F is again a subfield. In particular, if F is an extension field of K and S is a subset of F, then the intersection of all subfields of F that contain both K and S is a subfield of F. This intersection is contained in any subfield that contains both K and S. This guarantees the existence of the field defined below.

6.1.4 Definition. *Let F be an extension field of K, and let $u_1, u_2, \ldots, u_n \in F$. The smallest subfield of F that contains K and $u_1, u_2, \ldots, u_n$ will be denoted by $K(u_1, u_2, \ldots, u_n)$. It is called the **extension field of K generated by** $u_1, u_2, \ldots, u_n$. Alternatively, $K(u_1, u_2, \ldots, u_n)$ is called the **extension field of K defined by adjoining** $u_1, u_2, \ldots, u_n$ to K.*

*If $F = K(u)$ for a single element $u \in F$, then F is said to be a **simple extension** of K.*

If F is an extension field of K, and $u_1, u_2, \ldots, u_n \in F$, then it is possible to construct $K(u_1, u_2, \ldots, u_n)$ by adjoining one element u_i at a time. That is, we first construct $K(u_1)$, and then consider the smallest subfield of F that contains $K(u_1)$ and u_2. This would be written as $K(u_1)(u_2)$, but it is clear from the definition of $K(u_1, u_2)$ that the two fields are equal. This procedure can be repeated to construct $K(u_1, u_2, \ldots, u_n)$. The next proposition describes the adjunction of a single element.

6.1.5 Proposition. *Let F be an extension field of K, and let $u \in F$.*

   (a) If u is algebraic over K, then $K(u) \cong K[x]/\langle p(x) \rangle$, where $p(x)$ is the minimal polynomial of u over K.

   (b) If u is transcendental over K, then $K(u) \cong K(x)$, where $K(x)$ is the quotient field of the integral domain $K[x]$.

Proof. Define $\phi_u : K[x] \to F$ by $\phi_u(f(x)) = f(u)$, for all polynomials $f(x) \in K[x]$. This defines a ring homomorphism, and $\ker(\phi_u)$ is the set of all polynomials $f(x)$ with $f(u) = 0$. The image of ϕ_u is a subring of F consisting of all elements of the form $a_0 + a_1u + \ldots + a_nu^n$, and it must be contained in every subring of F that contains K and u. In particular, the image of ϕ_u must be contained in $K(u)$.

   (a) If u is algebraic over K, then $\ker(\phi_u) = \langle p(x) \rangle$, for the minimal polynomial $p(x)$ of u over K. In this case the fundamental homomorphism theorem for rings implies that the image of ϕ_u is isomorphic to $K[x]/\langle p(x) \rangle$, which is a field since $p(x)$ is irreducible. But then the image of ϕ_u must in fact be equal to $K(u)$, since the image is a subfield containing K and u.

   (b) If u is transcendental over K, then $\ker(\phi_u) = \{0\}$, and so the image of ϕ_u is isomorphic to $K[x]$. Since F is a field, by Theorem 5.4.6 there exists an isomorphism θ from the field of quotients of $K[x]$ into F. Since every element of the image of θ is a quotient of elements that belong to $K(u)$, it follows that this image must be contained in $K(u)$. Then since $\theta(Q(K[x])) = \theta(K(x))$ is a field that contains u, it must be equal to $K(u)$. $\square$

To help understand the field $K(u)$, it may be useful to approach its construction from a more elementary point of view. Given an extension field F of K and an element $u \in F$, any subfield that contains K and u must be closed under sums and products, so it must contain all elements of the form $a_0 + a_1u + \ldots + a_nu^n$, where $a_i \in K$ for $0 \leq i \leq n$. Furthermore, since it must be closed under division, it must contain all elements of the form

$$\frac{a_0 + a_1u + \ldots + a_nu^n}{b_0 + b_1u + \ldots + b_mu^m}$$

such that the denominator is nonzero. The set of all such quotients can be shown to be a subfield of F, and so it must be equal to $K(u)$.

If u is algebraic of degree n over K, let the minimal polynomial of u over K be $p(x) = c_0 + c_1x + \ldots + c_nx^n$. Since $c_0 + c_1u + \ldots + c_nu^n = 0$, we can solve for u^n and obtain a formula that allows us to reduce any expression of the form $a_0 + a_1u + \ldots + a_mu^m$ to one involving only $u, u^2, \ldots, u^{n-1}$ and elements of K. Given any expression of the form $a_0 + a_1u + \ldots + a_{n-1}u^{n-1}$, we let $f(x)$ be the corresponding polynomial $a_0 + a_1x + \ldots + a_{n-1}x^{n-1}$. If $f(x)$ is nonzero, then it must be relatively prime to $p(x)$ since $p(x)$ is irreducible and $\deg(f(x)) < \deg(p(x))$. Thus there exist polynomials $g(x)$ and $q(x)$ such that $f(x)g(x) + p(x)q(x) = 1$. Substituting u gives $f(u)g(u) = 1$ since $p(u) = 0$, and so $g(u) = 1/f(u)$. Thus the denominators can be eliminated in our description of $K(u)$. We conclude that when u is algebraic over K of degree n, each element of $K(u)$ has the form $a_0 + a_1u + \ldots + a_{n-1}u^{n-1}$ for elements $a_0, a_1, \ldots, a_{n-1} \in K$.

Example 6.1.4 ($Q(\sqrt{2}) \cong Q[x]/\langle x^2 - 2 \rangle$).

In the light of Example 6.1.1 and Proposition 6.1.5 (a), we can improve the result in Example 5.2.10, since we now know that the kernel of the evaluation mapping $\phi : Q[x] \to R$ defined by $\sqrt{2}$ is the ideal $\langle x^2 - 2 \rangle$. $\square$

Example 6.1.5 (Computations in $Q(\sqrt[3]{2})$).

We can let u be any root of the polynomial $x^3 - 2$, since the computations will be the same as for $u = \sqrt[3]{2}$. The extension field $Q(u)$ of Q is isomorphic to the factor ring $Q[x]/\langle x^3 - 2 \rangle$, and so computations can be done in either field. For example, let us compute $(1 + u^2)^{-1}$. In $Q(u)$ we can set up the equation $(1 + u^2)(a + bu + cu^2) = 1$. Using the identities $u^3 = 2$ and $u^4 = 2u$ to multiply out the left-hand side, we obtain the equations $a + 2b = 1$, $b + 2c = 0$, and $a + c = 0$. These lead to the solution $a = 1/5$, $b = 2/5$, and $c = -1/5$.

On the other hand, if we let $\langle x^3 - 2 \rangle = I$, then to find the multiplicative inverse of the coset $1 + x^2 + I$ in $\mathbf{Q}[x]/I$, we can use the Euclidean algorithm to solve for

$$\gcd(x^2 + 1, x^3 - 2) .$$

We obtain $x^3 - 2 = x(x^2 + 1) - (x + 2)$ and then $x^2 + 1 = (x - 2)(x + 2) + 5$. Solving for the linear combinations that give the greatest common divisor yields the equation

$$1 = \left(-\frac{1}{5}x^2 + \frac{2}{5}x + \frac{1}{5} \right)(x^2 + 1) + \left(\frac{1}{5}x - \frac{2}{5} \right)(x^3 - 2) .$$

Thus $1 + I = (1 + x^2 + I)\left(\frac{1}{2} + \frac{2}{5}x - \frac{1}{5}x^2 + I \right)$ and so we can use the isomorphism to obtain the same answer we got previously:

$$\left(1 + u^2 \right)^{-1} = \frac{1}{5} + \frac{2}{5}u - \frac{1}{5}u^2 . \quad \square$$

Because of its importance in our current discussion, we recall the statement and proof of Kronecker's theorem. We could now add the following information to the statement of the theorem: if $f(x)$ is irreducible, then it is the minimal polynomial for u over K.

4.3.8. Theorem (Kronecker) *Let K be a field, and let $f(x)$ be any nonconstant polynomial in $K[x]$. Then there exists an extension field F of K and an element $u \in F$ such that $f(u) = 0$.*

Proof. Recall that the extension field F is constructed as $K[x]/\langle p(x) \rangle$, where $p(x)$ is an irreducible factor of $f(x)$. Then K is viewed as isomorphic to the subfield consisting of all cosets of the form $a + \langle p(x) \rangle$, where $a \in K$. The element u is the coset determined by x, and it follows that $f(u) = 0$. $\square$

EXERCISES: SECTION 6.1

1. Show that the following complex numbers are algebraic over $\mathbf{Q}$.
    †(a) $\sqrt{2}$
     (b) $\sqrt{n}$, for $n \in \mathbf{Z}^+$
    †(c) $\sqrt{3} + \sqrt{5}$
     (d) $\sqrt{2 + \sqrt{3}}$
    †(e) $(-1 + \sqrt{3}i)/2$
     (f) $\sqrt[3]{2} + \sqrt{2}$

$1^3 + 3(1)^2 (2) + 3 (1)(2)^2 + (2)^3$

2. Let F be an extension field of K, and let u be a nonzero element of F that is algebraic over K. Show that u^{-1} is also algebraic over K.

3. Suppose that u is algebraic over the field K, and that $a \in K$. Show that $u + a$ is algebraic over K, find its minimal polynomial over K, and show that the degree of $u + a$ over K is equal to the degree of u over K.

4. Show that $\sqrt{3} \notin \mathbf{Q}(\sqrt{2})$.

5. (a) Show that $f(x) = x^3 + 3x + 3$ is irreducible over $\mathbf{Q}$.

†(b) Let u be a root of $f(x)$. Express u^{-1} and $(1 + u)^{-1}$ in the form $a + bu + cu^2$, where $a, b, c \in \mathbf{Q}$.

6. Show that the intersection of any collection of subfields of a given field is again a subfield.

7. Let $F = K(u)$, where u is transcendental over the field K. If E is a field such that $K \subset E \subseteq F$, then show that u is algebraic over E.

8. Let F be an extension field of K.

(a) Show that F is a vector space over K.

(b) Let $u \in F$. Show that u is algebraic over K if and only if the subspace spanned by $\{1, u, u^2, \ldots\}$ is a field.

9. Let F be an extension field of K. If $u \in F$ is transcendental over K, then show that every element of $K(u)$ that is not in K is also transcendental over K.

10. Let u and r be positive real numbers, with $u \neq 1$. It follows from a famous theorem of Gelfand and Schneider that if r is irrational and both u and r are algebraic over $\mathbf{Q}$, then u^r must be transcendental over $\mathbf{Q}$. You may use this result to show that the following numbers are transcendental over $\mathbf{Q}$.

(a) $\sqrt[3]{7}^{\sqrt{5}}$

(b) $\sqrt[3]{7}^{\sqrt{5}} + 7$

11. Show that there exist irrational numbers $a, b \in \mathbf{R}$ such that a^b is rational.

12. Assuming that π is transcendental over $\mathbf{Q}$, prove that either $\pi + e$ or $\pi \cdot e$ is irrational.

6.2 Finite and Algebraic Extensions

If F is an extension field of K, then multiplication in F defines a scalar multiplication, if we consider the elements of K as scalars and the elements of F as vectors. It is easy to check that the necessary axioms hold for this scalar multiplication, and since F is an abelian group under addition, we see that F is a vector space over K. This fact is worth formally recording as a proposition.

6.2.1 Proposition. *If F is an extension field of K, then F is a vector space over K.*

Knowing that an extension field is a vector space over the base field allows us to make use of the concept of the dimension of a vector space. (If you need to review results on dimension, see Section A.7 of the appendix.)

$K \subseteq F$

6.2.2 Proposition. *Let F be an extension field of K and let $u \in F$ be an element algebraic over K. If the minimal polynomial of u over K has degree n, then $K(u)$ is an n-dimensional vector space over K.*

Proof. Let $p(x) = c_0 + c_1 x + \ldots + c_n x^n$ be the minimal polynomial of u over K. We will show that the set $\mathcal{B} = \{1, u, u^2, \ldots, u^{n-1}\}$ is a basis for $K(u)$ over K. By Proposition 6.1.5, $K(u) \cong K[x]/\langle p(x) \rangle$, and since each coset of $K[x]/\langle p(x) \rangle$ contains a unique representative of degree less than n, it follows from this isomorphism that each element of $K(u)$ can be represented uniquely in the form $a_0 1 + a_1 u + \ldots + a_{n-1} u^{n-1}$. Thus $\mathcal{B}$ spans $K(u)$, and the uniqueness of representations implies that $\mathcal{B}$ is also a linearly independent set of vectors. □

6.2.3 Definition. *Let F be an extension field of K. If the dimension of F as a vector space over K is finite, then F is said to be a **finite** extension of K.*

*The dimension of F as a vector space over K is called the **degree** of F over K, and is denoted by $[F : K]$.*

In the next proposition, by using the notion of the degree of an extension, we are able to give a useful characterization of algebraic elements, which implies, in particular, that every element of a finite extension must be algebraic. After working through Example 6.1.3, the observant reader will have asked the question of whether the sum of two algebraic elements is always algebraic. The machinery that we are developing will let us get a handle on this not so innocent question.

6.2.4 Proposition. *Let F be an extension field of K and let $u \in F$. The following conditions are equivalent:*

    (1) u is algebraic over K;

    (2) $K(u)$ is a finite extension of K;

    (3) u belongs to a finite extension of K.

Proof. It is clear that (1) implies (2) and (2) implies (3). To prove (3) implies (1), suppose that $u \in E$, for a field E with $K \subseteq E$ and $[E : K] = n$. The set $\{1, u, u^2, \ldots, u^n\}$ contains $n + 1$ elements, and these cannot be linearly independent in an n-dimensional vector space. Thus there exists a relation $a_0 + a_1 u + \ldots + a_n u^n = 0$ with scalars $a_i \in K$ that are not all zero. This shows that u is a root of a nonzero polynomial in $K[x]$. □

Counting arguments often provide very useful tools. In case we have extension fields $K \subseteq E \subseteq F$, we can consider the degree of E over K and the degree of F over E. The next theorem shows that there is a very simple relationship between these two degrees and the degree of F over K. Theorem 6.2.5 will play a very important role in our study of extension fields.

6.2.5 Theorem. *Let E be a finite extension of K and let F be a finite extension of E. Then F is a finite extension of K, and*

$$[F : K] = [F : E][E : K] .$$

Proof. Let $[F : E] = n$ and let $[E : K] = m$. Let $u_1, u_2, \ldots, u_n$ be a basis for F over E and let $v_1, v_2, \ldots, v_m$ be a basis for E over K. We claim that the set $\mathcal{B}$ of nm products $u_i v_j$ (where $1 \le i \le n$ and $1 \le j \le m$) is a basis for F over K.

We must first show that $\mathcal{B}$ spans F over K. If u is any element of F, then $u = \sum_{i=1}^{n} a_i u_i$ for elements $a_i \in E$. For each element a_i we have $a_i = \sum_{j=1}^{m} c_{ij} v_j$, where $c_{ij} \in K$. Substituting gives $u = \sum_{i=1}^{n} \sum_{j=1}^{m} c_{ij} v_j u_i$, and so $\mathcal{B}$ spans F over K.

To show that $\mathcal{B}$ is a linearly independent set, suppose that $\sum_{i,j} c_{ij} v_j u_i = 0$ for some linear combination of the elements of $\mathcal{B}$, with coefficients in K. This expression can be written as $\sum_{i=1}^{n} \left(\sum_{j=1}^{m} c_{ij} v_j \right) u_i$. Since the elements $u_1, u_2, \ldots, u_n$ form a basis for F over E, each of the coefficients $\sum_{j=1}^{m} c_{ij} v_j$ (which belong to E) must be zero. Then since the elements $v_1, v_2, \ldots, v_m$ form a basis for E over K, for each i we must have $c_{ij} = 0$ for all j. $\square$

As illustrated by the next example, the central idea used in the proof of Theorem 6.2.5 is often useful in finding a basis for a field extension.

Example 6.2.1 ($[\mathbf{Q}(\sqrt{3}, \sqrt{2}) : \mathbf{Q}] = 4$).

Consider the fields $\mathbf{Q} \subseteq \mathbf{Q}(\sqrt{2}) \subseteq \mathbf{Q}(\sqrt{3}, \sqrt{2})$. We showed in Example 6.1.1 that $[\mathbf{Q}(\sqrt{2}) : \mathbf{Q}] = 2$, so to apply Theorem 6.2.5 we only need to find $[\mathbf{Q}(\sqrt{3}, \sqrt{2}) : \mathbf{Q}(\sqrt{2})]$. To compute $[\mathbf{Q}(\sqrt{2}, \sqrt{3}) : \mathbf{Q}(\sqrt{2})]$, we need to show that $x^2 - 3$ is irreducible over $\mathbf{Q}(\sqrt{2})$, in which case it will be the minimal polynomial of $\sqrt{3}$ over $\mathbf{Q}(\sqrt{2})$. The roots $\pm\sqrt{3}$ of the polynomial do not belong to $\mathbf{Q}(\sqrt{2})$ (see Exercise 4 of Section 6.1). The desired conclusion now follows from Theorem 6.2.5.

To find a basis for $\mathbf{Q}(\sqrt{3}, \sqrt{2})$ over $\mathbf{Q}$, we note that $\{1, \sqrt{2}\}$ is a basis for $\mathbf{Q}(\sqrt{2})$ over $\mathbf{Q}$, and $\{1, \sqrt{3}\}$ is a basis for $\mathbf{Q}(\sqrt{3}, \sqrt{2})$ over $\mathbf{Q}(\sqrt{2})$. Taking all possible products of these basis elements gives us the basis $\{1, \sqrt{2}, \sqrt{3}, \sqrt{6}\}$ for $\mathbf{Q}(\sqrt{3}, \sqrt{2})$ over $\mathbf{Q}$. $\square$

Example 6.2.2 ($[\mathbf{Q}(\sqrt{2} + \sqrt{3}) : \mathbf{Q}] = 4$).

In Example 6.1.3 we showed that $\sqrt{2} + \sqrt{3}$ has degree 4 over $\mathbf{Q}$ by showing that it has the minimal polynomial $x^4 - 10x^2 + 1$. The previous theorem can be used to give an alternate proof, using the fact that $\mathbf{Q}(\sqrt{2} + \sqrt{3}) = \mathbf{Q}(\sqrt{2}, \sqrt{3})$.

To show that the two field extensions are equal, we first observe that $\mathbf{Q}(\sqrt{2} + \sqrt{3}) \subseteq \mathbf{Q}(\sqrt{2}, \sqrt{3})$, since we have $\sqrt{2} + \sqrt{3} \in \mathbf{Q}(\sqrt{2}, \sqrt{3})$ and the field $\mathbf{Q}(\sqrt{2} + \sqrt{3})$ is defined by the property that it contains $\sqrt{2} + \sqrt{3}$ and is contained in any extension field that contains $\mathbf{Q}$ and $\sqrt{2} + \sqrt{3}$. On the other hand, $(\sqrt{3} - \sqrt{2})(\sqrt{3} + \sqrt{2}) = 1$, and so $\sqrt{3} - \sqrt{2} \in \mathbf{Q}(\sqrt{3} + \sqrt{2})$ since it is the multiplicative inverse of $\sqrt{3} + \sqrt{2}$. Because $\sqrt{3} = ((\sqrt{3} + \sqrt{2}) + (\sqrt{3} - \sqrt{2}))/2$, it follows that $\sqrt{3}, \sqrt{2} \in \mathbf{Q}(\sqrt{2} + \sqrt{3})$, and so we also have $\mathbf{Q}(\sqrt{2}, \sqrt{3}) \subseteq \mathbf{Q}(\sqrt{2} + \sqrt{3})$.

It now follows from Example 6.2.1 that $\sqrt{2} + \sqrt{3}$ has degree 4 over $\mathbf{Q}$. This argument using degrees shows that any monic polynomial of degree 4 that has $\sqrt{2} + \sqrt{3}$ as a root must be its minimal polynomial.

In contrast to the basis found in Example 6.2.1, we note that the basis for $\mathbf{Q}(\sqrt{2} + \sqrt{3})$ over $\mathbf{Q}$ produced by Proposition 6.2.2 would consist of the powers of $\sqrt{2} + \sqrt{3}$. This yields the basis $\{1, \sqrt{2} + \sqrt{3}, 5 + \sqrt{6}, 11\sqrt{2} + 9\sqrt{3}\}$. $\square$

6.2.6 Corollary. *Let F be a finite extension of K. If $u \in F$, then the degree of u over K is a divisor of $[F : K]$.*

Proof. If $u \in F$, then $[F : K] = [F : K(u)][K(u) : K]$. $\square$

Example 6.2.3 (Irreducible polynomials over R have degree 1 or 2).

Using Corollary 6.2.6 we can give another proof of Theorem 4.4.12. Suppose that $f(x) \in \mathbf{R}[x]$ is irreducible and has positive degree. By the fundamental theorem of algebra there is a root u of $f(x)$ in $\mathbf{C}$, and its degree is a divisor of $[\mathbf{C} : \mathbf{R}] = 2$. This implies that $f(x)$ has degree 1 or 2. $\square$

6.2.7 Corollary. *Let F be an extension field of K, with algebraic elements*

$$u_1, u_2, \ldots, u_n \in F .$$

Then the degree of $K(u_1, u_2, \ldots, u_n)$ over K is at most the product of the degrees of u_i over K, for $1 \le i \le n$.

Proof. We give a proof by induction on n. By Proposition 6.2.2, the result is true for $n = 1$. If the result is assumed to be true for the case $n - 1$, then let $E = K(u_1, u_2, \ldots, u_{n-1})$. Since

$$K(u_1, u_2, \ldots, u_{n-1}, u_n) = E(u_n) ,$$

the desired conclusion will follow from the equality

$$[E(u_n) : K] = [E(u_n) : E][E : K] ,$$

if we can show that $[E(u_n) : E]$ is at most the degree of u_n over K. Now the minimal polynomial of u_n over K is in particular a polynomial over E, and thus the minimal polynomial of u_n over E must be a divisor of it. Applying Proposition 6.2.2 completes the proof.   □

Example 6.2.4 ($[\mathbf{Q}(\sqrt[3]{2}, \sqrt{2}) : \mathbf{Q}] = 6$).

Since $\sqrt{2}$ has degree 2 over $\mathbf{Q}$, and $\sqrt[3]{2}$ has degree 3 over $\mathbf{Q}$, it follows from Corollary 6.2.7 that $[\mathbf{Q}(\sqrt[3]{2}, \sqrt{2}) : \mathbf{Q}] \le 6$. On the other hand, Corollary 6.2.6 implies that $[\mathbf{Q}(\sqrt[3]{2}, \sqrt{2}) : \mathbf{Q}]$ is divisible by both 2 and 3, so we must have $[\mathbf{Q}(\sqrt[3]{2}, \sqrt{2}) : \mathbf{Q}] = 6$.   □

It is important to note that the inequality in Corollary 6.2.7 can be a strict inequality, as illustrated by the following example.

Example 6.2.5 ($[\mathbf{Q}(\sqrt[4]{2}, \sqrt{2}) : \mathbf{Q}] = 4$).

Since $\sqrt{2} = (\sqrt[4]{2})^2 \in \mathbf{Q}(\sqrt[4]{2})$, to show that $[\mathbf{Q}(\sqrt[4]{2}, \sqrt{2}) : \mathbf{Q}] = 4]$ we only need to note that $\sqrt[4]{2}$ has degree 4 over $\mathbf{Q}$.   □

6.2.8 Corollary. *Let F be an extension field of K. The set of all elements of F that are algebraic over K forms a subfield of F.*

Proof. If u, v are algebraic elements of F, then $K(u, v) \supseteq K(u) \supseteq K$, and since $[K(u, v) : K(u)]$ and $[K(u) : K]$ are finite, it follows from Theorem 6.2.5 that $[K(u, v) : K]$ is finite. Since $u + v$, $u - v$, and uv all belong to $K(u, v)$, these elements are algebraic by Proposition 6.2.4. The same argument applies to u/v, if $v \neq 0$.   □

6.2.9 Definition. *An extension field F of K is said to be over K if each element of F is algebraic over K.*

6.2.10 Proposition. *Every finite extension is an algebraic extension.*

Proof. This follows immediately from Proposition 6.2.4.   □

The following example shows that an algebraic extension need not be a finite extension.

Example 6.2.6 (Algebraic numbers).

Let $\overline{\mathbf{Q}}$ be the set of complex numbers $u \in \mathbf{C}$ such that u is algebraic over $\mathbf{Q}$. Then $\overline{\mathbf{Q}}$ is a subfield of $\mathbf{C}$ by Corollary 6.2.8, called the **field of algebraic numbers**. By definition, $\overline{Q}$ is an algebraic extension of $\mathbf{Q}$.

On the other hand, we showed in Corollary 4.4.7 that for any prime p the polynomial $1 + x + \ldots + x^{p-2} + x^{p-1}$ is irreducible over $\mathbf{Q}$. The roots of this polynomial exist in $\mathbf{C}$ (they are the primitive pth roots of unity) and thus $\overline{\mathbf{Q}}$ contains algebraic numbers of arbitrarily large degree over $\mathbf{Q}$, which shows that it cannot have finite degree over $\mathbf{Q}$. Thus $\overline{\mathbf{Q}}$ is an algebraic extension of $\mathbf{Q}$, but not a finite extension. $\square$

6.2.11 Proposition. *Let F be an algebraic extension of E and let E be an algebraic extension of K. Then F is an algebraic extension of K.*

Proof. If F is algebraic over E, then any element $u \in F$ must satisfy some nonzero polynomial $f(x) = a_0 + a_1 x + \ldots + a_n x^n$ over E. Since E is not necessarily a finite extension of K, we consider the smaller extension $K(a_0, a_1, \ldots, a_n)$, which *is* a finite extension of K since each element a_i is algebraic over K, for $0 \leq i \leq n$. Because u is algebraic over $K(a_0, a_1, \ldots, a_n)$, it follows that $K(a_0, a_1, \ldots, a_n, u)$ is a finite extension of K, and thus u is algebraic over K by Proposition 6.2.4. $\square$

The following diagram illustrates the proof of Proposition 6.2.11.

Figure 6.2.1:

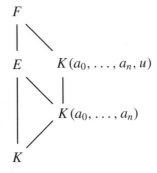

EXERCISES: SECTION 6.2

1. Find the degree and a basis for each of the given field extensions.

†(a) $\mathbf{Q}(\sqrt{3})$ over $\mathbf{Q}$

(b) $\mathbf{Q}(\sqrt{3}, \sqrt{7})$ over $\mathbf{Q}$

†(c) $\mathbf{Q}(\sqrt{3} + \sqrt{7})$ over $\mathbf{Q}$

(d) $\mathbf{Q}(\sqrt{2}, \sqrt[3]{2})$ over $\mathbf{Q}$

†(e) $\mathbf{Q}(\sqrt{2} + \sqrt[3]{2})$ over $\mathbf{Q}$

(f) $\mathbf{Q}(\omega)$ over $\mathbf{Q}$, where $\omega = (-1 + \sqrt{3}i)/2$

2. Find the degree and a basis for each of the given field extensions.

(a) $\mathbf{Q}(\sqrt{3}, \sqrt{21})$ over $\mathbf{Q}(\sqrt{7})$

(b) $\mathbf{Q}(\sqrt{3} + \sqrt{7})$ over $\mathbf{Q}(\sqrt{7})$

(c) $\mathbf{Q}(\sqrt{3}, \sqrt{7})$ over $\mathbf{Q}(\sqrt{3} + \sqrt{7})$

3. Find the degree of $\mathbf{Q}(\sqrt[3]{2}, \sqrt[4]{5})$ over $\mathbf{Q}$.

4. Let F be a finite extension of K such that $[F : K] = p$, a prime number. If $u \in F$ but $u \notin K$, show that $F = K(u)$.

5. Let $f(x)$ be an irreducible polynomial in $K[x]$. Show that if F is an extension field of K such that $\deg(f(x))$ is relatively prime to $[F : K]$, then $f(x)$ is irreducible in $F[x]$.

6. Let $K \subseteq E \subseteq F$ be fields. Prove that if F is algebraic over K, then F is algebraic over E and E is algebraic over K.

7. Let $F \supseteq K$ be fields, and let R be a ring such that $F \supseteq R \supseteq K$. If F is an algebraic extension of K, show that R is a field. What happens if we do not assume that F is algebraic over K?

8.†Determine $[\mathbf{Q}(\sqrt{n}) : \mathbf{Q}]$ for all $n \in \mathbf{Z}^+$.

9. For any positive integers a, b, show that $\mathbf{Q}(\sqrt{a} + \sqrt{b}) = \mathbf{Q}(\sqrt{a}, \sqrt{b})$.

10. Let F be an extension field of K. Let $a \in F$ be algebraic over K, and let $t \in F$ be transcendental over K. Show that $a + t$ is transcendental over K.

11. Let F be an algebraic extension of K, and let S be a subset of F such that $S \supseteq K$, S is a vector space over K, and $s^n \in S$ for all $s \in S$ and all positive integers n. Prove that if $\operatorname{char}(K) \neq 2$, then S is a subfield of F.

(The result is false in characteristic 2, and all of the tools necessary to construct a counterexample are at hand, but the counterexample is not an easy one.)

6.3 Geometric Constructions

In this section we will exploit what we have learned about the degree of a finite extension defined in successive steps. We will show the impossibility of several geometric constructions which were first investigated by the ancient Greeks. One of the most elementary constructions taught in high-school geometry is how to bisect an angle. In this section, the word "construction" will be assumed to mean a geometric construction using only a straightedge and compass. It is important to note that the straightedge is not a ruler that can measure arbitrary lengths. Furthermore, according to classical Greek practice a compass is "collapsible". While you can draw a circle with given center, passing through a given point, you cannot use the compass to transfer lengths.

There is no general method for trisecting an angle, since we will show that a $20°$ angle cannot be constructed, and thus it is impossible for any general method to successfully trisect a $60°$ angle. Secondly, given a circle, it is impossible (in general) to construct a square with the same area as that of the circle. (This is known as "squaring the circle.") Finally, given a cube, it is not generally possible to construct a cube with double the volume of the given cube.

We will take as given a line segment that will be defined to be one unit in length. Using this line segment, lengths corresponding to all positive rational numbers can be constructed. Since the constructions we will allow must involve only a straightedge and compass, they will be limited to the following: (i) constructing a line through two points whose coordinates are known, (ii) constructing a circle with center at a point with known coordinates and passing though a point with known coordinates, and (iii) finding the points of intersection of given lines and circles. The reader should review several constructions that we will need: that of a line parallel to a given line and passing through a point not on the given line, and that of a line perpendicular to a given line and passing through a given point.

6.3.1 Definition. *The real number a is said to be a **constructible number** if it is possible to construct a line segment of length $|a|$ by using only a straightedge and compass.*

One issue we need to address is how to transfer a length from one line to another. Figure 6.3.1 illustrates the case in which we copy a line segment joining the constructible points A and B to a second line ℓ. Given a constructible point C on ℓ, we first construct a line through C parallel to the line determined by A and B. Then we construct a line through B parallel to the line determined by A and C. With C as center, we can construct a circle through E, and its intersection with the line ℓ produces a length $|\overline{CD}|$ equal to the length $|\overline{AB}|$. We can transfer a length on a line to another position on the line by first copying it to a second line, and then using the above procedure.

The next proposition implies that all rational numbers are constructible.

Figure 6.3.1:

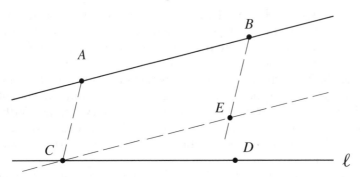

6.3.2 Proposition. *The set of all constructible real numbers is a subfield of the field of all real numbers.*

Proof. Let a, b be constructible real numbers, which we may assume to be positive. We must show that $a \pm b$ and ab are constructible, and that a/b is constructible, provided $b \neq 0$. Assuming $a > b$, using the method outlined in the discussion preceding the proposition we can copy the line segment whose length is b to the line containing the segment whose length is a, and then it is clear how to construct $a + b$ and $a - b$. Furthermore, given positive constructible numbers y, z, w, by choosing any angle α we can easily construct a triangle with two sides of length z and w, as in Figure 6.3.2. Using the length y, we can construct a line parallel to the third side of the triangle, giving us two similar triangles with sides of length x, y, z, w that satisfy the relation $x/y = z/w$. (The diagram presumes that $y > w$, and can easily be modified if not.) To construct $x = ab$, choose $y = a$, $z = b$, and $w = 1$. To construct $x = a/b$, choose $y = 1$, $z = a$, and $w = b$. $\square$

Figure 6.3.2:

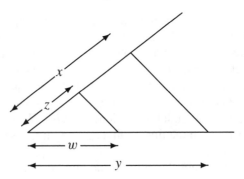

We now need further information about how a real number can actually be constructed. We will obtain it by considering intermediate extension fields between **Q** and the field of all constructible numbers.

6.3.3 Definition. *Let F be a subfield of* **R**.

*A straight line with an equation of the form $ax + by + c = 0$, for elements $a, b, c \in F$, is called a **line over** F.*

*Any circle with an equation of the form $x^2 + y^2 + ax + by + c = 0$, for elements $a, b, c \in F$, is called a **circle over** F.*

We note that a line over $F \subset \mathbf{R}$ may contain some points whose coordinates do not belong to F. For example, the point $(\sqrt{2}/2, \sqrt{2}/2)$ belongs to the line $x - y = 0$ over **Q** and to the circle $x^2 + y^2 - 1 = 0$ over **Q**.

6.3.4 Lemma. *Let F be a subfield of* **R**.

(a) *Any straight line joining two points whose coordinates belong to F is a line over F.*

(b) *Any circle such that its radius and the coordinates of its center belong to F is a circle over F.*

Proof. (a) Let (a_1, b_1) and (a_2, b_2) be points in $\mathbf{R}^2$ with $a_1, b_1, a_2, b_2 \in F$. The line $x - a_1 = 0$ takes care of the case $a_1 = a_2$. Otherwise, the two-point form

$$y - b_1 = \frac{b_2 - b_1}{a_2 - a_1}(x - a_1)$$

of the equation of a line determines an equation of the form we need.

(b) If $r \in F$ and $(a, b) \in \mathbf{R}^2$ with $a, b \in F$, then the equation of the circle with radius r and center (a, b) is

$$(x - a)^2 + (y - b)^2 = r^2 \, ,$$

and expanding it gives the form we need, since F is a field. $\square$

A word of caution may be in order, concerning the next result. Lemma 6.3.5 does *not* say that two circles or a line and a circle must necessarily have a point of intersection.

6.3.5 Lemma. *Let F be a subfield of* **R**. *The points of intersection of lines over F and circles over F belong to the field $F(\sqrt{u})$, for some $u \in F$.*

Proof. Given two lines over F, we can find their point of intersection by using elementary row operations on the associated matrix. Since we use only field operations, the coordinates of the point of intersection must still belong to F. Given two circles in F, subtracting one equation from the other reduces the question of their points of intersection to the question of the points of intersection of a line over F and a circle over F. To complete the proof, assume that we are given the equation of a line over F and the equation of a circle over F. In the equation of the line, one of x or y must have a nonzero coefficient, say y. Then we can solve for y and substitute into the equation of the given circle, yielding a quadratic equation in x. Unless the line and circle have no intersection in $\mathbf{R}^2$, consider the solution produced by using the quadratic formula. Since F is closed under the operations of $\mathbf{R}$, each number involved in the solution, including the term u under the radical, is again an element of F. Substituting back into the equation of the line shows that the coordinates of the points of intersection belong to $F(\sqrt{u})$. □

6.3.6 Theorem. *The real number u is constructible if and only if there exists a finite set $u_1, u_2, \ldots, u_n$ of real numbers such that*

(i) $u_1^2 \in \mathbf{Q}$,

(ii) $u_i^2 \in \mathbf{Q}(u_1, \ldots, u_{i-1})$, *for $i = 2, \ldots, n$, and*

(iii) $u \in \mathbf{Q}(u_1, \ldots, u_n)$.

Proof. If u is constructible, then the construction can be done in a finite number of steps. Starting with $\mathbf{Q}$, the first step must consist of finding an intersection of lines or circles over $\mathbf{Q}$, so either this can be done in $\mathbf{Q}$, or else by Lemma 6.3.5 we obtain an extension of the form $\mathbf{Q}(\sqrt{v_1})$, for some $v_1 \in \mathbf{Q}$. We may assume that the next step in the construction involves lines and circles over $\mathbf{Q}(\sqrt{v_1})$, and so the points we obtain have coordinates in $\mathbf{Q}(\sqrt{v_1}, \sqrt{v_2})$ for some $v_2 \in \mathbf{Q}(\sqrt{v_1})$. Continuing in this manner allows us to obtain u as an element of a field of the required form $\mathbf{Q}(u_1, u_2, \ldots, u_n)$.

To show the converse, it suffices to show that if F is any subfield of the field of constructible numbers, then $\sqrt{u}$ is constructible for all $u \in F$, since this implies that every element of $F(\sqrt{u})$ is constructible. Given $u \in F$, we can construct a circle of diameter $1 + u$. Then, as in Figure 6.3.3, we can construct a perpendicular line on the diameter, at a distance of 1 from its end. If x is the length on this line between the diameter and the intersection with the circle, then we have constructed similar triangles that yield the proportion $x/1 = u/x$. Thus $x = \sqrt{u}$, and the proof is complete. □

6.3.7 Corollary. *If u is a constructible real number, then u is algebraic over $\mathbf{Q}$, and the degree of its minimal polynomial over $\mathbf{Q}$ is a power of 2.*

Proof. Assume that u is a constructible real number. Then by Theorem 6.3.6, u belongs to a field $F \subseteq \mathbf{R}$ of the form $F = \mathbf{Q}(u_1, u_2, \ldots, u_n)$, where (i) $u_1^2 \in \mathbf{Q}$ and

Figure 6.3.3:

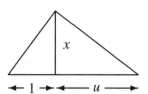

(ii) $u_i^2 \in \mathbf{Q}(u_1, \ldots, u_{i-1})$, for $i = 2, \ldots, n$. By Theorem 6.2.5, the degree of F over $\mathbf{Q}$ is a power of 2, since $[\mathbf{Q}(u_1) : \mathbf{Q}] \le 2$ and

$$[\mathbf{Q}(u_1, \ldots, u_i) : \mathbf{Q}(u_1, \ldots, u_{i-1})] \le 2 \, ,$$

for $1 < i \le n$. The desired conclusion also follows from Theorem 6.2.5, since

$$[F : \mathbf{Q}] = [F : \mathbf{Q}(u)][\mathbf{Q}(u) : \mathbf{Q}] \, ,$$

and thus $[\mathbf{Q}(u) : \mathbf{Q}]$ is a divisor of a power of 2. $\square$

6.3.8 Lemma. *For any angle θ, the following trigonometric identities hold.*
   (a) $2 \cos^2 \theta - \cos(2\theta) - 1 = 0$
   (b) $4 \cos^3 \theta - 3 \cos \theta - \cos(3\theta) = 0$

Proof. (a) We recall the trigonometric formulas involving the sum of two angles:

$$\sin(\alpha + \beta) = \sin \alpha \cos \beta + \cos \alpha \sin \beta \quad \text{and} \quad \cos(\alpha + \beta) = \cos \alpha \cos \beta - \sin \alpha \sin \beta \, .$$

From these we obtain $\sin(2\theta) = 2 \sin \theta \cos \theta$ and $\cos(2\theta) = \cos^2 \theta - \sin^2 \theta$.
   Using the identity $\sin^2 \theta = 1 - \cos^2 \theta$, we have

$$\begin{aligned}
\cos(2\theta) &= \cos^2 \theta - \sin^2 \theta \\
&= \cos^2 \theta - (1 - \cos^2 \theta) \\
&= 2 \cos^2 \theta - 1 \, .
\end{aligned}$$

It follows that $2 \cos^2 \theta - \cos(2\theta) - 1 = 0$.
   (b) Using the above identities, let $\alpha = 2\theta$ and $\beta = \theta$ in the formula for the cosine of the sum of two angles. Then we have

$$\begin{aligned}
\cos(3\theta) &= \cos(2\theta + \theta) \\
&= \cos(2\theta) \cos \theta - \sin(2\theta) \sin \theta \\
&= (2 \cos^2 \theta - 1) \cos \theta - (2 \sin \theta \cos \theta) \sin \theta \\
&= 2 \cos^3 \theta - \cos \theta - 2 \cos \theta \sin^2 \theta \\
&= 2 \cos^3 \theta - \cos \theta - 2 \cos \theta (1 - \cos^2 \theta) \\
&= 4 \cos^3 \theta - 3 \cos \theta \, .
\end{aligned}$$

It follows immediately that $4\cos^3\theta - 3\cos\theta - \cos(3\theta) = 0$. $\square$

6.3.9 Theorem. *It is impossible to find a general construction for trisecting an angle, duplicating a cube, or squaring a circle.*

Proof. If a $60°$ angle (which is of course constructible) could be trisected, then it would be possible to construct a $20°$ angle, and so $u = \cos 20°$ would be a constructible real number. For $\theta = 20°$ we have $\cos 3\theta = 1/2$, so by Lemma 6.3.8 we have $4\cos^3\theta - 3\cos\theta - 1/2 = 0$. Multiplying by 2 shows that $u = \cos\theta$ is a root of the polynomial $8x^3 - 6x - 1$. Using Proposition 4.4.1 it is easy to check that this polynomial has no rational roots, so it is irreducible over $\mathbf{Q}$, and therefore the minimal polynomial of u over $\mathbf{Q}$ has degree 3. Since the degree of u over $\mathbf{Q}$ is not a power of 2, it cannot be constructible.

To construct a cube with double the volume of the unit cube requires the construction of a cube of volume 2. This requires constructing $\sqrt[3]{2}$, which is impossible since $\sqrt[3]{2}$ has degree 3 over $\mathbf{Q}$.

Finally, constructing a square with the same area as a circle of radius 1 requires that $\sqrt{\pi}$ be constructible. This is not true since π is not algebraic over $\mathbf{Q}$. For a proof of this deep fact we refer the student to a book by I. Niven called *Irrational Numbers*. Thus we have (almost) completed the proof. $\square$

EXERCISES: SECTION 6.3

1. Show that the roots of the polynomial $8x^3 - 6x - 1$ used in Theorem 6.3.9 are $u_1 = \cos\frac{\pi}{9}$, $u_2 = \cos\frac{5\pi}{9}$, and $u_3 = \cos\frac{7\pi}{9}$.

2. Use the identity $4\cos^3\theta - 3\cos\theta - \cos(3\theta) = 0$ to show that the roots of the polynomial $x^3 - 3x + 1$ are $u_1 = 2\cos\frac{2\pi}{9}$, $u_2 = 2\cos\frac{4\pi}{9}$, and $u_3 = 2\cos\frac{8\pi}{9}$.

3. In this exercise we outline how to construct a regular pentagon. Let $\zeta = \cos(2\pi/5) + i\sin(2\pi/5)$.

   (a) Show that ζ is a primitive fifth root of unity.

   (b) Show that $\left(\zeta + \zeta^{-1}\right)^2 + \left(\zeta + \zeta^{-1}\right) - 1 = 0$.

   (c) Show that $\zeta + \zeta^{-1} = \left(-1 + \sqrt{5}\right)/2$.

   (d) Show that $\cos(2\pi/5) = \left(-1 + \sqrt{5}\right)/4$ and that $\sin(2\pi/5) = \left(\sqrt{10 + 2\sqrt{5}}\right)/4$.

   (e) Conclude that a regular pentagon is constructible.

4. Prove that a regular heptagon is not constructible.

   Hint: Let $\zeta = \cos(2\pi/7) + i\sin(2\pi/7)$. Show that $[\mathbf{Q}(\zeta) : \mathbf{Q}]$ is not a power of 2.

6.4 Splitting Fields

We will be interested, ultimately, in the question of determining when a given polynomial equation is solvable by radicals. The answer to this question involves a comparison between the field generated by the coefficients of the polynomial and the field generated by the roots of the polynomial. This comparison must be done in some field that contains all roots of the polynomial. Over such a field the given polynomial can be factored (or "split") into a product of linear factors. Our task in this section is to study the existence and uniqueness of such fields. Recall that we have seen in Section 6.1 that given any field and any polynomial over that field there exists an extension field in which the polynomial has a root. Now we simply need to iterate this procedure, to obtain all roots of the polynomial.

6.4.1 Definition. *Let K be a field and let $f(x) = a_0 + a_1x + \ldots + a_nx^n$ be a polynomial in $K[x]$ of degree $n > 0$. An extension field F of K is called a **splitting field for $f(x)$ over K** if there exist elements $r_1, r_2, \ldots, r_n \in F$ such that*
   (i) $f(x) = a_n(x - r_1)(x - r_2) \cdots (x - r_n)$, *and*
   (ii) $F = K(r_1, r_2, \ldots, r_n)$.

The elements $r_1, r_2, \ldots, r_n$ are roots of $f(x)$, and so F is obtained by adjoining to K a complete set of roots of $f(x)$. We say that $f(x)$ **splits** over the field E if E contains the splitting field of F.

The proof of the next theorem parallels that of Corollary 4.3.9. But we now know about the degree of an extension, and so we can give more information about the construction by keeping track of the degrees of the extensions that we use.

6.4.2 Theorem. *Let $f(x) \in K[x]$ be a polynomial of degree $n > 0$. Then there exists a splitting field F for $f(x)$ over K, with $[F : K] \leq n!$.*

Proof. The proof is by induction on the degree of $f(x)$. If $\deg(f(x)) = 1$, then K itself is a splitting field. Assume that $\deg(f(x)) = n > 1$ and that the theorem is true for any polynomial $g(x)$ with $1 \leq \deg(g(x)) < n$ over any field K. Let $p(x)$ be an irreducible factor of $f(x)$. By Kronecker's theorem there exists an extension field E of K in which $p(x)$ has a root r. We now consider the field $K(r)$. Over this field $f(x)$ factors as

$$f(x) = p(x)q(x) = (x - r)g(x)$$

for some polynomial $g(x) \in K(r)[x]$ of degree $n - 1$. Thus by the induction hypothesis there exists a splitting field F of $g(x)$ over $K(r)$, say $F = K(r)(r_1, r_2, \ldots, r_{n-1})$, with $[F : K(r)] \leq (n - 1)!$. Then $F = K(r, r_1, r_2, \ldots, r_{n-1})$ and it is clear that $f(x)$ splits over F. Finally,

$$[F : K] = [F : K(r)][K(r) : K] \leq (n - 1)! \cdot n = n! \,,$$

which completes the proof.   □

Example 6.4.1 (Splitting field for $x^2 + 1$ over Q).

If we consider $x^2 + 1$ as a polynomial with rational coefficients, then to obtain a splitting field we only need to adjoin i to $\mathbf{Q}$. Thus $\mathbf{Q}(i)$ is a splitting field for $x^2 + 1$ over $\mathbf{Q}$.   □

Example 6.4.2 (Splitting field for $x^3 - 2$ over Q).

We must adjoin to $\mathbf{Q}$ all solutions of the equation $x^3 = 2$. It is obvious that $x = \sqrt[3]{2}$ is a solution. But in addition, $\omega\sqrt[3]{2}$ is also a solution for any complex number ω such that $\omega^3 = 1$. Since we have

$$x^3 - 1 = (x - 1)(x^2 + x + 1)$$

we can let ω be a root of $x^2 + x + 1$; that is

$$\omega = \frac{-1 + \sqrt{-3}}{2} = -\frac{1}{2} + \frac{\sqrt{3}}{2}i \ .$$

Then $\sqrt[3]{2}$, $\omega\sqrt[3]{2}$, and $\omega^2\sqrt[3]{2}$ are the roots of $x^3 - 2$. Thus we obtain the splitting field as $\mathbf{Q}(\sqrt[3]{2}, \omega)$.

We claim that the degree of the splitting field over $\mathbf{Q}$ is 6. Since $x^3 - 2$ is irreducible over $\mathbf{Q}$, we have $[\mathbf{Q}(\sqrt[3]{2}) : \mathbf{Q}] = 3$. The polynomial $x^2 + x + 1$ is irreducible over $\mathbf{Q}$ and stays irreducible over $\mathbf{Q}(\sqrt[3]{2})$ since it has no root in that field. The degree of $x^2 + x + 1$ is not a divisor of $[\mathbf{Q}(\sqrt[3]{2}) : \mathbf{Q}]$. Since ω is a root of $x^2 + x + 1$, this implies that $[\mathbf{Q}(\sqrt[3]{2}, \omega) : \mathbf{Q}(\sqrt[3]{2})] = 2$.   □

Example 6.4.3 (Splitting fields for $x^2 + 1$ over R).

Recall that our standard construction for a splitting field of the polynomial $x^2 + 1$ over $\mathbf{R}$ is to consider $\mathbf{R}[x]/\langle x^2 + 1 \rangle$. (See Example 4.3.2 and Corollary 4.4.10.) Then the field $\mathbf{R}$ is identified with the cosets of the form $a + \langle x^2 + 1 \rangle$, for $a \in \mathbf{R}$.

Another familiar construction is to simply use ordered pairs of the form $a + bi$, where $a, b \in \mathbf{R}$ and the "imaginary" number i is a square root of -1. This can be done rigorously by using the set of all ordered pairs (a, b) such that $a, b \in \mathbf{R}$, with componentwise addition and multiplication $(a, b) \cdot (c, d) = (ac - bd, ad + bc)$. In this case the field $\mathbf{R}$ is identified with ordered pairs of the form $(a, 0)$ and i is identified with the ordered pair $(0, 1)$. We thus obtain $\mathbf{C}$ as the splitting field for $x^2 + 1$ over $\mathbf{R}$.

Another construction of this splitting field uses the set of 2×2 matrices over $\mathbf{R}$. We first identify $\mathbf{R}$ with the set of scalar matrices. Let

$$F = \left\{ \begin{bmatrix} a & b \\ -b & a \end{bmatrix} \;\middle|\; a, b \in \mathbf{R} \right\} .$$

It can be shown that the function $\phi : \mathbf{C} \to F$ defined by $\phi(a + bi) = \begin{bmatrix} a & b \\ -b & a \end{bmatrix}$ is an isomorphism. We have thus adjoined to the set of scalar matrices the matrix $\begin{bmatrix} 0 & 1 \\ -1 & 0 \end{bmatrix}$, which satisfies the polynomial $x^2 + 1$. $\square$

Example 6.4.4 (Alternate proof of Kronecker's theorem).

The technique of the previous example can be extended to give another proof of Kronecker's theorem, which guarantees the existence of roots. If $p(x)$ is a monic irreducible polynomial of degree n over the field K, we first identify K with the field of $n \times n$ scalar matrices over K. There exists an $n \times n$ matrix C with $p(C) = 0$, called the *companion matrix* of the polynomial. (Refer to Exercise 8 for further details.)

The set of all matrices of the form $a_0 I + a_1 C + \ldots + a_{n-1} C^{n-1}$, such that $a_0, a_1, \ldots, a_{n-1} \in K$, then defines an extension field F of K in which $p(x)$ has a root. The fact that $p(C) = 0$ guarantees that C^k can be expressed as element of F, for $k \geq n$, and so F is closed under multiplication. For any nonzero element $q(C) \in F$, we see that $q(x)$ is relatively prime to $p(x)$, since $\deg(q(x)) < n$. Thus there exist $f(x), g(x) \in K[x]$ with $p(x)f(x) + q(x)g(x) = 1$. Replacing 1 with the identity matrix I and substituting $x = C$ yields $q(C)g(C) = I$, which completes the proof that F is a field. $\square$

The previous examples show that splitting fields can be constructed in a variety of ways. One would hope that there would be some uniqueness involved. In fact, the next results show that splitting fields are unique up to isomorphism. This provides the necessary basis for our study of solvability by radicals. We will also use this fact in Section 6.5 to show that any two finite fields with the same number of elements must be isomorphic.

6.4.3 Lemma. *Let $\theta : K \to L$ be an isomorphism of fields. Let F be an extension field of K such that $F = K(u)$ for an algebraic element $u \in F$. Let $p(x)$ be the minimal polynomial of u over K. Let $q(x)$ be the image of $p(x)$ under θ. If v is any root of $q(x)$ in an extension field of L and $E = L(v)$, then there is a unique way to extend θ to an isomorphism $\phi : F \to E$ such that $\phi(u) = v$ and $\phi(a) = \theta(a)$ for all $a \in K$.*

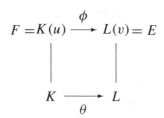

Proof. If $p(x)$ has degree n, then elements of $K(u)$ have the form $a_0 + a_1 u + \ldots +$ $a_{n-1}u^{n-1}$ for elements $a_0, a_1, \ldots, a_{n-1} \in K$. Therefore the required isomorphism $\phi : K(u) \to L(v)$ must have the form

$$\phi(a_0 + a_1 u + \ldots + a_{n-1}u^{n-1}) = \theta(a_0) + \theta(a_1)v + \ldots + \theta(a_{n-1})v^{n-1} .$$

We could simply show by direct computation that this function is an isomorphism. However, it seems to be easier to show that ϕ is a composite of functions that we already know to be isomorphisms.

Let $\iota : L \to L[x]$ be the natural inclusion mapping, with $\iota(a) = a$ for all $a \in L$. By Proposition 5.2.7, the ring homomorphism $\iota\theta$ can be extended uniquely to a ring homomorphism $\widehat{\theta} : K[x] \to L[x]$ such that $\widehat{\theta}(x) = x$. Since θ is an isomorphism, it is clear that $\widehat{\theta}$ is also an isomorphism. By assumption, $\widehat{\theta}(p(x)) = q(x)$, and so $\widehat{\theta}$ maps the ideal $\langle p(x) \rangle$ generated by $p(x)$ onto the ideal $\langle q(x) \rangle$ generated by $q(x)$. Example 5.3.5 shows that $\widehat{\theta}$ induces an isomorphism $\overline{\theta} : K[x]/\langle p(x) \rangle \to$ $L[x]/\langle q(x) \rangle$. Let $\eta : K[x]/\langle p(x) \rangle \to K(u)$ and $\epsilon : L[x]/\langle q(x) \rangle \to L(v)$ be the isomorphisms defined by evaluation at u and v, respectively. Then $\phi = \epsilon\overline{\theta}\eta^{-1}$ defines the required isomorphism from $K(u)$ onto $L(v)$. $\square$

6.4.4 Lemma. *Let F be a splitting field for the polynomial $f(x) \in K[x]$. If $\theta : K \to L$ is a field isomorphism that maps $f(x)$ to $g(x) \in L[x]$ and E is a splitting field for $g(x)$ over L, then there exists an isomorphism $\phi : F \to E$ such that $\phi(a) = \theta(a)$ for all $a \in K$.*

Proof. The proof uses induction on the degree n of $f(x)$. If $f(x)$ has degree one, then $F = K$ and $E = L$, so there is nothing to prove. We now assume that the result holds for all polynomials of positive degree less than n and for all fields K. Let $p(x)$ be an irreducible factor of $f(x)$, which maps to the irreducible factor $q(x)$ of $g(x)$. All roots of $p(x)$ belong to F, so we may choose one, say u, which gives $K \subseteq K(u) \subseteq F$. Similarly, we may choose a root v of $q(x)$ in E, which gives $L \subseteq L(v) \subseteq E$. By Lemma 6.4.3 there exists an isomorphism $\widehat{\theta} : K(u) \to L(v)$ such that $\widehat{\theta}(u) = v$ and $\widehat{\theta}(a) = \theta(a)$ for all $a \in K$. If we write $f(x) = (x - u)s(x)$ and $g(x) = (x - v)t(x)$, then the polynomial $s(x)$ has degree less than n, the extension F is a splitting field for $s(x)$ over $K(u)$, the polynomial $s(x)$ is mapped by $\widehat{\theta}$ to $t(x)$, and the extension E is a splitting field for $t(x)$ over $L(v)$. Thus the induction hypothesis can be applied, and so there exists an isomorphism $\phi : F \to E$

such that $\phi(w) = \widehat{\theta}(w)$ for all $w \in K(u)$. In particular, $\phi(a) = \widehat{\theta}(a) = \theta(a)$ for all $a \in K$, and the proof is complete. $\square$

The following theorem, which shows that splitting fields are unique "up to isomorphism", is a special case of Lemma 6.4.4. In the induction argument in the proof of Lemma 6.4.4 we needed to change the base field. That made it necessary to use as an induction hypothesis the more general statement of that lemma.

6.4.5 Theorem. *Let $f(x)$ be a polynomial of positive degree over the field K. If E and F are splitting fields of $f(x)$ over K, then there exists an isomorphism $\phi : F \rightarrow E$ such that $\phi(a) = a$ for all $a \in K$.*

EXERCISES: SECTION 6.4

1. Determine the splitting fields in **C** for the following polynomials (over **Q**).

   †(a) $x^2 - 2$

   (b) $x^2 + 3$

   †(c) $x^4 + x^2 - 6$

   (d) $x^3 - 5$

2. Determine the splitting fields in **C** for the following polynomials (over **Q**).

   (a) $x^3 - 1$

   (b) $x^4 - 1$

   (c) $x^3 + 3x^2 + 3x - 4$

3. Determine the splitting fields over $\mathbf{Z}_2$ for the following polynomials.

   †(a) $x^2 + x + 1$

   (b) $x^2 + 1$

   †(c) $x^3 + x + 1$

   (d) $x^3 + x^2 + 1$

4. Let p be a prime number. Determine the splitting field in **C** for $x^p - 1$ (over **Q**).

5. †Determine the splitting field for $x^p - x$ over $\mathbf{Z}_p$.

6. Determine the splitting field for $x^9 - x$ over $\mathbf{Z}_3$.

   Hint: See Exercise 13 of Section 4.2.

7. Prove that if F is an extension field of K of degree 2, then F is the splitting field over K for some polynomial.

8. Let K be a field. For a monic polynomial $f(x) = a_0 + a_1 x + \ldots + a_{n-1} x^{n-1} + x^n$ in $K[x]$, the following matrix C is called the **companion matrix** of $f(x)$:

$$\begin{bmatrix} 0 & 1 & 0 & \cdots & 0 \\ 0 & 0 & 1 & \cdots & 0 \\ 0 & 0 & 0 & \cdots & 0 \\ \vdots & \vdots & \vdots & & \vdots \\ 0 & 0 & 0 & \cdots & 1 \\ -a_0 & -a_1 & -a_2 & \cdots & -a_{n-1} \end{bmatrix}.$$

This exercise outlines a proof that $f(C) = 0$. (That is, $a_0 I + a_1 C + \ldots + a_{n-1} C^{n-1} + C^n = 0$, where I is the $n \times n$ identity matrix.) Let $\mathbf{v}_1 = (1, 0, \ldots, 0)$, $\mathbf{v}_2 = (0, 1, \ldots, 0)$, ..., $\mathbf{v}_n = (0, 0, \ldots, 1)$ be the standard basis vectors for K^n.

(a) Show that $\mathbf{v}_i C = \mathbf{v}_{i+1}$ for $i = 1, \ldots, n-1$, and $\mathbf{v}_n C = \sum_{j=1}^{n} -a_{j-1} \mathbf{v}_j$.

(b) Find similar expressions for $\mathbf{v}_1 C^2, \ldots, \mathbf{v}_1 C^{n-1}, \mathbf{v}_1 C^n$, and show that $\mathbf{v}_1 f(C) = \mathbf{0}$.

(c) Show that $\mathbf{v}_i f(C) = \mathbf{0}$, for $i = 2, \ldots, n$ and conclude that $f(C) = 0$.

Hint: Use the fact that $C^i f(C) = f(C) C^i$ for $i = 1, \ldots, n$.

9. Let K be a field, let $f(x) = a_0 + a_1 x + \ldots + a_{n-1} x^{n-1} + x^n \in F[x]$, and let C be the companion matrix of $f(x)$, as defined in Exercise 8. Show that the set $R = \{b_0 I + b_1 C + \ldots + b_{n-1} C^{n-1} \mid b_i \in F \text{ for } i = 0, \ldots, n-1\}$ is a commutative ring isomorphic to the ring $F[x]/\langle f(x) \rangle$.

10. Strengthen Theorem 6.4.2 by proving that under the conditions of the theorem there exists a splitting field F for $f(x)$ over K for which $[F : K]$ is a divisor of $n!$.

11. Let K be a field, and let F be an extension field of K. Let $\phi : F \to F$ be an automorphism of F such that $\phi(a) = a$, for all $a \in K$. Show that for any polynomial $f(x) \in K[x]$, and any root $u \in F$ of $f(x)$, the image $\phi(u)$ must be a root of $f(x)$.

12. Use Exercise 11 to show that there are only two automorphisms of the field $\mathbf{Q}(i)$: the identity automorphism, and the one defined by $\phi(a + bi) = a - bi$, for all $a, b \in \mathbf{Q}$.

13. Use Exercise 11 to show that there are at most four distinct automorphisms of the field $\mathbf{Q}(\sqrt{2}, \sqrt{3})$.

14. (a) Show that the splitting field of $x^4 - 2$ over $\mathbf{Q}$ is $\mathbf{Q}(\sqrt[4]{2}, i)$.

(b) Show that $\mathbf{Q}(\sqrt[4]{2}, i)$ is also the splitting field of $x^4 + 2$ over $\mathbf{Q}$.

15. Use Exercise 11 to show that there are at most eight distinct automorphisms of the splitting field $\mathbf{Q}(\sqrt[4]{2}, i)$ of $x^4 - 2$ over $\mathbf{Q}$.

6.5   Finite Fields

We first met finite fields in Chapter 1, in studying $\mathbf{Z}_p$, where p is a prime number. With the field theory we have developed it is now possible to give a complete description of the structure of all finite fields. We note that Proposition 5.2.11 implies that any finite field has prime characteristic.

6.5.1 Proposition. *Let F be a finite field of characteristic p. Then F has p^n elements, for some positive integer n.*

Proof. Recall that if F has characteristic p, then the ring homomorphism $\phi : \mathbf{Z} \to F$ defined by $\phi(n) = n \cdot 1$ for all $n \in \mathbf{Z}$ has kernel $p\mathbf{Z}$, and thus the image of ϕ is a subfield K of F isomorphic to $\mathbf{Z}_p$. Since F is finite, it must certainly have finite dimension as a vector space over K, say $[F : K] = n$. If $v_1, v_2, \ldots, v_n$ is a basis for F over K, then each element of F has the form $a_1 v_1 + a_2 v_2 + \ldots + a_n v_n$ for elements $a_1, a_2, \ldots, a_n \in K$. Thus to define an element of F there are n coefficients a_i, and for each coefficient there are p choices, since K has only p elements. Therefore the total number of elements in F is p^n. $\square$

If F is any field, then the smallest subfield of F that contains the identity element 1 is called the **prime subfield** of F. As noted in the proof of Proposition 6.5.1, if F is a finite field, then its prime subfield is isomorphic to $\mathbf{Z}_p$, where $p = \mathrm{char}(F)$.

6.5.2 Theorem. *Let F be a finite field with p^n elements. Then F is the splitting field of the polynomial $x^{p^n} - x$ over the prime subfield of F.*

Proof. Since F has p^n elements, the order of the multiplicative group $F^\times$ of nonzero elements of F is $p^n - 1$. By Corollary 3.2.11 (b), we have $x^{p^n - 1} = 1$ for all $0 \neq x \in F$, and so $x^{p^n} = x$ for all $x \in F$. Since the polynomial $f(x) = x^{p^n} - x$ can have at most p^n roots in any field, we see that F must contain all of its roots. Hence, since F is generated by these roots, it is a splitting field of $f(x)$ over its prime subfield. $\square$

Example 6.5.1 (Wilson's theorem).

The field $\mathbf{Z}_p$ is the splitting field of $x^p - x$, so we have

$$x^p - x = x(x - 1)(x - 2) \cdots (x - (p - 1)) \ .$$

Thus

$$x^{p-1} - 1 = (x - 1)(x - 2) \cdots (x - (p - 1)) \ ,$$

and substituting $x = p$ gives $p^{p-1} - 1 = (p - 1)(p - 2) \cdots (1)$. Hence $-1 = (p - 1)!$ in $\mathbf{Z}_p$, or, equivalently, $(p - 1)! \equiv -1 \pmod{p}$, which is Wilson's theorem. (See Exercise 27 of Section 1.4.) $\square$

6.5.3 Corollary. *Two finite fields are isomorphic if and only if they have the same number of elements.*

Proof. Let F and E be finite fields with p^n elements, containing prime subfields K and L, respectively. Then $K \cong \mathbf{Z}_p \cong L$ and so $F \cong E$ by Lemma 6.4.4, since by Theorem 6.5.2 both F and E are splitting fields of $x^{p^n} - x$ over K and L, respectively. $\square$

6.5.4 Lemma. *Let F be a field of prime characteristic p, and let $n \in \mathbf{Z}^+$.*

(a) $(a \pm b)^{p^n} = a^{p^n} \pm b^{p^n}$ *for all* $a, b \in F$.

(b) $\{a \in F \mid a^{p^n} = a\}$ *is a subfield of F.*

Proof. (a) With the exception of the coefficients of x^p and y^p, each binomial coefficient $(p!)/(k!(p-k)!)$ in the expansion of $(x \pm y)^p$ contains p in the numerator but not the denominator, because p is prime. Since $\operatorname{char}(F) = p$, this implies that $(x \pm y)^p = x^p \pm y^p$, for all $x, y \in F$. Applying this formula inductively to $(a \pm b)^{p^k}$ for $k \leq n$ shows that $(a \pm b)^{p^n} = a^{p^n} \pm b^{p^n}$.

(b) Let $E = \{a \in F \mid a^{p^n} = a\}$. It follows immediately from part (a) that E is closed under addition and subtraction. Since $(ab)^{p^n} = a^{p^n} b^{p^n}$, it is clear that E is closed under multiplication. To complete the proof of part (b), we only need to observe that if $a \in E$ is nonzero, then $(a^{-1})^{p^n} = (a^{p^n})^{-1} = a^{-1}$, and so $a^{-1} \in E$. $\square$

The next proposition gives a complete characterization of the subfields of any finite field. It shows, for example, that in a field with 8 elements the only proper subfield is its prime subfield. On the other hand, a field with 16 elements must have precisely two proper subfields, with 2 and 4 elements, respectively. The proof uses the multiplication of the field in a critical way, since group theory only tells us that subgroups have an order that is a divisor of the total number of elements.

We will need the fact that if m, n are positive integers with $m \mid n$, then $x^m - 1$ is a divisor of $x^n - 1$ in $\mathbf{Z}[x]$. (See the stronger result in Exercise 8.) If $n = mq$, for $q \in \mathbf{Z}$, then Lemma 4.1.8 implies that $y - 1$ is a factor of $y^q - 1$, and substituting $y = x^m$ shows that $x^m - 1 \mid x^n - 1$.

6.5.5 Proposition. *Let F be a field with p^n elements. Each subfield of F has p^m elements for some divisor m of n. Conversely, for each positive divisor m of n there exists a unique subfield of F with p^m elements.*

Proof. Let K be the prime subfield of F. Any subfield E of F must have p^m elements, where $m = [E : K]$. Then $m \mid n$ since $n = [F : K] = [F : E][E : K]$.

Conversely, suppose that $m \mid n$ for some $m > 0$. Then $p^m - 1$ is a divisor of $p^n - 1$, and so $g(x) = x^{p^m - 1} - 1$ is a divisor of $f(x) = x^{p^n - 1} - 1$. Since F is the splitting field of $xf(x)$ over K, with distinct roots, it must contain all p^m distinct roots of $xg(x)$. By Lemma 6.5.4, these roots form a subfield of F. Furthermore, any other subfield with p^m elements must be a splitting field of $xg(x)$, and so it must consist of precisely the same elements. $\square$

6.5.6 Lemma. *Let F be a field of characteristic p. If n is a positive integer not divisible by p, then the polynomial $x^n - 1$ has no repeated roots in any extension field of F.*

Proof. Let c be a root of $x^n - 1$ in an extension E of F. A direct computation shows that we must have the factorization

$$x^n - 1 = (x - c)(x^{n-1} + cx^{n-2} + c^2 x^{n-3} + \ldots + c^{n-2} x + c^{n-1}) \,.$$

With the notation $x^n - 1 = (x - c) f(x)$, we only need to show that $f(c) \neq 0$. Since $f(x)$ has n terms, we have $f(c) = nc^{n-1}$, and then $f(c) \neq 0$ since $p \nmid n$. $\square$

6.5.7 Theorem. *For each prime p and each positive integer n, there exists a field with p^n elements.*

Proof. Let F be the splitting field of $f(x) = x^{p^n} - x$ over the field $\mathbf{Z}_p$. Since $x^{p^n} - x = x(x^{p^n - 1} - 1)$ and $p^n - 1$ is not divisible by p, Lemma 6.5.6 implies that $f(x)$ has distinct roots. By Lemma 6.5.4, the set of all roots of $f(x)$ is a subfield of F, and so we conclude that F must consist of precisely the roots of $f(x)$, of which there are exactly p^n elements. $\square$

Any finite field F of characteristic p contains an isomorphic copy of $\mathbf{Z}_p$ as its prime subfield. Theorem 6.5.2 shows that F is a splitting field, so it is "unique up to isomorphism". Thus we can think of the field constructed in Theorem 6.5.7 as "the" field of order p^n.

6.5.8 Definition. *Let p be a prime number and let $n \in \mathbf{Z}^+$. The field with p^n elements is called the **Galois field of order** p^n, denoted by $\mathrm{GF}(p^n)$.*

For a prime number p, we now have two different notations for the set of congruence classes of integers modulo p, and these can be used interchangeably. We will generally use the notation $\mathrm{GF}(p)$ for the prime subfield of $\mathrm{GF}(p^n)$, particularly when we want to emphasize the field structure. We will retain the notation $\mathbf{Z}_p$ when considering only the group structure, and in most cases we will retain the notation $\mathbf{Z}_p[x]$ for the polynomial ring with coefficients in $\mathbf{Z}_p$.

Example 6.5.2 (GF(2^2)).

The polynomial $x^2 + x + 1$ is irreducible over $\mathbf{Z}_2$, so $\mathbf{Z}_2[x]/\langle x^2 + x + 1\rangle$ provides a model for GF(2^2). The cosets $0+\langle x^2 + x + 1\rangle$ and $1+\langle x^2 + x + 1\rangle$ form the prime subfield GF(p). Note that since the multiplicative group of nonzero elements has order 3, it is cyclic. Either $x + \langle x^2 + x + 1\rangle$ or $1 + x + \langle x^2 + x + 1\rangle$ is a generator for GF(2^2)$^\times$. Addition and multiplication tables for this field can be found in Table 4.3.1 in Section 4.3. □

Our next goal is to show that the multiplicative group $F^\times$ of any finite field F is cyclic. In fact, without any additional effort, we can prove a more general theorem that does not require the field to be finite. We need the following lemma, whose proof is taken from that of Proposition 3.5.9.

6.5.9 Lemma. *Let G be a finite abelian group. If $a \in G$ is an element of maximal order in G, then the order of every element of G is a divisor of the order of a.*

Proof. Let a be an element of maximal order in G, and let x be any element of G different from the identity. If $o(x) \nmid o(a)$, then in the prime factorizations of the respective orders there must exist a prime p that occurs to a higher power in $o(x)$ than in $o(a)$. Let $o(a) = p^\alpha n$ and $o(x) = p^\beta m$, where $\alpha < \beta$ and $p \nmid n$, $p \nmid m$. Now $o(a^{p^\alpha}) = n$ and $o(x^m) = p^\beta$, and so the orders are relatively prime since $p \nmid n$. It follows that the order of the product $a^{p^\alpha} x^m$ is equal to np^β, which is greater than $o(a)$, a contradiction. □

6.5.10 Theorem. *Any finite subgroup of the multiplicative group of a field is cyclic.*

Proof. Let F be a field, and let H be a finite subgroup of $F^\times$. Let a be an element of H of maximal order, with $o(a) = m$. By Lemma 6.5.9, each element of H satisfies the polynomial $x^m - 1$. Since F is a field, there are at most m roots of this polynomial, and so $|H| \leq m$. This implies that $o(a) = |H|$, and so H is cyclic. □

Example 6.5.3 (GF(2^3)).

The polynomial $x^3 + x + 1$ is irreducible over $\mathbf{Z}_2$, since it has no roots in $\mathbf{Z}_2$, so $\mathbf{Z}_2[x]/\langle x^3 + x + 1\rangle$ provides a model for GF(2^3). If we let $u = x + \langle x^3 + x + 1\rangle$, then we can describe GF(2^3) as the set of elements 0, 1, u, $1 + u$, u^2, $1 + u^2$, $u + u^2$, and $1 + u + u^2$. The elements are added like polynomials, but multiplication involves numerous substitutions, using the identity $u^3 = 1 + u$.

It is much more convenient to make use of the fact that the multiplicative group GF(2^3)$^\times$ is cyclic. With the given representation, the element u is a generator, so the correspondences $1 + u = u^3$, $1 + u^2 = u^6$, $u + u^2 = u^4$, and $1 + u + u^2 = u^5$ allow multiplication to be done in the exponential form. For example, we have $(1 + u^2)(u + u^2) = u^6 u^4 = u^7 u^3 = u^3 = 1 + u$. Note the analogy with real logarithms.

As in the proof of Kronecker's theorem, u is a root of the polynomial $x^3 + x + 1$ since $u^3 + u + 1 = 0$. Using Lemma 6.5.4, it can be shown that the mapping $\phi : \mathrm{GF}(2^3) \to \mathrm{GF}(2^3)$ defined by $\phi(x) = x^2$ is an automorphism of $\mathrm{GF}(2^3)$. (See Exercise 12.) Since $\phi(0) = 0$ and $\phi(1) = 1$, it must map any root of $x^3 + x + 1$ to a root of $x^3 + x + 1$. (See Exercise 11 of Section 6.4.) Therefore u, u^2, and $u^4 = u + u^2$ are roots of $x^3 + x + 1$, showing that $\mathrm{GF}(2^3)$ is the splitting field for $x^3 + x + 1$ over $\mathrm{GF}(3)$. □

Example 6.5.4 ($\mathbf{Z}_2[x]/\langle x^3 + x^2 + 1 \rangle \cong \mathbf{Z}_2[x]/\langle x^3 + x + 1 \rangle$).

Corollary 6.5.3 implies that the field $F_2 = \mathbf{Z}_2[x]/\langle x^3 + x^2 + 1 \rangle$ is isomorphic to $F_1 = \mathbf{Z}_2[x]/\langle x^3 + x + 1 \rangle$. It is instructive to actually construct an isomorphism. Let $w = x + \langle x^3 + x^2 + 1 \rangle$, so that w is a root of $x^3 + x^2 + 1$ in F_2. It follows from Lemma 6.4.3 that to construct an isomorphism from F_2 to F_1 we only need to find a root v of $x^3 + x^2 + 1$ in F_1, and then map w to v. The lemma then tells us how to extend this mapping.

In Example 6.5.3 we studied F_1, and let $u = x + \langle x^3 + x + 1 \rangle$. We saw that u, u^2, and $u + u^2$ are roots of $x^3 + x + 1$. Since F_1 is a splitting field over $\mathrm{GF}(2)$ for $x^8 - x = x(x - 1)(x^3 + x + 1)(x^3 + x^2 + 1)$, it follows (by process of elimination) that the roots of $x^3 + x^2 + 1$ must be $1 + u$, $1 + u^2$, and $1 + u + u^2$. Thus if we choose $v = 1 + u$, we then map an element $aw^2 + bw + c \in F_2$ to $av^2 + bv + c \in F_1$.

In this particular case we can give another proof that uses earlier results. By Example 5.2.2, the mapping $\theta : \mathbf{Z}_2[x] \to \mathbf{Z}_2[x]$ defined by $\theta(f(x)) = f(x+1)$ is an automorphism. Since $\theta(x^3 + x^2 + 1) = (x+1)^3 + (x+1)^2 + 1 = x^3 + x + 1$, it follows from Example 5.3.5 that $F_2 \cong F_1$. □

Recall that an extension field F of K is called a *simple* extension of K if $F = K(u)$ for some $u \in F$.

6.5.11 Theorem. *Any finite field is a simple extension of its prime subfield.*

Proof. Let F be a finite field with prime subfield K. By Theorem 6.5.10, the multiplicative group $F^\times$ is cyclic. It is clear that $F = K(u)$ for any generator u of $F^\times$. □

6.5.12 Corollary. *For each positive integer n there exists an irreducible polynomial of degree n over* GF(p).

Proof. By Theorem 6.5.11, if GF(p^n) is obtained by adjoining u to GF(p), then the minimal polynomial $p(x)$ of u over GF(p) must be an irreducible polynomial of degree n. □

EXERCISES: SECTION 6.5

1. Give addition and multiplication tables for the finite field GF(2^3), as described in Example 6.5.3.

2. Give addition and multiplication tables for the finite field GF(3^2), and find a generator for the cyclic group of nonzero elements under multiplication.

3.†Find a generator for the cyclic group of nonzero elements of GF(2^4).

4. Find the splitting fields over GF(3) for the following polynomials.
   (a) $x^4 + 2$
   (b) $x^4 - 2$
   Hint: See Exercise 13 of Section 4.2.

5. Show that $x^3 - x - 1$ and $x^3 - x + 1$ are irreducible over GF(3). Construct their splitting fields and explicitly exhibit the isomorphism between these splitting fields.

6. Show that $x^3 - x^2 + 1$ is irreducible over GF(3). Construct its splitting field and explicitly exhibit the isomorphism between this field and the splitting field of $x^3 - x + 1$ over GF(3).

7. Show that if $g(x)$ is irreducible over GF(p) and $g(x) \mid (x^{p^m} - x)$, then deg($g(x)$) is a divisor of m.

8. Let m, n be positive integers with gcd(m, n) = d. Show that, over any field, the greatest common divisor of $x^m - 1$ and $x^n - 1$ is $x^d - 1$.

9.†If E and F are subfields of GF(p^n) with p^e and p^f elements respectively, how many elements does $E \cap F$ contain? Prove your claim.

10. Let p be an odd prime.
    (a) Show that the set S of squares in GF(p^n) contains $(p^n + 1)/2$ elements.
    (b) Given $a \in$ GF(p^n), let $T = \{a - x \mid x \in S\}$. Show that $T \cap S \neq \emptyset$.
    (c) Show that every element of GF(p^n) is a sum of two squares.
    (d) What can be said about GF(2^n)?

11. Show that $x^p - x + a$ is irreducible over GF(p) for all nonzero elements $a \in$ GF(p).

12. Define the function $\phi :$ GF(2^3) $\to$ GF(2^3) by $\phi(x) = x^2$, for all $x \in$ GF(2^3).

    (a) Show that ϕ is an isomorphism.

    (b) Choose an irreducible polynomial $p(x)$ to represent GF(2^3) as $\mathbf{Z}_2[x]/ \langle p(x)\rangle$, and give an explicit computation of ϕ, ϕ^2, and ϕ^3.

6.6 Irreducible Polynomials over Finite Fields

In this section, our task is to study the irreducible polynomials over a finite field. The following theorem is the key result, and comes from our earlier results on the structure of finite fields.

6.6.1 Theorem. *Let $F =$ GF(q), where $q = p^n$. The monic irreducible factors of $x^{q^m} - x$ in $F[x]$ are precisely the monic irreducible polynomials in $F[x]$ whose degree is a divisor of m.*

Proof. The splitting field for $f(x) = x^{q^m} - x$ over F is GF(q^m), so the degree of any root of an irreducible factor of $f(x)$ must be a divisor of $m = [$GF(q^m) : GF(q)$]$. Thus the degree of any irreducible factor is a divisor of m.

On the other hand, let $p(x) \in F[x]$ be any irreducible polynomial of degree k such that $k|m$. Adjoining a root u of $p(x)$ gives a field $F(u)$ with q^k elements, which must be isomorphic to a subfield of GF(q^m) since $k|m$. Since $p(x)$ is still the minimal polynomial of the image of u in GF(q^m), it follows that $p(x)$ is a factor of $x^{q^m} - x$. $\square$

Let $f(x)$ be an irreducible polynomial over the finite field K, and suppose that $f(x)$ has a root in the extension F, with $|F| = p^n$. Then the elements of F are the roots of $x^{p^n} - x$, and $f(x) \mid x^{p^n} - x$, so there are several consequences. First, since $x^{p^n} - x$ has no repeated roots the same condition holds for $f(x)$. Secondly, $f(x)$ splits over F since F contains all the roots of $f(x)$. These remarks prove the following corollary.

6.6.2 Corollary. *Let K be a finite field, and let $f(x) \in K[x]$ be an irreducible polynomial. If F is an extension field of K that contains a root u of $f(x)$, then $K(u)$ is a splitting field for $f(x)$ over K.*

Our next goal is to find a way to simply count the number of monic irreducible polynomials of a given degree over a given finite field.

Example 6.6.1.

We recall that the Euler φ-function counts the number of positive integers less than n and relatively prime to n. In Corollary 3.5.6 we proved that if the prime factorization of n is $n = p_1^{\alpha_1} \cdots p_k^{\alpha_k}$, then

$$\varphi(n) = n\left(1 - \frac{1}{p_1}\right) \cdots \left(1 - \frac{1}{p_k}\right).$$

For example, since $12 = 2^2 \cdot 3$, we have

$$\varphi(12) = 12\left(1 - \frac{1}{2}\right)\left(1 - \frac{1}{3}\right) = 12\left(\frac{1}{1} + \frac{-1}{2} + \frac{-1}{3} + \frac{1}{6}\right).$$

The last term above uses the factors 1, 2, 3, and 6 of 12, while the remaining factors 4 and 12 are omitted. The omission of terms which are divisible by powers of primes provides one motivation for the definition of the function μ in Definition 6.6.3. □

Convention: In each of the notations $\quad \sum_{d|n}, \quad \prod_{d|n}, \quad$ and $\quad \prod_{p|n},$ we will assume that $d|n$ refers to the positive divisors of n, and that $p|n$ will only be used to refer to the positive prime divisors of n.

6.6.3 Definition. *If d is a positive integer, we define the **Möbius** function $\mu(d)$ as follows:*

$\mu(1) = 1;$

$\mu(d) = 1$ *if d has an even number of prime factors (each occurring only once);*

$\mu(d) = -1$ *if d has an odd number of prime factors (each occurring only once);*

$\mu(d) = 0$ *if d is divisible by the square of a prime.*

Our first proposition follows easily from the definition, and the proof will be omitted.

6.6.4 Proposition. *If $m, n \in \mathbf{Z}^+$ and $(m, n) = 1$, then $\mu(mn) = \mu(m)\mu(n)$.*

The property in the previous proposition is important enough to merit a definition. If R is a commutative ring, then a function $f : \mathbf{Z}^+ \to R$ is said to be a **multiplicative** function if $f(mn) = f(m)f(n)$, whenever $(m, n) = 1$.

6.6.5 Proposition. *Let R be a commutative ring, and let $f : \mathbf{Z}^+ \to R$ be a multiplicative function. If $F : \mathbf{Z}^+ \to R$ is defined by*

$$F(n) = \sum_{d|n} f(d), \quad \text{for all } n \in \mathbf{Z}^+,$$

then F is a multiplicative function.

Proof. If $\gcd(m, n) = 1$, then

$$F(mn) = \sum_{d|mn} f(d) = \sum_{a|m} \sum_{b|n} f(ab) = \sum_{a|m} \sum_{b|n} f(a) f(b)$$
$$= \left(\sum_{a|m} f(a) \right) \left(\sum_{b|n} f(b) \right) = F(m) F(n) ,$$

and so the function F is multiplicative. $\square$

6.6.6 Proposition. *For any positive integer n,*

$$\sum_{d|n} \mu(d) = \begin{cases} 1 & \text{if } n = 1 \\ 0 & \text{if } n > 1 \end{cases}$$

Proof. Since $\mu(1) = 1$, the case $n = 1$ is trivial. In the case of a prime power p^m, we have

$$\sum_{d|p^m} \mu(d) = \mu(1) + \mu(p) + \mu(p^2) + \ldots + \mu(p^m) = \mu(1) + \mu(p) = 1 - 1 = 0 .$$

It follows form Proposition 6.6.5 that the given sum is a multiplicative function, so the existence of even one prime divisor of n implies that the sum is zero. Thus $\sum_{d|n} \mu(d) = 0$ if $n > 1$. $\square$

6.6.7 Theorem (Möbius Inversion Formula). *Let R be a commutative ring, and let $f : \mathbf{Z}^+ \to R$ be any function. If the function $F : \mathbf{Z}^+ \to R$ is defined by*

$$F(n) = \sum_{d|n} f(d) \quad \text{for all } n \in \mathbf{Z}^+ ,$$

then

$$f(m) = \sum_{n|m} \mu(m/n) F(n) \quad \text{for all } m \in \mathbf{Z}^+ .$$

Proof. We have

$$\sum_{n|m} \mu(m/n) F(n) = \sum_{n|m} \mu(m/n) \left(\sum_{d|n} f(d) \right) = \sum_{n|m} \sum_{d|n} \mu(m/n) f(d)$$
$$= \sum_{ij|m} \mu(i) f(j) = \sum_{j|m} \left(\sum_{i|\frac{m}{j}} \mu(i) \right) f(j) = f(m) .$$

The last equality holds since we have $\sum_{i|\frac{m}{j}} \mu(i) = 0$ unless $\frac{m}{j} = 1$, or $j = m$, in which case $\sum_{i|\frac{m}{j}} \mu(i) = 1$. $\square$

There is also a multiplicative form of the Möbius inversion formula, whose proof is similar to that for the additive form. Note that in the special case $R = \mathbf{R}$, setting $f(d) = \log g(d)$ and $F(n) = \log f(n)$ reduces the multiplicative form to the additive one.

6.6.8 Theorem (Möbius Inversion Formula (2)). *Let R be a commutative ring, and let $g : \mathbf{Z}^+ \to R$ be any function. If the function $G : \mathbf{Z}^+ \to R$ is defined by*

$$G(n) = \prod_{d|n} g(d) \quad \text{for all } n \in \mathbf{Z}^+ ,$$

then

$$g(m) = \prod_{n|m} G(n)^{\mu(m/n)} \quad \text{for all } m \in \mathbf{Z}^+ .$$

Proof. We have

$$
\prod_{n|m} G(n)^{\mu(m/n)} \;=\; \prod_{n|m} \left(\prod_{d|n} g(d) \right)^{\mu(m/n)} \;=\; \prod_{n|m} \prod_{d|n} g(d)^{\mu(m/n)}
$$
$$
=\; \prod_{ij|m} g(j)^{\mu(i)} \;=\; \prod_{j|m} g(j)^{\sum_{i|\frac{m}{j}} \mu(i)} \;=\; g(m) .
$$

The last equality holds since we have $\sum_{i|\frac{m}{j}} \mu(i) = 0$ unless $\frac{m}{j} = 1$, or $j = m$, in which case $\sum_{i|\frac{m}{j}} \mu(i) = 1$. $\square$

6.6.9 Corollary. *For any positive integer n, $\varphi(n) = n \sum_{d|n} \frac{\mu(d)}{d}$.*

Proof. We first show that the formula $n = \sum_{d|n} \varphi(d)$ can be proved using facts from group theory. Let $G = \langle a \rangle$ be a cyclic group of order n, and recall that Corollary 3.5.4 shows that the subgroups of G are $\langle a^d \rangle$ where $d|n$. Furthermore, the corollary shows that G has $\varphi(n)$ generators, and that $\langle a^d \rangle$ has $\varphi(n/d)$ generators, since its order is n/d. Because every element of G generates one and only one cyclic subgroup of G, we have

$$n = \sum_{d|n} \varphi(n/d) = \sum_{d|n} \varphi(d) .$$

Applying the Möbius inversion formula, with $f(n) = \varphi(n)$ and $F(n) = n$, we obtain $\varphi(n) = \sum_{d|n} \mu(n/d) \cdot d$. Finally, if we interchange d and n/d, for the positive divisors of n, we obtain the desired formula. $\square$

With these tools in hand, we now turn to the problem of counting the number of monic irreducible polynomials of a given degree over a finite field. For example, the results of Exercise 12 of Section 4.2 show that over GF(2) there is only one monic irreducible polynomial of degree 2. There are two monic irreducible polynomials of degree 3, three of degree 4, and six of degree 5. Exercise 14 of Section 4.2 shows that over GF(p) there are $(p^2 - p)/2$ monic irreducible polynomials of degree 2.

6.6.10 Definition. *The number of monic irreducible polynomials of degree m over the finite field GF(q), where q is a prime power, will be denoted by $\mathrm{I}_q(m)$.*

The following formula for $I_q(m)$ is due to Gauss.

6.6.11 Theorem. *For any prime power q and any positive integer m,*

$$I_q(m) = \frac{1}{m} \sum_{d|m} \mu(m/d)q^d .$$

Proof. Theorem 6.6.1 shows that over $GF(q)$ the polynomial $x^{q^m} - x$ is the product of all monic irreducible polynomials whose degree is a divisor of m. Comparing degrees, we see that $q^m = \sum_{d|m} d \cdot I_q(d)$. Using the Möbius inversion formula with $f(d) = d \cdot I_q(d)$ and $F(m) = q^m$, we obtain $mI_q(m) = \sum_{d|m} \mu(m/d)q^d$. □

6.6.12 Corollary. *For all positive integers m and all prime powers q we have* $I_q(m) \geq 1$.

Proof. Since $\mu(d) \geq -1$ for $d > 1$ and $\mu(1) = 1$, we have

$$\begin{aligned}
I_q(m) &= \frac{1}{m} \sum_{d|m} \mu(m/d)q^d \\
&\geq \frac{1}{m} \left(q^m - q^{m-1} - q^{m-2} - \ldots - q \right) \\
&\geq \frac{1}{m} .
\end{aligned}$$

This completes the proof. □

The above corollary gives another proof that for each prime power p^m there exists a finite field with p^m elements, since $I_p(m) \geq 1$.

EXERCISES: SECTION 6.6

1. Verify Theorem 6.6.1 in the special case of $x^{16} - x$ over $GF(2)$, by multiplying out the appropriate irreducible polynomials from the list given in the answer to Exercise 12 of Section 4.2.

2. Use Theorem 6.6.1 to show that over $GF(2)$ the polynomial $x^{32} + x$ factors as a product of the terms x, $x + 1$, $x^5 + x^2 + 1$, $x^5 + x^3 + 1$, $x^5 + x^4 + x^3 + x + 1$, $x^5 + x^4 + x^2 + x + 1$, $x^5 + x^3 + x^2 + x + 1$, and $x^5 + x^4 + x^3 + x^2 + 1$.

3. Let F be a field of characteristic p, with prime subfield $K = GF(p)$. Show that if $u \in F$ is a root of a polynomial $f(x) \in K[x]$, then u^p is also a root of $f(x)$.

4. Let u be a primitive element of GF(p^m), and let $M^{(i)}(x)$ be the minimal polynomial of u^i over GF(p). Show that every element of the form u^{ip^k} is also a root of $M^{(i)}(x)$.

5. Let GF(2^6) be represented by $\mathbf{Z}_2[x]/\langle x^6 + x + 1 \rangle$, and let u be any primitive element of GF(2^6). Show that GF(2^3) = $\{0, 1, u^9, u^{18}, u^{27}, u^{36}, u^{45}, u^{54}\}$.

6. Let F be a field, and let n be a positive integer. An element $\zeta \in F$ is called a **primitive nth root of unity** if it has order n in the multiplicative group $F^\times$. Show that no field of characteristic $p > 0$ contains a primitive pth root of unity.

7. Let $n \in \mathbf{Z}^+$, and define $\tau(n)$ to be the number of divisors of n.

   (a) Show that τ is a multiplicative function.

   (b) Show that if $n = p_1^{\alpha_1} p_2^{\alpha_2} \cdots p_k^{\alpha_k}$, then $\tau(n) = (\alpha_1 + 1)(\alpha_2 + 1) \cdots (\alpha_k + 1)$.

   (c) Show that $\tau(n)$ is odd if and only if n is a square.

   (d) Show that $\sum_{d|n} \tau(d)\mu(n/d) = 1$.

8. Let $n \in \mathbf{Z}^+$ and define $\sigma(n) = \sum_{d|n} d$, the sum of the positive divisors of n.

   (a) Show that σ is a multiplicative function.

   (b) Show that if $n = p_1^{\alpha_1} p_2^{\alpha_2} \cdots p_k^{\alpha_k}$, then $\sigma(n) = \prod_{i=1}^k \left((p_i^{\alpha_i+1} - 1)/(p_i - 1) \right)$.

   (c) Show that $\sigma(n)$ is odd if and only if n is a square or two times a square.

   (d) Show that $\sum_{d|n} \sigma(d)\mu(n/d) = n$.

9. A positive integer n is called *perfect* if it is equal to the sum of its proper positive divisors. Thus n is perfect if and only if $\sigma(n) = 2n$. Prove that n is an even perfect number if and only if $n = 2^{p-1}(2^p - 1)$, where p and $2^p - 1$ are prime numbers.

   Note: Prime numbers of the form $2^p - 1$, for p prime, are called **Mersenne** primes. It is not known whether there are infinitely many Mersenne primes. It is also not known whether there are any odd perfect numbers.

10. Let D be an integral domain. Show that if $f : \mathbf{Z}^+ \to D$ is a nonzero multiplicative function, then $\sum_{d|n} \mu(d) f(d) = \prod_{p|n} (1 - f(p))$, for all $n \in \mathbf{Z}^+$, where the product is taken over all prime divisors p of n.

11. Let R be a commutative ring. Let $\mathcal{R}$ be the set of all functions $f : \mathbf{Z}^+ \to R$. For $f, g \in \mathcal{R}$ define $f + g$ by ordinary addition of functions: $(f+g)(n) = f(n)+g(n)$, for all $n \in \mathbf{Z}^+$. Define a product $*$ on $\mathcal{R}$ as follows:

$$(f * g)(n) = \sum_{d|n} f(d)g(n/d), \quad \text{for all } n \in \mathbf{Z}^+.$$

The product $*$ is called the **convolution product** of the functions f and g. Define $\epsilon : \mathbf{Z}^+ \to R$ by $\epsilon(1) = 1$ and $\epsilon(n) = 0$, for all $n > 1$.

   (a) Show that $\mathcal{R}$ is a commutative ring under the operations $+$ and $*$, with identity ϵ.

   (b) Show that $f \in \mathcal{R}$ has a multiplicative inverse if and only if $f(1)$ is invertible in R.

(c) Show that if $f, g \in \mathcal{R}$ are multiplicative functions, then so is $f * g$.

(d) Show that if R is an integral domain, then the set of nonzero multiplicative functions in $\mathcal{R}$ is a subgroup of $\mathcal{R}^\times$, the group of units of $\mathcal{R}$.

(e) Define $\mu_R \in \mathcal{R}$ as in Definition 6.6.3, with the understanding that 0 and 1 are (respectively) the additive and multiplicative identities of R. Let $\nu \in \mathcal{R}$ be defined by $\nu(n) = 1$, for all $n \in \mathbf{Z}^+$. Show that $\mu_R * \nu = \epsilon$, and that $\mu_R * \nu * f = f$, for all $f \in \mathcal{R}$.

Note: The formula $f = \mu_R * \nu * f$ is a generalized Möbius inversion formula.

12. Let R be a commutative ring. Let $f : \mathbf{Z}^+ \to R$ be any function, and let $F : \mathbf{Z}^+ \to R$ be defined by $F(n) = \sum_{d|n} f(d)$, for all $n \in \mathbf{Z}^+$. Show that if F is a multiplicative function, then so is f.

6.7 Quadratic Reciprocity

Gauss's law of quadratic reciprocity is one of the gems of eighteenth and nineteenth century mathematics. This remarkable result was first conjectured by Euler, and again later (independently) by Adrien Marie Legendre (1752–1833). An incomplete proof was given by Lagrange, and the first complete proof was finally found by Gauss in 1796. His proof was first published in his book *Disquisitiones Arithmeticae* in 1801. By 1818, Gauss had found his sixth proof of the result. It is evident that he considered this work on quadratic reciprocity one of his most important and favorite contributions to our subject. The quadratic reciprocity law, when used in conjunction with other results such as the Chinese remainder theorem, allows us to determine precisely which quadratic congruences are solvable.

6.7.1 Definition. *Let n be a positive integer, and let a be an integer such that $n \nmid a$. Then a is called a **quadratic residue modulo** n if the congruence*

$$x^2 \equiv a \ (mod \ n)$$

*is solvable, and a **quadratic nonresidue** otherwise.*

When n is a prime, we write $\left(\dfrac{a}{n}\right) = 1$ if a is a quadratic residue modulo n and $\left(\dfrac{a}{n}\right) = -1$ if a is a quadratic nonresidue modulo n.

The symbol $\left(\dfrac{a}{n}\right)$ is called the **Legendre symbol**. The reason for choosing this notation will become apparent later in the discussion.

We begin with several easy observations about the Legendre symbol, when the modulus p is prime. We note that if $a \equiv b$ (mod p), then $\left(\dfrac{a}{p}\right) = \left(\dfrac{b}{p}\right)$ and so the symbol $\left(\dfrac{x}{p}\right)$ is well-defined for nonzero elements $x \in \mathbf{Z}_p$. This abuse of notation should not cause any confusion. It is routine to check that $\left(\dfrac{ab}{p}\right) = \left(\dfrac{a}{p}\right)\left(\dfrac{b}{p}\right)$ for all $a, b \in \mathbf{Z}$ such that $p \nmid a$ and $p \nmid b$. Another easily verified fact is that $\left(\dfrac{1}{p}\right) = 1$.

If p and q are distinct odd primes, the law of quadratic reciprocity allows us to evaluate the symbol $\left(\dfrac{p}{q}\right)$ provided we already know the value of the symbol $\left(\dfrac{q}{p}\right)$. We will establish this beautiful result using our work on finite fields.

Let p be an odd prime. The set $Q = \left\{ 1^2, 2^2, \ldots, \left(\dfrac{p-1}{2}\right)^2 \right\}$ is a set of quadratic residues modulo p. If a and b are positive integers that are less than or equal to $(p-1)/2$ and $a^2 \equiv b^2$ (mod p), then $a \equiv b$ (mod p), and so $a = b$ since $a \equiv -b$ (mod p) cannot occur when $2 \leq a + b \leq p - 1$. Since $a^2 \equiv (-a)^2 \equiv (p-a)^2$ (mod p), every quadratic residue modulo p is congruent to one and only one element of Q. Thus there are $(p-1)/2$ quadratic residues modulo p and $(p-1)/2$ quadratic nonresidues modulo p.

Before stating our main theorem, we prove the following important proposition. Recall that if p prime, then the Galois field with p elements can be denoted by GF(p), as well as by $\mathbf{Z}_p$. We will use the notation GF(p) throughout this section, and we need to recall the fact that GF(p)$^\times$ is a cyclic group. (Theorem 6.5.10 states that the multiplicative group of nonzero elements of any finite field is cyclic.)

6.7.2 Proposition (Euler's Criterion). *If p is an odd prime, and if $a \in \mathbf{Z}$ with $p \nmid a$, then*

$$\left(\frac{a}{p}\right) \equiv a^{(p-1)/2} \ (mod \ p) \ .$$

Proof. Let $[w]_p$ be a generator of the multiplicative group GF(p)$^\times$. Then $a \equiv w^j$ (mod p) for some $0 \leq j \leq p - 2$. Now $\left(\dfrac{a}{p}\right) = 1$ if and only if j is even, and so $\left(\dfrac{a}{p}\right) = (-1)^j$. Since -1 has order two in the cyclic group GF(p)$^\times$, we have

$$-1 \equiv w^{(p-1)/2} \ (mod \ p) \ ,$$

and so

$$\left(\frac{a}{p}\right) \equiv (-1)^j \equiv w^{(p-1)j/2} \equiv a^{(p-1)/2} \ (mod \ p) \ .$$

This completes the proof.   □

In the course of the proof of Theorem 6.7.3, we will abuse our notation and write j for the congruence class $[j]_q$ in GF(q). This will simplify our formulas and should not confuse the careful reader.

6.7.3 Theorem (Quadratic Reciprocity). *Let p, q be distinct odd primes. Then*

$$\left(\frac{p}{q}\right)\left(\frac{q}{p}\right) = (-1)^{(p-1)(q-1)/4} \ .$$

Proof. Let K be an extension field of GF(q) that contains a root $\omega \neq 1$ of the polynomial $x^p - 1$. Let G be the multiplicative group GF(p)$^\times$. Since $\omega^p = 1$, for each $t \in G$ the expression ω^t is well-defined.

We define the Gauss sum for each $a \in G$ as

$$g(a) = \sum_{t \in G} \left(\frac{t}{p}\right) \omega^{at} \ .$$

We note that $g(a) \in K$. As shown by the following argument, $g(a)$ is completely determined by $g(1)$ and $\left(\frac{a}{p}\right)$. Since $\left(\frac{a^{-1}}{p}\right) = \left(\frac{a}{p}\right)$, we have

$$
\begin{aligned}
\left(\frac{a}{p}\right) \cdot g(1) &= \left(\frac{a^{-1}}{p}\right) \cdot g(1) = \left(\frac{a^{-1}}{p}\right) \sum_{t \in G} \left(\frac{t}{p}\right) \omega^t \\
&= \sum_{t \in G} \left(\frac{a^{-1}t}{p}\right) \omega^t = \sum_{t \in G} \left(\frac{t}{p}\right) \omega^{at} = g(a) \ .
\end{aligned}
$$

The next to last equality holds because multiplication by a defines a permutation of G, and thus the sum over all $t \in G$ is the same as the sum over all elements $at \in G$.

Since the characteristic of K is q, we have

$$
\begin{aligned}
g(1)^q &= \left[\sum_{t \in G} \left(\frac{t}{p}\right) \omega^t\right]^q = \sum_{t \in G} \left(\frac{t}{p}\right)^q \omega^{qt} \\
&= \sum_{t \in G} \left(\frac{t}{p}\right) \omega^{qt} = g(q) = \left(\frac{q}{p}\right) g(1) \ .
\end{aligned}
$$

We have the following useful expression for $g(1)^2$.

$$
\begin{aligned}
g(1)^2 &= \left[\sum_{t \in G} \left(\frac{t}{p}\right) \omega^t\right]\left[\sum_{s \in G} \left(\frac{s}{p}\right) \omega^s\right] = \sum_{t,s \in G} \left(\frac{t}{p}\right)\left(\frac{s}{p}\right) \omega^{t+s} \\
&= \sum_{t,s \in G} \left(\frac{ts}{p}\right) \omega^{t+s} = \sum_{u \in \mathrm{GF}(p)} \left[\sum_{t \in G, \ t \neq u} \left(\frac{t(u-t)}{p}\right) \omega^u\right]
\end{aligned}
$$

Next, we check that

$$\left(\frac{t(u-t)}{p}\right) = \left(\frac{-1}{p}\right)\left(\frac{t^2}{p}\right)\left(\frac{1-t^{-1}u}{p}\right) = (-1)^{(p-1)/2}\left(\frac{1-t^{-1}u}{p}\right).$$

Let $d_u = \sum_{t\in G,\, t\neq u}\left(\dfrac{1-t^{-1}u}{p}\right)$. Then if $u=0$, we have $d_0 = \sum_{t\in G}\left(\dfrac{1}{p}\right) = p-1$.

On the other hand, if $u \neq 0$, then $1-t^{-1}u$ runs over $G - \{1\}$ as t runs over G, with $t \neq u$, and since there are an equal number of quadratic residues and nonresidues in G, we have

$$d_u = \sum_{t\in G,\, t\neq u}\left(\frac{1-t^{-1}u}{p}\right) = \sum_{s\in G}\left(\frac{s}{p}\right) - \left(\frac{1}{p}\right) = 0 - \left(\frac{1}{p}\right) = -1.$$

From our previous expression for $g(1)^2$, we have

$$
\begin{aligned}
g(1)^2 &= \sum_{u\in GF(p)}\left[\sum_{t\in G,\, t\neq u}\left(\frac{t(u-t)}{p}\right)\omega^u\right]\\
&= (-1)^{(p-1)/2}\sum_{u\in GF(p)} d_u\omega^u\\
&= (-1)^{(p-1)/2}\left[(p-1)\omega^0 + \sum_{u\in G}(-1)\omega^u\right]\\
&= (-1)^{(p-1)/2}\left[p\omega^0 + (-1)\sum_{u\in GF(p)}\omega^u\right].
\end{aligned}
$$

Since $\omega \neq 1$ is a root of $x^p - 1$, we have

$$0 = \omega^{p-1} + \omega^{p-2} + \ldots + \omega + 1 = \sum_{u\in GF(p)}\omega^u,$$

and so $g(1)^2 = (-1)^{(p-1)/2}p$. Because $g(1) \neq 0$ in K, the equation $g(1)^q = \left(\dfrac{q}{p}\right)g(1)$ implies that $g(1)^{q-1} = \left(\dfrac{q}{p}\right)$. Combining $g(1)^2 = (-1)^{(p-1)/2}p$ and $g(1)^{q-1} = \left(\dfrac{q}{p}\right)$ and another application of Euler's criterion, we get

$$
\begin{aligned}
\left(\frac{p}{q}\right)\left(\frac{q}{p}\right) &= \left(\frac{p}{q}\right)g(1)^{q-1} = \left(\frac{p}{q}\right)[g(1)^2]^{(q-1)/2}\\
&= \left(\frac{p}{q}\right)[(-1)^{(p-1)/2}p]^{(q-1)/2} = \left(\frac{p}{q}\right)p^{(q-1)/2}(-1)^{(p-1)(q-1)/4}\\
&= \left(\frac{p}{q}\right)\left(\frac{p}{q}\right)(-1)^{(p-1)(q-1)/4} = (-1)^{(p-1)(q-1)/4}
\end{aligned}
$$

as required. $\square$

The next theorem provides a pair of useful supplements to the law of quadratic reciprocity.

6.7.4 Theorem. *Let p be an odd prime. Then*

(i) $\left(\dfrac{-1}{p}\right) = (-1)^{(p-1)/2}$, *and*

(ii) $\left(\dfrac{2}{p}\right) = (-1)^k$, *where* $k = \dfrac{p^2 - 1}{8}$.

Proof. Part (i) follows immediately from Euler's criterion.

In proving part (ii), we first note that $8 \mid (p^2 - 1)$ because p is odd, and

$$(-1)^{(p^2-1)/8} = \begin{cases} 1 & \text{if } p \equiv \pm 1 \ (\text{mod } 8) \\ -1 & \text{if } p \equiv \pm 5 \ (\text{mod } 8) \ . \end{cases}$$

Let ω be a root of $x^8 - 1$ that is not a root of $x^4 - 1$ in some extension field K of GF(p). Then since $\omega^4 \neq 1$, we have $\omega^4 = -1$, and so $\omega^2 = -\omega^{-2}$ and $\omega^5 = -\omega$. Now let $\rho = \omega + \omega^{-1}$. Thus $\rho^2 = \omega^2 + 2 + \omega^{-2} = 2$. Since $\rho^p = \omega^p + \omega^{-p}$, we have $\rho^p = \rho$ if $p \equiv \pm 1 \ (\text{mod } 8)$ and $\rho^p = \omega^5 + \omega^{-5} = -(\omega + \omega^{-1}) = -\rho$ if $p \equiv \pm 5 \ (\text{mod } 8)$. Thus

$$\left(\dfrac{2}{p}\right) = 2^{(p-1)/2} = (\rho^2)^{(p-1)/2} = \rho^{p-1} = \begin{cases} 1 & \text{if } p \equiv \pm 1 \ (\text{mod } 8) \\ -1 & \text{if } p \equiv \pm 5 \ (\text{mod } 8) \ . \end{cases}$$

Hence $\left(\dfrac{2}{p}\right) = (-1)^k$ for $k = (p^2 - 1)/8$, as required. □

Example 6.7.1.

Is the congruence $x^2 \equiv 3 \ (\text{mod } 47)$ solvable? To answer this question we need to determine $\left(\dfrac{3}{47}\right)$. In the statement of the quadratic reciprocity law we can multiply both sides of the equation by $\left(\dfrac{q}{p}\right)$. This gives us the following form of the law, where p and q are odd primes.

$$\left(\dfrac{p}{q}\right) = \left(\dfrac{q}{p}\right)(-1)^{(p-1)(q-1)/4}$$

Using the above form of the quadratic reciprocity law, we have

$$\left(\dfrac{3}{47}\right) = \left(\dfrac{47}{3}\right)(-1)^{(47-1)(3-1)/4} = \left(\dfrac{2}{3}\right)(-1)^{23} = (-1)(-1) = 1$$

since $47 \equiv 2 \ (\text{mod } 3)$ and since it is easy to check that $x^2 \equiv 2 \ (\text{mod } 3)$ is not solvable. Since $\left(\dfrac{3}{47}\right) = 1$, we know that $x^2 \equiv 3 \ (\text{mod } 47)$ is solvable. □

Example 6.7.2.

For the prime numbers 461 and 773 we have the following computation:

$$\left(\frac{461}{773}\right) = \left(\frac{773}{461}\right)(-1)^{(773-1)(461-1)/4} = \left(\frac{312}{461}\right)$$

$$= \left(\frac{8\cdot 3\cdot 13}{461}\right) = \left(\frac{2}{461}\right)^3\left(\frac{3}{461}\right)\left(\frac{13}{461}\right).$$

We first note that $\left(\dfrac{2}{461}\right) = (-1)^3 = -1$, since $461 \equiv 5 \pmod 8$. We

have $\left(\dfrac{3}{461}\right) = \left(\dfrac{461}{3}\right)(-1)^{(461-1)(3-1)/4} = \left(\dfrac{2}{3}\right) = -1$. Next, we have

$\left(\dfrac{13}{461}\right) = \left(\dfrac{461}{13}\right)(-1)^{(461-1)(13-1)/4} = \left(\dfrac{6}{13}\right) = \left(\dfrac{2}{13}\right)\left(\dfrac{3}{13}\right)$, and since

$13 \equiv 5 \pmod 8$, we have $\left(\dfrac{2}{13}\right) = -1$. Thus we have

$$\left(\frac{461}{773}\right) = \left(\frac{2}{461}\right)^3\left(\frac{3}{461}\right)\left(\frac{13}{461}\right) = (-1)(-1)(-1)\left(\frac{3}{13}\right)$$

$$= (-1)\left(\frac{13}{3}\right)(-1)^{(13-1)(3-1)/4} = (-1)\left(\frac{1}{3}\right) = -1.\quad\square$$

Example 6.7.3.

In this example we determine the value of $\left(\dfrac{3}{p}\right)$, when p is an odd prime
different from 3. By the quadratic reciprocity law we have

$$\left(\frac{3}{p}\right) = \left(\frac{p}{3}\right)(-1)^{(p-1)(3-1)/4} = \left(\frac{p}{3}\right)(-1)^{(p-1)/2}.$$

We have either $p \equiv 1 \pmod 3$, in which case $\left(\dfrac{p}{3}\right) = 1$, or $p \equiv 2 \pmod 3$,
in which case $\left(\dfrac{p}{3}\right) = -1$. Furthermore, we have either $p \equiv 1 \pmod 4$, in
which case $(p-1)/2$ is even, or $p \equiv 3 \pmod 4$, in which case $(p-1)/2$ is
odd.

The assumptions on p show that it is relatively prime to 12, so p is congruent
to one of $1, 5, 7, 11$ modulo 12. If $p \equiv 1 \pmod{12}$, then $p \equiv 1 \pmod 3$
and $p \equiv 1 \pmod 4$, so $\left(\dfrac{3}{p}\right) = (1)(1) = 1$. If $p \equiv 5 \pmod{12}$, then

$p \equiv 2 \pmod 3$ and $p \equiv 1 \pmod 4$, so $\left(\dfrac{3}{p}\right) = (-1)(1) = -1$. If $p \equiv$ 7 $\pmod{12}$, then $p \equiv 1 \pmod 3$ and $p \equiv 3 \pmod 4$, so $\left(\dfrac{3}{p}\right) = (1)(-1) =$ -1. If $p \equiv 11 \pmod{12}$, then $p \equiv 2 \pmod 3$ and $p \equiv 3 \pmod 4$, so $\left(\dfrac{3}{p}\right) = (-1)(-1) = 1$.

To summarize, we have

$$\left(\frac{3}{p}\right) = \begin{cases} 1 & \text{if } p \equiv \pm 1 \pmod{12} \\ -1 & \text{if } p \equiv \pm 5 \pmod{12} \, . \end{cases} \qquad \square$$

EXERCISES: SECTION 6.7

1. Prove that $\left(\dfrac{ab}{p}\right) = \left(\dfrac{a}{p}\right)\left(\dfrac{b}{p}\right)$ for all $a, b \in \mathbf{Z}$ such that $p \nmid a$ and $p \nmid b$.

2.†Compute the following values of the Legendre symbol.

(a) $\left(\dfrac{231}{997}\right)$

(b) $\left(\dfrac{783}{997}\right)$

3. Is the congruence $x^2 \equiv 180873 \pmod{997}$ solvable?

4. Determine the value of $\left(\dfrac{r}{p}\right)$ for the indicated values of r, where p is an odd prime subject to the indicated conditions.

†(a) $r = 5$, $p \neq 5$

(b) $r = 6$, $p \neq 3$

†(c) $r = 7$, $p \neq 7$

(d) $r = 11$, $p \neq 11$

(e) $r = 13$, $p \neq 13$

5. If a is a quadratic nonresidue of each of the odd primes p and q, is the congruence $x^2 \equiv a \pmod{pq}$ solvable?

6. If p and q are odd primes and $p = q + 4t$ for some $t \in \mathbf{Z}$, prove that $\left(\dfrac{t}{p}\right) = \left(\dfrac{t}{q}\right)$.

7. If p and q are distinct odd primes, prove that

$$\left(\frac{p}{q}\right)\left(\frac{q}{p}\right) = \begin{cases} -1 & \text{if } p \equiv q \equiv 3 \pmod 4 \\ 1 & \text{otherwise} \end{cases} .$$

8. Prove that there are infinitely many primes of the form $4m + 1$.

Notes

In Theorem 6.3.6 we gave a characterization of constructible numbers. We were
then able to use the characterization to show that it is impossible to find a general
construction for trisecting an angle, duplicating a cube, or squaring a circle.

A related problem is that of constructing (using only straightedge and compass)
a regular polygon with n sides. The modern solution uses the methods of Galois
theory and is beyond the scope of our book, but we can give a small part of the story
at this point. The ancient Greeks already knew how to construct regular polygons
of three, four, five, and six sides. The problem of constructing a regular pentagon
can be translated into the problem of finding the complex roots of the polynomial
$x^5 - 1$. It is certainly possible to give a trigonometric solution as a complex number
of the form $\cos(2\pi/5) + i\,\sin(2\pi/5)$, but we need to know that this can be expressed
in terms of rational numbers and their square roots. By results in Section 4.4 we
can give the factorization

$$x^5 - 1 = (x - 1)(x^4 + x^3 + x^2 + x + 1)$$

of $x^5 - 1$ into factors irreducible over $\mathbf{Q}$. The primitive fifth root of unity that we
need to construct is a root of the equation $x^4 + x^3 + x^2 + x + 1 = 0$, and since
$x^5 = 1$, we can rewrite it in the form

$$x^{-1} + x^{-2} + x^2 + x + 1 = (x^2 + 1 + x^{-2}) + (x + x^{-1}) = 0 .$$

Substituting $y = x + x^{-1}$ yields the equation

$$y^2 + y - 1 = 0$$

and from its solution we can find a solution (by radicals) of $x^5 - 1 = 0$.

When he was eighteen, Gauss discovered that the regular 17-gon is constructible.
He published the solution in *Disquisitiones Arithmeticae* as a special case of the
general solution of the "cyclotomic equation" $x^n - 1 = 0$. He proved that a regular
n-gon is constructible with straightedge and compass only if n has the form $n =
2^\alpha p_2 \cdots p_k$, where the numbers p_i are odd primes of the form $2^{2^n} + 1$. The converse
is also true, as was proved almost forty years later (see Section 8.5 for a proof).
We should note that not all numbers of the form $2^{2^n} + 1$ are prime, as had been
conjectured by Fermat. Although $2^1 + 1 = 3, 2^2 + 1 = 5, 2^4 + 1 = 17, 2^8 + 1 = 257$,
and $2^{16} + 1 = 65537$ are all prime, Euler showed that $2^{32} + 1 = 641 \cdot 6700417$.
The general solution of the cyclotomic equation, and his proof of the fundamental
theorem of algebra, represent the most important contributions that Gauss made to
the theory of algebraic equations.

Chapter 7

STRUCTURE OF GROUPS

Our ultimate goal is to attach a group (called the Galois group) to any polynomial equation and show that the equation is solvable by radicals if and only if the group has a certain structure. (A group with this structure is called "solvable.") This requires a much more detailed knowledge of the structure of groups than we have already acquired. To show (in Chapter 8) that there exist polynomial equations over **C** of degree 5 that cannot be solved by radicals, our approach will be to find an equation whose Galois group is isomorphic to S_5, and then show that S_5 does not have the required property. This demands further study of permutation groups, in particular of A_n, which we do in Section 7.7.

Lagrange's theorem states that if H is a subgroup of a finite group G, then the order of H is a divisor of the order of G. The converse is false (A_4 provides a counterexample), but *does* hold in certain classes of groups. In a cyclic group of order n, the converse is true by Proposition 3.5.3: for any divisor m of n, there exists a subgroup of order m. In Section 7.5 we will determine the structure of each finite abelian group (originally done by Kronecker in 1870), and using this characterization we will be able to show that the converse of Lagrange's theorem holds for such groups. We will also be able to obtain some partial results in this direction, for arbitrary finite groups. The Sylow theorems (proved in Section 7.4) state that if the order of the group is divisible by a power of a prime, then there exists a subgroup whose order is the given prime power. These theorems were proved in 1872 by M.L. Sylow (1832–1918) in the context of permutation groups, and then in 1887 Georg Frobenius (1849–1917) published a new proof based on the axiomatic definition of a group.

A finite group G is *solvable* if and only if it has a sequence of subgroups

$$G = N_0 \supseteq N_1 \supseteq \ldots \supseteq N_{k-1} \supseteq N_k = \{e\}$$

such that each subgroup is normal in the previous one and each of the factor groups N_{i-1}/N_i is cyclic of prime order. We study such chains of subgroups in Section 7.6. One way to study the structure of groups is to find suitable "building blocks" and then determine how they can be put together to construct groups. The appropriate building blocks are simple groups, which have no nontrivial normal subgroups. For any finite group it is possible to find a sequence of subgroups of the type given above, in which each factor group is simple (but not necessarily simple and abelian, as is the case for solvable groups).

The problem, then, is to determine the structure of each finite simple group and to solve the "extension problem." That is, given a group G with normal subgroup N such that the structure of N as well as that of G/N are known, what is the structure of G? The extension problem is still open, but the classification of finite simple groups is generally accepted as complete.

An abelian group is simple if and only if it is cyclic of prime order. We will show in Section 7.7 that the alternating group on n elements is simple, if $n \geq 5$. We can easily describe one other family of finite simple groups. Let F be any finite field, and let $G = \mathrm{GL}_n(F)$, the group of invertible $n \times n$ matrices over F. Then G has a normal subgroup $N = \mathrm{SL}_n(F)$, the subgroup of all matrices of determinant 1. (Note that N is normal because it is the kernel of the determinant mapping.) The center of N, which we denote by Z, may be nontrivial, in which case N is not simple. However, the factor group N/Z is simple except for the cases $n = 2$ and $F = \mathrm{GF}(2)$ or $\mathrm{GF}(3)$.

William Burnside (1852–1927) states in the second edition of his text *Theory of Groups of Finite Order* (published in 1911) that his research "suggests inevitably that simple groups of odd order do not exist." He had shown that the order of a simple finite group of odd order (nonabelian, of course) must have at least seven prime factors, and then he had checked all orders up to $40,000$. This was finally shown to be true in 1963, by Walter Feit and John Thompson, in a 255-page paper that proved that all groups of odd order are solvable. This sparked a great deal of interest in the problem of classifying all finite simple groups, and the classification was finally completed in 1981. (The work required the efforts of many people, and is still being checked and simplified.) There are a number of infinite families of finite simple groups in addition to those mentioned above. In addition there are 26 "sporadic" ones, which do not fit into the other classes. The largest of these is known as the "monster," and has approximately 10^{54} elements.

7.1 Isomorphism Theorems; Automorphisms

We need to recall the fundamental homomorphism theorem for groups. If G_1 and G_2 are groups, and $\phi : G_1 \to G_2$ is a group homomorphism, then $\ker(\phi)$ is a normal subgroup of G_1, $\phi(G_1)$ is a subgroup of G_2, and the factor group $G_1/\ker(\phi)$ is isomorphic to the image $\phi(G_1)$. We will exploit this theorem in proving two isomorphism theorems. Refer to Figure 7.1.1 for the first isomorphism theorem and Figure 7.1.2 for the second isomorphism theorem.

7.1.1 Theorem (First Isomorphism Theorem). *Let G be a group, let N be a normal subgroup of G, and let H be any subgroup of G. Then HN is a subgroup of G, $H \cap N$ is a normal subgroup of H, and*

$$(HN)/N \cong H/(H \cap N) \, .$$

Proof. Define $\phi : H \to G/N$ by $\phi(h) = hN$, for all $h \in H$. Since ϕ is the restriction of the natural projection $\pi : G \to G/N$, it is a group homomorphism. Recall that the set $HN = \{hn \mid h \in H, \ n \in N\}$ is a subgroup of G by Proposition 3.3.2.

$$
\begin{aligned}
\phi(H) &= \{gN \in G/N \mid gN = hN \text{ for some } h \in H\} \\
&= \{gN \in G/N \mid g \in HN\} \\
&= HN/N
\end{aligned}
$$

Finally, $\ker(\phi) = H \cap N$, and so $H \cap N$ is a normal subgroup of H. By the fundamental homomorphism theorem, we have $\phi(H) \cong H/\ker(\phi)$. $\quad\square$

Figure 7.1.1:

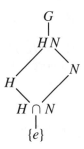

Example 7.1.1.

Let G be the dihedral group D_8, given by elements a of order 8 and b of order 2, with $ba = a^{-1}b$. Let $N = \langle a^2 \rangle$, and let $H = \{e, a^4, b, a^4b\}$. Then N is normal in G (see Exercise 18 in Section 3.7), $HN = \{e, a^2, a^4, a^6, b, a^2b, a^4b, a^6b\}$, and $H \cap N = \{e, a^4\}$. Thus N has two cosets in HN, and $H \cap N$ has two cosets in H, so it is clear that HN/N must be isomorphic to $H/H \cap N$. $\quad\square$

Example 7.1.2.

Let $G = \mathrm{GL}_2(\mathbf{Q})$, let $N = \mathrm{SL}_2(\mathbf{Q})$, and let H be the set of diagonal matrices in G. Since N is the kernel of the determinant mapping from $\mathrm{GL}_2(\mathbf{Q})$ into $\mathbf{Q}^\times$, it is a normal subgroup of G, and it is easy to check that H is a subgroup of G. Then $H \cap N$ is the set of diagonal matrices of determinant 1, and $HN = G$, since any element of G (with determinant d) can be expressed in the form

$$\begin{bmatrix} a_{11} & a_{12} \\ a_{21} & a_{22} \end{bmatrix} = \begin{bmatrix} d & 0 \\ 0 & 1 \end{bmatrix} \begin{bmatrix} a_{11}/d & a_{12}/d \\ a_{21} & a_{22} \end{bmatrix}.$$

It follows from the first isomorphism theorem that $GL_2(\mathbf{Q}) / SL_2(\mathbf{Q}) \cong H/(H \cap N)$. $\square$

7.1.2 Theorem (Second Isomorphism Theorem). *Let G be a group with normal subgroups N and H such that $N \subseteq H$. Then H/N is a normal subgroup of G/N, and*

$$(G/N) / (H/N) \cong G/H .$$

Proof. By H/N we mean the set of all cosets of the form hN, where $h \in H$. Define $\phi : G/N \to G/H$ by $\phi(aN) = aH$ for all $a \in G$. Then ϕ is well-defined since if $aN = bN$ for $a, b \in G$, we have $b^{-1}a \in N$. Since $N \subseteq H$, this implies $b^{-1}a \in H$, and so $aH = bH$. It is clear that ϕ maps G/N onto G/H. To show that ϕ is a homomorphism we only need to note that

$$\phi(aNbN) = \phi(abN) = abH = aHbH = \phi(aN)\phi(bN) .$$

Finally, $\ker(\phi) = \{aN \mid aH = H\} = H/N$, and so H/N is a normal subgroup of G/N. The fundamental homomorphism theorem for groups implies that $(G/N) / \ker(\phi) \cong G/H$, the desired result. $\square$

Figure 7.1.2:

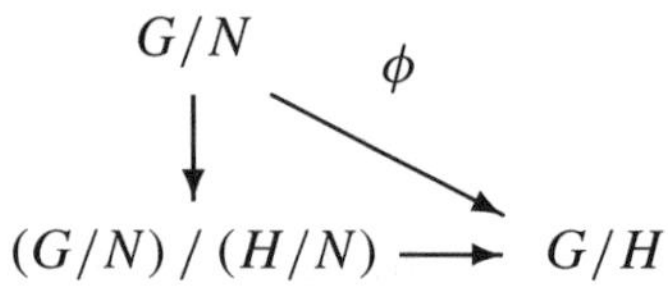

$$G/N \qquad \phi$$

$$(G/N) / (H/N) \longrightarrow G/H$$

Example 7.1.3 ($\mathbf{Z}_n/m\mathbf{Z}_n \cong \mathbf{Z}_m$ if $m|n$).

We have already proved this directly, in Example 3.8.10, but it also follows immediately from the second isomorphism theorem. Let $G = \mathbf{Z}$, let m, n be positive integers with $m|n$, let $N = n\mathbf{Z}$, and let $H = m\mathbf{Z}$. Then $N \subseteq H$, and so by the second isomorphism theorem we have $(G/N) / (H/N) \cong G/H$.

That is, $(\mathbf{Z}/n\mathbf{Z}) \,/\, (m\mathbf{Z}/n\mathbf{Z}) \cong \mathbf{Z}/m\mathbf{Z}$. In our standard notation, this is written as $\mathbf{Z}_n/m\mathbf{Z}_n \cong \mathbf{Z}_m$. Of course, since $\mathbf{Z}_n$ is cyclic, every subgroup of $\mathbf{Z}_n$ has the form $m\mathbf{Z}_n$, for some positive divisor m of n, and so we have characterized all factor groups of $\mathbf{Z}_n$. $\square$

Example 7.1.4.

Let $G = D_8$, and let N and H be the subgroups defined in Example 7.1.1. Then N and $H \cap N$ are normal in G, and $H \cap N \subseteq N$. It follows from the second isomorphism theorem that $(G/(H \cap N)) \,/\, (N/(H \cap N)) \cong G/N$. $\square$

The next theorem will be crucial in proving later theorems which describe the structure of a finite abelian group. It is also useful in proving, for example, that $D_6 \cong S_3 \times \mathbf{Z}_2$, since we would only need to find normal subgroups of D_6 isomorphic to S_3 and $\mathbf{Z}_2$ which satisfy the conditions of the theorem.

7.1.3 Theorem. *Let G be a group with normal subgroups H, K such that $HK = G$ and $H \cap K = \{e\}$. Then $G \cong H \times K$.*

Proof. We claim that $\phi : H \times K \to G$ defined by $\phi(h, k) = hk$, for all $(h, k) \in H \times K$ is a homomorphism. First, for all $(h_1, k_1), (h_2, k_2) \in H \times K$ we have

$$\phi((h_1, k_1)(h_2, k_2)) = \phi((h_1 h_2, k_1 k_2)) = h_1 h_2 k_1 k_2 \ .$$

To show that this is equal to

$$\phi((h_1, k_1))\phi((h_2, k_2)) = h_1 k_1 h_2 k_2$$

it suffices to show that $h_2 k_1 = k_1 h_2$. For any elements $h \in H$ and $k \in K$, we have $hkh^{-1}k^{-1} \in H \cap K$ since $hkh^{-1}, k^{-1} \in K$ and $h, kh^{-1}k^{-1} \in H$. By assumption $H \cap K = \{e\}$, and so $hkh^{-1}k^{-1} = e$, or $hk = kh$. We have now verified our claim that ϕ is a homomorphism.

Since $HK = G$, it is clear that ϕ is onto. Finally, if $\phi((h, k)) = e$ for $(h, k) \in H \times K$, then $hk = e$ implies $h = k^{-1} \in H \cap K$, and so $h = e$ and $k = e$, which shows that $\ker(\phi)$ is trivial and hence ϕ is one-to-one. $\square$

We have been using the definition that a subgroup H of a group G is normal if $ghg^{-1} \in H$ for all $h \in H$ and $g \in G$. We now introduce a more sophisticated point of view using the notion of an inner automorphism. The more general notion of an automorphism of a group is also extremely important.

7.1.4 Proposition. *Let G be a group and let $a \in G$. The function $i_a : G \to G$ defined by $i_a(x) = axa^{-1}$ for all $x \in G$ is an isomorphism.*

Proof. If $x, y \in G$, then

$$i_a(xy) = a(xy)a^{-1} = (axa^{-1})(aya^{-1}) = i_a(x)i_a(y) \, ,$$

and so i_a is a homomorphism. If $i_a(x) = e$, then $axa^{-1} = e$, so $x = e$ and i_a is one-to-one since its kernel is trivial. Given $y \in G$, we have $y = i_a(a^{-1}ya)$, and so i_a is also an onto mapping. $\square$

7.1.5 Definition. *Let G be a group. An isomorphism from G onto G is called an* **automorphism** *of G.*

An automorphism of G of the form i_a, *for some* $a \in G$, *where* $i_a(x) = axa^{-1}$ *for all* $x \in G$, *is called an* **inner automorphism** *of G.*

The set of all automorphisms of G will be denoted by Aut(G) *and the set of all inner automorphisms of G will be denoted by* Inn(G).

The condition that a subgroup H of G is normal can be expressed by saying that $i_a(h) \in H$ for all $h \in H$ and all $a \in G$. Equivalently, H is normal if and only if $i_a(H) \subseteq H$ for all $a \in G$, and we can also express this by saying that H is invariant under all inner automorphisms of G.

7.1.6 Proposition. *Let G be a group. Then* Aut(G) *is a group under composition of functions, and* Inn(G) *is a normal subgroup of* Aut(G).

Proof. Composition of functions is always associative. We already know that the composite of two isomorphisms is again an isomorphism, and that the inverse of an isomorphism is an isomorphism, so it follows immediately that Aut(G) is a group.

For any elements $a, b \in G$, we have

$$i_a i_b(x) = a(bxb^{-1})a^{-1} = (ab)x(ab)^{-1} = i_{ab}(x)$$

for all $x \in G$, and so this yields the formula $i_a i_b = i_{ab}$. It follows easily that i_e is the identity mapping and $(i_a)^{-1} = i_{a^{-1}}$, so Inn(G) is a subgroup of Aut(G). To show that it is normal, let $\beta \in$ Aut(G) and let $i_a \in$ Inn(G). For $x \in G$, we have

$$
\begin{aligned}
\beta i_a \beta^{-1}(x) &= \beta(a(\beta^{-1}(x))a^{-1}) \\
&= (\beta(a))(\beta\beta^{-1}(x))(\beta(a^{-1})) \\
&= (\beta(a))(x)(\beta(a))^{-1} = bxb^{-1} \\
&= i_b(x)
\end{aligned}
$$

for the element $b = \beta(a)$. Thus $\beta i_a \beta^{-1} \in$ Inn(G), and so Inn(G) is a normal subgroup of Aut(G). $\square$

7.1.7 Definition. *For any group G, the subset*

$$Z(G) = \{x \in G \mid xg = gx \ \text{for all} \ g \in G\}$$

*is called the **center** of G.*

7.1.8 Proposition. *For any group G, the center $Z(G)$ is a normal subgroup, and*

$$G/Z(G) \cong \text{Inn}(G) \ .$$

Proof. Define $\phi : G \rightarrow \text{Inn}(G)$ by $\phi(a) = i_a$, for all $a \in G$. Then

$$\phi(ab) = i_{ab} = i_a i_b = \phi(a)\phi(b)$$

and we have defined a homomorphism. Since ϕ is onto by the definition of $\text{Inn}(G)$, we only need to compute $\ker(\phi)$. If i_a is the identity mapping, then for all $x \in G$ we have $axa^{-1} = x$, or $ax = xa$, so the kernel of ϕ is the center $Z(G)$. It follows that $Z(G)$ is normal, and that $G/Z(G) \cong \text{Inn}(G)$. $\square$

Example 7.1.5 ($\text{Aut}(\mathbf{Z}) \cong \mathbf{Z}_2$ and $\text{Inn}(\mathbf{Z}) = \{1\}$).

To compute $\text{Aut}(\mathbf{Z})$ and $\text{Inn}(\mathbf{Z})$ we first observe that all inner automorphisms of an abelian group are trivial (equal to the identity mapping). Next, we observe that any isomorphism between cyclic groups maps generators to generators, so if $\alpha \in \text{Aut}(\mathbf{Z})$, then $\alpha(1) = \pm 1$. Thus there are two possible automorphisms, with the formulas $\alpha(n) = n$ or $\alpha(n) = -n$, for all $n \in \mathbf{Z}$. $\square$

Example 7.1.6 ($\text{Aut}(\mathbf{Z}_n) \cong \mathbf{Z}_n^\times$).

The computation of $\text{Aut}(\mathbf{Z}_n)$ is similar to that of $\text{Aut}(\mathbf{Z})$. For any automorphism α of $\mathbf{Z}_n$, let $\alpha([1]) = [a]$. Since $[1]$ is a generator, $[a]$ must also be a generator, and thus $\gcd(a, n) = 1$. Then α must be given by the formula $\alpha([m]) = [am]$, for all $[m] \in \mathbf{Z}_n$. Since the composition of such functions corresponds to multiplying the coefficients, it follows that $\text{Aut}(\mathbf{Z}_n) \cong \mathbf{Z}_n^\times$, where $\mathbf{Z}_n^\times$ is the multiplicative group of units of $\mathbf{Z}_n$. $\square$

EXERCISES: SECTION 7.1

1. In $G = \mathbf{Z}_{32}^\times$ find cyclic subgroups H of order 2 and K of order 8 with $HK = G$ and $H \cap K = \{e\}$. Conclude that $\mathbf{Z}_{32}^\times \cong \mathbf{Z}_2 \times \mathbf{Z}_8$.

2. Prove that $D_6 \cong S_3 \times \mathbf{Z}_2$.

3.†Determine Aut$(\mathbf{Z}_2 \times \mathbf{Z}_2)$.

4. Let G be a finite abelian group of order n, and let m be a positive integer with $(n, m) = 1$. Show that $\phi : G \to G$ defined by $\phi(g) = g^m$ for all $g \in G$ belongs to Aut(G).

5.†Let $\phi : G \to G$ be the function defined by $\phi(g) = g^{-1}$ for all $g \in G$. Find conditions on G such that ϕ is an automorphism.

6. Show that for $G = S_3$, Inn$(G) \cong G$.

7.†Determine Aut(S_3).

8. For groups G_1 and G_2, determine the center of $G_1 \times G_2$.

9. Show that $G/Z(G)$ cannot be a nontrivial cyclic group. (That is, if $G/Z(G)$ is cyclic, then G must be abelian, and hence $Z(G) = G$.)

10. Describe the centers $Z(D_n)$ of the dihedral groups D_n, for all integers $n \geq 3$.

11. In the group GL$_2(\mathbf{C})$ of all invertible 2×2 matrices with complex entries, let Q be the following set of matrices (the quaternion group, defined in Example 3.3.7):

$$\pm \begin{bmatrix} 1 & 0 \\ 0 & 1 \end{bmatrix}, \quad \pm \begin{bmatrix} i & 0 \\ 0 & -i \end{bmatrix}, \quad \pm \begin{bmatrix} 0 & 1 \\ -1 & 0 \end{bmatrix}, \quad \pm \begin{bmatrix} 0 & i \\ i & 0 \end{bmatrix}.$$

(a) Show that Q is not isomorphic to D_4.

(b) Find the center $Z(Q)$ of Q.

12. Let F_{20} be the subgroup of GL$_2(\mathbf{Z}_5)$ consisting of all matrices of the form $\begin{bmatrix} m & n \\ 0 & 1 \end{bmatrix}$, such that $m, n \in \mathbf{Z}_5$ and $m \neq 0$, as defined in Exercise 23 of Section 3.8. This group will be called the **Frobenius** group of degree 5. Find the center of F_{20}.

13. Show that the Frobenius group F_{20} defined in Exercise 12 can be defined by generators and relations as follows. Let $a = \begin{bmatrix} 1 & 1 \\ 0 & 1 \end{bmatrix}$ and $b = \begin{bmatrix} 2 & 0 \\ 0 & 1 \end{bmatrix}$.

(a) Show that $o(a) = 5$, $o(b) = 4$, and $ba = a^2b$.

(b) Show that each element of F_{20} can be expressed in the form $a^i b^j$ for $0 \leq i \leq 4$ and $0 \leq j \leq 3$.

14. Let G be the subgroup of GL$_2(\mathbf{R})$ consisting of all matrices $\begin{bmatrix} a_{11} & a_{12} \\ a_{21} & a_{22} \end{bmatrix}$ such that $a_{21} = 0$ and $a_{22} = 1$. (See Exercises 10 and 11 of Section 3.1.)

(a) Let N be the set of matrices in G with $a_{11} = 1$. Show that N is a normal subgroup of G.

(b) Let $a = \begin{bmatrix} 1 & 1 \\ 0 & 1 \end{bmatrix}$ and $b = \begin{bmatrix} 2 & 0 \\ 0 & 1 \end{bmatrix}$. Show that if $H = \langle a \rangle$, then bHb^{-1} is a proper subset of H. Conclude that H is not normal in any subgroup that contains b.

15. Give another proof of Theorem 7.1.1 by constructing an isomorphism from $(HN)/N$ onto $H/(H \cap N)$.

(In our proof we constructed an isomorphism from $H/(H \cap N)$ onto $(HN)/N$. The point is that it may be much easier to define a function in one direction than the other.)

7.2   Conjugacy

If G is a group with a subgroup H that is not normal in G, then there must exist at least one element $a \in G$ such that $aHa^{-1} \neq H$. The set aHa^{-1} is a subgroup of G, since it is the image of H under the inner automorphism i_a of G. It is important to study the subgroups related to H in this way. It is also important to study elements of the form aha^{-1}, for $h \in H$, since H is normal if and only if it contains all such elements.

7.2.1 Definition. *Let G be a group, and let $x, y \in G$. The element y is said to be a **conjugate** of the element x if there exists an element $a \in G$ such that $y = axa^{-1}$.*

*If H and K are subgroups of G, then K is said to be a **conjugate subgroup of** H if there exists $a \in G$ such that $K = aHa^{-1}$.*

In an abelian group, elements or subgroups are only conjugate to themselves. More generally, an element x of a group G has no conjugates other than itself if and only if $axa^{-1} = x$ for all $a \in G$, and this holds if and only if x is a member of the center $Z(G)$.

Recall Proposition 3.8.8, which states that a subgroup $H \subseteq G$ is normal if and only if $aH = Ha$, for all $a \in G$. As a slight modification of this result, we see that H is normal in G if and only if $aHa^{-1} = H$, for all $a \in G$. Thus H is a normal subgroup if and only if it has no conjugate subgroups other than itself.

We can exploit the fact that a conjugate axa^{-1} of the element $x \in G$ is the image of x under the inner automorphism i_a to obtain valuable information. Since an automorphism preserves orders of elements, each conjugate of x must have the same order as x. Since an automorphism preserves inverses, if y is conjugate to x, then y^{-1} is conjugate to x^{-1}.

7.2.2 Proposition.

(a) *Conjugacy of elements defines an equivalence relation on any group G.*

(b) *Conjugacy of subgroups defines an equivalence relation on the set of all subgroups of G.*

Proof. For $x, y \in G$, write $x \sim y$ if y is a conjugate of x. Then for all $x \in G$, we have $x \sim x$ since $x = exe^{-1}$. If $x \sim y$, then $y = axa^{-1}$ for some $a \in G$, and it follows that $x = byb^{-1}$ for $b = a^{-1}$, which shows that $y \sim x$. Finally, if $z \in G$ and $x \sim y$, $y \sim z$, then there exist $a, b \in G$ with $z = aya^{-1}$ and $y = bxb^{-1}$. Thus $z = abxb^{-1}a^{-1}$, so $z = (ab)x(ab)^{-1}$ and we have $x \sim z$.

A similar proof shows that conjugacy of subgroups defines an equivalence relation on the set of all subgroups of G. $\square$

The equivalence classes under conjugacy of elements are called the **conjugacy classes** of G. Thus the conjugacy class of $x \in G$ is

$$\{g \in G \mid \text{ there exists } a \in G \text{ with } g = axa^{-1}\} .$$

Note that the conjugacy class of x is $\{x\}$ if and only if x commutes with each elements of G. We say that a conjugacy class is **trivial** if it contains only one member.

Example 7.2.1 (Conjugacy in S_3).

If the group S_3 is given by generators a, b of order 3 and order 2, respectively, and $ba = a^2b$, then we can find the conjugacy classes as follows. Since a does not commute with b, we know that there is more than one element in the conjugacy class of a, and we can compute $bab^{-1} = a^2bb^{-1} = a^2$, so that a^2 is conjugate to a. We know that the subgroup $\{e, a, a^2\}$ is normal, so a cannot be conjugate to any element outside of the subgroup.

To find the conjugates of b, we have

$$aba^{-1} = aba^2 = aa^2ba = ba = a^2b ,$$

which shows that a^2b is conjugate to b. Conjugating b by a^2b gives ab, and so the conjugacy classes of S_3 are the following:

$$\{e\}, \quad \{a, a^2\}, \quad \{b, ab, a^2b\} .$$

Now that we know which elements of S_3 are conjugate, it is easy to see which subgroups of S_3 are conjugate. The subgroup $\{e, a, a^2\}$ is normal, so it is conjugate only to itself. The subgroups $\{e, b\}$, $\{e, ab\}$, and $\{e, a^2\}$ are conjugate to each other. (See Figure 3.6.5 for a diagram of the subgroups.) $\square$

Let x be an element of a group G. The computations in the preceding example show that we need to find an answer to the general question of when elements $a, b \in G$ determine the same conjugate of x. Since $axa^{-1} = bxb^{-1}$ if and only if $(b^{-1}a)x = x(b^{-1}a)$, it turns out that we need to find the elements that commute with x. Then the cosets of this subgroup correspond to distinct conjugates of x.

7.2.3 Definition. *Let G be a group. For any element $x \in G$, the set*

$$\{a \in G \mid axa^{-1} = x\}$$

*is called the **centralizer** of x in G, denoted by $C(x)$.*

　For any subgroup H of G, the set

$$\{a \in G \mid aHa^{-1} = H\}$$

*is called the **normalizer** of H in G, denoted by $N(H)$.*

Using the above definition, note that $C(x) = G$ if and only if $x \in Z(G)$, and $N(H) = G$ if and only if H is a normal subgroup of G.

7.2.4 Proposition. *Let G be a group.*
　(a) *If $x \in G$, then $C(x)$ is a subgroup of G.*
　(b) *If H is a subgroup of G, then $N(H)$ is a subgroup of G.*

Proof. (a) If $a, b \in C(x)$, then $(ab)x(ab)^{-1} = a(bxb^{-1})a^{-1} = axa^{-1} = x$, and so $ab \in C(x)$. Furthermore, $a^{-1} \in C(x)$, since $axa^{-1} = x$ implies that $x = a^{-1}xa$, and thus $a^{-1}x(a^{-1})^{-1} = x$. Finally, it is clear that $e \in C(x)$.
　(b) The proof is similar to part (a), and is left as an exercise.　□

If the conjugacy classes of G are known, then it is possible to tell whether or not a subgroup H is normal by checking that each conjugacy class lies either entirely inside of H or entirely outside of H. Equivalently, H is normal if and only if it is a union of some of the conjugacy classes of G. We note that the normalizer of H is in fact the largest subgroup of G in which H is normal.

The next proposition explains why the centralizer of an element is so useful in finding its conjugates. There is a similar result for conjugate subgroups, stated in Exercise 2.

7.2.5 Proposition. *Let x be an element of the group G. Then the elements of the conjugacy class of x are in one-to-one correspondence with the left cosets of the centralizer $C(x)$ of x in G.*

Proof. For $a, b \in G$, we have $axa^{-1} = bxb^{-1}$ if and only if $(b^{-1}a)x(b^{-1}a)^{-1} = x$, or equivalently, if and only if $b^{-1}a \in C(x)$. It follows from Proposition 3.8.1 that a and b determine the same conjugate of x if and only if a and b belong to the same left coset of $C(x)$.　□

Example 7.2.2 (Conjugacy in D_4).

We now compute the conjugacy class and centralizer of each element in the dihedral group D_4, described by the generators a and b of order 4 and order 2, respectively, with $ba = a^{-1}b$.

It follows from the relation $ba^i = a^{-i}b$ that an element of the form a^i commutes with b if and only if $a^i = a^{-i}$, and this holds if and only if $i = 2$. On the other hand, the same relation can be used to show that a^2 commutes with every other element of D_4. Furthermore, a similar computation shows that any element of the form $a^j b$ fails to commute with a, and so $Z(D_4) = \{e, a^2\}$. Thus the conjugacy classes with exactly one element are $\{e\}$ and $\{a^2\}$.

It is always true that $\langle x \rangle \subseteq C(x)$, since any power of x must commute with x. Thus $\{e, a, a^2, a^3\} \subseteq C(a)$, and we must have equality since $a \notin Z(D_4)$ implies $C(a) \neq D_4$. The conjugacy class of a is $\{a, a^3\}$. Similarly, $b \in C(b)$, and we also have $a^2 \in C(b)$ since a^2 commutes with b. Again, $C(b) \neq D_4$, and so we may conclude that $C(b) = \{e, b, a^2, a^2b\}$, and then we can easily show that the conjugacy class of b is $\{b, a^2b\}$ by conjugating b by any element not in $C(b)$. Remember that the conjugates of b correspond to the left cosets of $C(b)$. Finally, a similar computation shows that $C(ab) = \{e, ab, a^2, a^3b\}$ and the conjugacy class of ab is $\{ab, a^3b\}$. Thus the complete list of conjugacy classes of D_4 is

$$\{e\}, \quad \{a^2\}, \quad \{a, a^3\}, \quad \{b, a^2b\}, \quad \{ab, a^3b\}. \quad \square$$

Example 7.2.3 (Conjugacy in S_n).

Writing the elements of S_3 in cyclic notation gives a clue to what happens in S_n. Using Example 7.2.1, the conjugacy classes of S_3 are $\{(1)\}$, $\{(1, 2, 3), (1, 3, 2)\}$ and $\{(1, 2), (1, 3), (2, 3)\}$.

We will show that two permutations are conjugate in S_n if and only if they have the same shape (i.e., the same number of disjoint cycles, of the same lengths). For example, in S_5 the conjugacy class of the identity permutation has only one element, and then in addition there is one conjugacy class for each of the following shapes (assuming the permutations are written in cyclic notation): (a, b), (a, b, c), $(a, b)(c, d)$, (a, b, c, d), $(a, b, c)(d, e)$, and (a, b, c, d, e). Thus, in particular, cycles of the same length are always conjugate.

Recall how a permutation $\sigma \in S_n$ is written in cyclic notation. Starting with a number i we construct the cycle $(i, \sigma(i), \sigma^2(i), \ldots)$, and continue with additional disjoint cycles as necessary. If $\tau \in S_n$, then to construct the cyclic representation of the conjugate $\tau \sigma \tau^{-1}$, we can start with the number $\tau(i)$ and then proceed as follows. We have $\tau \sigma \tau^{-1}(\tau(i)) = \tau(\sigma(i))$ as the next entry of the cycle, showing that if σ maps i to j, then $\tau \sigma \tau^{-1}$ maps $\tau(i)$ to $\tau(j)$. Thus if $\sigma = (a_1, \ldots, a_k)$, then $\tau \sigma \tau^{-1}$ is the cycle $(\tau(a_1), \ldots, \tau(a_k))$. The cycles of $\tau \sigma \tau^{-1}$ are found by simply applying τ to the entries of the cycles of σ, resulting in precisely the same cycle structure.

On the other hand, if σ and ρ have the same shape, then a simple substitution can be made in which the entries of σ are replaced by the corresponding entries of ρ. For the permutation τ that is defined by this substitution, we have $\tau \sigma \tau^{-1} = \rho$, showing that ρ is conjugate to σ. $\square$

The equation in the following theorem is called the **conjugacy class equation** of the group G. Recall that the number of left cosets of a subgroup H of a group G is called the index of H in G, and is denoted by $[G : H]$. We will see that a great deal of information can be obtained simply by counting the elements in G according to its conjugacy classes.

7.2.6 Theorem (Class Equation). *Let G be a finite group. Then*

$$|G| = |Z(G)| + \sum [G : C(x)]$$

where the sum ranges over one element x from each nontrivial conjugacy class.

Proof. Since conjugacy defines an equivalence relation, the conjugacy classes partition G. By Proposition 7.2.5, the number of elements in the conjugacy class of the element x is the index $[G : C(x)]$ of the centralizer of x. The conjugacy classes containing only one element can be grouped together, and this yields the form of the class equation given in the statement of the theorem. $\square$

Example 7.2.4 (Class equation of D_4).

It follows from the calculations in Example 7.2.2 that the class equation for D_4 is $8 = 2 + (2 + 2 + 2)$. $\square$

Example 7.2.5 (Class equation of S_4).

To find the class equation of S_4, we can apply Example 7.2.3, which shows that the conjugacy classes are determined by the possible shapes of elements in S_4.

To construct a transposition (a, b), we choose 2 of 4 elements, and this can be done in $\binom{4}{2} = 6$ ways. To construct a 3-cycle, we choose 3 of 4 elements, but then each choice can be arranged in two different ways, giving a total of $2\binom{4}{3} = 8$ cycles of the form (a, b, c). To construct a 4-cycle, we fix 1 as the first element, and then there are $3! = 6$ ways to complete the cycle

$(1, a, b, c)$. Finally, there are 3 permutations of the form $(a, b)(c, d)$. Thus the class equation for S_4 is

$$24 = 1 + (6 + 8 + 6 + 3) \, .$$

Any normal subgroup must contain the identity, and cannot contain only part of a conjugacy class. It follows from the class equation for S_4 that $12 = 1 + 8 + 3, 4 = 1 + 3$, and 1 are the only possible orders for a proper normal subgroup, since these are the only proper divisors of 24 that can be written as 1 plus a subfamily of 6, 8, 6, 3. In this case the class equation also determines the elements that must belong to such subgroups, and so we conclude that the only proper nontrivial normal subgroups of S_4 are A_4 and the subgroup $V = \{(1), (1, 2)(3, 4), (1, 3)(2, 4), (1, 4)(2, 3)\}$. □

We next consider some applications to groups of prime power order.

7.2.7 Definition. *A group of order p^n, with p a prime number and $n \geq 1$, is called a p-group. Such groups are said to have **prime power order**.*

7.2.8 Theorem (Burnside). *Let p be a prime number. The center of any p-group is nontrivial.*

Proof. Let G be a p-group. In the conjugacy class equation of G, the order of G is by definition divisible by p, and the terms $[G : C(x)]$ are all divisible by p since $x \notin Z(G)$ implies $[G : C(x)] > 1$. Remember that $[G : C(x)]$ is a divisor of $|G|$ and hence is a power of p. This implies that $|Z(G)|$ is divisible by p. □

7.2.9 Corollary. *Any group of order p^2 (where p is prime) is abelian.*

Proof. If $|G| = p^2$ and $Z(G) \neq G$, then let $a \in G - Z(G)$. Then $C(a)$ is a subgroup containing both a and $Z(G)$, with $|Z(G)| \geq p$ by the previous theorem. This shows that $C(a) = G$, a contradiction. Thus $Z(G) = G$, and so G is abelian. □

Augustin Cauchy (1789–1857) was one of the first (along with Lagrange and Ruffini) to investigate permutation groups, and the next theorem bears his name. The following proof makes use only of the conjugacy class equation. The next section contains a stronger statement of the theorem, together with a shorter proof, but the motivation for that proof is not as transparent.

7.2.10 Theorem (Cauchy). *If G is a finite group and p is a prime divisor of the order of G, then G contains an element of order p.*

Proof. Let $|G| = n$. The proof proceeds by induction on n. We start the induction with the observation that we certainly know that the theorem holds for $n = 1$. We may assume that the theorem holds for all groups of order less than n, and suppose that $p \mid n$. Consider the class equation

$$n = |Z(G)| + \sum[G : C(x)] .$$

Case 1. For each $x \notin Z(G)$, p is a divisor of $[G : C(x)]$.
As in the proof of Theorem 7.2.8, in this case $p \mid |Z(G)|$. Now if $Z(G) \neq G$, then the induction hypothesis shows that $Z(G)$ contains an element of order p and we are done. Thus we may assume that $Z(G) = G$, and so G is abelian. Let $a \in G$, with $a \neq e$, and consider $H = \langle a \rangle$. If $H = G$, then G is cyclic and the theorem holds, so we may assume that $|H| < |G|$. If p is a divisor of k, where $k = |H|$, then H has an element of order p by the induction hypothesis and we are done.

Thus we may assume that p is not a divisor of k, and hence must be a divisor of $|G/H|$. But then, since $|G/H| < n$, the group G/H must contain a coset of order p, say bH. Therefore $(bH)^p = H$, or, equivalently, $b^p \in H$. If $c = b^k$, then $c^p = (b^k)^p = (b^p)^k$, and this must give the identity element since $b^p \in H$ and $k = |H|$. If $c = e$, then $b^k = e$, which in turn implies $(bH)^k = H$ in G/H. Since bH has order p, we then have $p|k$, a contradiction. This shows that c is an element of order p.

Case 2. For some $x \notin Z(G)$, p is not a divisor of $[G : C(x)]$.
In this case, for the given element x, it follows that p is a divisor of $|C(x)|$ since p is a divisor of $|G| = |C(x)| \cdot [G : C(x)]$. Then we are done since $|C(x)| < n$ and the induction hypothesis implies that $C(x)$ contains an element of order p. $\square$

EXERCISES: SECTION 7.2

1. Let H be a subgroup of the group G. Prove that $N(H)$ is a subgroup of G.

2. Let H be a subgroup of the group G. Prove that the subgroups of G that are conjugate to H are in one-to-one correspondence with the left cosets of $N(H)$ in G.

3. Let G be a group with subgroups H and K such that $H \subseteq K$. Show that H is a normal subgroup of K if and only if $K \subseteq N(H)$.

4. Let p be a prime number, and let C be a cyclic subgroup of order p in S_p. Compute the order of $N(C)$.

5. Let G be a group, let H be a subgroup of G, and let $a \in G$. Show that there exists a subgroup K of G such that K is conjugate to H and $aH = Ka$.

6. Let G be a group, let $x, y \in G$, and let $n \in \mathbf{Z}$. Show that y is a conjugate of x^n if and only if y is the nth power of a conjugate of x.

7. Find the conjugate subgroups of D_4. (See Figure 3.6.6.)

8.†Find the conjugacy classes of D_5.

9. Describe the conjugacy classes of S_5 by listing the types of elements and the number of each type in each class.

10.†Find the conjugacy classes of A_4.

   Note: Two elements may be conjugate in S_4 but not in A_4.

11. Find the conjugacy classes of the quaternion group Q defined in Example 3.3.7.

12. Write out the conjugacy class equations for the following groups.

   †(a) A_4

   (b) S_5

13. Let the dihedral group D_n be given by elements a of order n and b of order 2, where $ba = a^{-1}b$. Show that a^m is conjugate to only itself and a^{-m}, and that $a^m b$ is conjugate to $a^{m+2k}b$, for any integer k.

14. Show that the Frobenius group F_{20} (defined in Exercise 12 of Section 7.1) is isomorphic to the subgroup of S_5 generated by the permutations $(1, 2, 3, 4, 5)$ and $(2, 3, 5, 4)$. Use this fact to help in finding the conjugacy classes of F_{20}, and its conjugacy class equation.

15. Show that if a group G has an element a which has precisely two conjugates, then G has a nontrivial proper normal subgroup.

16. Show that for each prime p there exists a nonabelian group of order p^3.

17. Let G be a nonabelian group of order p^3, for a prime number p. Show that $Z(G)$ must have order p.

18. Determine the conjugacy classes of the alternating group A_5, and use this information to show that A_5 is a simple group.

7.3   Groups Acting on Sets

Recall that if S is a set and G is a subgroup of the group $\mathrm{Sym}(S)$ of all permutations of S, then G is called a group of permutations. Historically, at first the theory of groups meant only the study of groups of permutations. The concept of an abstract group was introduced in studying properties that do not depend on the underlying set. Cayley's theorem states that every abstract group is isomorphic to a group of permutations, so the abstraction really just provides a different point of view. Still, when studying abstract groups, it is often important to be able to relate them to groups in which direct computations can actually be done. Since the appropriate way to relate the algebraic structures of two groups is via a homomorphism, we should study homomorphisms into groups of permutations.

If $\phi : G \rightarrow \text{Sym}(S)$ is a group homomorphism, suppose that $g \in G$ and $\phi(g) = \sigma$. Then for any element $x \in S$, $\sigma(x)$ is another element of S, say y, and it makes sense to think of g as "acting" on x to produce y. In many cases it is easier to think of G as "acting" on the set S instead of thinking in terms of a homomorphism. For any group acting on a set, we will be able to obtain a very useful formula that generalizes the conjugacy class equation.

7.3.1 Definition. *Let G be a group and let S be a set. A multiplication of elements of S by elements of G (defined by a function from $G \times S \rightarrow S$) is called a **group action** of G on S provided for each $x \in S$:*

(i) $a(bx) = (ab)x$ *for all $a, b \in G$, and*

(ii) $ex = x$ *for the identity element e of G.*

Our usual notation for the action of $a \in G$ on $x \in S$ is ax, but there will be occasions when it will be preferable to write $a \cdot x$ or $a * x$.

Example 7.3.1.

It is clear that any group G of permutations of the set S determines a group action of G on S. $\square$

Example 7.3.2.

If H is a subgroup of the group G, then H acts on the set G by using the group multiplication defined on G. $\square$

Example 7.3.3.

Let G be the multiplicative group $F^\times$ of nonzero elements of a field F. If V is any vector space over F, then scalar multiplication defines an action of G on V. The two conditions that must be satisfied are the only two vector space axioms that deal exclusively with scalar multiplication. $\square$

Example 7.3.4.

Let V be the vector space over the field F of all column vectors with n entries, and let G be any subgroup of the general linear group $\text{GL}_n(F)$ of all invertible $n \times n$ matrices over F. The standard multiplication of (column) vectors by matrices defines a group action of G on V. $\square$

The point of view of the next proposition will be useful in giving some more interesting examples. It explains the introductory statements hinting at the relationship
between group actions and representations of abstract groups via homomorphisms
into groups of permutations.

7.3.2 Proposition. *Let G be a group and let S be a set. Any group homomorphism
from G into the group* $\mathrm{Sym}(S)$ *of all permutations of S defines an action of G on S.
Conversely, every action of G on S arises in this way.*

Proof. Let $\phi : G \rightarrow \mathrm{Sym}(S)$ be a homomorphism. For $a \in G$, it is convenient to
let the permutation $\phi(a)$ be denoted by λ_a. Since ϕ is a homomorphism, we have
the formula

$$\lambda_a \lambda_b = \phi(a)\phi(b) = \phi(ab) = \lambda_{ab}$$

for all $a, b \in G$. For $x \in S$ and $a \in G$, we define $ax = \lambda_a(x)$. Since λ_e is the
identity permutation, and

$$a(bx) = \lambda_a(\lambda_b(x)) = \lambda_{ab}(x) = (ab)x$$

we have defined a group action.

Conversely, suppose that the group G acts on S. For each $a \in G$ define a
function $\lambda_a : S \rightarrow S$ by setting $\lambda_a(x) = ax$, for all $x \in S$. Then λ_a is one-to-one
since $\lambda_a(x_1) = \lambda_a(x_2)$ implies that $ax_1 = ax_2$, so multiplying by a^{-1} and using
the defining properties of the group action we obtain $x_1 = x_2$. Given $y \in S$, the
element $a^{-1}y$ is a solution to the equation $\lambda_a(x) = y$, and thus λ_a is onto. It is
not hard to show that $\lambda_a \lambda_b = \lambda_{ab}$, and this in turn can be used to show that the
function $\phi : G \rightarrow \mathrm{Sym}(S)$ defined by $\phi(a) = \lambda_a$ for each $a \in G$ is a group
homomorphism. $\square$

7.3.3 Definition. *Let G be a group acting on the set S. For each element $x \in S$, the
set*

$$Gx = \{s \in S \mid s = ax \text{ for some } a \in G\}$$

*is called the **orbit** of x under G, and the set*

$$G_x = \{a \in G \mid ax = x\}$$

*is called the **stabilizer** of x in G. The set*

$$S^G = \{x \in S \mid ax = x \text{ for all } a \in G\}$$

*is called the **subset of S fixed by** G.*

Example 7.3.5.

A subgroup $H \subseteq G$ has a natural action on the entire group G, as seen in Example 7.3.2. The orbit of an element $g \in G$ is the right coset Hg. The stabilizer H_g of g is just $\{e\}$. If H is nontrivial, then the fixed subset G^H equals $\{g \in G \mid hg = g \text{ for all } h \in H\}$, and this must be the empty set. $\square$

Example 7.3.6.

For any group G, the homomorphism $\phi : G \rightarrow \text{Aut}(G)$ defined by $\phi(a) = i_a$, where i_a is the inner automorphism defined by a, gives a group action of G on itself. The orbit Gg of an element g is just its conjugacy class, and the stabilizer G_g is just the centralizer of g in G. The fixed subset

$$\{g \in G \mid aga^{-1} = g \text{ for all } a \in G\}$$

is the center $Z(G)$. $\square$

Example 7.3.7.

This example is closely related to Example 7.3.5. Let G be a group, and let S be the set of all subgroups of G. If $a \in G$ and H is a subgroup of G, define $a * H = aHa^{-1}$. The fact that $a(bHb^{-1})a^{-1} = abH(ab)^{-1}$ shows that $a * (b * H) = (ab) * H$ for all $H \in S$ and all $a, b \in G$. Since $e * H = H$ for all subgroups H of G, the multiplication $*$ defines a group action of G on S. (We have introduced the $*$ here to avoid confusion with the usual coset notation.)

The orbit $G * H$ of a subgroup H is the set of all subgroups conjugate to H. The stabilizer of H in G is just the normalizer of H. Finally, the fixed subset S^G is the set of normal subgroups of G.

To illustrate the many possibilities, in a closely related example we can let H be any subgroup and let S be the set of all subgroups of G that are conjugate to H. If K is any subgroup of G, then for any $k \in K$ and $J \in S$, the subgroup $k * J = kJk^{-1}$ is still a member of S since it is conjugate to H, and so the $*$ operation defines an action of K on S. With this operation the stabilizer of $J \in S$ is $K \cap N(J)$, and the subset of S left fixed by the action is the set of conjugates J of H for which $K \subseteq N(J)$. $\square$

Example 7.3.8.

Let H be a proper subgroup of G, and let S denote the set of left cosets of H. We define a group action of G on S as follows: for $a, x \in G$ define $a \cdot (xH) = (ax)H$. To show that this multiplication yields a group action, let $xH \in S$ and $a, b \in G$. Then $a \cdot (b \cdot xH) = a \cdot bxH = a(bx)H = (ab)xH = (ab) \cdot xH$ and $e \cdot xH = (ex)H = xH$.

We note that the orbit of each coset xH is all of S, and thus $S^G = \emptyset$. The stabilizer of xH is the subgroup xHx^{-1}. To see this, we first observe that if $xhx^{-1} \in xHx^{-1}$, then $(xhx^{-1} \cdot xH = xhH = xH$. On the other hand, if $a \in G$ and $a \cdot xH = xH$, then $ax \in xH$, so $ax = xh$ for some $h \in H$, which shows that $a = xhx^{-1} \in xHx^{-1}$. $\square$

7.3.4 Proposition. *Let G be a group that acts on the set S, and let $x \in S$.*

(a) *The stabilizer G_x of x in G is a subgroup of G.*

(b) *There is a one-to-one correspondence between the elements of the orbit Gx of x under G and the left cosets of G_x in G.*

Proof. (a) If $a, b \in G_x$, then $(ab)x = a(bx) = ax = x$, and so $ab \in G_x$. Furthermore, $a^{-1}x = a^{-1}(ax) = (a^{-1}a)x$, and then $ex = x$ shows that $a^{-1} \in G_x$, as well as showing that $G_x \neq \emptyset$.

(b) For $a, b \in G$ we have $ax = bx$ if and only if $b^{-1}ax = x$, which occurs if and only if $b^{-1}a \in G_x$. Since this is equivalent to the condition that $aG_x = bG_x$, the function that assigns to the left coset aG_x the element ax in the orbit of x is well-defined and one-to-one. This function is clearly onto, completing the proof. $\square$

Applying the above proposition to Example 7.3.6 shows that the normalizer $N(H)$ of a subgroup H is a subgroup of G. Furthermore, the number of distinct subgroups conjugate to H is equal to $[G : N(H)]$.

7.3.5 Proposition. *Let G be a finite group acting on the set S.*

(a) *The orbits of S (under the action of G) partition S.*

(b) *For any $x \in S$, $|Gx| = [G : G_x]$.*

Proof. (a) For $x, y \in S$ define $x \sim y$ if there exists $a \in G$ such that $x = ay$. This defines an equivalence relation on S, since, to begin with, for all $x \in S$, $x = ex$ implies $x \sim x$. If $x \sim y$, then there exists $a \in G$ with $x = ay$, and then $y = a^{-1}x$ implies that $y \sim x$. If $x \sim y$ and $y \sim z$ for $x, y, z \in S$, then there exist $a, b \in G$ such that $x = ay$ and $y = bz$. Therefore $x = (ab)z$ and $x \sim z$. The equivalence classes of $\sim$ are precisely the orbits Gx.

(b) This follows immediately from Proposition 7.3.4. $\square$

7.3.6 Theorem. *Let G be a finite group acting on the finite set S. Then*

$$|S| = |S^G| + \sum_\Gamma [G : G_x] \,,$$

where Γ is a set of representatives of the orbits Gx for which $|Gx| > 1$.

Proof. Since the orbits Gx partition S, we have $|S| = \sum |Gx|$. The equation we need to verify simply collects together the orbits with only one element and counts their members as $|S^G|$. $\square$

If we apply Theorem 7.3.6 to Example 7.3.5, we obtain the class equation of Theorem 7.2.6.

7.3.7 Lemma. *Let G be a finite p-group acting on the finite set S. Then*

$$|S| \equiv |S^G| \ (mod \ p) \,.$$

Proof. Assume that $|G| = p^n$ for some integer n. We simply reduce the equation in Theorem 7.3.6 modulo p. Each term $[G : G_x] > 1$ in the sum $\sum [G : G_x]$ must be a divisor of $|G| = p^n$, and so for some $\alpha \geq 1$ we have $[G : G_x] = p^\alpha \equiv 0 \ (mod \ p)$. $\square$

We now apply the theory of groups acting on sets to give a proof of a stronger version of Cauchy's theorem.

7.3.8 Theorem (Cauchy). *If G is a finite group and p is a prime divisor of $|G|$, then the number of solutions in G of the equation $x^p = e$ is a multiple of p. In particular, G has an element of order p.*

Proof. Let $|G| = n$ and let S be the set of all p-tuples $(x_1, x_2, \ldots, x_p)$ such that $x_i \in G$ for each i, and $x_1 x_2 \cdots x_p = e$. The entry x_p is determined by the first $p - 1$ entries, since $x_p = (x_1 x_2 \cdots x_{p-1})^{-1}$, and so $|S| = n^{p-1} \equiv 0 \ (mod \ p)$, since p is a divisor of n. The motivation for considering this set is that the elements $x \in G$ satisfying $x^p = e$ are precisely the elements such that $(x_1, x_2, \ldots, x_p) \in S$ for $x_1 = x_2 = \cdots = x_p = x$.

Let C be the cyclic subgroup of the permutation group S_p generated by the cycle $\sigma = (1, 2, \ldots, p)$. Then C acts on S by simply permuting indices. That is,

$$\sigma(x_1, x_2, \ldots, x_p) = (x_2, x_3, \ldots, x_1)$$

The product extends to powers of σ in the obvious way. Note that the action produces elements of S, since if $yz = e$, then $zy = e$. We have already observed that the fixed subset S^C consists of p-tuples $(x_1, x_2, \ldots, x_p) \in S$ such that $x_1 = x_2 = \cdots =$

$x_p = x$ and $x^p = e$. Note that S^C is nonempty since it contains $(e, e, \ldots, e)$. The result follows from the previous lemma since

$$|S^C| \equiv |S| \equiv 0 \pmod{p}$$

and $|S^C| \neq 0$.   □

If G is a p-group, then by Lagrange's theorem, the order of each element of G is a power of p. The standard definition of a p-group, allowing its usage for infinite groups, is that G is a p-group if each element of G has an order that is some power of p.

EXERCISES: SECTION 7.3

1. Let G be a group acting on the set S, and let $\phi : G \to \text{Sym}(S)$ be the group homomorphism defined in Proposition 7.3.2. Show that $\ker(\phi) = \cap_{x \in S} G_x$.

2. Let H be a subgroup of G, and let S denote the set of left cosets of H. Define a group action of G on S by setting $a \cdot (xH) = axH$, for all $a, x \in G$ (see Example 7.3.8).

   (a) Let $\phi : G \to \text{Sym}(S)$ be the homomorphism that corresponds to the group action defined above. Show that $\ker(\phi)$ is the largest normal subgroup of G that is contained in H.

   (b) Assume that G is finite and let $[G : H] = n$. Show that if $n!$ is not divisible by $|G|$, then H must contain a nontrivial normal subgroup of G.

3. Let H and K be subgroups of the group G, and let S be the set of left cosets of K. Define a group action of H on S by setting $a \cdot (xH) = axK$, for all $a \in H$ and $x \in G$. By considering the orbit of K under this action, show that $|HK| = \dfrac{|H| |K|}{|H \cap K|}$.

4. Let G be a group of order 21, which acts on the set S.

   (a) Show that if $|S| = 8$, then $S^G \neq \emptyset$.

   (b) For what other integers n between 1 and 100 can you prove that if $|S| = n$, then $S^G \neq \emptyset$?

   (c) For the remaining integers between 1 and 100, show that there *is* a set S with $|S| = n$ and $S^G = \emptyset$.

5. Let G be a group of order 28. Use Exercise 2 to show that G has a normal subgroup of order 7. Show that if G also has a normal subgroup of order 4, then it must be an abelian group.

6. Let G be any non-abelian group of order 6. By Cauchy's theorem, G has an element, say a, of order 2. Let $H = \langle a \rangle$, and let S be the set of left cosets of H.

   (a) Show that H is not normal in G.

   Hint: If H is normal, then $H \subseteq Z(G)$, and it can then be shown that G is abelian.

   (b) Use Exercise 2 and part (a) to show that G must be isomorphic to $\mathrm{Sym}(S)$. Thus any non-abelian group of order 6 is isomorphic to S_3.

7. Let G be a p-group, with $|G| = p^n$. Show that G has a normal subgroup of order p^m for each integer $0 < m < n$.

8. Let G be a p-group with proper subgroup H. Show that there exists an element $a \in G - H$ such that $a^{-1}Ha = H$.

9. Let G be a p-group, with $|G| = p^n$. Show that any subgroup of order p^{n-1} must be normal in G.

10. Let G be a group acting on a set S. Prove that $S^G = \{x \in S \mid G_x = G\}$ and that $S^G = \{x \in S \mid Gx = \{x\}\}$.

11. Prove that if G is a finite p-group acting on a finite set S with $p \nmid |S|$, then G has at least one orbit which contains only one element.

12. If G is a finite group of order n and p is the least prime such that $p|n$, show that any subgroup of index p is normal in G.

13. Let G be a group acting on a set S. We say that G acts **transitively** on S if for each pair x, y of elements of S there exists an element $g \in G$ such that $gx = y$.

    (a) Show that the symmetric group S_n acts transitively on the set $\{1, 2, \ldots, n\}$.

    (b) Show that if $n \neq 2$, then the alternating group A_n acts transitively on the set $\{1, 2, \ldots, n\}$.

    (c) Show that if V is an n-dimensional vector space over the field F, and S is the set of nonzero vectors in V, then $\mathrm{GL}_n(F)$ acts transitively on S.

    (d) Show that if G acts transitively on S, then $[G : G_x] = |S|$ for all $x \in S$. Show that if $|S| > 1$, then $S^G = \emptyset$.

14. If G is a subgroup of the symmetric group S_n, then G is called a **transitive** subgroup if it acts transitively on the set $\{1, 2, \ldots, n\}$.

    (a) Show that if p is a prime number, and G is a transitive subgroup of S_p, then G must contain a cycle of length p.

    (b) Given an example in S_4 of a transitive subgroup that does not contain a cycle of length 4.

7.4   The Sylow Theorems

Lagrange's theorem shows that for any finite group the order of a subgroup is a divisor of the order of the group. The converse is not true. For example, the alternating group A_4 has order 12, but has no subgroup of order 6. Cauchy's theorem gives a weak version of the converse (for prime divisors). The major results in this direction are due to Sylow.

7.4.1 Theorem (First Sylow Theorem). *Let G be a finite group. If p is a prime such that p^α is a divisor of $|G|$ for some $\alpha \geq 0$, then G contains a subgroup of order p^α.*

Proof. We will use induction on $n = |G|$. The theorem is certainly true for $n = 1$, and so we assume that it holds for all groups of order less than n. Consider the class equation

$$|G| = |Z(G)| + \sum [G : C(x)] ,$$

where the sum ranges over one entry from each nontrivial conjugacy class. We will consider two cases, depending on whether or not each term in the summation $\sum [G : C(x)]$ is divisible by p.

   Case 1. For each $x \notin Z(G)$, p is a divisor of $[G : C(x)]$.
In this case the class equation shows that p must be a divisor of $|Z(G)|$, and so $Z(G)$ contains an element a of order p by Cauchy's theorem. Then $\langle a \rangle$ is a normal subgroup of G since $a \in Z(G)$, and so by the induction hypothesis, $G/\langle a \rangle$ contains a subgroup of order $p^{\alpha-1}$, since $p^{\alpha-1}$ is a divisor of $|G/\langle a \rangle|$. The inverse image in G of this subgroup has order p^α since each coset of $\langle a \rangle$ contains p elements.

   Case 2. For some $x \notin Z(G)$, p is not a divisor of $[G : C(x)]$.
Since p^α is a divisor of $n = |C(x)| \cdot [G : C(x)]$, it follows that p^α is a divisor of $|C(x)|$. But then the induction hypothesis can be applied to $C(x)$, since $x \notin Z(G)$ implies $|C(x)| < |G|$, and so $C(x)$ contains a subgroup of order p^α. $\square$

7.4.2 Definition. *Let G be a finite group, and let p be a prime number. A subgroup P of G is called a **Sylow p-subgroup** of G if $|P| = p^\alpha$ for some integer $\alpha \geq 1$ such that p^α is a divisor of $|G|$ but $p^{\alpha+1}$ is not.*

The cyclic group $\mathbf{Z}_6$ has unique subgroups of order 2 and order 3, and these are the Sylow p-subgroups for $p = 2$ and $p = 3$. More generally, if $n = p_1^{\alpha_1} \cdots p_m^{\alpha_m}$ is the prime factorization of n, then for $1 \leq i \leq m$ the group $\mathbf{Z}_n$ has an element of order $p_i^{\alpha_i}$, and the subgroup it generates is the unique Sylow p_i-subgroup of $\mathbf{Z}_n$.

   In the symmetric group S_3 there are three subgroups of order 2, so we do not have uniqueness. Nevertheless, at least they are conjugate. The unique subgroup of order 3 is normal, and it turns out to be true in general that there is a unique Sylow p-subgroup if and only if there is a normal Sylow p-subgroup.

7.4.3 Lemma. *Let G be a finite group with* $|G| = mp^\alpha$, *where* $\alpha \geq 1$ *and m is not divisible by p. If P is a normal Sylow p-subgroup, then P contains every p-subgroup of G.*

Proof. Suppose that $a \in H$ for a p-subgroup H of G. Since P is a normal subgroup of G, we may consider the coset aP as an element of the factor group G/P. On the one hand, the order of a is a power of p since it belongs to a subgroup whose order is a power of p. The order of the coset aP must be a divisor of the order of a, so it is also a power of p. On the other hand, the order of aP must be a divisor of $[G : P]$, but by assumption $[G : P]$ is not divisible by p. This is a contradiction unless $aP = P$, so $a \in P$, and we have shown that $H \subseteq P$. $\square$

7.4.4 Theorem (Second and Third Sylow Theorems). *Let G be a finite group of order n, and let p be a prime number.*

(a) *All Sylow p-subgroups of G are conjugate, any p-subgroup of G is contained in a Sylow p-subgroup, and any maximal p-subgroup is a Sylow p-subgroup.*

(b) *Let* $n = mp^\alpha$, *with* $\gcd(m, p) = 1$, *and let k be the number of Sylow p-subgroups of G. Then* $k|m$ *and* $k \equiv 1 \pmod{p}$.

Proof. Let P be a Sylow p-subgroup of G with $|P| = p^\alpha$, let S be the set of all conjugates of P, and let P act on S by conjugation. If $Q \in S$ is left fixed by the action of P, then $P \subseteq N(Q)$. Since $|Q| = p^\alpha$, p is not a divisor of $[G : Q]$ and hence p is not a divisor of $[N(Q) : Q]$. Thus the hypothesis of Lemma 7.4.3 is satisfied by Q in $N(Q)$, since Q is normal in $N(Q)$. It follows that $P \subseteq Q$, so $P = Q$ since $|P| = |Q|$. Therefore the only member of S left fixed by the action of P is P itself, so $|S^P| = 1$, and then Lemma 7.3.7 shows that $|S| \equiv 1 \pmod{p}$.

Next let Q be any maximal p-subgroup. (That is, let Q be any p-subgroup that is not contained in any larger p-subgroup of G.) Let Q act on S by conjugation. Now $|S| \equiv 1 \pmod{p}$ implies by Lemma 7.3.7 that $|S^Q| \equiv 1 \pmod{p}$. In particular, some conjugate K of P must be left fixed by Q. Then $Q \subseteq N(K)$, and as before it follows from Lemma 7.4.3 that $Q \subseteq K$. But then since Q is a maximal p-subgroup, we must have $Q = K$. This shows that Q is conjugate to P. This implies not only that all Sylow p-subgroups are conjugate, but that any maximal p-subgroup is a Sylow p-subgroup. It is clear that any p-subgroup is contained in a maximal p-subgroup, so we have proved part (a).

Since we now know that S is the set of all Sylow p-subgroups of G, we have $k \equiv 1 \pmod{p}$. Finally, $k = [G : N(P)]$, since this is the number of conjugates of P. Since $P \subseteq N(P)$ we see that $k|m$ because $[G : N(P)]$ is a divisor of $[G : P]$. $\square$

Example 7.4.1.

To give a simple application of the Sylow theorems, we will show that any group of order 100 has a normal subgroup of order 25. We simply note that the number of Sylow 5-subgroups must be congruent to 1 modulo 5 and also a divisor of 4. The only possibility is that there is just one such subgroup (of order 25), which must then be normal.   □

Example 7.4.2.

As a slightly less straightforward example, we will show that any group of order 30 must have a nontrivial normal subgroup. The number of Sylow 3-subgroups must be congruent to 1 modulo 3 and a divisor of 10, so it must be either 1 or 10. The number of Sylow 5-subgroups must be congruent to 1 modulo 5 and a divisor of 6, so it must be either 1 or 6. Any Sylow 3-subgroup must have order 3, so the intersection of two distinct such subgroups must be trivial. Therefore ten Sylow 3-subgroups would yield twenty elements of order 3. Similarly, six Sylow 5-subgroups would yield twenty-four elements of order 5. Together, this would simply give too many elements for the group, so we conclude that there must be either one Sylow 3-subgroup or one Sylow 5-subgroup, showing the existence of a nontrivial normal subgroup.   □

As further applications of the structure theorems we have proved, we can obtain the following information about the structure of groups of certain types. The amount of work it takes to get even such limited results should make the student appreciate the difficulty of determining the structure of groups. Note that Proposition 7.4.5 is a special case of Proposition 7.4.6 (b). Its proof is of interest since it requires considerably less machinery than the proof of Proposition 7.4.6.

7.4.5 Proposition. *Let $p > 2$ be a prime, and let G be a group of order $2p$. Then G is either cyclic or isomorphic to the dihedral group D_p of order $2p$.*

Proof. By Cauchy's theorem, G contains an element a of order p and an element b of order 2. The cyclic subgroup $\langle a \rangle$ has index 2 in G, and so it must be a normal subgroup. Thus conjugating a by b gives $bab = a^n$ for some n. Then $a = b(bab)b = ba^n b = a^{n^2}$, and so $n^2 \equiv 1 \pmod{p}$. It follows that $n \equiv \pm 1 \pmod{p}$, and thus $bab = a$ or else $bab = a^{-1}$. In the first case a and b commute, and so ab has order $\text{lcm}(2, p) = 2p$ and G is cyclic. In the second case, we obtain $ba = a^{-1}b$ (or $ba = a^{p-1}b$), the familiar equation that defines D_p.   □

7.4.6 Proposition. *Let G be a group of order pq, where p > q are primes.*

(a) *If q is not a divisor of p − 1, then G is cyclic.*

(b) *If q is a divisor of p − 1, then either G is cyclic or else G is generated by two elements a and b satisfying the following equations:*

$$a^p = e, \qquad b^q = e, \qquad ba = a^n b$$

where n $\not\equiv$ 1 (mod p) but $n^q \equiv 1$ (mod p).

Proof. The number of Sylow p-subgroups is a divisor of q, so it must be either 1 or q. In the latter case it could not also be congruent to 1 modulo p, since $p > q$. Thus the Sylow p-subgroup is cyclic and normal, say $\langle a \rangle$. There exists an element b of order q, and since $\langle b \rangle$ is a Sylow q-subgroup, there are two cases. If the number of Sylow q-subgroups is 1, then $\langle b \rangle$ is a normal subgroup, and ab has order pq, showing that G is cyclic. (The intersection of Sylow subgroups for different primes is always trivial. Since both are normal subgroups, the element $aba^{-1}b^{-1}$ belongs to both subgroups and hence must be equal to e, showing that $ab = ba$.) In the second case, since $\langle a \rangle$ is normal, we have $bab^{-1} \in \langle a \rangle$, and so $ba = a^n b$ for some n. We can assume n is not congruent to 1 modulo p, since that would imply $ba = ab$, covered in the previous case. Conjugating repeatedly gives $b^q a b^{-q} = a^{n^q}$, or simply $a = a^{n^q}$, which shows that $n^q \equiv 1$ (mod p) since a has order p. $\square$

EXERCISES: SECTION 7.4

1. Let G be a finite abelian group, and let p be a prime divisor of $|G|$. Show that the Sylow p-subgroup of G consists of e and all elements whose order is a power of p.

2. Let G be a finite group, and let p be a prime divisor of $|Z(G)|$. Show that each Sylow p-subgroup of G contains the Sylow p-subgroup of $Z(G)$.

3.† In S_4 find a Sylow 2-subgroup and a Sylow 3-subgroup.

4. Find all Sylow subgroups of D_5 and D_6.

5. Find all Sylow subgroups of D_n, for the case in which $n = p^k$ is a prime power.

6.† Find all Sylow 3-subgroups of S_4 and show explicitly how they are conjugate.

7. Show that A_4 has no subgroup of order 6.

8. This exercise classifies all subgroups of S_4.

   (a) Show that any proper, nontrivial subgroup of S_4 is isomorphic to one of the following groups: $\mathbf{Z}_2$, $\mathbf{Z}_3$, $\mathbf{Z}_4$, $\mathbf{Z}_2 \times \mathbf{Z}_2$, S_3, D_4, or A_4.

(b) For each of the groups in part (a), determine all subgroups of S_4 that are isomorphic to the given group, determine which of these are conjugate, and determine their normalizers in S_4.

9. Show that there is no simple group of order 148.

10. Show that there is no simple group of order 56.

11. Let G be a group of order p^2q, where p and q are distinct primes. Show that G must contain a proper nontrivial normal subgroup.

12. Show that there is no simple group of order 48.

13. Show that there is no simple group of order 36.

14. Let G be a finite group in which each Sylow subgroup is normal. Prove that G is isomorphic to the direct product of its Sylow subgroups.

7.5   Finite Abelian Groups

If $m > 1$ and $n > 1$ are relatively prime numbers, and $\phi : \mathbf{Z}_{mn} \to \mathbf{Z}_m \times \mathbf{Z}_n$ is defined by $\phi([x]_{mn}) = ([x]_m, [x]_n)$, then ϕ is an isomorphism. The statement that ϕ is onto is precisely the statement of the Chinese remainder theorem. Another proof is to observe that the second group is cyclic since the order of the element $([1], [1])$ is $\text{lcm}[m, n] = mn$. Applying this result repeatedly, we can show that for any $n > 1$, the cyclic group $\mathbf{Z}_n$ is isomorphic to a direct product of cyclic groups of prime power order, where the prime powers are those in the prime factorization of n. (See Theorem 3.5.5 of Section 3.5.) The goal of this section is to prove a much more general result: any finite abelian group is isomorphic to a direct product of cyclic groups of prime power order.

Since all of the groups under discussion in this section are abelian, we will use additive notation in the results leading up to, and including, the fundamental structure theorem. In additive notation, Proposition 7.1.3 states that if H and K are subgroups of an abelian group G such that $H \cap K = \{0\}$ and $H + K = G$, then $G \cong H \times K$. In this case each element of G can be written uniquely in the form $h + k$, where $h \in H$ and $k \in K$. We generalize this in the next definition.

7.5.1 Definition. *Let $H_1, \ldots, H_n$ be subgroups of the abelian group G. If each element $g \in G$ can be written uniquely in the form $g = h_1 + \cdots + h_n$, with $h_i \in H_i$ for all i, then G is called the direct sum of the subgroups $H_1, \ldots, H_n$, and we write $G = H_1 \oplus \cdots \oplus H_n$.*

7.5.2 Proposition. *If H and K are subgroups of an abelian group G such that $H \cap K = \{0\}$ and $H + K = G$, then $G = H \oplus K$.*

Proof. It is clear that each element of G can be written in the form $h + k$, for some $h \in H$ and $k \in K$. Now suppose that $h_1 + k_1 = h_2 + k_2$, with $h_1, h_2 \in H$ and $k_1, k_2 \in K$. Then $h_1 - h_2 = k_2 - k_1 \in H \cap K$, so $h_1 - h_2 = 0 = k_2 - k_1$, and thus $h_1 = h_2$ and $k_1 = k_2$. This shows that the sums are unique. $\square$

Our first step in describing the structure of a finite abelian group G is to show that G is a direct sum of subgroups of prime power order. This reduces the study to that of abelian p-groups.

7.5.3 Theorem. *A finite abelian group is the direct sum of its Sylow p-subgroups.*

Proof. Let G be a finite abelian group, with $|G| = np^\alpha$, where $p \nmid n$. Let $H_1 = \{a \in G \mid p^\alpha a = 0\}$ and let $K_1 = \{a \in G \mid na = 0\}$. Since G is abelian, both are subgroups, and H_1 is the Sylow p-subgroup of G.

We will show that (i) $H_1 \cap K_1 = \{0\}$ and (ii) $H_1 + K_1 = G$. Then we can decompose K_1 in a similar fashion, etc., to get $G = H_1 \oplus H_2 \oplus \cdots \oplus H_k$, where each subgroup H_i is a Sylow p-subgroup for some prime p.

To prove (i), we simply observe that if $a \in H_1 \cap K_1$, then the order of a is a common divisor of p^α and n, which implies that $a = 0$. To prove (ii), let $a \in G$. Then the order k of a is a divisor of $p^\alpha n$, and so $k = p^\beta m$, where $m|n$, $\beta \le \alpha$, and $p \nmid m$. Since $\gcd(p^\beta, m) = 1$, there exist $r, s \in \mathbf{Z}$ with $sm + rp^\beta = 1$. Then $a = s(ma) + r(p^\beta a)$, and $a \in H_1 + K_1$ since $ma \in H_1$ and $p^\beta a \in K_1$. The last statement follows from the fact that $p^\alpha(ma) = 0$ and $n(p^\beta a) = 0$ since $p^\alpha m$ and np^β are multiples of the order of a. $\square$

7.5.4 Lemma. *Let G be a finite abelian p-group, and let $a \in G$ be an element whose order is maximal in G. Then each coset of $\langle a \rangle$ contains an element d such that $\langle d \rangle \cap \langle a \rangle = \{0\}$.*

Proof. The outline of the proof is this: for $b \in G$, let γ be the smallest nonnegative integer such that $p^\gamma b \in \langle a \rangle$. We seek $x \in \langle a \rangle$ such that $p^\gamma x = p^\gamma b$, and then let $d = b - x$. The details go as follows.

Let $o(a) = p^\alpha$. Given $b \in G$, let $o(b) = p^\beta$, and let p^γ be the order of the coset $b + \langle a \rangle$ in the factor group $G/\langle a \rangle$. Note that $\gamma \le \beta \le \alpha$ since a is an element of maximal order in G.

Since $p^\gamma b \in \langle a \rangle$, we have $p^\gamma b = p^\delta qa$ for some nonnegative integers δ, q such that $p \nmid q$. Then qa is a generator for $\langle a \rangle$, since q is relatively prime to $o(a)$, and hence $o(qa) = p^\alpha$. We have $p^{\beta-\gamma} p^\delta (qa) = p^{\beta-\gamma} p^\gamma b = 0$, so $o(a) \mid p^{\beta-\gamma} p^\delta$ and thus $\alpha \le \beta - \gamma + \delta$. It follows that $\gamma \le \delta - (\alpha - \beta) \le \delta$.

Let $x = p^{\delta-\gamma} qa$, and set $d = b - x$. Note that $p^\gamma x = p^\gamma b$. Then $d \in b + \langle a \rangle$, as required. To show that $\langle d \rangle \cap \langle a \rangle = \{0\}$, suppose that $nd \in \langle a \rangle$, for some $n \in \mathbf{Z}$. Then $nb - nx = nd \in \langle a \rangle$, and so $nb \in \langle a \rangle$ since $x \in \langle a \rangle$. This implies that $n(b + \langle a \rangle) = 0 + \langle a \rangle$ in $G/\langle a \rangle$, and so $p^\gamma \mid n$ since p^γ is the order of $b + \langle a \rangle$. But then $nd = 0$ since $p^\gamma d = p^\gamma(b - x) = 0$. $\square$

7.5.5 Lemma. *Let G be a finite abelian p-group. If $\langle a \rangle$ is a cyclic subgroup of G of maximal order, then there exists a subgroup H with $G = \langle a \rangle \oplus H$.*

Proof. The outline of the proof is to factor out $\langle a \rangle$ and use induction to decompose $G / \langle a \rangle$ into a direct sum of cyclic subgroups. Then Lemma 7.5.4 can be used to choose the right preimages of the generators of $G / \langle a \rangle$ to generate the complement H of $\langle a \rangle$.

We use induction on the order of G. If $|G|$ is prime, then G is cyclic and there is nothing to prove. Consequently, we may assume that the statement of the theorem holds for all groups of order less than $|G| = p^\alpha$. If G is cyclic, then we are done. If not, let $\langle a \rangle$ be a cyclic subgroup of G of maximal order, and use the induction hypothesis repeatedly to write $G / \langle a \rangle$ as a direct sum $H_1 \oplus H_2 \oplus \cdots \oplus H_n$ of cyclic subgroups.

We next use Lemma 7.5.4 to choose, for each i, a coset $a_i + \langle a \rangle$ that corresponds to a generator of H_i such that $\langle a_i \rangle \cap \langle a \rangle = \{0\}$. We claim that $G = \langle a \rangle \oplus H$ for the smallest subgroup $H = \langle a_1, a_2, \ldots, a_n \rangle$ that contains $a_1, a_2, \ldots, a_n$.

First, if $g \in \langle a \rangle \cap \langle a_1, \ldots, a_n \rangle$, then $g = m_1 a_1 + \cdots + m_n a_n \in \langle a \rangle$ for some integers $m_1, \ldots, m_n$. Thus $g + \langle a \rangle = m_1 a_1 + \cdots + m_n a_n + \langle a \rangle = \langle a \rangle$, and since $G / \langle a \rangle$ is a direct sum, this implies that $m_i a_i + \langle a \rangle = \langle a \rangle$ for each i. But then $m_i a_i \in \langle a \rangle$, and so $m_i a_i = 0$ since $\langle a_i \rangle \cap \langle a \rangle = \{0\}$. Thus $g = 0$.

Next, given $g \in G$, express the coset $g + \langle a \rangle$ as $m_1 a_1 + \cdots + m_n a_n + \langle a \rangle$ for integers $m_1, \ldots, m_n$. Then $g \in g + \langle a \rangle$, and so $g = ma + m_1 a_1 + \cdots + m_n a_n$ for some integer m.

Thus we have shown that $\langle a \rangle \cap H = \{0\}$ and $G = \langle a \rangle + H$, so $G = \langle a \rangle \oplus H$.   □

7.5.6 Theorem (Fundamental Theorem of Finite Abelian Groups). *Any finite abelian group is isomorphic to a direct product of cyclic groups of prime power order. Any two such decompositions have the same number of factors of each order.*

Proof. We can use Theorem 7.5.3 to decompose any finite abelian group G into a direct sum of p-groups, and then we can use Lemma 7.5.5 to write each of these groups as a direct sum of cyclic subgroups.

Uniqueness is shown by induction on $|G|$. It is enough to prove the uniqueness for a given p-group. Suppose that

$$\mathbf{Z}_{p^{\alpha_1}} \oplus \mathbf{Z}_{p^{\alpha_2}} \oplus \cdots \oplus \mathbf{Z}_{p^{\alpha_n}} = \mathbf{Z}_{p^{\beta_1}} \oplus \mathbf{Z}_{p^{\beta_2}} \oplus \cdots \oplus \mathbf{Z}_{p^{\beta_m}}$$

where $\alpha_1 \geq \alpha_2 \geq \ldots \geq \alpha_n$ and $\beta_1 \geq \beta_2 \geq \ldots \geq \beta_m$. Consider the subgroups in which each element has been multiplied by p. By induction, $\alpha_1 - 1 = \beta_1 - 1, \ldots$, which gives $\alpha_1 = \beta_1, \ldots$, with the possible exception of the α_i's and β_j's that equal 1. But the groups have the same order, and this determines that each has the same number of factors isomorphic to $\mathbf{Z}_p$.   □

Example 7.5.1.

We will find all finite abelian groups of order 72. The first step is to find the prime factorization: $72 = 2^3 3^2$. There are three possible groups of order 8: $\mathbf{Z}_8$, $\mathbf{Z}_4 \times \mathbf{Z}_2$, and $\mathbf{Z}_2 \times \mathbf{Z}_2 \times \mathbf{Z}_2$. There are two possible groups of order 9: $\mathbf{Z}_9$ and $\mathbf{Z}_3 \times \mathbf{Z}_3$. This gives us the following possible groups:

$$\mathbf{Z}_8 \times \mathbf{Z}_9 \qquad \mathbf{Z}_4 \times \mathbf{Z}_2 \times \mathbf{Z}_9 \qquad \mathbf{Z}_2 \times \mathbf{Z}_2 \times \mathbf{Z}_2 \times \mathbf{Z}_9$$
$$\mathbf{Z}_8 \times \mathbf{Z}_3 \times \mathbf{Z}_3 \qquad \mathbf{Z}_4 \times \mathbf{Z}_2 \times \mathbf{Z}_3 \times \mathbf{Z}_3 \qquad \mathbf{Z}_2 \times \mathbf{Z}_2 \times \mathbf{Z}_2 \times \mathbf{Z}_3 \times \mathbf{Z}_3 \, .$$

Example 7.5.2.

There is another way to describe the possible abelian groups of order 72. We can combine the highest powers of each prime by using the fact that $\mathbf{Z}_m \times \mathbf{Z}_n \cong \mathbf{Z}_{mn}$ if $(m, n) = 1$. Then we have the groups in the following form:

$$\mathbf{Z}_{72} \qquad \mathbf{Z}_{36} \times \mathbf{Z}_2 \qquad \mathbf{Z}_{18} \times \mathbf{Z}_2 \times \mathbf{Z}_2$$
$$\mathbf{Z}_{24} \times \mathbf{Z}_3 \qquad \mathbf{Z}_{12} \times \mathbf{Z}_6 \qquad \mathbf{Z}_6 \times \mathbf{Z}_6 \times \mathbf{Z}_2 \, .$$

Note we have arranged the cyclic factors in such a way that the order of each factor is a divisor of the order of the preceding one. $\quad\square$

7.5.7 Proposition. *Let G be a finite abelian group. Then G is isomorphic to a direct product of cyclic groups $\mathbf{Z}_{n_1} \times \mathbf{Z}_{n_2} \times \cdots \times \mathbf{Z}_{n_k}$ such that $n_i \mid n_{i-1}$ for $i = 2, 3, \ldots, k$.*

Proof. We will use induction on the number of prime divisors of $|G|$. If G is a p-group, then we only need to arrange its factors in decreasing order of size, since the divisibility condition automatically follows. If $|G|$ has more than one prime factor, let H_p denote its p-Sylow subgroup, and let K_p be a subgroup with $G \cong H_p \times K_p$. Then the induction hypothesis may be applied to K_p to give $K_p \cong \mathbf{Z}_{n_1} \times \mathbf{Z}_{n_2} \times \cdots$. Furthermore, $H_p \cong \mathbf{Z}_{p^{\alpha_1}} \times \mathbf{Z}_{p^{\alpha_2}} \times \cdots$. Since n_1 and p^{α_1} are relatively prime, the subgroup $\mathbf{Z}_{p^{\alpha_1}} \times \mathbf{Z}_{n_1}$ is cyclic. Similarly, we may combine successive factors of H_p with factors of K_p, and in doing so we maintain the necessary divisibility relations. $\quad\square$

7.5.8 Corollary. *Let G be a finite abelian group. If $a \in G$ is an element of maximal order in G, then the order of every element of G is a divisor of the order of a.*

Proof. Let G be isomorphic to a direct product of cyclic groups $\mathbf{Z}_{n_1} \times \mathbf{Z}_{n_2} \times \cdots \times \mathbf{Z}_{n_k}$ such that $n_i \mid n_{i-1}$ for $i = 2, 3, \ldots, k$. Recall that the order of an element in a direct product is the least common multiple of the orders of its components. Thus the largest possible order of an element of G is n_1. Furthermore, it is clear that the order of any element must be a divisor of n_1. $\quad\square$

If p is prime, then $\mathbf{Z}_p$ is a field, and Theorem 6.5.10 implies that $\mathbf{Z}_p^\times$ is a cyclic group. Of the composite numbers up to 20, it can be checked that $\mathbf{Z}_n^\times$ is cyclic for $n = 4, 6, 9, 10, 14, 18$, while $\mathbf{Z}_n^\times$ fails to be cyclic for $n = 8, 12, 15, 16, 20$. We leave it to the reader to make a conjecture as to when $\mathbf{Z}_n^\times$ is cyclic. (Results in this section will provide the answer.)

When studying the multiplicative group $\mathbf{Z}_n^\times$, it is natural to use multiplicative notation. We begin our description of $\mathbf{Z}_n^\times$ with an elementary number-theoretic lemma.

7.5.9 Lemma. *Let p be a prime number, and let k, a, b be integers.*

(a) *If $1 \leq k \leq p - 1$, then p is a divisor of the binomial coefficient $\binom{p}{k}$.*

(b) *If $k \geq 1$ and $a \equiv b \pmod{p^k}$, then $a^p \equiv b^p \pmod{p^{k+1}}$.*

(c) *If $k \geq 2$ and p is an odd prime, then*

$$(1 + ap)^{p^{k-2}} \equiv 1 + ap^{k-1} \pmod{p^k} \, .$$

(d) *If p is an odd prime and $p \nmid a$, then*

$$(1 + ap)^{p^{k-1}} \equiv 1 \pmod{p^k} \quad \text{and} \quad (1 + ap)^{p^{k-2}} \not\equiv 1 \pmod{p^k} \, .$$

Proof. (a) Since $\binom{p}{k} = p!/k!(p - k)!$ and p is prime, no factor in the denominator cancels the factor p in the numerator.

(b) Since $a = b + qp^k$ for some $q \in \mathbf{Z}$, we have

$$a^p = (b + qp^k)^p = b^p + pb^{p-1} \cdot qp^k + mp^{2k} = b^p + (b^{p-1}q + mp^{k-1})p^{k+1}$$

for some $m \in \mathbf{Z}$. Therefore $a^p \equiv b^p \pmod{p^{k+1}}$.

(c) The proof is by induction, starting with $k = 2$. This case is obvious since $(1 + ap)^1 \equiv 1 + ap^1 \pmod{p^2}$. Now assume that the result holds for k. By part (b) we have

$$\left((1 + ap)^{p^{k-2}}\right)^p \equiv (1 + ap^{k-1})^p \pmod{p^{k+1}} \, .$$

Then

$$(1 + ap)^{p^{k-1}} \equiv \sum_{j=0}^{p} \binom{p}{j}(ap^{k-1})^j \equiv 1 + ap^k \pmod{p^{k+1}}$$

since $(k - 1)j \geq k + 1$ for $j \geq 3$ and $2(k - 1) \geq k$ and $p \mid \binom{p}{2}$. Thus the result holds for $k + 1$.

(d) By part (c) we have $(1+ap)^{p^{k-1}} \equiv 1+ap^k \pmod{p^{k+1}}$, and so $(1+ap)^{p^{k-1}} \equiv 1 \pmod{p^k}$. On the other hand, $(1 + ap)^{p^{k-2}} \equiv 1 + ap^{k-1} \pmod{p^k}$ by part (c), and $1 + ap^{k-1} \not\equiv 1 \pmod{p^k}$ since $p \nmid a$ $\square$

7.5.10 Theorem. *Let p be an odd prime, and let k be a positive integer. Then* $\mathbf{Z}_{p^k}^{\times}$ *is a cyclic group.*

Proof. We have already noted that $\mathbf{Z}_p^{\times}$ is cyclic, so we choose a generator $[a]$ of $\mathbf{Z}_p^{\times}$. Then $[a+p] = [a]$, and so $[a+p]$ is also a generator. If $a^{p-1} \equiv 1 \pmod{p^2}$, then

$$(a+p)^{p-1} \equiv a^{p-1} + (p-1)a^{p-2} \cdot p \equiv 1 + p(p-1)a^{p-2} \pmod{p^2} .$$

Since $p \nmid a$, we have $(a+p)^{p-1} \not\equiv 1 \pmod{p^2}$. Thus without loss of generality we may assume that $a^{p-1} \equiv 1 \pmod p$, but $a^{p-1} \not\equiv 1 \pmod{p^2}$.

We will show that $[a]$ is an element of order $\varphi(p^k)$ in the group $\mathbf{Z}_{p^k}^{\times}$. Thus we must show that $a^n \equiv 1 \pmod{p^k}$ implies that $\varphi(p^k) \mid n$. Since $a^{p-1} \equiv 1 \pmod p$, there exists $q \in \mathbf{Z}$ such that $a^{p-1} = 1 + qp$, and then since $a^{p-1} \not\equiv 1 \pmod{p^2}$, we must have $p \nmid q$. By Lemma 7.5.9 (d) we have

$$(1+qp)^{p^{k-1}} \equiv 1 \pmod{p^k} \quad \text{and} \quad (1+qp)^{p^{k-2}} \not\equiv 1 \pmod{p^k} ,$$

which implies that the order of $[1+qp]$ in $\mathbf{Z}_{p^k}^{\times}$ is p^{k-1}. Since $[1+qp]^n = [a^{p-1}]^n = [1]$, we must have $p^{k-1} \mid n$. Set $n = mp^{k-1}$. Since $a^n = (a^{p^{k-1}})^m \equiv a^m \pmod p$ and $a^n \equiv 1 \pmod p$, we conclude that $a^m \equiv 1 \pmod p$. Since the order of $[a]$ in $\mathbf{Z}_p^{\times}$ is $p-1$, we see that $p-1 \mid m$. Thus $\varphi(p^k) = p^{k-1}(p-1)$ is a divisor of n, and our proof is complete. $\square$

We next consider the prime $p = 2$. Of course, the group $\mathbf{Z}_2^{\times}$ is trivially cyclic, and $\mathbf{Z}_4^{\times}$ is cyclic of order of 2.

7.5.11 Lemma. *Let k be an integer with $k \geq 2$.*

(a) *In* $\mathbf{Z}_{2^k}^{\times}$, *the element* $[5]$ *has order* 2^{k-2}.

(b) *In* $\mathbf{Z}_{2^k}^{\times}$, *the elements* $[\pm 5^n]$ *are distinct, as n ranges over nonnegative integers less than* 2^{k-2}.

Proof. (a) For $k = 2$ we have $5 \equiv 1 \pmod 4$, and so we assume that $k \geq 3$. It suffices to show that $5^{2^{k-3}} \equiv 1 + 2^{k-1} \pmod{2^k}$, since then $5^{2^{k-3}} \not\equiv 1 \pmod{2^k}$, but $5^{2^{k-2}} = (5^{2^{k-3}})^2 \equiv (1 + 2^{k-1})^2 \equiv 1 \pmod{2^k}$. We will give a proof by induction. The case $k = 3$ is clear. Assume that the result holds for k. Then by Lemma 7.5.9 (b) we have $5^{2^{k-2}} = (5^{2^{k-3}})^2 \equiv (1 + 2^{k-1})^2 \pmod{2^{k+1}}$. But then $(1 + 2^{k-1})^2 = 1 + 2^k + 2^{2k-2} \equiv 1 + 2^k \pmod{2^{k+1}}$ since $2k - 2 \geq k + 1$ for $k \geq 3$.

(b) Assume that m, n are nonnegative integers less than 2^{k-2} such that $\pm 5^m \equiv \pm 5^n \pmod{2^k}$, with $m \geq n$. Since $5 \equiv 1 \pmod 4$, reducing modulo 4 shows that the signs must be the same. Then $5^m \equiv 5^n \pmod{2^k}$ implies that $5^{m-n} \equiv 1 \pmod{2^k}$, because we can multiply by 5^{-n} since $\gcd(5, 2) = 1$. By part (a) of the lemma, $m - n \equiv 0 \pmod{2^{k-2}}$, and therefore $m = n$. $\square$

7.5.12 Theorem. *If $k \geq 3$, then $\mathbf{Z}_{2^k}^{\times}$ is isomorphic to the direct product of a cyclic group of order 2 and a cyclic group of order 2^{k-2}.*

Proof. In the group $G = \mathbf{Z}_{2^k}^{\times}$, the subgroup H generated by $[-1]$ has order 2, and if $k \geq 3$, then Lemma 7.5.11 (a) shows that the subgroup K generated by $[5]$ has order 2^{k-2}. Since the order of G is $\varphi(p^k) = 2^{k-1}$, Lemma 7.5.11 (b) implies that $G = \{[\pm 5^n] \mid 0 \leq n < 2^{k-2}\}$. Thus $G = HK$ and $H \cap K = \{1\}$, so Theorem 7.1.3 implies that G is isomorphic to $H \times K$. □

7.5.13 Corollary. *The group $\mathbf{Z}_n^{\times}$ is cyclic if and only if n is of the form 2, 4, p^k, or $2p^k$ for an odd prime p.*

Proof. Assume that n has the prime decomposition $n = p_1^{\alpha_1} p_2^{\alpha_2} \cdots p_m^{\alpha_m}$, and suppose that $\mathbf{Z}_n^{\times}$ is cyclic. In Example 5.2.13 we showed that

$$\mathbf{Z}_n^{\times} \cong \mathbf{Z}_{p_1^{\alpha_1}}^{\times} \times \mathbf{Z}_{p_2^{\alpha_2}}^{\times} \times \cdots \times \mathbf{Z}_{p_m^{\alpha_m}}^{\times} .$$

Each of the nontrivial component groups in this direct product has even order, so the assumption that $\mathbf{Z}_n^{\times}$ is cyclic implies that there is only one nontrivial component. (If not, an element of maximal order in $\mathbf{Z}_n^{\times}$ would have order equal to the least common multiple of the orders of the groups on the right hand side, which is less than $|\mathbf{Z}_n^{\times}|$.) Since $\mathbf{Z}_2^{\times}$ is trivial, this implies that n has the form $2p^k$ for an odd prime p, or the form p^k. In the latter case, if $p = 2$, then k must be 1 or 2 by Theorem 7.5.12.

Conversely, if n has the stated form, then since $\mathbf{Z}_2^{\times}$ is trivial, Theorem 7.5.10 implies that $\mathbf{Z}_n^{\times}$ is cyclic. □

In elementary number theory, an integer g is called a **primitive root** for the modulus n if $\mathbf{Z}_n^{\times}$ is a cyclic group and $[g]_n$ is a generator for $\mathbf{Z}_n^{\times}$. Corollary 7.5.13 determines which moduli n have primitive roots. The proof of Theorem 7.5.10 shows how to find a generator for $\mathbf{Z}_{p^k}^{\times}$.

EXERCISES: SECTION 7.5

1. Give a representative of each isomorphism class of abelian groups of order 64.

2. Using both the form of Theorem 7.5.6 and that of Proposition 7.5.7, list all nonisomorphic abelian groups of order 5^6.

3. Using both the form of Theorem 7.5.6 and that of Proposition 7.5.7, list all nonisomorphic abelian groups of the following orders.

   (a) order 108

   (b) order 200

   (c) order 900

4. Write each of the following groups as a direct product of cyclic groups of prime power order.

†(a) $\mathbf{Z}_{20}^{\times}$

(b) $\mathbf{Z}_{54}^{\times}$

†(c) $\mathbf{Z}_{70}^{\times}$

(d) $\mathbf{Z}_{180}^{\times}$

5. Show that the following conditions are equivalent, for any positive integer n:

(i) all abelian groups of order n are isomorphic;

(ii) all abelian groups of order n are cyclic;

(iii) the integer n is not divisible by the square of any prime number.

6. Show that $H = \{([m]_4, [n]_4) \mid m \equiv n \pmod 2\}$ is a subgroup of $\mathbf{Z}_4 \times \mathbf{Z}_4$. Write H as a direct sum of cyclic groups of prime power order.

7. Prove that if p is a prime and $\mathbf{Z}_{p^\alpha} \cong G_1 \times G_2$, then either $G_1 \cong \mathbf{Z}_{p^\alpha}$ or $G_2 \cong \mathbf{Z}_{p^\alpha}$.

8. Let p be an odd prime, and let g be a primitive root modulo p. Prove that $g^{(p-1)/2} \equiv -1 \pmod p$.

9. Prove that if the modulus n has a primitive root, then it has exactly $\phi(\phi(n))$ pairwise incongruent primitive roots.

10. Let a, b be positive integers, and let $d = \gcd(a, b)$ and $m = \operatorname{lcm}[a, b]$. In the group $G = \mathbf{Z}_a \times \mathbf{Z}_b$, let M be the subgroup generated by $(1, 1)$. Find a subgroup H of G with $G = M \oplus H$ (compare Lemma 7.5.5).

We now have enough information to complete a classification of groups of order less than 12. This can be done by using the fundamental theorem of finite abelian groups and Proposition 7.4.5 and 7.4.6. However, the case of order 8 requires some additional analysis.

11. Let G be a nonabelian group of order 8.

(a) Prove that G must have an element of order 4, but none of order 8.

(b) Let a be an element of order 4, and let $N = \langle a \rangle$. Show that there exists an element b such that $G = N \cup Nb$.

(c) Show that either $b^2 = e$ or $b^2 = a^2$. (Since N is normal, consider the order of Nb in G/N.)

(d) Show that bab^{-1} has order 4 and must be equal to a^3.

(e) Conclude that either $G \cong D_4$ or else G is determined by the equations $a^4 = e$, $ba = a^3 b$, $b^2 = a^2$. Review Example 3.3.7 (the quaternion group) to verify that the second case can occur.

12. Determine (up to isomorphism) all groups of order less than 12.

7.6   Solvable Groups

We are now ready to study groups arising from equations that are solvable by radicals. (We will not be able to give a proof of this correspondence until after we develop some new ideas in the next chapter.) Groups in the class are simply said to be *solvable*. It is obvious from the following definition that any abelian group is solvable.

7.6.1 Definition. *The group G is said to be **solvable** if there exists a finite chain of subgroups $G = N_0 \supseteq N_1 \supseteq \ldots \supseteq N_n$ such that*

   (i) *N_i is a normal subgroup in N_{i-1} for $i = 1, 2, \ldots, n$,*
   (ii) *N_{i-1}/N_i is abelian for $i = 1, 2, \ldots, n$, and*
   (iii) *$N_n = \{e\}$.*

Example 7.6.1 (S_3 is solvable).

> Let $G = S_3$, the group of all permutations on three elements (the smallest nonabelian group). For the descending chain of subgroups $N_0 = G, N_1 = A_3$, and $N_2 = \{(1)\}$ we have $N_0/N_1 \cong \mathbf{Z}_2$ and $N_1/N_2 \cong \mathbf{Z}_3$. Recall that A_3 is the set $\{(1), (1, 2, 3), (1, 3, 2)\}$ of all even permutations of S_3. This shows that S_3 is a solvable group.   □

Example 7.6.2 (S_4 is solvable).

> Let $G = S_4$, and let $N_0 = G$, $N_1 = A_4$. Since A_4 has index 2 in S_4, we must have $N_0/N_1 \cong \mathbf{Z}_2$. Let N_3 be the trivial subgroup $\{(1)\}$, and let
>
> $$N_2 = \{(1), (1, 2)(3, 4), (1, 3)(2, 4), (1, 4)(2, 3)\} \, .$$
>
> Then N_2 is a subgroup of G since it is closed under multiplication. Moreover, it is a normal subgroup of both G and N_1, since conjugating an element of N_2 by any element of G must yield an element that has the same cycle structure, and the elements of N_2 are the only permutations in S_4 that can be expressed as products of disjoint transpositions. Since $[N_1 : N_2] = 3$, we have $N_1/N_2 \cong \mathbf{Z}_3$. For all $\sigma \in N_2$, we have $\sigma^2 = (1)$, and so $N_2/N_3 \cong N_2$ is isomorphic to the Klein four-group $\mathbf{Z}_2 \times \mathbf{Z}_2$. Thus each factor group in the given descending chain of subgroups is abelian, and this shows that S_4 is a solvable group.   □

In Example 7.6.2 we could have added another term at the bottom of the descending chain of subgroups, by letting $N_3 = \{(1), (1, 2)(3, 4)\}$ and $N_4 = \{(1)\}$.

Then each factor group N_i/N_{i+1} would have been isomorphic to a cyclic group. This can always be done, as the next proposition shows (for finite groups).

7.6.2 Proposition. *Let G be a finite group. Then G is solvable if and only if there exists a finite chain of subgroups $G = N_0 \supseteq N_1 \supseteq \ldots \supseteq N_n$ such that*

   (i) N_i *is a normal subgroup in N_{i-1} for $i = 1, 2, \ldots, n$,*

   (ii) N_{i-1}/N_i *is cyclic of prime order for $i = 1, 2, \ldots, n$, and*

   (iii) $N_n = \{e\}$.

Proof. Assume that G is solvable, with a chain of subgroups $N_0 \supseteq N_1 \supseteq \ldots \supseteq N_n$ satisfying the conditions of Definition 7.6.1. By omitting all unnecessary terms, we can assume that each factor group is nontrivial. If some factor group N_i/N_{i+1} is not cyclic of prime order, then let p be a prime number such that p divides $|N_i/N_{i+1}|$. By Cauchy's theorem there exists an element aN_{i+1} of N_i/N_{i+1} of order p. Recalling that subgroups of N_i/N_{i+1} correspond to subgroups of N_i that contain N_{i+1}, we let H be the inverse image in N_i of $\langle aN_{i+1} \rangle$. Thus we have $N_i \supseteq H \supseteq N_{i+1}$, and H is a normal subgroup of N_i since $\langle aN_{i+1} \rangle$ is a normal subgroup of the abelian group N_i/N_{i+1}. Furthermore, N_{i+1} is normal in H since N_{i+1} is normal in $N_i \supseteq H$. Since N_i/H is a homomorphic image of the abelian group N_i/N_{i+1}, it is abelian, and $H/N_{i+1} \cong \langle aN_{i+1} \rangle$ is a cyclic group of order p. We next consider the descending chain of subgroups constructed from the original chain by adding H. We can apply the same procedure repeatedly, ultimately arriving at a descending chain of subgroups, each normal in the previous one, such that all factors are cyclic of prime order.

   The converse is obvious. $\square$

7.6.3 Theorem. *Let p be a prime number. Any finite p-group is solvable.*

Proof. Let G be any group of order p^m. First let C_0 be the trivial subgroup. The center $Z(G) = \{x \in G \mid xg = gx \text{ for all } g \in G\}$ of G is nontrivial, so we let $C_1 = Z(G)$. It follows from the definition of $Z(G)$ that C_1 is abelian, and also that it is normal in G. Since the factor group G/C_1 is defined, it also has nontrivial center $Z(G/C_1)$ since its order is again a power of p. Let C_2 be the subgroup of G that contains C_1 and corresponds to $Z(G/C_1)$. Since normal subgroups correspond to normal subgroups, we see that C_2 is normal in G. Furthermore, $C_2/C_1 \cong Z(G/C_1)$, and so this factor is abelian. We can continue this procedure until we obtain $C_n = G$ for some n. Then we have constructed a chain $G = C_n \supseteq \ldots \supseteq C_1 \supseteq C_0$ satisfying the conditions of Definition 7.6.1, and so G is solvable. $\square$

7.6.4 Definition. *Let G be a group. An element $g \in G$ is called a **commutator** if $g = aba^{-1}b^{-1}$ for elements $a, b \in G$.*

   *The smallest subgroup that contains all commutators of G is called the **commutator subgroup** or **derived subgroup** of G, and is denoted by G'.*

We note that the commutators themselves do not necessarily form a subgroup.

7.6.5 Proposition. *Let G be a group with commutator subgroup G'.*

(a) *The subgroup G' is normal in G, and the factor group G/G' is abelian.*

(b) *If N is any normal subgroup of G, then the factor group G/N is abelian if and only if $G' \subseteq N$.*

Proof. (a) Let $x \in G'$ and let $g \in G$. Then $gxg^{-1}x^{-1} \in G'$, and since $x \in G'$, we must have $gxg^{-1} = gxg^{-1}x^{-1}x \in G'$.

The factor group G/G' must be abelian since for any cosets aG', bG' we have

$$aG'bG'a^{-1}G'b^{-1}G' = aba^{-1}b^{-1}G' = G' .$$

Thus $aG'bG' = bG'aG'$.

(b) Let N be a normal subgroup of G. If $N \supseteq G'$, then G/N is a homomorphic image of G/G' and must be abelian. Conversely, suppose that G/N is abelian. Then $aNbN = bNaN$ for all $a, b \in G$, or simply $aba^{-1}b^{-1}N = N$, showing that every commutator of G belongs to N. This implies that $G' \subseteq N$. □

7.6.6 Definition. *Let G be a group. The subgroup $(G')'$ is called the **second derived subgroup** of G. We define $G^{(k)}$ inductively as $(G^{(k-1)})'$, and call it the **kth derived subgroup** of G.*

We note that the kth derived subgroup of G is always normal in G (see Exercise 1). In fact, it can be shown to be invariant under all automorphisms of G. Our reason for considering the commutator subgroups is to develop the following criterion for solvability.

7.6.7 Theorem. *A group G is solvable if and only if $G^{(n)} = \{e\}$ for some positive integer n.*

Proof. First assume that G is solvable and that $G = N_0 \supseteq N_1 \supseteq \ldots \supseteq N_n = \{e\}$ is a chain of subgroups such that N_i/N_{i+1} is abelian. Since G/N_1 is abelian, we have $G' \subseteq N_1$. Then we must have $G^{(2)} = (G')' \subseteq (N_1)' \subset N_2$ since N_1/N_2 is abelian. In general, $G^{(k)} \subseteq N_k$, and so $G^{(n)} = \{e\}$.

Conversely, if $G^{(n)} = \{e\}$, then in the descending chain $G \supseteq G' \supseteq \ldots \supseteq G^{(n)} = \{e\}$, each subgroup is normal in G and each factor $G^{(i)}/G^{(i+1)}$ is abelian, showing that G is solvable. □

7.6.8 Corollary. *Let G be a group.*

(a) *If G is solvable, then so is any subgroup or homomorphic image of G.*

(b) *If N is a normal subgroup of G such that both N and G/N are solvable, then G is solvable.*

Proof. (a) Assume that G is solvable, with $G^{(n)} = \{e\}$, and let H be a subgroup of G. Since $H' \subseteq G'$, it follows inductively that $H^{(n)} \subseteq G^{(n)} = \{e\}$.

To show that any homomorphic image of G is solvable, it suffices to show that G/N is solvable for any normal subgroup N of G. Commutators $aba^{-1}b^{-1}$ of G correspond directly to commutators $aNbNa^{-1}Nb^{-1}N = aba^{-1}b^{-1}N$ of G/N, and so the kth derived subgroup of G/N is the projection of the kth derived subgroup of G onto G/N. It is then obvious that G/N is solvable.

(b) Assume that N is a normal subgroup of G such that N and G/N are solvable. Then $(G/N)^{(n)} = \{eN\}$ for some positive integer n, and the correspondence between commutators of G/N and G that we observed in the proof of part (a) shows that $G^{(n)} \subseteq N$. But then $N^{(k)} = \{e\}$ for some positive integer k, and so $G^{(n+k)} = (G^{(n)})^{(k)} \subseteq N^{(k)} = \{e\}$. Thus G is solvable. $\square$

We have seen several methods of determining whether or not a given group is solvable. The methods involved sequences of subgroups, determined in various ways. This raises the question of uniqueness of such sequences. Theorem 7.6.10 shows that if we have a sequence that cannot be lengthened, then there is a certain amount of uniqueness, which we can illustrate with the following example. In $\mathbf{Z}_6$ we have the following two descending chains of subgroups:

$$\mathbf{Z}_6 \supset 3\mathbf{Z}_6 \supset \{0\} \quad \text{and} \quad \mathbf{Z}_6 \supset 2\mathbf{Z}_6 \supset \{0\} .$$

In the first chain we have $\mathbf{Z}_6/3\mathbf{Z}_6 \cong \mathbf{Z}_3$ and $3\mathbf{Z}_6 \cong \mathbf{Z}_2$. On the other hand, in the second chain we have $\mathbf{Z}_6/2\mathbf{Z}_6 \cong \mathbf{Z}_2$ and $2\mathbf{Z}_6 \cong \mathbf{Z}_3$. At least we have the same factor groups, even though they occur in a different order.

7.6.9 Definition. *Let G be a group. A chain of subgroups $G = N_0 \supseteq N_1 \supseteq \ldots \supseteq N_n$ such that*

(i) *N_i is a normal subgroup in N_{i-1} for $i = 1, 2, \ldots, n$,*

(ii) *N_{i-1}/N_i is simple for $i = 1, 2, \ldots, n$, and*

(iii) *$N_n = \{e\}$*

*is called a **composition series** for G.*

*The factor groups N_{i-1}/N_i are called the **composition factors** determined by the series. The number n is called the **length** of the series.*

Note that any finite group G has at least one composition series. Let $N_0 = G$ and then let N_1 be a maximal proper normal subgroup of G. (That is, let N_1 be a proper normal subgroup of G that is not contained in any strictly larger proper normal subgroup of G.) To continue the series, let N_2 be a maximal proper normal subgroup of N_1, etc. Since G is finite, the sequence must terminate at the trivial subgroup after at most a finite number of steps.

Example 7.6.3 (Composition series for S_4).

In Example 7.6.2 we constructed a descending chain $S_4 = N_0 \supseteq N_1 \supseteq N_2 \supseteq$ $\{(1)\}$, where $N_1 = A_4$ and N_2 is isomorphic to the Klein four-group. This is not a composition series for S_4 since N_2 is not a simple group. If we refine the series by including the subgroup $N_3 = \{(1), (1, 2)(3, 4)\}$, and if we let $N_4 = \{(1)\}$, then

$$S_4 = N_0 \supseteq N_1 \supseteq N_2 \supseteq N_3 \supseteq N_4$$

is a composition series of length 4, since $N_0/N_1 \cong \mathbf{Z}_2$, $N_1/N_2 \cong \mathbf{Z}_3$, $N_2/N_3 \cong \mathbf{Z}_2$, and $N_3/N_4 \cong \mathbf{Z}_2$. □

7.6.10 Theorem (Jordan-Hölder). *Any two composition series for a finite group have the same length. Furthermore, there exists a one-to-one correspondence between composition factors of the two composition series under which corresponding composition factors are isomorphic.*

Proof. The proof uses induction on the length of a composition series for the group G. That is, we will show that if G is a finite group with a composition series

$$G \supseteq N_1 \supseteq \ldots \supseteq N_k = \{e\}$$

of length k, then any other composition series

$$G \supseteq H_1 \supseteq \ldots \supseteq H_m = \{e\}$$

for G must have $m = k$ and there must exist a permutation $\sigma \in S_k$ such that $N_{i-1}/N_i \cong H_{\sigma(i)-1}/H_{\sigma(i)}$ for $i = 1, 2, \ldots, k$. If $k = 1$, then G must be simple and so there is only one possible composition series.

Assume that G has a composition series of length k, as above. In addition, assume that the induction hypothesis is satisfied for all groups with a composition series of length less than k, and assume that G has another composition series of length m, as above. If $N_1 = H_1$, then we can apply the induction hypothesis to the composition series

$$N_1 \supseteq N_2 \supseteq \ldots \supseteq N_k = \{e\}$$

and thus obtain the result for G. If $H_1 \neq N_1$, then let

$$N_1 \cap H_1 \supseteq K_3 \supseteq \ldots \supseteq K_n = \{e\}$$

be a composition series. This gives the diagram in Figure 7.6.1.

Since N_1 and H_1 are normal in G, so is their intersection. Furthermore, $N_1 H_1$ is normal in G, and so it must be equal to G since it contains both N_1 and H_1,

Figure 7.6.1:

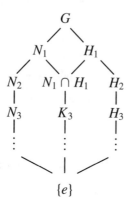

which are maximal normal subgroups. Applying the first isomorphism theorem (Theorem 7.1.1) gives us the following isomorphisms:

$$N_1 / (N_1 \cap H_1) \cong (N_1 H_1) / H_1 = G/H_1$$

and

$$H_1 / (N_1 \cap H_1) \cong (N_1 H_1) / N_1 = G/N_1 .$$

This implies that $N_1 \cap H_1$ is a maximal normal subgroup of both N_1 and H_1, so we have the following four composition series for G:

$$G \supseteq N_1 \supseteq N_2 \supseteq \ldots \supseteq N_k = \{e\} ,$$

$$G \supseteq N_1 \supseteq N_1 \cap H_1 \supseteq \ldots \supseteq K_n = \{e\} ,$$

$$G \supseteq H_1 \supseteq N_1 \cap H_1 \supseteq \ldots \supseteq K_n = \{e\} ,$$

$$G \supseteq H_1 \supseteq H_2 \supseteq \ldots \supseteq H_m = \{e\} .$$

The first two composition series have N_1 as the first term, so we must have $n = k$, and isomorphic composition factors. The last two composition series have H_1 as the first term, so again we must have $m = n$, and isomorphic composition factors. The isomorphisms given above show that the middle two composition series have isomorphic composition factors, and so by transitivity the first and last composition series have the same length and isomorphic composition factors. □

EXERCISES: SECTION 7.6

1. Let G be a group and let N be a normal subgroup of G. For $a, b \in G$, let $[a, b]$ denote the commutator $aba^{-1}b^{-1}$.

   (a) Show that $g[a, b]g^{-1} = [gag^{-1}, gbg^{-1}]$, for all $g \in G$.

   (b) Show that N' is a normal subgroup of G.

2. Prove that an abelian group has a composition series if and only if it is finite.

3.†Give an example of two groups G_1 and G_2 that have the same composition factors, but are not isomorphic.

4. Find a composition series for D_{15}.

5. Let G be the Frobenius group F_{20} of matrices of the form $\begin{bmatrix} m & n \\ 0 & 1 \end{bmatrix}$ such that $n \in \mathbf{Z}_5$ and $m \in \mathbf{Z}_5^\times$.

   (a) Find a composition series for G.

   (b) Find the descending series of commutator subgroups of G.

6. Prove that if $G_1 \times G_2 \times \cdots \times G_s \cong H_1 \times H_2 \times \cdots \times H_t$, where each of the groups G_i and H_i is a finite simple group, then $s = t$ and there exists $\sigma \in S_t$ such that $G_j \cong H_{\sigma(j)}$ for $j = 1, \ldots, t$.

7. Let p and q be primes, not necessarily distinct. A famous theorem of Burnside states that any group of order $p^n q^m$ is solvable for all $n, m \in \mathbf{Z}^+$.

   (a) Show that any group of order pq is solvable.

   (b) Show that any group of order $p^2 q$ is solvable.

   (c) Show that any group of order $p^n q$ is solvable if $p > q$.

8. Let G be a group. A subgroup H of G is called a **characteristic** subgroup if $\phi(H) \subseteq H$ for all $\phi \in \text{Aut}(G)$.

   (a) Prove that any characteristic subgroup is normal.

   (b) Prove that if H is a normal subgroup of G, and K is a characteristic subgroup of H, then K is normal in G.

   (c) Prove that the center of any group is a characteristic subgroup.

   (d) Prove that the commutator subgroup is always a characteristic subgroup.

   (e) Prove that if G is finite, then any normal Sylow p-subgroup of G is a characteristic subgroup.

   (f) Prove that the product of two characteristic subgroups is a characteristic subgroup.

9. Prove that a finite solvable group of order ≥ 2 must contain a nontrivial normal abelian subgroup.

10. Prove that if G is a finite group which is not solvable, then G must contain a nontrivial normal subgroup N such that $N' = N$.

7.7 Simple Groups

Cyclic groups of prime order form the most elementary class of simple groups. In the introduction to Chapter 7 we presented, without proof, another class of finite simple groups, constructed from the group of invertible $n \times n$ matrices over a finite field. In this section we will return to this class (in the special case $n = 2$), after considering the class of alternating groups. The fact that A_n is simple, for $n \geq 5$, will play a crucial role in Chapter 8.

7.7.1 Lemma. *If $n \geq 3$, then every permutation in A_n can be expressed as a product of 3-cycles.*

Proof. The product of any two transpositions must have one of the following forms (where different letters represent distinct positive integers):

$$
\begin{aligned}
(a, b)(a, b) &= (1) = (a, b, c)(a, b, c)(a, b, c) \,, \\
(a, b)(b, c) &= (a, b, c) \,, \\
(a, b)(c, d) &= (a, b, c)(b, c, d) \,.
\end{aligned}
$$

Since any element of A_n is a product of an even number of transpositions, this shows that any element of A_n can be expressed as a product of 3-cycles. $\square$

7.7.2 Theorem. *The symmetric group S_n is not solvable for $n \geq 5$.*

Proof. We first show that the derived subgroup $(A_n)'$ is equal to A_n. Let (a, b, c) be any 3-cycle in A_n. Since $n \geq 5$, we can choose $d, f \in \{1, 2, \ldots, n\}$ different from a, b, c. Then

$$(a, b, c) = (a, b, d)(a, c, f)(a, d, b)(a, f, c)$$

and we have shown that any 3-cycle is a commutator. Together with Lemma 7.7.1, this shows that any element of A_n is a product of commutators of elements in A_n.

Finally, we have $(S_n)' \subseteq A_n$ since the factor group S_n/A_n is abelian. In the other direction, we have $A_n = (A_n)' \subseteq (S_n)'$, and so $(S_n)' = A_n$. It follows that $(S_n)^{(k)} = A_n$ for all $k \geq 1$, and thus S_n is not solvable. $\square$

7.7.3 Lemma. *If $n \geq 4$, then no proper normal subgroup of A_n contains a 3-cycle.*

Proof. Let N be a normal subgroup of A_n that contains a 3-cycle (a, b, c). Note that N must also contain the square (a, c, b) of (a, b, c). Conjugating (a, c, b) by the even permutation $(a, b)(c, x)$, we obtain (a, b, x), which must belong to N since N is normal in A_n. By repeating this argument we can obtain any 3-cycle (x, y, z), and thus N contains all 3-cycles. By Lemma 7.7.1, we have $N = A_n$. $\square$

7.7.4 Theorem. *The alternating group A_n is simple if $n \geq 5$.*

Proof. Let N be a nontrivial normal subgroup of A_n. Using the previous lemma, we only need to show that N contains a 3-cycle. Let $\sigma \in N$, and assume that σ is written as a product of disjoint cycles. If σ is itself a 3-cycle, then we are done.

If σ contains a cycle of length ≥ 4, say $\sigma = (a, b, c, d, \ldots) \cdots$, then let $\tau = (b, c, d)$. Since N is a normal subgroup, σ^{-1} and $\tau \sigma \tau^{-1}$ must belong to N, and so $\sigma^{-1} \tau \sigma \tau^{-1} \in N$. A direct computation shows that $\sigma^{-1} \tau \sigma \tau^{-1} = (a, b, d)$, and thus N contains a 3-cycle.

If σ contains a 3-cycle but no longer cycle, then either $\sigma = (a, b, c)(d, f, g) \cdots$ or $\sigma = (a, b, c)(d, f) \cdots$. Let $\tau = (b, c, d)$. In the first case, $\sigma^{-1} \tau \sigma \tau^{-1} = (a, b, d, c, g)$, and in the second case, $\sigma^{-1} \tau \sigma \tau^{-1} = (a, b, d, c, f)$. Thus, by the previous argument, N must again contain a 3-cycle.

Finally, if σ consists of only transpositions, then either $\sigma = (a, b)(c, d)$ or $\sigma = (a, b)(c, d) \cdots$. The second case reduces to the first, since $\sigma^{-1} \tau \sigma \tau^{-1} = (a, d)(b, c)$, again using $\tau = (b, c, d)$. In the first case, since $n \geq 5$, there must be a fifth element, say f. For the permutation $\rho = (c, d, f)$, we have $\sigma^{-1} \rho \sigma \rho^{-1} = (c, d, f)$. Thus N must contain a 3-cycle, completing the proof. $\square$

We will now consider another infinite family of simple groups. Recall that for a field F the set of all invertible $n \times n$ matrices with entries in F is called the general linear group $\mathrm{GL}_n(F)$. If $A, B \in \mathrm{GL}_n(F)$, then $\det(AB) = \det(A) \det(B)$, and so the determinant defines a homomorphism from $\mathrm{GL}_n(F)$ into the multiplicative group $F^\times$ of nonzero elements of F. The kernel of this homomorphism is a normal subgroup consisting of the set of all matrices with determinant 1.

7.7.5 Definition. *Let F be a field. The set of all $n \times n$ matrices with entries in F and determinant 1 is called the **special linear group** over F, and is denoted by $\mathrm{SL}_n(F)$.*

*The group $\mathrm{SL}_n(F)$ modulo its center is called the **projective special linear group** and is denoted by $\mathrm{PSL}_n(F)$.*

7.7.6 Proposition. *For any field F, the center of $\mathrm{SL}_n(F)$ is the set of nonzero scalar matrices with determinant 1.*

Proof. Let I be the $n \times n$ identity matrix, and let e_{ij} denote the matrix having 1 as the ijth entry and zeros elsewhere. If $i \neq j$, then $I + e_{ij} \in \mathrm{SL}_n(F)$, and so for any matrix A (with entries $\{a_{ij}\}$) in the center of $\mathrm{SL}_n(F)$ we must have $(I + e_{ij})A(I + e_{ij})^{-1} = A$. Considering the iith entry of this matrix equation leads to the equation $a_{ii} + a_{ji} = a_{ii}$, and so A must be a diagonal matrix. With this assumption, considering the ijth entry of the matrix equation yields the equation $a_{jj} - a_{ii} = 0$, showing that A must be a scalar matrix. Since $A \in \mathrm{SL}_n(F)$, we must have $\det(A) = 1$. $\square$

To find a family of finite simple groups, we will examine the structure of $GL_n(F)$, for any finite field F. The proof of Proposition 7.7.6 shows that the center of $GL_n(F)$ is the set of all scalar matrices with nonzero determinant, and so it contains the center of $SL_n(F)$, which we will denote by Z. Thus Z is a normal subgroup of $GL_n(F)$, and so we have a chain of normal subgroups

$$\{I\} \subseteq Z \subseteq SL_n(F) \subseteq GL_n(F)$$

where $GL_n(F)/SL_n(F) \cong F^\times$ and Z is a finite abelian group. By Theorem 6.5.10 we know that $F^\times$ is cyclic, and Theorem 7.5.6 gives a complete description of all finite abelian groups. Thus the interesting part of a composition series for $GL_n(F)$ is the factor $SL_n(F)/Z = PSL_n(F)$. With two exceptions, when $n = 2$ and either $|F| = 2$ or $|F| = 3$, the groups $PSL_n(F)$ are simple. We will prove only a special case of this result, for $n = 2$.

To determine the order of $GL_2(F)$, we consider the number of ways in which we can construct an invertible 2×2 matrix A with entries in F. The first column of A must be nonzero, and then the second column cannot be a multiple of the first column. If $|F| = q$, then there are $q^2 - 1$ ways in which to choose a nonzero column. Next, there are q multiples of this column that we cannot choose for the second column, so the number of choices for the second column is $q^2 - q$, and thus the total number of choices for the matrix is $(q^2 - 1)(q^2 - q)$. Therefore

$$| GL_2(F) | = (q^2 - 1)(q^2 - q) .$$

Since there are $q - 1$ cosets of $SL_2(F)$ in $GL_2(F)$, by Lagrange's theorem we have

$$| SL_2(F) | = (q^2 - 1)q .$$

If $\operatorname{char}(F) = 2$, then the center Z of $SL_2(F)$ is trivial. In any other characteristic, the entries in a matrix in Z must satisfy $x^2 = 1$, and so since F contains exactly two solutions of this equation, we have $Z = \{\pm I\}$, where I is the 2×2 identity matrix.

We begin by looking at the exceptional cases $PSL_2(F)$, with $|F| = 2$ and $|F| = 3$.

Example 7.7.1 ($PSL_2(F) \cong S_3$ if $|F| = 2$).

If $|F| = 2$, then $PSL_2(F) = SL_2(F) = GL_2(F)$. We have shown in Example 3.4.5 that $GL_2(F)$ is isomorphic to the symmetric group S_3. □

Example 7.7.2 ($PSL_2(F) \cong A_4$ if $|F| = 3$).

Assume that F is a field with 3 elements. We will sketch a proof of the fact that $PSL_2(F)$ is isomorphic to the alternating group A_4, showing that it is not simple. (The details are left as an exercise.)

To simplify the notation, let $G = \mathrm{SL}_2(F)$. We first define the subgroup

$$H = \left\{ \begin{bmatrix} 1 & 0 \\ 0 & 1 \end{bmatrix}, \begin{bmatrix} 1 & 1 \\ 0 & 1 \end{bmatrix}, \begin{bmatrix} 1 & -1 \\ 0 & 1 \end{bmatrix} \right\}.$$

The product HZ of H and the center Z is a subgroup of order 6, consisting of the elements of H and their negatives. We let G act on the left cosets of HZ by defining $a(xHZ) = axHZ$, for all $a, x \in G$. (See Example 7.3.8.) This action defines a group homomorphism ϕ from G into the symmetric group S_4. It can be shown the action of each element of G produces an even permutation, and that $\ker(\phi) = Z$. Thus we have $\mathrm{PSL}_2(F) = \mathrm{SL}_2(F)/Z \cong A_4$. $\square$

7.7.7 Lemma. *Let F be any field. Then $\mathrm{SL}_2(F)$ is generated by elements of the form* $\begin{bmatrix} 1 & u \\ 0 & 1 \end{bmatrix}$ *and* $\begin{bmatrix} 1 & 0 \\ v & 1 \end{bmatrix}$.

Proof. If $a_{21} \neq 0$, then

$$\begin{bmatrix} a_{11} & a_{12} \\ a_{21} & a_{22} \end{bmatrix} = \begin{bmatrix} 1 & x \\ 0 & 1 \end{bmatrix} \begin{bmatrix} 1 & 0 \\ a_{21} & 1 \end{bmatrix} \begin{bmatrix} 1 & y \\ 0 & 1 \end{bmatrix}$$

for $x = (a_{11} - 1)a_{21}^{-1}$ and $y = (a_{22} - 1)a_{21}^{-1}$. This depends on the identity

$$y + a_{21}xy + x = a_{12},$$

which follows from the computation

$$\begin{aligned} a_{21}(y + a_{21}xy + x) &= (a_{22} - 1) + (a_{11} - 1)(a_{22} - 1) + (a_{11} - 1) \\ &= a_{11}a_{22} - 1 = a_{21}a_{12}. \end{aligned}$$

Similarly, if $a_{12} \neq 0$, then for $x = (a_{11} - 1)a_{12}^{-1}$ and $y = (a_{22} - 1)a_{12}^{-1}$ we have

$$\begin{bmatrix} a_{11} & a_{12} \\ a_{21} & a_{22} \end{bmatrix} = \begin{bmatrix} 1 & 0 \\ y & 1 \end{bmatrix} \begin{bmatrix} 1 & a_{12} \\ 0 & 1 \end{bmatrix} \begin{bmatrix} 1 & 0 \\ x & 1 \end{bmatrix}.$$

If $a_{21} = a_{12} = 0$, then $a_{11} \neq 0$, and

$$\begin{bmatrix} 1 & 0 \\ 1 & 1 \end{bmatrix} \begin{bmatrix} a_{11} & 0 \\ 0 & a_{22} \end{bmatrix} = \begin{bmatrix} a_{11} & 0 \\ a_{11} & a_{22} \end{bmatrix},$$

so it follows from the first case that

$$\begin{bmatrix} a_{11} & 0 \\ 0 & a_{22} \end{bmatrix} = \begin{bmatrix} 1 & 0 \\ -1 & 1 \end{bmatrix} \begin{bmatrix} 1 & x \\ 0 & 1 \end{bmatrix} \begin{bmatrix} 1 & 0 \\ a_{11} & 1 \end{bmatrix} \begin{bmatrix} 1 & y \\ 0 & 1 \end{bmatrix}$$

for $x = (a_{11} - 1)a_{11}^{-1}$ and $y = (a_{22} - 1)a_{11}^{-1}$. $\square$

7.7.8 Lemma. *Let F be any finite field, and let N be a normal subgroup of $\mathrm{SL}_2(F)$. If N contains an element of the form $\begin{bmatrix} 1 & a \\ 0 & 1 \end{bmatrix}$ with $a \neq 0$, then $N = \mathrm{SL}_2(F)$.*

Proof. Assume that F is a finite field with $|F| = q$, and that N is a normal subgroup of $\mathrm{SL}_2(F)$. We let

$$U = \left\{ \begin{bmatrix} 1 & u \\ 0 & 1 \end{bmatrix} \, \middle| \, u \in F \right\} \quad \text{and} \quad V = \left\{ \begin{bmatrix} 1 & 0 \\ v & 1 \end{bmatrix} \, \middle| \, v \in F \right\}$$

and assume that N contains a matrix $\begin{bmatrix} 1 & a \\ 0 & 1 \end{bmatrix}$ in the subgroup U with $a \neq 0$. Note that $|U| = |V| = q$. We will use the following equalities to show that N contains a set of generators for $\mathrm{SL}_2(F)$.

$$\begin{bmatrix} 1 & x^2 a \\ 0 & 1 \end{bmatrix} = \begin{bmatrix} x & 0 \\ 0 & x^{-1} \end{bmatrix} \begin{bmatrix} 1 & a \\ 0 & 1 \end{bmatrix} \begin{bmatrix} x & 0 \\ 0 & x^{-1} \end{bmatrix}^{-1}$$

$$\begin{bmatrix} 1 & 0 \\ -x^2 a & 1 \end{bmatrix} = \begin{bmatrix} 0 & -x^{-1} \\ x & 0 \end{bmatrix} \begin{bmatrix} 1 & a \\ 0 & 1 \end{bmatrix} \begin{bmatrix} 0 & -x^{-1} \\ x & 0 \end{bmatrix}^{-1}$$

The mapping $\phi : F^\times \to F^\times$ defined by $\phi(x) = x^2$ is a group homomorphism, and $\ker(\phi)$ consists of the solutions in F of the equation $x^2 = 1$. Thus either $\ker(\phi) = \{1\}$ and every element of F is a square, or $\ker(\phi) = \{\pm 1\}$ and exactly half of the nonzero elements of F are squares. Since $a \neq 0$, the set $\{x^2 a \mid x \in F^\times\}$ has either $q - 1$ or $(q - 1)/2$ elements. Thus $U \cap N$ has at least $1 + (q - 1)/2$ elements, so $U \cap N = U$ since $|U \cap N| > |U|/2$. We conclude that $U \subseteq N$, and a similar argument shows that $V \subseteq N$, so $N = \mathrm{SL}_2(F)$ by Lemma 7.7.7. $\quad\square$

Before proving the next theorem, we note one of the facts that we will use. If N is a normal subgroup of the group G and $a \in N$, then $xax^{-1}a^{-1} \in N$ for all $x \in G$.

7.7.9 Theorem. *Let F be any finite field with $|F| > 3$. Then the projective special linear group $\mathrm{PSL}_2(F)$ is a simple group.*

Proof. Let F be any finite field with $|F| > 3$. Any normal subgroup of $\mathrm{PSL}_2(F)$ corresponds to a normal subgroup of $\mathrm{SL}_2(F)$ that contains the center Z of $\mathrm{SL}_2(F)$, so it suffices to prove the following assertion: if N is a normal subgroup of $\mathrm{SL}_2(F)$ that properly contains Z, then $N = \mathrm{SL}_2(F)$. Assume that the normal subgroup N is given, and that the matrix $A = \begin{bmatrix} a_{11} & a_{12} \\ a_{21} & a_{22} \end{bmatrix}$ belongs to $N - Z$. We will show that N contains a matrix $\begin{bmatrix} 1 & a \\ 0 & 1 \end{bmatrix}$, with $a \neq 0$, and then it follows from Lemma 7.7.8 that $N = \mathrm{SL}_2(F)$.

First assume that $a_{21} = 0$. Then $a_{11}a_{22} = 1$, and the following matrix belongs to N.

$$\begin{bmatrix} 1 & 1-a_{11}^2 \\ 0 & 1 \end{bmatrix} = \begin{bmatrix} 1 & 1 \\ 0 & 1 \end{bmatrix} \begin{bmatrix} a_{11} & a_{12} \\ 0 & a_{22} \end{bmatrix} \begin{bmatrix} 1 & 1 \\ 0 & 1 \end{bmatrix}^{-1} \begin{bmatrix} a_{11} & a_{12} \\ 0 & a_{22} \end{bmatrix}^{-1}$$

If $1 - a_{11}^2 \neq 0$, then we are done. If not, there are two possibilities. If $a_{11} = 1$, then $a_{22} = 1$ and $a_{12} \neq 0$ since $A \notin Z$, and thus A itself has the required form. If $a_{11} = -1 \neq 1$, then $\mathrm{char}(F) \neq 2$, $a_{22} = -1$, and $a_{12} \neq 0$. In this case A^2 has the required form since $-2a_{12} \neq 0$.

Now assume that $a_{21} \neq 0$.

$$\begin{bmatrix} 1 & x \\ 0 & 1 \end{bmatrix} \begin{bmatrix} a_{11} & a_{12} \\ a_{21} & a_{22} \end{bmatrix} \begin{bmatrix} 1 & x \\ 0 & 1 \end{bmatrix}^{-1} = \begin{bmatrix} a_{11} + a_{21}x & * \\ a_{21} & -a_{21}x + a_{22} \end{bmatrix}$$

If we choose $x = -a_{21}^{-1}a_{11}$, then $a_{11} + a_{21}x = 0$, and we have constructed a matrix $B \in N$ with $B = \begin{bmatrix} 0 & b_{12} \\ b_{21} & b_{22} \end{bmatrix}$, where $b_{21} = a_{21} \neq 0$, $b_{12} = -a_{21}^{-1}$, and $b_{22} = a_{11} + a_{22}$. Then the following matrix belongs to N.

$$\begin{bmatrix} x^2 & (1-x^2)b_{12}b_{22} \\ 0 & x^{-2} \end{bmatrix} = \begin{bmatrix} x & 0 \\ 0 & x^{-1} \end{bmatrix} \begin{bmatrix} 0 & b_{12} \\ b_{21} & b_{22} \end{bmatrix}^{-1} \begin{bmatrix} x & 0 \\ 0 & x^{-1} \end{bmatrix}^{-1} \begin{bmatrix} 0 & b_{12} \\ b_{21} & b_{22} \end{bmatrix}$$

If this matrix is not in Z, then we have reduced the problem to the first case. We only need to know that F contains an element x such that $x^2 \neq \pm 1$. If $|F| > 5$, then F contains at most 4 solutions of $x^4 = 1$, and we are done. If $|F| = 4$, then by Theorem 6.5.2 the nonzero elements of F are roots of $x^3 - 1$, and so again we can find an element x that does not satisfy $x^4 = 1$. This leaves only the case $|F| = 5$.

In the exceptional case $F = \mathrm{GF}(5)$, if $b_{22} = 0$, then since $b_{21}b_{12} = -1$ we have

$$\begin{bmatrix} 1 & 0 \\ 2b_{21} & 1 \end{bmatrix} \begin{bmatrix} 0 & b_{12} \\ b_{21} & 0 \end{bmatrix} \begin{bmatrix} 1 & 0 \\ 2b_{21} & 1 \end{bmatrix}^{-1} = \begin{bmatrix} 2 & b_{12} \\ 0 & -2 \end{bmatrix}.$$

This reduces the problem to the first case of the general proof. If $b_{22} \neq 0$ then we can choose $x = 2$, to obtain the matrix

$$\begin{bmatrix} x^2 & (1-x^2)b_{12}b_{22} \\ 0 & x^{-2} \end{bmatrix} = \begin{bmatrix} -1 & 2b_{12}b_{22} \\ 0 & -1 \end{bmatrix},$$

whose square $\begin{bmatrix} 1 & b_{12}b_{22} \\ 0 & 1 \end{bmatrix}$ is in N and has the required form. This completes the proof. $\square$

EXERCISES: SECTION 7.7

1. Let G be a group of order $2m$, where m is odd. Show that G is not simple.

2.†Find all normal subgroups of S_n, for $n \geq 5$.

3. (a) Let G be a group with a subgroup H of index n. Show that there is a homomorphism $\phi : G \to S_n$ for which $\ker(\phi) = \cap_{g \in G} gHg^{-1}$.

   Hint: Let G act on the set of left cosets of H by defining $a(gH) = agH$, for all $a, g \in G$.

   (b) Prove that if G is a simple group that contains a subgroup of index n, then G can be embedded in S_n.

   (c) Prove that if an infinite group contains a subgroup of finite index, then it is not simple.

4. (a) Let S be the set $\{1, 2, \ldots\}$ of positive integers, and let

   $$G = \{\sigma \in \text{Sym}(S) \mid \sigma(j) = j \text{ for all but finitely many } j\}.$$

   Show that G is a subgroup of $\text{Sym}(S)$.

   (b) Let A_∞ be the subgroup of G generated by all 3-cycles (a, b, c). Show that A_∞ is a simple group.

   (c) Show that every finite group can be embedded in A_∞.

5. Show that A_∞ is the only proper nontrivial normal subgroup of the group G defined in Exercise 4.

6. Let G be a finite group containing a subgroup H of index p, where p is the smallest prime divisor of $|G|$. Prove that H is normal in G, and hence G is not simple if $|G| > p$.

7. Let H be a subgroup of G and let S be the set of all subgroups conjugate to H. Define an action of G on S by letting $a(gHg^{-1}) = (ag)H(ag)^{-1}$, for $a, g \in G$. Show that the corresponding homomorphism $\phi : G \to \text{Sym}(S)$ has kernel $\cap_{g \in G} gN(H)g^{-1}$.

8. Let G be an infinite group containing an element (not equal to the identity) which has only finitely many conjugates. Use Exercise 7 to prove that G is not simple.

9. Prove that (up to isomorphism) the only simple group of order 60 is A_5.

10. Prove that if F is a finite field with $|F| > 3$, then the group $SL_2(F)$ is equal to its commutator subgroup.

11.†Let F be a finite field, with $|F| = q$. Determine the orders of the groups $GL_n(F)$, $SL_n(F)$, and $PSL_n(F)$.

12. Let F be a finite field, with $\text{char}(F) = p$. Show that $\left\{ \begin{bmatrix} 1 & x \\ 0 & 1 \end{bmatrix} \,\middle|\, x \in F \right\}$ is a Sylow p-subgroup of $SL_2(F)$.

13. Provide the details in Example 7.7.2 to complete the proof that if $|F| = 3$, then $\mathrm{PSL}_2(F)$ is isomorphic to the alternating group A_4.

14. Prove that if F is a field with $|F| = 5$, then $\mathrm{PSL}_2(F)$ is isomorphic to the alternating group A_5.

Chapter 8

GALOIS THEORY

A polynomial equation with real coefficients can be solved by radicals, provided it has degree less than five. The solutions for cubic and quartic equations were discovered in the sixteenth century (see the notes in Chapter 4). From that time until the beginning of the nineteenth century, some of the best mathematicians of the period (such as Euler and Lagrange) attempted to find a similar solution by radicals for equations of degree five. All attempts ended in failure, and finally a paper was published by Ruffini in 1798, in which he asserted that no method of solution could be found. The proof was not well received at the time, and even further elaborations of the ideas, published in 1802 and 1813, were not regarded as constituting a proof. Ruffini's proof is based on the assumption that the radicals necessary in the solution of the quintic can all be expressed as rational functions of the roots of the equation, and this was only proved later by Abel. It is generally agreed that the first fully correct proof of the insolvability of the quintic was published by Abel in 1826.

Abel attacked the general problem of when a polynomial equation could be solved by radicals. His papers inspired Galois to formulate a theory of solvability of equations involving the structures we now know as groups and fields. Galois worked with fields to which all roots of the given equation had been adjoined. He then considered the set of all permutations of these roots that leave the coefficient field unchanged. The permutations form a group, called the Galois group of the equation. From the modern point of view, the permutations of the roots can be extended to automorphisms of the field, and form a group under composition of functions. Then an equation is solvable by radicals if and only if its Galois group is solvable. Thus to show that there exists an equation that is not solvable by radicals, it is enough to find an equation whose Galois group is S_5.

In 1829, at the age of seventeen, Galois presented two papers on the solution of algebraic equations to the Académie des Sciences de Paris. Both were sent to Cauchy, who lost them. In 1830, Galois presented another paper to the Académie, which was this time given to Joseph Fourier (1768–1830), who died before reading it. A third, revised version, was submitted in 1831. This manuscript was reviewed carefully, but was not understood. It was only published in 1846 by Liouville, fourteen years after the death of Galois. Galois had been involved in the 1830 revolution and had been imprisoned twice. In May of 1832 he was forced to accept a duel (the circumstances and motives are unclear), and certain that he would be killed, he spent the night before the duel writing a long letter to his friend Auguste Chevalier (1809–1868) explaining the basic ideas of his research. This work involved other areas in addition to what we now call Galois theory, and included the construction of finite fields, which are now called Galois fields (see Section 6.5).

In our modern terminology, we let F be the splitting field of a polynomial over the field K that has no repeated roots. The fundamental theorem of Galois theory (in Section 8.3) shows that there is a correspondence between the normal subgroups of the Galois group and the intermediate fields between K and F that are splitting fields for some polynomial. This gives the necessary connection between the structure of the Galois group and the successive adjunctions that must be made in order to solve the equation by radicals.

8.1   The Galois Group of a Polynomial

We begin by reviewing some facts about automorphisms. Recall that an automorphism ϕ of a field F is a one-to-one correspondence $\phi : F \to F$ such that $\phi(1) = 1$, and for all $a, b \in F$,

$$\phi(a + b) = \phi(a) + \phi(b) \qquad \text{and} \qquad \phi(ab) = \phi(a)\phi(b) .$$

That is, ϕ is an automorphism of the additive group of F, and since $a \neq 0$ implies $\phi(a) \neq 0$, it is also an automorphism when restricted to the multiplicative group $F^\times$. We use the notation $\text{Aut}(F)$ for the group of all automorphisms of F.

Recall that the smallest subfield containing the identity element 1 is called the *prime subfield* of F. If F has characteristic zero, then the prime subfield is isomorphic to $\mathbf{Q}$, and consists of all elements of the form

$$\{(n \cdot 1)(m \cdot 1)^{-1} \mid n, m \in \mathbf{Z}, m \neq 0\} .$$

If F has characteristic p (p a prime), then the prime subfield is isomorphic to $\text{GF}(p)$, and consists of all elements of the form

$$\{n \cdot 1 \mid n = 0, 1, \ldots, p - 1\} .$$

For any automorphism ϕ of F, we have

$$\phi(n \cdot 1) = n \cdot \phi(1) = n \cdot 1$$

for any $n \in \mathbf{Z}$. Furthermore, if $n \cdot 1 \neq 0$, then

$$\phi((n \cdot 1)^{-1}) = (\phi(n \cdot 1))^{-1} = (n \cdot 1)^{-1} .$$

This shows that any automorphism of F must leave the prime subfield of F fixed.

To study solvability by radicals of a polynomial equation $f(x) = 0$, we let K be the field generated by the coefficients of $f(x)$, and let F be a splitting field for $f(x)$ over K. (We know, by Kronecker's theorem, that we can always find roots. The question is whether or not the roots have a particular form.) Galois considered permutations of the roots that leave the coefficient field fixed. The modern approach is to consider the automorphisms determined by these permutations.

8.1.1 Proposition. *Let F be an extension field of K. The set of all automorphisms $\phi : F \to F$ such that $\phi(a) = a$ for all $a \in K$ is a group under composition of functions.*

Proof. We only need to show that the given set is a subgroup of $\mathrm{Aut}(F)$. It certainly contains the identity function. If $\phi, \theta \in \mathrm{Aut}(F)$ and $\phi(a) = a$, $\theta(a) = a$ for all $a \in K$, then $\phi\theta(a) = \phi(\theta(a)) = \phi(a) = a$ and $\phi^{-1}(a) = \phi^{-1}\phi(a) = a$ for all $a \in K$. $\square$

8.1.2 Definition. *Let F be an extension field of K. The set*

$$\{\theta \in \mathrm{Aut}(F) \mid \theta(a) = a \ \text{for all} \ a \in K\}$$

*is called the **Galois group** of F over K, denoted by $\mathrm{Gal}(F/K)$.*

8.1.3 Definition. *Let K be a field, let $f(x) \in K[x]$, and let F be a splitting field for $f(x)$ over K. Then $\mathrm{Gal}(F/K)$ is called the **Galois group of** $f(x)$ **over** K, or the **Galois group of the equation** $f(x) = 0$ **over** K.*

8.1.4 Proposition. *Let F be an extension field of K, and let $f(x) \in K[x]$. Then any element of $\mathrm{Gal}(F/K)$ defines a permutation of the roots of $f(x)$ that lie in F.*

Proof. Let $f(x) = a_0 + a_1 x + \ldots + a_n x^n$, where $a_i \in K$ for $i = 0, 1, \ldots, n$. If $u \in F$ with $f(u) = 0$ and $\theta \in \mathrm{Gal}(F/K)$, then we have

$$
\begin{aligned}
\theta(f(u)) &= \theta(a_0 + a_1 u + \ldots + a_n u^n) \\
&= \theta(a_0) + \theta(a_1 u) + \ldots + \theta(a_n u^n) \\
&= \theta(a_0) + \theta(a_1)\theta(u) + \ldots + \theta(a_n)(\theta(u))^n
\end{aligned}
$$

since θ preserves sums and products. Finally, since $\theta(a_i) = a_i$ for $i = 0, 1, \ldots, n$, we have

$$\theta(f(u)) = a_0 + a_1\theta(u) + \ldots + a_n(\theta(u))^n \ .$$

Since $f(u) = 0$, we must have $\theta(f(u)) = 0$, and thus

$$a_0 + a_1\theta(u) + \ldots + a_n(\theta(u))^n = 0 \ ,$$

showing that $f(\theta(u)) = 0$.

Thus θ maps roots of $f(x)$ to roots of $f(x)$. Since there are only finitely many roots and θ is one-to-one, θ must define a permutation of those roots of $f(x)$ that lie in F. $\square$

At this point in the development of the theory we have enough information to do some easy calculations. These computations will help you understand the definitions, but you should be aware that later results such as Proposition 8.4.2 and Theorem 8.4.3 provide much more powerful techniques, so that it is not always necessary to do the detailed analysis of automorphisms that is presented in the following examples.

Example 8.1.1 ($\mathrm{Gal}(\mathbf{Q}(\sqrt[3]{2}\,)/\mathbf{Q})$ **is trivial**).

In general, if $F = K(u_1, \ldots, u_n)$, then any automorphism $\theta \in \mathrm{Gal}(F/K)$ is completely determined by the values $\theta(u_1), \ldots, \theta(u_n)$. If we let $\theta \in \mathrm{Gal}(\mathbf{Q}(\sqrt[3]{2})/\mathbf{Q})$, then θ is completely determined by $\theta(\sqrt[3]{2})$. Now $\sqrt[3]{2}$ is a root of the polynomial $x^3 - 2 \in \mathbf{Q}[x]$, so by Proposition 8.1.4, θ must map $\sqrt[3]{2}$ into a root of $x^3 - 2$. The other two roots are $\omega\sqrt[3]{2}$ and $\omega^2\sqrt[3]{2}$, where $\omega = (-1 + \sqrt{3}i)/2$ is a complex cube root of unity. Since these values are not real numbers, they cannot belong to $\mathbf{Q}(\sqrt[3]{2})$, and so we must have $\theta(\sqrt[3]{2}) = \sqrt[3]{2}$. This shows that $\mathrm{Gal}(\mathbf{Q}(\sqrt[3]{2})/\mathbf{Q})$ is the trivial group consisting only of the identity automorphism. Note also that any automorphism of $\mathbf{Q}(\sqrt[3]{2})$ fixes $\mathbf{Q}$, so in fact $\mathrm{Aut}(\mathbf{Q}(\sqrt[3]{2}))$ is also trivial. $\square$

Example 8.1.2 ($\mathrm{Gal}(\mathbf{Q}(\sqrt{2} + \sqrt{3})/\mathbf{Q}) \cong \mathbf{Z}_2 \times \mathbf{Z}_2$).

The field extension $\mathbf{Q}(\sqrt{2} + \sqrt{3})$ is easier to work with when expressed as $\mathbf{Q}(\sqrt{2}, \sqrt{3})$. (We showed in Example 6.2.2 that these two fields are the same.)

Let θ be any automorphism in $\mathrm{Gal}(\mathbf{Q}(\sqrt{2}, \sqrt{3})/\mathbf{Q})$. The roots of $x^2 - 2$ are $\pm\sqrt{2}$ and the roots of $x^2 - 3$ are $\pm\sqrt{3}$, and so we must have $\theta(\sqrt{2}) = \pm\sqrt{2}$ and $\theta(\sqrt{3}) = \pm\sqrt{3}$. Since $\{1, \sqrt{3}\}$ is a basis for $\mathbf{Q}(\sqrt{3})$ over $\mathbf{Q}$ and $\{1, \sqrt{2}\}$ is a basis for $\mathbf{Q}(\sqrt{2}, \sqrt{3})$ over $\mathbf{Q}(\sqrt{3})$, recall that $\{1, \sqrt{2}, \sqrt{3}, \sqrt{2}\sqrt{3}\}$ is a basis for $\mathbf{Q}(\sqrt{2}, \sqrt{3})$ over $\mathbf{Q}$. As soon as we know the action of θ on $\sqrt{2}$ and $\sqrt{3}$, its action on $\sqrt{6} = \sqrt{2}\sqrt{3}$ is determined. This gives a total of four possibilities for θ, which we have labeled with subscripts:

$$
\begin{aligned}
\theta_1(a + b\sqrt{2} + c\sqrt{3} + d\sqrt{6}) &= a + b\sqrt{2} + c\sqrt{3} + d\sqrt{6}\,, \\
\theta_2(a + b\sqrt{2} + c\sqrt{3} + d\sqrt{6}) &= a - b\sqrt{2} + c\sqrt{3} - d\sqrt{6}\,, \\
\theta_3(a + b\sqrt{2} + c\sqrt{3} + d\sqrt{6}) &= a + b\sqrt{2} - c\sqrt{3} - d\sqrt{6}\,, \\
\theta_4(a + b\sqrt{2} + c\sqrt{3} + d\sqrt{6}) &= a - b\sqrt{2} - c\sqrt{3} + d\sqrt{6}\,.
\end{aligned}
$$

It is left as an exercise to show that each of these functions defines an automorphism. In each case, repeating the automorphism gives the identity mapping. Thus $\theta^2 = e$ for all $\theta \in \text{Gal}(\mathbf{Q}(\sqrt{2}, \sqrt{3})/\mathbf{Q})$, which shows that this Galois group must be isomorphic to the Klein four-group $\mathbf{Z}_2 \times \mathbf{Z}_2$.

We note that $\mathbf{Q}(\sqrt{2}, \sqrt{3})$ is the splitting field of $(x^2 - 2)(x^2 - 3)$ over $\mathbf{Q}$. Theorem 8.1.6 will show that the number of elements in the Galois group must be equal to $[\mathbf{Q}(\sqrt{2}, \sqrt{3}) : \mathbf{Q}]$. This degree is 4, as was shown in Example 6.1.3. Since we found at most four distinct mappings that carry roots to roots, they must in fact all be automorphisms. Thus if we utilize Theorem 8.1.6, we do not actually have to go through the details of showing that the mappings preserve addition and multiplication. □

8.1.5 Lemma. *Let $f(x) \in K[x]$ be a polynomial of positive degree such that each irreducible factor has no repeated roots, and let F be a splitting field for $f(x)$ over K. If $\phi : K \to L$ is a field isomorphism that maps $f(x)$ to $g(x) \in L[x]$ and E is a splitting field for $g(x)$ over L, then there exist exactly $[F : K]$ isomorphisms $\theta : F \to E$ such that $\theta(a) = \phi(a)$ for all $a \in K$.*

Proof. The proof uses induction on the degree of $f(x)$. We follow the proof of Lemma 6.4.4 exactly, except that now we keep careful track of the number of possible isomorphisms. If $f(x)$ has degree 1, then $F = K$ and $E = L$, so there is nothing to prove. We now assume that the result holds for all polynomials of degree less than n and for all fields K. Let $p(x)$ be an irreducible factor of $f(x)$ of degree d, which maps to the irreducible factor $q(x)$ of $g(x)$. All roots of $p(x)$ belong to F, so we may choose one, say u, which gives $K \subseteq K(u) \subseteq F$. Since $f(x)$ has no repeated roots, the same is true of $q(x)$, so we may choose any one, say v, of the d roots of $q(x)$ in E, which gives $L \subseteq L(v) \subseteq E$. By Lemma 6.4.3 there exist d isomorphisms $\phi' : K(u) \to L(v)$ (one for each root v) such that $\phi'(u) = v$ and $\phi'(a) = \phi(a)$ for all $a \in K$. If we write $f(x) = (x - u)s(x)$ and $g(x) = (x - v)t(x)$, then the polynomial $s(x)$ has degree less than n, the extension F is a splitting field for $s(x)$ over $K(u)$, and the extension E is a splitting field for $t(x)$ over $L(v)$. Thus the induction hypothesis can be applied, and so there exist $[F : K(u)]$ isomorphisms $\theta : F \to E$ such that $\theta(x) = \phi'(x)$ for all $x \in K(u)$. In particular, $\theta(a) = \phi'(a) = \phi(a)$ for all $a \in K$. Thus we have precisely $[F : K] = [F : K(u)][K(u) : K]$ extensions of the original isomorphism ϕ, and the proof is complete. □

8.1.6 Theorem. *Let K be a field, let $f(x)$ be a polynomial of positive degree in $K[x]$, and let F be a splitting field for $f(x)$ over K. If no irreducible factor of $f(x)$ has repeated roots, then*

$$|\text{Gal}(F/K)| = [F : K] .$$

Proof. This is an immediate consequence of the preceding lemma.   □

Lemma 6.5.4 shows that if F is a field of characteristic p, then $(a+b)^p = a^p + b^p$, for all $a, b \in F$. Thus the mapping $\phi : F \to F$ defined by $\phi(x) = x^p$, for all $x \in F$, is a ring homomorphism, since it is clear that ϕ also respects multiplication and maps 1 to 1. Since F is a field, we have $\ker(\phi) = (0)$. If F is finite, then ϕ must also be onto, and thus ϕ is an automorphism of F.

8.1.7 Definition. *Let F be a finite field with* $\text{char}(F) = p$. *The automorphism* $\phi : F \to F$ *defined by* $\phi(x) = x^p$, *for all* $x \in F$, *is called the* **Frobenius automorphism** *of F.*

8.1.8 Theorem. *Let K be a finite field with* $|K| = p^r$, *where* $p = \text{char}(F)$, *let F be an extension field of K with* $[F : K] = m$, *and let ϕ be the Frobenius automorphism of F. Then* $\text{Gal}(F/K)$ *is a cyclic group of order m, generated by* ϕ^r.

Proof. Since F is a finite extension of K of degree m, it has $(p^r)^m$ elements. If we let $n = rm$, then it follows from Theorem 6.5.2 that F is the splitting field of the polynomial $x^{p^n} - x$ over its prime subfield, and hence over K. Since $f(x)$ has no repeated roots, we may apply Theorem 8.1.6 to conclude that $|\text{Gal}(F/K)| = m$.

Define $\theta : F \to F$ by $\theta(x) = x^{p^r}$. We note that $\theta = \phi^r$, and so θ is an automorphism of F. Furthermore, $\theta \in \text{Gal}(F/K)$ since K is the splitting field of $x^{p^r} - x$ over its prime subfield, and thus θ leaves the elements of K fixed. To compute the order of θ in $\text{Gal}(F/K)$, we first note that θ^m is the identity automorphism since $\theta^m(x) = x^{p^{rm}} = x^{p^n} = x$ for all $x \in F$. Furthermore, θ^s does not equal the identity for any $1 \le s < m$, since this would imply that $x^{p^{rs}} = x$ for all $x \in F$, and this equation cannot have p^n roots. Thus $\theta = \phi^r$ is a generator for $\text{Gal}(F/K)$.   □

Example 8.1.3 ($\text{Gal}(\text{GF}(2^4)/\text{GF}(2^2)) \cong \mathbf{Z}_2$).

Since $[\text{GF}(2^4) : \text{GF}(2^2)] = 2$, it follows immediately from Theorem 8.1.8 that $\text{Gal}(\text{GF}(2^4)/\text{GF}(2^2)) \cong \mathbf{Z}_2$. The Galois group is generated by ϕ^2, where ϕ is the Frobenius automorphism of $\text{GF}(2^4)$. Thus the generator is defined by $\phi^2(x) = x^4$, for all $x \in \text{GF}(2^4)$.

It is interesting to look at this result in the context of a particular description of $\text{GF}(2^4)$. Consider the field $\text{GF}(2)[x]/ < x^4 + x + 1 >$. (Exercise 12 of Section 4.2 shows that $x^4 + x + 1$ is irreducible over $\text{GF}(2)$.) If we let $u = x + < x^4 + x + 1 >$, then Exercise 3 of Section 6.5 shows that u is a generator for the multiplicative group of the field. The four element subfield is the set of elements left fixed by ϕ^2, so it is the set of elements $a \in \text{GF}(2^4)$ satisfying $a^4 = a$. This includes 0, 1, and the elements of order 3. Since $u^5 = u \cdot u^4 = u^2 + u$ and $u^{10} = (u^2 + u)^2 = u^4 + u^2 = u^2 + u + 1$, we have $\{0, 1, u^2 + u, u^2 + u + 1\} \cong \text{GF}(2^2)$.   □

The first step in computing the Galois group of a polynomial $f(x)$ over a field K is to find the splitting field for the polynomial. If $f(x)$ has no repeated roots, then Theorem 8.1.6 shows that the order of the Galois group is the same as the degree of the splitting field over K. In Section 8.2 we will address the question of repeated roots, and we will show that there is nothing to worry about over finite fields or fields of characteristic zero, since in these cases no irreducible polynomial has repeated roots.

Our next goal is a deeper study of splitting fields. In Section 8.3 we will be able to give the following characterization.

The following conditions are equivalent for an extension field F of a field K:

(1) F is the splitting field over K of a polynomial with no repeated roots;

(2) there is a finite group G of automorphisms of F such that $a \in K$ if and only if $\theta(a) = a$ for all $\theta \in G$;

(3) F is a finite extension of K; the minimal polynomial over K of any element in F has no repeated roots; and if $p(x) \in K[x]$ is irreducible and has a root in F, then it splits over F.

If F is the splitting field of a polynomial $f(x)$ over K, then certain subfields between K and F may be splitting fields for other polynomials. The fundamental theorem of Galois theory (Theorem 8.3.8) explains the connection between subfields of F and subgroups of $\mathrm{Gal}(F/K)$. It shows that the intermediate splitting fields correspond to normal subgroups, and this plays a crucial role in Section 8.4, where we will investigate the connection between solvability by radicals and solvable groups.

EXERCISES: SECTION 8.1

1. Show that $\mathrm{Gal}(\mathrm{GF}(2^2)/\mathrm{GF}(2)) \cong \mathbf{Z}_2$.

2.† Find a basis for $\mathrm{GF}(2^3)$, and then write out an explicit formula for each of the elements of $\mathrm{Gal}(\mathrm{GF}(2^3)/\mathrm{GF}(2))$.

3. Prove by a direct computation that the function θ_2 in Example 8.1.2 preserves products.

4.† In Example 8.1.2, find $\{x \in \mathbf{Q}(\sqrt{2} + \sqrt{3}) \mid \theta_2(x) = x\}$ and show that it is a subfield of $\mathbf{Q}(\sqrt{2} + \sqrt{3})$.

5. Show that the Galois group of $x^3 - 1$ over $\mathbf{Q}$ is cyclic of order 2.

6. Show that the Galois group of $(x^2 - 2)(x^2 + 2)$ over $\mathbf{Q}$ is isomorphic to $\mathbf{Z}_2 \times \mathbf{Z}_2$.

7. Let E and F be two splitting fields of a polynomial over the field K. We already know that $E \cong F$. Prove that $\mathrm{Gal}(E/K) \cong \mathrm{Gal}(F/K)$.

8.2   Multiplicity of Roots

In the previous section, we showed that the order of the Galois group of a polynomial with no repeated roots is equal to the degree of its splitting field over the base field. If we want to compute the Galois group of a polynomial $f(x)$ over the field K, we can factor $f(x)$ into a product

$$f(x) = p_1(x)^{\alpha_1} p_2(x)^{\alpha_2} \cdots p_n(x)^{\alpha_n}$$

of distinct irreducible factors. The splitting field F of $f(x)$ over K is the same as the splitting field of

$$g(x) = p_1(x) p_2(x) \cdots p_n(x)$$

over K. Note that distinct irreducible polynomials cannot have roots in common. (If $\gcd(p(x), q(x)) = 1$, then there exist $a(x), b(x)$ such that $a(x)p(x) + b(x)q(x) = 1$, showing that there can be no common roots.) Thus $g(x)$ has no repeated roots if and only if each of its irreducible factors has no repeated roots. We will show in this section that over fields of characteristic zero, and over finite fields, irreducible polynomials have no repeated roots. It follows that in these two situations, when computing a Galois group we can always reduce to the case of a polynomial with no repeated roots. The first thing that we need to do in this section is to develop methods to determine whether or not a polynomial has repeated roots.

8.2.1 Definition. *Let $f(x)$ be a polynomial in $K[x]$, and let F be a splitting field for $f(x)$ over K. If $f(x)$ has the factorization*

$$f(x) = (x - r_1)^{m_1} (x - r_2)^{m_2} \cdots (x - r_t)^{m_t}$$

*over F, then we say that the root r_i has **multiplicity** m_i.*
    *If $m_i = 1$, then r_i is called a **simple** root.*

   In Proposition 4.2.11 we showed that a polynomial $f(x)$ over **R** has no repeated factors if and only if it has no factors in common with its derivative $f'(x)$. We can extend this result to polynomials over any field and use it to check for multiple roots.

8.2.2 Definition. *Let $f(x) \in K[x]$, with $f(x) = \sum_{k=0}^{t} a_k x^k$. The **formal derivative** $f'(x)$ of $f(x)$ is defined by the formula*

$$f'(x) = \sum_{k=1}^{t} k a_k x^{k-1},$$

where $k a_k$ denotes the sum of a_k added to itself k times.

It is not difficult to show from this definition that the standard differentiation formulas hold. The next proposition gives a test for multiple roots, which can be carried out over K without actually finding the roots.

8.2.3 Proposition. *The polynomial $f(x) \in K[x]$ has no multiple roots if and only if $\gcd(f(x), f'(x)) = 1$.*

Proof. Let F be a splitting field for $f(x)$ over K. In using the Euclidean algorithm to find the greatest common divisor of $f(x)$ and $f'(x)$, we make use of the division algorithm. We can do the necessary computations in both $K[x]$ and $F[x]$. But the quotients and remainders that occur over $K[x]$ also serve as the appropriate quotients and remainders over $F[x]$. Using the fact that these answers must be unique over $F[x]$, it follows that it makes no difference if we do the computations in the Euclidean algorithm over the splitting field F.

If $f(x)$ has a root r of multiplicity $m > 1$ in F, then we can write $f(x) = (x - r)^m g(x)$ for some polynomial $g(x) \in F[x]$. Thus

$$f'(x) = m(x - r)^{m-1} g(x) + (x - r)^m g'(x) ,$$

and so $(x - r)^{m-1}$ is a common divisor of $f(x)$ and $f'(x)$, which shows that $\gcd(f(x), f'(x)) \neq 1$.

On the other hand, if $f(x)$ has no multiple roots, then we can write $f(x) = (x - r_1)(x - r_2) \cdots (x - r_t)$. Applying the product rule, we find that $f'(x)$ is a sum of terms, each of which contains all but one of the linear factors of $f(x)$. It follows that each of the linear factors of $f(x)$ is a divisor of all but one of the terms in $f'(x)$, and so $\gcd(f(x), f'(x)) = 1$. $\square$

8.2.4 Proposition. *Let $f(x)$ be an irreducible polynomial over the field K. Then $f(x)$ has no multiple roots unless $\mathrm{char}(K) = p \neq 0$ and $f(x)$ has the form $f(x) = a_0 + a_1 x^p + a_2 x^{2p} + \ldots + a_n x^{np}$.*

Proof. Using the previous proposition, the only case in which $f(x)$ has a multiple root is if $\gcd(f(x), f'(x)) \neq 1$. Since $f(x)$ is irreducible and $\deg(f'(x)) < \deg(f(x))$, the only way this can happen is if $f'(x)$ is the zero polynomial. This is impossible over a field of characteristic zero. However, over a field of characteristic $p > 0$, it is possible if every coefficient of $f'(x)$ is a multiple of p. Thus $f(x)$ has no multiple roots unless it has the form specified in the statement of the proposition. $\square$

8.2.5 Definition. *A polynomial $f(x)$ over the field K is called **separable** if its irreducible factors have only simple roots.*

*An algebraic extension field F of K is called **separable over** K if the minimal polynomial of each element of F is separable.*

*The field F is called **perfect** if every polynomial over F is separable.*

8.2.6 Theorem. *Any field of characteristic zero is perfect. A field of characteristic $p > 0$ is perfect if and only if each of its elements has a pth root.*

Proof. Proposition 8.2.4 implies that any field of characteristic zero is perfect, and so we consider the case of a field F of characteristic $p > 0$ and a polynomial $f(x)$ irreducible over F.

If $f(x)$ has a multiple root, then it has the form $f(x) = \sum_{i=0}^{n} a_i (x^p)^i$. If each element of F has a pth root, then for each i we may let b_i be the pth root of a_i. Thus

$$f(x) = \sum_{i=0}^{n} (b_i)^p (x^p)^i = \sum_{i=0}^{n} (b_i x^i)^p = \left(\sum_{i=0}^{n} b_i x^i \right)^p$$

and so $f(x)$ is reducible, a contradiction.

To prove the converse, suppose that there exists an element $a \in F$ that has no pth root. Consider the polynomial $f(x) = x^p - a$ and let $g(x)$ be an irreducible factor of $f(x)$. Next, consider the extension field $F(b)$, where b is a root of $g(x)$. Then $b^p = a$, and so over $F(b)$ the polynomial $f(x)$ has the form $f(x) = x^p - b^p = (x - b)^p$. But then $g(x)$ must have the form $g(x) = (x - b)^k$ with $1 < k \leq p$ since $b \notin F$. Thus we have produced an irreducible polynomial with repeated roots. $\square$

8.2.7 Corollary. *Any finite field is perfect.*

Proof. Let F be a finite field of characteristic p, and let ϕ be the Frobenius automorphism of F, defined by $\phi(x) = x^p$, for all $x \in F$. Since ϕ maps F onto F, it follows that every element of F has a pth root. $\square$

It can be shown that if p is a prime number, and $K = \mathrm{GF}(p)$, then in the field $K(x)$ of rational functions over K, the element x has no pth root (see Exercise 8). Therefore this rational function field is not perfect.

In the final result of this section, we will use some of the ideas we have developed to investigate the structure of finite extensions. We began by studying field extensions of the form $F = K(u)$. If F is a finite field, then we showed in Theorem 6.5.10 that the multiplicative group $F^\times$ is cyclic. If the generator of this group is a, then it is easy to see that $F = K(a)$ for any subfield K. We now show that any finite separable extension has this form. Recall Definition 6.1.4: the extension field F of K is called a simple extension if there exists an element $u \in F$ such that $F = K(u)$. In this case, u is called a **primitive** element.

8.2.8 Theorem (Primitive element theorem). *Let F be a finite extension of the field K. If F is separable over K, then it is a simple extension of K.*

Proof. Let F be a finite separable extension of K. If K is a finite field, then F is also a finite field. As remarked above, we must have $F = K(u)$ for any generator u of the cyclic group $F^\times$.

If we can prove the result in case $F = K(u_1, u_2)$, then it is clear how to extend it by induction to the case $F = K(u_1, u_2, \ldots, u_n)$. Thus we may assume that K is infinite and $F = K(u, v)$ for elements $u, v \in F$.

Let $f(x)$ and $g(x)$ be the minimal polynomials of u and v over K, and assume their degrees to be m and n, respectively. Let E be an extension of F over which both $f(x)$ and $g(x)$ split. Since F is separable, the roots $u = u_1, u_2, \ldots, u_m$ and $v = v_1, v_2, \ldots, v_n$ of $f(x)$ and $g(x)$ are distinct. If $j \neq 1$, then the equation $u_i + v_j x = u + vx$ has a unique solution $x = (u - u_i)/(v_j - v)$ in E. Therefore, since K is infinite, there must exist an element $a \in K$ such that $u + av \neq u_i + av_j$ for all i and all $j \neq 1$.

We will show that $F = K(t)$ for $t = u + av$. It is clear that $K(t) \subseteq K(u, v) = F$. If we can show that $v \in K(t)$, then it will follow easily that $u \in K(t)$, giving us the desired equality. Our strategy is to show that the minimal polynomial $p(x)$ of v over $K(t)$ has degree 1, which will force v to belong to $K(t)$.

Let $h(x) = f(t - ax)$. This polynomial has coefficients in $K(t)$ and has v as a root since $h(v) = f(t - av) = f(u) = 0$. Since we also have $g(v) = 0$, it follows that $p(x)$ is a common divisor of $h(x)$ and $g(x)$. Now we consider all three polynomials $p(x)$, $h(x)$, and $g(x)$ over the extension field E. Since $t \neq u_i + av_j$, it follows that $t - av_j \neq u_i$ for all i and all $j \neq 1$. Thus v_j is not a root of $h(x)$, for $j = 2, 3, \ldots, n$. Since $g(x)$ splits over E and $x - v_j$ is not a divisor of $h(x)$ for $j = 2, 3, \ldots, n$, we can conclude that over E we have $\gcd(h(x), g(x)) = x - v$. But $p(x)$ is a common divisor of $h(x)$ and $g(x)$ over E as well as over $K(t)$, and so $p(x)$ must be linear. $\square$

EXERCISES: SECTION 8.2

1. Let $p(x)$ be an irreducible polynomial of degree n over a finite field K. Show that its Galois group over K is cyclic of order n.

2. Find the Galois group of $x^4 - 2$ over GF(3).

3. Find the Galois group of $x^4 + 2$ over GF(3).

4. Show that any algebraic extension of a perfect field is perfect.

5. Show that if $F \supseteq E \supseteq K$ are fields and F is separable over K, then F is separable over E.

6. Show that the product rule holds for the derivative defined in Definition 8.2.2.

7.†Let $\omega = (-1 + \sqrt{3}i)/2$ (a primitive cube root of unity). Find a primitive element for the extension $\mathbf{Q}(\omega, \sqrt[3]{2})$ of $\mathbf{Q}$.

8. Let p be a prime number, and let $K = \mathrm{GF}(p)$. Show that in the field $K(x)$ of rational functions over K, the element x has no pth root.

9. Let F be a field of characteristic $p \neq 0$, and let $a \in F$. Show that in $F[x]$ the polynomial $x^p - a$ is either irreducible or a pth power.

8.3   The Fundamental Theorem of Galois Theory

In this section we study the connection between subgroups of $\text{Gal}(F/K)$ and fields between K and F. This is a critical step in proving that a polynomial is solvable by radicals if and only if its Galois group is solvable.

8.3.1 Proposition. *Let F be a field, and let G be a subgroup of $\text{Aut}(F)$. Then $\{a \in F \mid \theta(a) = a$ for all $\theta \in G\}$ is a subfield of F.*

Proof. If a and b are elements of F that are left fixed by all automorphisms in G, then for any $\theta \in G$ we have

$$\theta(a \pm b) = \theta(a) \pm \theta(b) = a \pm b$$

and

$$\theta(ab) = \theta(a)\theta(b) = ab \ .$$

For $a \neq 0$ we have $\theta(a^{-1}) = (\theta(a))^{-1} = a^{-1}$, and thus we have shown that the given set is a subfield. $\square$

8.3.2 Definition. *Let F be a field, and let G be a subgroup of $\text{Aut}(F)$. Then $\{a \in F \mid \theta(a) = a$ for all $\theta \in G\}$ is called the G-**fixed subfield** of F, or the G-**invariant subfield** of F, and is denoted by F^G.*

For example, let $F = \mathbf{C}$, and let $G = \{1_{\mathbf{C}}, \theta\}$, where $\theta : \mathbf{C} \to \mathbf{C}$ is defined by complex conjugation. A complex number is real if and only if it is left unchanged by conjugation, and so $F^G = \mathbf{R}$.

8.3.3 Proposition. *If F is the splitting field over K of a separable polynomial and $G = \text{Gal}(F/K)$, then $F^G = K$.*

Proof. Let $E = F^G$, so that we have $K \subseteq E \subseteq F$. It is clear that F is a splitting field over E as well as over K, and that $G = \text{Gal}(F/E)$. By Theorem 8.1.6 we have both $|G| = [F : E]$ and $|G| = [F : K]$. This implies that $[E : K] = 1$, and so $E = K$. $\square$

8.3.4 Lemma (Artin). *Let G be a finite group of automorphisms of the field F, and let $K = F^G$. Then $[F : K] \leq |G|$.*

Proof. Let $G = \{\theta_1, \theta_2, \ldots, \theta_n\}$, with the identity element of G denoted by θ_1. Suppose that there exist $n + 1$ elements $\{u_1, u_2, \ldots, u_{n+1}\}$ of F which are linearly independent over K. We next consider the following system of equations:

$$\theta_1(u_1)x_1 + \theta_1(u_2)x_2 + \ldots + \theta_1(u_{n+1})x_{n+1} = 0,$$
$$\theta_2(u_1)x_1 + \theta_2(u_2)x_2 + \ldots + \theta_2(u_{n+1})x_{n+1} = 0,$$
$$\cdot \quad \cdot \quad \cdot \quad \cdot \quad \cdot$$
$$\theta_n(u_1)x_1 + \theta_n(u_2)x_2 + \ldots + \theta_n(u_{n+1})x_{n+1} = 0.$$

There are n equations corresponding to the n elements of G and $n + 1$ unknowns corresponding to the $n + 1$ linearly independent elements of F, and so it follows from the elementary theory of systems of linear equations that there exists a nontrivial solution $(a_1, a_2, \ldots, a_{n+1})$ in F. Among all such solutions, choose one with the smallest number of nonzero terms. By relabeling the indices on the x's and u's, we may assume that $a_1 \neq 0$. Dividing each of the solutions by a_1 still gives us a set of solutions, and so we may assume that $a_1 = 1$.

Since θ_1 is the identity, the first equation is $u_1 x_1 + u_2 x_2 + \ldots + u_{n+1} x_{n+1} = 0$. Not all of the elements a_i can belong to K, since this would contradict the fact that $u_1, u_2, \ldots, u_{n+1}$ are assumed to be linearly independent over K. We can again relabel the indices on $x_2, \ldots, x_n$ and $u_2, \ldots, u_n$ to guarantee that $a_2 \notin K$. Since $K = F^G$, this means that a_2 is not invariant under all automorphisms in G, say $\theta_i(a_2) \neq a_2$. If we apply θ_i to each of the equations in the system, we do not change the system, since G is a group and multiplying each element by θ_i merely permutes the elements. On the other hand, since θ_i is an automorphism, this leads to a second solution $(1, \theta_i(a_2), \ldots, \theta_i(a_{n+1}))$. Subtracting the second solution from the first gives a nontrivial solution (since $a_2 - \theta_i(a_2) \neq 0$) that has fewer nonzero entries. This is a contradiction, completing the proof. $\square$

Theorem 8.3.6 will show that the following property holds for the splitting field of a separable polynomial. Let F be the splitting field of $f(x)$ over K, and assume that $f(x)$ has no repeated roots. Then if $p(x) \in K[x]$ is the minimal polynomial of any element of F, it follows that $p(x)$ splits into linear factors in $F[x]$. This is equivalent to saying that F contains a splitting field for any polynomial in $K[x]$ that has a root in F. It is convenient to make the following definition, in which the choice of the term normal is justified by part (c) of Theorem 8.3.8.

8.3.5 Definition. *Let F be an algebraic extension of the field K. Then F is said to be a **normal** extension of K if every irreducible polynomial in $K[x]$ that contains a root in F is a product of linear factors in $F[x]$.*

8.3.6 Theorem. *The following conditions are equivalent for an extension field F of K:*

> **(1)** *F is the splitting field over K of a separable polynomial;*
> **(2)** $K = F^G$ *for some finite group G of automorphisms of F;*
> **(3)** *F is a finite, normal, separable extension of K.*

Proof. (1) implies (2): If F is the splitting field over K of a separable polynomial, then by Proposition 8.3.3 we must have $K = F^G$ for $G = \mathrm{Gal}(F/K)$. The Galois group is finite by Theorem 8.1.6.

(2) implies (3): Assume that $K = F^G$ for some finite group G of automorphisms of F. Then Artin's lemma shows that $[F : K] \leq |G|$, and so F is a finite extension of K. Let $f(x)$ be a monic irreducible polynomial in $K[x]$ that has a root r in F. Since G is a finite group, the set $\{\theta(r) \mid \theta \in G\}$ is finite, say with distinct elements $r_1 = r, r_2, \ldots, r_m$. These elements are roots of $f(x)$, since G is a group of automorphisms of F that fixes the coefficients of $f(x)$, so $\deg(f(x)) \geq m$. Let

$$h(x) = (x - r_1)(x - r_2) \cdots (x - r_m) .$$

It follows from our choice of the elements r_i that applying any automorphism $\theta \in G$ yields $\theta(r_i) = r_j$, for some j. Therefore applying $\theta \in G$ to the polynomial $h(x)$ simply permutes the factors, showing that $\theta(h(x)) = h(x)$. Using the fact that θ is an automorphism and equating corresponding coefficients shows that every coefficient of $h(x)$ must be left fixed by G, so by assumption the coefficients of $h(x)$ belong to $K = F^G$. Now $f(x)$ is the minimal polynomial of r over K, and $h(r) = 0$, so $f(x) \mid h(x)$. Since $f(x)$ and $h(x)$ are monic and $\deg(f(x)) \geq \deg(h(x))$, we must have $f(x) = h(x)$. In particular, we have shown that $f(x)$ has distinct roots, all belonging to F. This implies that F is both normal and separable over K.

(3) implies (1): If F is a finite separable extension of K, then by Theorem 8.2.8 it is a simple extension, say $F = K(u)$. If F is a normal extension of K and u has the minimal polynomial $f(x)$ over K, then F contains all of the roots of $f(x)$, and so F is a splitting field for $f(x)$. □

8.3.7 Corollary. *If F is an extension field of K such that $K = F^G$ for some finite group G of automorphisms of F, then $G = \mathrm{Gal}(F/K)$.*

Proof. By assumption $K = F^G$, so G is a subgroup of $\mathrm{Gal}(F/K)$, and the result follows since

$$[F : K] \leq |G| \leq |\mathrm{Gal}(F/K)| = [F : K] .$$

The first inequality comes from Artin's lemma, and the last equality is implied by Theorems 8.1.6 and 8.3.6. □

Example 8.3.1 $(\mathrm{Gal}(\mathrm{GF}(p^n)/\mathrm{GF}(p)))$.

In Corollary 8.1.8 we showed that $\mathrm{Gal}(\mathrm{GF}(p^n)/\mathrm{GF}(p))$ is cyclic of order n, generated by the Frobenius automorphism ϕ defined by $\phi(x) = x^p$, for all $x \in \mathrm{GF}(p^n)$.

In Section 6.5 we showed that the subfields of $\mathrm{GF}(p^n)$ are of the form $\mathrm{GF}(p^r)$, for divisors r of n, and then Corollary 8.1.8 shows that $\mathrm{Gal}(\mathrm{GF}(p^n)/\mathrm{GF}(p^r))$ has order m, where $n = mr$, and is generated by ϕ^r. Note that as the value of r is increased, to give a larger subfield, the power of ϕ is increased, and so the corresponding subgroup it defines is smaller. It happens very generally that there is an order-reversing correspondence between subfields and subgroups of the Galois group, and this is proved in the following fundamental theorem of Galois theory.

The Galois group of $\mathrm{GF}(p^r)$ is also generated by the Frobenius automorphism, when restricted to the smaller field. Is there a connection between $\mathrm{Gal}(\mathrm{GF}(p^n)/\mathrm{GF}(p))$ and $\mathrm{Gal}(\mathrm{GF}(p^r)/\mathrm{GF}(p))$? Since any power of ϕ^r leaves elements of $\mathrm{GF}(p^r)$ fixed, for any integer we have $\phi^t(x) = \phi^{t+rs}(x)$ for all $x \in \mathrm{GF}(p^r)$. This suggests that the cosets of $\langle \phi^r \rangle$ in $\langle \phi \rangle$ might correspond to nontrivial automorphisms in $\mathrm{Gal}(\mathrm{GF}(p^r)/\mathrm{GF}(p))$, and indeed this is the case, as will be shown in Theorem 8.3.8. $\square$

8.3.8 Theorem (Fundamental Theorem of Galois Theory). *Let F be the splitting field of a separable polynomial over the field K, and let $G = \mathrm{Gal}(F/K)$.*

(a) *There is a one-to-one order-reversing correspondence between subgroups of G and subfields of F that contain K:*

    (i) *If H is a subgroup of G, then the corresponding subfield is F^H, and*

$$H = \mathrm{Gal}(F/F^H) \ .$$

    (ii) *If E is a subfield of F that contains K, then the corresponding subgroup of G is $\mathrm{Gal}(F/E)$, and*

$$E = F^{\mathrm{Gal}(F/E)} \ .$$

(b) *For any subgroup H of G, we have*

$$[F : F^H] = |H| \quad \text{and} \quad [F^H : K] = [G : H] \ .$$

(c) *Under the above correspondence, the subgroup H is normal if and only if the subfield $E = F^H$ is a normal extension of K. In this case,*

$$\mathrm{Gal}(E/K) \cong \mathrm{Gal}(F/K) / \mathrm{Gal}(F/E) \ .$$

Proof.   (a) If we verify that (i) and (ii) hold, then it is clear that the mapping that assigns to each subgroup H of G the subfield F^H has an inverse mapping, and so it defines a one-to-one correspondence. We will first show that the one-to-one correspondence reverses the natural ordering in the diagrams of subgroups and subfields. If $H_1 \subseteq H_2$ are subgroups of G, then it is clear that the subfield left fixed by H_2 is contained in the subfield left fixed by H_1. On the other hand, if $E_1 \subseteq E_2$, then it is clear that $\mathrm{Gal}(F/E_2) \subseteq \mathrm{Gal}(F/E_1)$.

Given the subgroup H, Corollary 8.3.7 implies that $H = \mathrm{Gal}(F/F^H)$. On the other hand, given a subfield E with $K \subseteq E \subseteq F$, it is clear from the initial assumption that F is also a splitting field over E, and then $E = F^{\mathrm{Gal}(F/E)}$ by Proposition 8.3.3.

(b) Given the subgroup H of G, since F is a splitting field over F^H and $H = \mathrm{Gal}(F/F^H)$, it follows from Theorem 8.1.6 that $[F : F^H] = |H|$. Since $[F : K] = |G|$ by Theorem 8.1.6, the desired equality $[F^H : K] = [G : H]$ follows from the two equalities $[F : K] = [F : F^H][F^H : K]$ and $|G| = |H| \cdot [G : H]$.

(c) Let E be a normal extension of K in F, and let ϕ be any element of G. If $u \in E$ with minimal polynomial $p(x)$, then $\phi(u)$ is also a root of $p(x)$, and so we must have $\phi(u) \in E$ since E is a normal extension of K. Thus for any $\theta \in \mathrm{Gal}(F/E)$, we have $\theta\phi(u) = \phi(u)$, and so $\phi^{-1}\theta\phi(u) = \phi^{-1}\phi(u) = u$. This implies that $\phi^{-1}\theta\phi \in \mathrm{Gal}(F/E)$, and therefore $\mathrm{Gal}(F/E)$ is a normal subgroup of G.

Conversely, let H be a normal subgroup of G, and let $E = F^H$. If $\phi \in G$ and $\theta \in H$, then by the normality of H we must have $\phi^{-1}\theta\phi = \theta'$ for some $\theta' \in H$, and then $\theta\phi = \phi\theta'$. For any $u \in E$ we have by definition $\theta'(u) = u$, and so $\theta(\phi(u)) = \phi(\theta'(u)) = \phi(u)$. This shows that $\phi(u) \in F^H$, and so ϕ maps E into E. Because ϕ leaves K fixed, it is a K-linear transformation and is one-to-one when restricted to E, so a dimension argument shows that it maps E onto E.

The above argument shows that the restriction mapping defines a function (which is easily seen to be a group homomorphism) from $G = \mathrm{Gal}(F/K)$ into $\mathrm{Gal}(E/K)$. By Theorem 8.2.8, F is a simple extension of E, and so Lemma 6.4.4 implies that any element of $\mathrm{Gal}(E/K)$ can be extended to an element of $\mathrm{Gal}(F/K)$. Thus the restriction mapping is onto, and since the kernel of this mapping is clearly $\mathrm{Gal}(F/E) = H$, by the fundamental homomorphism theorem we must have $\mathrm{Gal}(E/K) \cong \mathrm{Gal}(F/K) / \mathrm{Gal}(F/E)$.

By part (b) the index of $\mathrm{Gal}(F/E)$ in G equals $[E : K]$, and so $|\mathrm{Gal}(E/K)| = [E : K]$. Since $|\mathrm{Gal}(E/K)|$ is finite, Theorem 8.3.6 implies that E is a normal separable extension of $E^{\mathrm{Gal}(E/K)}$. But then $[E : E^{\mathrm{Gal}(E/K)}] = [E : K]$, and so $E^{\mathrm{Gal}(E/K)} = K$, which finally shows that E is a normal extension of K. $\square$

In the statement of the fundamental theorem we could have simply said that normal subgroups correspond to normal extensions. In the proof we noted that if E is a normal extension of K, then $\phi(E) \subseteq E$ for all $\phi \in \mathrm{Gal}(F/K)$. In the context of the fundamental theorem, we say that two intermediate subfields E_1 and E_2 are **conjugate** if there exists $\phi \in \mathrm{Gal}(F/K)$ such that $\phi(E_1) = E_2$. We now show that the subfields conjugate to an intermediate subfield E correspond to the subgroups conjugate to $\text{Gal}(F/E)$. Thus E is a normal extension if and only if it is conjugate only to itself.

8.3.9 Proposition. *Let F be the splitting field of a separable polynomial over the field K, and let E be a subfield such that $K \subseteq E \subseteq F$, with $H = \text{Gal}(F/E)$. If $\phi \in \text{Gal}(F/K)$, then $\text{Gal}(F/\phi(E)) = \phi H \phi^{-1}$.*

Proof. Since $\phi \in \text{Gal}(F/K)$, for any $\theta \in \text{Gal}(F/K)$ we have

$$
\begin{aligned}
\theta \in \text{Gal}(F/\phi(E)) \quad &\text{if and only if} \quad \theta\phi(x) = \phi(x) \ \text{ for all } \ x \in E \\
&\text{if and only if} \quad \phi^{-1}\theta\phi(x) = x \ \text{ for all } \ x \in E \\
&\text{if and only if} \quad \phi^{-1}\theta\phi \in \text{Gal}(F/E) = H \\
&\text{if and only if} \quad \theta \in \phi H \phi^{-1} \ .
\end{aligned}
$$

This shows that $\text{Gal}(F/\phi(E)) = \phi H \phi^{-1}$ and completes the proof. $\quad\square$

Example 8.3.2 (Galois group of $x^3 - 2$ over Q).

In this example we will compute the Galois group G of the polynomial $p(x) = x^3 - 2$ over **Q**. We will also illustrate the fundamental theorem by investigating the subfields of its splitting field. The roots of the polynomial are $\sqrt[3]{2}$, $\omega\sqrt[3]{2}$, and $\omega^2\sqrt[3]{2}$, where $\omega = (-1 + \sqrt{3}i)/2$ is a primitive cube root of unity. Thus the splitting field of $p(x)$ over **Q** can be constructed as $\mathbf{Q}(\omega, \sqrt[3]{2})$. Since $x^3 - 2$ is irreducible over **Q**, adjoining $\sqrt[3]{2}$ gives an extension of degree 3, which is contained in **R**. The minimal polynomial of ω over **Q** is $x^2 + x + 1$, and it is irreducible since its roots are ω and ω^2, which are not real. This implies that $[\mathbf{Q}(\omega, \sqrt[3]{2}) : \mathbf{Q}] = 6$, and so $|G| = 6$. The set $\{1, \sqrt[3]{2}, \sqrt[3]{4}, \omega, \omega\sqrt[3]{2}, \omega\sqrt[3]{4}\}$ is a basis for $\mathbf{Q}(\omega, \sqrt[3]{2})$ over **Q**.

Note that any automorphism $\phi \in G$ is completely determined by its values on $\sqrt[3]{2}$ and ω. Since ϕ must preserve roots of polynomials, the only possibilities are $\phi(\sqrt[3]{2}) = \sqrt[3]{2}$, $\omega\sqrt[3]{2}$, or $\omega^2\sqrt[3]{2}$, and $\phi(\omega) = \omega$, or ω^2. This gives six possible functions, and so they must determine the six possible elements of G. Let α be defined by $\alpha(\sqrt[3]{2}) = \omega\sqrt[3]{2}$ and $\alpha(\omega) = \omega$, and let β be defined by $\beta(\sqrt[3]{2}) = \sqrt[3]{2}$ and $\beta(\omega) = \omega^2$. Since ω^2 is the complex conjugate of ω, complex conjugation defines an automorphism of $\mathbf{Q}(\sqrt[3]{2}, \omega)$ that has the properties required of β, so evidently we have defined $\beta(x) = \overline{x}$, for all $x \in \mathbf{Q}(\sqrt[3]{2}, \omega)$. A direct computation shows that $\alpha^2(\sqrt[3]{2}) = \omega^2\sqrt[3]{2}$ and then $\alpha^3(\sqrt[3]{2}) = \sqrt[3]{2}$, so α has order 3. Similarly, β has order 2. Furthermore, $\beta\alpha(\sqrt[3]{2}) = \beta(\omega\sqrt[3]{2}) = \omega^2\sqrt[3]{2}$ and $\beta\alpha(\omega) = \beta(\omega) = \omega^2$. On the other hand, $\alpha^2\beta(\sqrt[3]{2}) = \alpha^2(\sqrt[3]{2}) = \omega^2\sqrt[3]{2}$ and $\alpha^2\beta(\omega) = \alpha^2(\omega^2) = \omega^2$. This shows that $\beta\alpha = \alpha^2\beta$, and it follows that G is isomorphic to S_3.

The subfield $\mathbf{Q}(\omega)$ is the splitting field of x^2+x+1 over $\mathbf{Q}$, and so it must be the fixed subfield of the only proper nontrivial normal subgroup $\{1, \alpha, \alpha^2\}$ of G. The subfield $\mathbf{Q}(\sqrt[3]{2})$ is the fixed subfield of the subgroup $H = \{1, \beta\}$, since $\beta(\sqrt[3]{2}) = \sqrt[3]{2}$. By Proposition 8.3.9, the subfield $\mathbf{Q}(\omega\sqrt[3]{2}) = \alpha(\mathbf{Q}(\sqrt[3]{2}))$ must be the fixed subfield of the subgroup $\alpha H \alpha^{-1} = \{1, \alpha^2\beta\}$. (A direct computation shows that $\alpha^2\beta(\omega\sqrt[3]{2}) = \omega\sqrt[3]{2}$.) The remaining subfield is $\mathbf{Q}(\omega^2\sqrt[3]{2}) = \alpha^2(\mathbf{Q}(\sqrt[3]{2}))$, and this suffices to determine the correspondence between subgroups and subfields. We give the respective diagrams in Figure 8.3.1. $\square$

Figure 8.3.1:

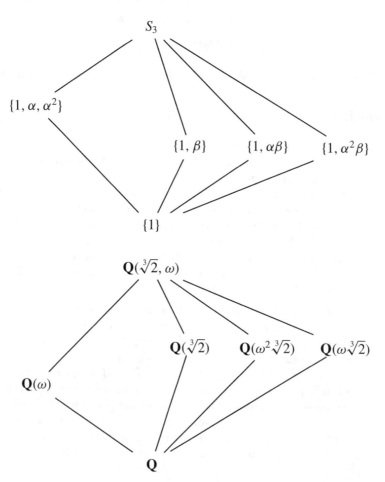

Example 8.3.3 (Galois group of $x^4 - 2$ over Q).

In the final example of this section we compute the Galois group G of $x^4 - 2$ over **Q** and investigate the subfields of its splitting field. We have $x^4 - 2 = (x^2 + \sqrt{2})(x^2 - \sqrt{2})$, and this factorization leads to the roots $\pm\sqrt[4]{2}, \pm i\sqrt[4]{2}$ of $x^4 - 2$. We can obtain the splitting field as $\mathbf{Q}(\sqrt[4]{2}, i)$ by first adjoining the root $\sqrt[4]{2}$ of $x^4 - 2$ to obtain $\mathbf{Q}(\sqrt[4]{2})$ and then adjoining the root i of the irreducible polynomial $x^2 + 1$. Since any element of the Galois group must simply permute the roots of $x^4 - 2$ and $x^2 + 1$, the eight possible permutations determine the Galois group, which has order 8.

Let α be the automorphism defined by $\alpha(\sqrt[4]{2}) = i\sqrt[4]{2}$ and $\alpha(i) = i$, and let β be defined by $\beta(\sqrt[4]{2}) = \sqrt[4]{2}$ and $\beta(i) = -i$. Then α represents a cyclic permutation of the roots of $x^4 - 2$ and has order 4, while β represents complex conjugation and has order 2. A direct computation shows that $\beta\alpha = \alpha^{-1}\beta$, and so $G \cong D_4$.

To compute the fixed subfields of the various subgroups of D_4, we first note that all powers of α leave i fixed, and so the fixed subfield corresponding to $\langle \alpha \rangle$ contains $\mathbf{Q}(i)$. We must have equality since the degree of the fixed subfield over **Q** is equal to the index of $\langle \alpha \rangle$ in D_4. Next, α^2 also leaves $\sqrt{2}$ fixed since $\alpha(\sqrt{2}) = \alpha(\sqrt[4]{2}\sqrt[4]{2}) = (i\sqrt[4]{2})(i\sqrt[4]{2}) = -\sqrt{2}$. A degree argument shows that the fixed subfield of $\langle \alpha^2 \rangle$ is $\mathbf{Q}(\sqrt{2}, i)$.

We next consider $\langle \alpha^2, \beta \rangle$ and its subgroups. Since $\sqrt{2}$ is a real number, it is left fixed by β, as well as by α^2. It follows that the subfield left fixed by $\langle \alpha^2, \beta \rangle$ is $\mathbf{Q}(\sqrt{2})$. Adjoining either $\sqrt[4]{2}$ or $i\sqrt[4]{2}$ to **Q** gives a subfield of degree 4 that contains $\mathbf{Q}(\sqrt{2})$, and so the corresponding subgroups must be contained in $\langle \alpha^2, \beta \rangle$. Because $\mathbf{Q}(\sqrt[4]{2})$ consists of real numbers, it is left fixed by β, and then $\mathbf{Q}(i\sqrt[4]{2})$ is the subfield fixed by $\langle \alpha^2\beta \rangle$.

To identify the subfield fixed by $\langle \alpha\beta \rangle$ requires some additional computations. As a basis for $\mathbf{Q}(\sqrt[4]{2}, i)$ over **Q** we will use the set

$$\{1, \sqrt[4]{2}, \sqrt{2}, \sqrt[4]{8}, i, i\sqrt[4]{2}, i\sqrt{2}, i\sqrt[4]{8}\} .$$

Using the definitions of α and β it can be shown that

$$\alpha\beta(\sqrt[4]{2}) = i\sqrt[4]{2}, \quad \alpha\beta(\sqrt{2}) = -\sqrt{2}, \quad \alpha\beta(\sqrt[4]{8}) = -i\sqrt[4]{8}, \quad \alpha\beta(i) = -i,$$

$$\alpha\beta(i\sqrt[4]{2}) = \sqrt[4]{2}, \quad \alpha\beta(i\sqrt{2}) = i\sqrt{2}, \quad \text{and} \quad \alpha\beta(i\sqrt[4]{8}) = -\sqrt[4]{8} .$$

It follows that an element

$$u = a_1 + a_2\sqrt[4]{2} + a_3\sqrt{2} + a_4\sqrt[4]{8} + a_5 i + a_6 i\sqrt[4]{2} + a_7 i\sqrt{2} + a_8 i\sqrt[4]{8}$$

is left fixed by $\alpha\beta$ if and only if $a_3 = a_5 = 0$, $a_6 = a_2$, and $a_8 = -a_4$. Thus

$$u = a_1 + a_2(\sqrt[4]{2} + i\sqrt[4]{2}) + a_7 i\sqrt{2} + a_4(\sqrt[4]{8} - i\sqrt[4]{8}) ,$$

and so the subfield fixed by $\langle \alpha\beta \rangle$ is $\mathbf{Q}(\sqrt[4]{2} + i\sqrt[4]{2})$. It is left as an exercise to verify that the subfield fixed by $\langle \alpha^3\beta \rangle$ is $\mathbf{Q}(\sqrt[4]{2} - i\sqrt[4]{2})$.

There are various ways in which to obtain the splitting field of $x^4 - 2$ by adjoining, at each stage, a root of a quadratic. These are reflected in the various chains of subfields. For example, we can first adjoin $\sqrt{2}$, to obtain the splitting field of $x^2 - 2$. Then we can adjoin the root $\sqrt[4]{2}$ of $x^2 - \sqrt{2}$, which does not produce a splitting field. Finally, after adjoining a root of $x^2 + 1$ we arrive at the splitting field $\mathbf{Q}(\sqrt[4]{2}, i)$. It may be useful to note that this field is also the splitting field of $x^4 + 2$ over $\mathbf{Q}$, which has roots $\pm\sqrt[4]{2}v$ and $\pm i\sqrt[4]{2}v$, where $v = \dfrac{\sqrt{2}}{2} + \dfrac{\sqrt{2}}{2}i$. Adjoining these roots produces the seemingly mysterious subfields $\mathbf{Q}(\sqrt[4]{2} + i\sqrt[4]{2})$ and $\mathbf{Q}(\sqrt[4]{2} - i\sqrt[4]{2})$.

The appropriate diagrams are given in Figure 8.3.2. $\square$

We end this section by proving the fundamental theorem of algebra. The proof uses algebraic techniques, although we do need to know that any polynomial over $\mathbf{R}$ of odd degree must have a root. If $f(x) \in \mathbf{R}[x]$ had odd degree, then $f(n)$ and $f(-n)$ have opposite signs for sufficiently large n, so $f(x)$ must have a root since it is a continuous function.

8.3.10 Theorem (Fundamental Theorem of Algebra). *Any nonconstant polynomial in* $\mathbf{C}[x]$ *has a root in* $\mathbf{C}$.

Proof. Let $f(x)$ be a nonconstant polynomial in $\mathbf{C}[x]$ and let L be a splitting field for $f(x)$ over $\mathbf{C}$. Let F be a splitting field for $(x^2 + 1)f(x)\overline{f(x)}$ over $\mathbf{R}$, where $\overline{f(x)}$ is the polynomial whose coefficients are obtained by taking the complex conjugates of the coefficients of $f(x)$. Since char$(\mathbf{R}) = 0$, we are in the situation of the fundamental theorem of Galois theory. We will use $G = \mathrm{Gal}(F/\mathbf{R})$ to show that $F = \mathbf{C}$. Then $L = \mathbf{C}$ since $F \supseteq L \supseteq \mathbf{C}$.

Let H be a Sylow 2-subgroup of G, and let $E = F^H$. Since H is a Sylow 2-subgroup, $[G : H]$ is odd, which implies by Theorem 8.3.8 that $[E : \mathbf{R}]$ is odd. By Theorem 8.2.8 there exists $u \in E$ such that $E = \mathbf{R}(u)$, and so the minimal polynomial of u over $\mathbf{R}$ has odd degree. The minimal polynomial is irreducible, so it can only have degree 1, since any polynomial of odd degree over $\mathbf{R}$ has a root in $\mathbf{R}$ (by our comment preceding the theorem), and this implies that $E = \mathbf{R}$. Now, since $F^H = \mathbf{R}$, we must have $H = G$, and so G is a 2-group.

The subgroup $G_1 = \mathrm{Gal}(F/\mathbf{C})$ of G is also a 2-group. If G_1 is not the trivial group, then the first Sylow theorem implies that it has a normal subgroup N of index 2. Since F is a normal extension of $\mathbf{C}$, we can again apply Theorem 8.3.8. If $K = F^N$, then $[K : \mathbf{C}] = [G_1 : N] = 2$, so $K = \mathbf{C}(v)$ for some $v \in K$ with a minimal polynomial that is a quadratic. We know that square roots exist in $\mathbf{C}$, so the quadratic formula is valid, which implies that no polynomial in $\mathbf{C}[x]$ of degree 2 is irreducible. Therefore G_1 must be the trivial group, showing that $F = \mathbf{C}$. $\square$

Figure 8.3.2:

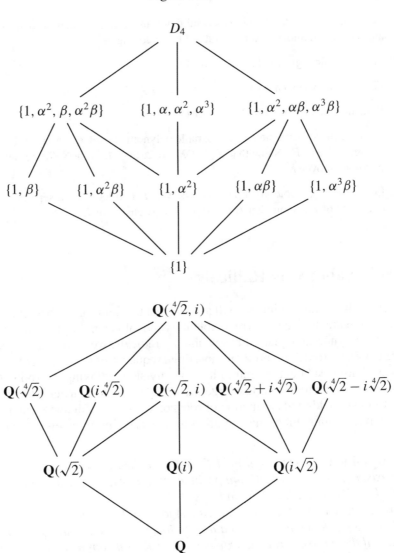

EXERCISES: SECTION 8.3

1. In Example 8.3.3 use a direct calculation to verify that the subfield fixed by $\langle \alpha^3 \beta \rangle$ is $\mathbf{Q}(\sqrt[4]{2} - i \sqrt[4]{2})$.

2. In Example 8.3.3 determine which subfields are conjugate, and in each case find an automorphism under which the subfields are conjugate.

3. Find the Galois group of $x^4 + 1$ over $\mathbf{Q}$.

4.†Find the Galois group of $x^4 - x^2 - 6$ over $\mathbf{Q}$.

5. Find the Galois group of $x^8 - 1$ over $\mathbf{Q}$.

6. Let F be the splitting field of a separable polynomial over K, and let E be a subfield between K and F. Show that if $[E : K] = 2$, then E is the splitting field of some polynomial over K.

7. Let E be a separable algebraic extension of F and let F be a separable algebraic extension of K. Show that E is a separable extension of K.

8.4   Solvability by Radicals

In most results in this section we will assume that the fields have characteristic zero, in order to guarantee that no irreducible polynomial has multiple roots. When we say that a polynomial equation is solvable by radicals, we mean that the solutions can be obtained from the coefficients in a finite sequence of steps, each of which may involve addition, subtraction, multiplication, division, or taking nth roots. Only the extraction of an nth root leads to a larger field, and so our formal definition is phrased in terms of subfields and adjunction of roots of $x^n - a$ for suitable elements a. The definition is reminiscent of the condition for constructibility given in Section 6.3.

8.4.1 Definition. *An extension field F of K is called a **radical extension** of K if there exist $u_1, u_2, \ldots, u_m \in F$ and positive integers $n_1, n_2, \ldots, n_m$ such that*
   (i) $F = K(u_1, u_2, \ldots, u_m)$, *and*
   (ii) $u_1^{n_1} \in K$ *and* $u_i^{n_i} \in K(u_1, \ldots, u_{i-1})$ *for* $i = 2, \ldots, m$.
   *For $f(x) \in K[x]$, the polynomial equation $f(x) = 0$ is said to be **solvable by radicals** if there exists a radical extension F of K that contains all roots of $f(x)$.*

We must first determine the structure of the Galois group of a polynomial of the form $x^n - a$. Then we will make use of the fundamental theorem of Galois theory to see what happens when we successively adjoin roots of such polynomials.

We first need to recall a result from group theory. We showed in Example 7.1.6 that $\mathrm{Aut}(\mathbf{Z}_n) \cong \mathbf{Z}_n^\times$. This follows from the fact that any automorphism of $\mathbf{Z}_n$ has the form $\phi_a([x]) = [ax]$ for all $[x] \in \mathbf{Z}_n$, where $\gcd(a, n) = 1$. It is easy to see that $\phi_a \phi_b = \phi_{ab}$, and it then follows immediately that $\mathrm{Aut}(\mathbf{Z}_n) \cong \mathbf{Z}_n^\times$.

8.4.2 Proposition. *Let F be the splitting field of $x^n - 1$ over a field K of characteristic zero. Then $\mathrm{Gal}(F/K)$ is an abelian group.*

Proof. Since $\mathrm{char}(K) = 0$, the polynomial $x^n - 1$ has n distinct roots, and it is easy to check that they form a subgroup C of $F^\times$, which is cyclic by Theorem 6.5.10. Every element of $\mathrm{Gal}(F/K)$ defines an automorphism of C, and using this observation it is apparent that $\mathrm{Gal}(F/K)$ is isomorphic to a subgroup of $\mathrm{Aut}(C)$. Since C is cyclic of order n, the remarks preceding the proposition show that $\mathrm{Gal}(F/K)$ is isomorphic to a subgroup of $\mathbf{Z}_n^\times$, and so it must be abelian. $\square$

Recall that the roots of the polynomial $x^n - 1$ are called the *nth roots of unity*. Any generator of the group of all nth roots of unity is called a *primitive nth root of unity.*

8.4.3 Theorem. *Let K be a field of characteristic zero that contains all nth roots of unity, let $a \in K$, and let F be the splitting field of $x^n - a$ over K. Then $\mathrm{Gal}(F/K)$ is a cyclic group whose order is a divisor of n.*

Proof. If u is any root of $x^n - a$ and ζ is a primitive nth root of unity, then all other roots of $x^n - a$ have the form $\zeta^i u$, for $i = 1, 2, \ldots, n - 1$. Thus $F = K(u)$, and so any element $\phi \in \mathrm{Gal}(F/K)$ is completely determined by its value on u, which must be another root, say $\phi(u) = \zeta^i u$. If $\theta \in \mathrm{Gal}(F/K)$ with $\theta(u) = \zeta^j u$, then we have $\phi\theta(u) = \phi(\zeta^j u) = \zeta^i \zeta^j u = \zeta^{i+j} u$. Assigning to $\phi \in \mathrm{Gal}(F/K)$ the exponent i of ζ in $\phi(u) = \zeta^i u$ defines a one-to-one homomorphism from $\mathrm{Gal}(F/K)$ into $\mathbf{Z}_n$. $\square$

8.4.4 Theorem. *Let p be a prime number, let K be a field that contains all pth roots of unity, and let F be an extension of K. If $[F : K] = |\mathrm{Gal}(F/K)| = p$, then $F = K(u)$ for some $u \in F$ such that $u^p \in K$.*

Proof. Let $G = \mathrm{Gal}(F/K)$. If $w \in F$ is not in K, then $F = K(w)$ since $[F : K]$ is prime. Let $C = \{\zeta_1, \zeta_2, \ldots, \zeta_p\} \subseteq K$ be the pth roots of unity, and let θ be a generator of G. Let $w_1 = w$ and $w_i = \theta(w_{i-1})$ for $1 < i \le p$. Equivalently, $w_i = \theta^{i-1}(w)$ for $1 < i \le p$. Since θ^p is the identity, we have $w_1 = \theta(w_p)$. For each i we let

$$w_1 + \zeta_i w_2 + \zeta_i^2 w_3 + \ldots + \zeta_i^{p-1} w_p = u_i \ .$$

Applying θ leaves the roots of unity fixed, and since $\zeta_i^p = 1$ we obtain

$$
\begin{aligned}
\theta(u_i) &= \theta(w_1) + \zeta_i \theta(w_2) + \ldots + \zeta_i^{p-1} \theta(w_p) \\
&= w_2 + \zeta_i w_3 + \ldots + \zeta_i^{p-2} w_p + \zeta_i^{p-1} w_1 \\
&= \zeta_i^{-1}(\zeta_i w_2 + \zeta_i^2 w_3 + \ldots + \zeta_i^{p-1} w_p + w_1) \\
&= \zeta_i^{-1} u_i
\end{aligned}
$$

and therefore we have

$$
\theta(u_i^p) = (\theta(u_i))^p = (\zeta_i^{-1} u_i)^p = \zeta_i^{-p} u_i^p = u_i^p \ .
$$

This shows that u_i^p is left fixed by the Galois group G, since θ is a generator of G, and it follows that $u_i^p \in K$.

Writing the definition of the u_i's in the matrix form

$$
\begin{bmatrix}
1 & \zeta_1 & \zeta_1^2 & \cdots & \zeta_1^{p-1} \\
1 & \zeta_2 & \zeta_2^2 & \cdots & \zeta_2^{p-1} \\
\vdots & \vdots & \vdots & & \vdots \\
1 & \zeta_p & \zeta_p^2 & \cdots & \zeta_p^{p-1}
\end{bmatrix}
\begin{bmatrix}
w_1 \\
w_2 \\
\vdots \\
w_p
\end{bmatrix}
=
\begin{bmatrix}
u_1 \\
u_2 \\
\vdots \\
u_p
\end{bmatrix}
$$

we see that the coefficient matrix has the form of a Vandermonde matrix, which is invertible since the ζ_i's are distinct. One way to see that the matrix must be invertible is to observe that it is the matrix used to solve for the coefficients of a polynomial $f(x) = a_0 + a_1 x + \ldots + a_{p-1} x^{p-1}$ over K such that $f(\zeta_i) = 0$ for all i. The only solution is the zero polynomial, since no other polynomial of degree $p - 1$ can have p roots. This means that it is possible to solve for $w = w_1$ as a linear combination (with coefficients in K) of the elements u_i. Thus if $u_i \in K$ for all i, then $w \in K$, a contradiction. We conclude that $u_j \notin K$ for some j, and so $F = K(u)$, for $u = u_j$, and $u^p \in K$, as we had previously shown. $\square$

8.4.5 Lemma. *Let K be a field of characteristic zero, and let E be a radical extension of K. Then there exists an extension F of E that is a normal radical extension of K.*

Proof. Let E be a radical extension of K with elements $u_1, u_2, \ldots, u_m \in E$ such that (i) $E = K(u_1, u_2, \ldots, u_m)$ and (ii) $u_1^{n_1} \in K$ and $u_i^{n_i} \in K(u_1, \ldots, u_{i-1})$ for $i = 2, \ldots, m$ and integers $n_1, n_2, \ldots, n_m$. Let F be the splitting field of the product $f(x)$ of the minimal polynomials of u_i over K, for $i = 1, 2, \ldots, n$. The proof that condition (2) implies condition (3) in Theorem 8.3.6 shows that in F each root of

$f(x)$ has the form $\theta(u_i)$ for some integer i and some automorphism $\theta \in \text{Gal}(F/K)$. For any $\theta \in \text{Gal}(F/K)$, we have $\theta(u_1)^{n_1} \in K$ and $\theta(u_i)^{n_i} \in K(\theta(u_1), \ldots, \theta(u_{i-1}))$ for $i = 2, \ldots, m$. Thus if $\text{Gal}(F/K) = \{\theta_1, \theta_2, \ldots, \theta_k\}$, then the elements $\{\theta_j(u_i)\}$ for $i = 1, \ldots, m$ and $j = 1, \ldots, k$ satisfy the conditions of Definition 8.4.1, showing that F is a radical extension of K. $\square$

8.4.6 Theorem. *Let $f(x)$ be a polynomial over a field K of characteristic zero. The equation $f(x) = 0$ is solvable by radicals if and only if the Galois group of $f(x)$ over K is solvable.*

Proof. We first assume that the equation $f(x) = 0$ is solvable by radicals. Let F be a radical extension of K that contains a splitting field E of $f(x)$ over K. By the previous lemma we may assume that F is a splitting field over K, with elements $u_1, u_2, \ldots, u_m \in F$ such that (i) $F = K(u_1, u_2, \ldots, u_m)$ and (ii) $u_1^{n_1} \in K$ and $u_i^{n_i} \in K(u_1, \ldots, u_{i-1})$ for $i = 2, \ldots, m$ and integers $n_1, n_2, \ldots, n_m$. Let n be the least common multiple of the exponents n_i. By adjoining a primitive nth root of unity ζ, we obtain a normal radical extension $F(\zeta)$ of $K(\zeta)$. Note that $K(\zeta)$ contains all of the n_ith roots of unity for $i = 1, \ldots, m$. It follows from the fundamental theorem of Galois theory that $\text{Gal}(E/K)$ is a factor group of $\text{Gal}(F(\zeta)/K)$. Since any factor of a solvable group is again solvable, it suffices to show that $\text{Gal}(F(\zeta)/K)$ is solvable.

Let $K(\zeta, u_1, \ldots, u_i) = F_i$ and let $\text{Gal}(F(\zeta)/K) = G$, $\text{Gal}(F(\zeta)/K(\zeta)) = N$, and $\text{Gal}(F(\zeta)/F_i) = N_i$, for $i = 1, 2, \ldots, m - 1$. Since F_{i-1} contains all n_ith roots of unity, F_i is the splitting field of $x^{n_i} - a_i$ over F_{i-1}, for some $a_i \in F_{i-1}$. Therefore N_i is a normal subgroup of N_{i-1} and $\text{Gal}(F_i/F_{i-1}) \cong N_{i-1}/N_i$ by the fundamental theorem. Furthermore, by Theorem 8.4.3, $\text{Gal}(F_i/F_{i-1})$ is cyclic. Finally, N is normal in $\text{Gal}(F(\zeta)/K)$, and $G/N \cong \text{Gal}(K(\zeta)/K)$ is abelian by Proposition 8.4.2. The descending chain of subgroups

$$G \supseteq N \supseteq N_1 \supseteq \ldots \supseteq N_m = \{e\}$$

shows that $G = \text{Gal}(F(\zeta)/K)$ is a solvable group.

To prove the converse, assume that the Galois group G of $f(x)$ over K is solvable, and let E be a splitting field for $f(x)$ over K. If $|G| = n$, let ζ be a primitive nth root of unity, and let $F = E(\zeta)$. We will show that $\phi(E) = E$ for any element ϕ of $\text{Gal}(E(\zeta)/K(\zeta))$. To see this, let $a \in E$, and let $g(x)$ be the minimal polynomial of a over K. Then $\phi(a)$ is a root of $g(x)$, and so $\phi(a) \in E$ since E is normal over K. The restriction of ϕ to E defines an element of $\text{Gal}(E/K) = G$. Thus $\text{Gal}(F/K(\zeta))$ is isomorphic to a subgroup of G and hence is solvable. By Proposition 7.6.2 there exists a finite chain of subgroups

$$\text{Gal}(F/K(\zeta)) \supset N_1 \supset \ldots \supset \{e\}$$

such that each subgroup is normal in the one above it and the factor groups N_i/N_{i+1} are cyclic of prime order p_i. By the fundamental theorem of Galois theory there is a corresponding ascending chain of subfields

$$K(\zeta) \subset F_1 \subset \ldots \subset F$$

with $N_i = \text{Gal}(F/F_i)$ and $\text{Gal}(F_{i+1}/F_i) \cong N_i/N_{i+1}$. Since $p_i|n$ and F_i contains a primitive nth root of unity, it contains all p_ith roots of unity, and we can apply Theorem 8.4.4 to show that $f(x)$ is solvable by radicals over $K(\zeta)$, and hence over K. □

This seems an appropriate point at which to remind the student that this fundamental result has motivated almost all of our work on groups and extension fields.

Theorem 7.7.2 shows that S_n is not solvable for $n \geq 5$, and so to give an example of a polynomial equation of degree n that is not solvable by radicals, we only need to find a polynomial of degree n whose Galois group over $\mathbf{Q}$ is S_n. We will give such an example of degree 5 over $\mathbf{Q}$, a special case of a more general construction, which we outline below.

Let m be a positive even integer, and let $n_1 < n_2 < \ldots < n_{k-2}$ be even integers, where k is odd and $k > 3$. Let

$$g(x) = (x^2 + m)(x - n_1)(x - n_2) \cdots (x - n_{k-2}) \ .$$

The polynomial $f(x) = g(x) - 2$ has exactly two nonreal roots in $\mathbf{C}$, if m is chosen large enough, and if k is prime, then its Galois group over $\mathbf{Q}$ is S_k (see Section 4.10 of Jacobson's *Basic Algebra I*). For degree 5, one of the simplest cases is

$$f(x) = (x^2 + 2)(x + 2)(x)(x - 2) - 2 = x^5 - 2x^3 - 8x - 2 \ .$$

To complete the proof that $f(x)$ has Galois group S_5, we need the following group theoretic lemma.

8.4.7 Lemma. *Any subgroup of S_5 that contains both a transposition and a cycle of length 5 must be equal to S_5 itself.*

Proof. By renaming the elements of S_5 we may assume that the given transposition is $(1, 2)$. We can then replace the cycle of length 5 with one of its powers to obtain $(1, 2, a, b, c)$, and then we can again rename the elements so that we may assume without loss of generality that we are given $(1, 2)$ and $(1, 2, 3, 4, 5)$.

We have $(1, 2)(1, 2, 3, 4, 5) = (2, 3, 4, 5)$, and conjugating $(1, 2)$ by powers of $(2, 3, 4, 5)$ gives $(1, 3)$, $(1, 4)$, and $(1, 5)$. Then it follows from the formula

$$(1, n)(1, m)(1, n) = (m, n)$$

that any subgroup of S_5 that contains the two given elements must contain every transposition. □

8.4.8 Theorem. *There exists a polynomial of degree 5 with rational coefficients that is not solvable by radicals.*

Proof. Let $f(x) = x^5 - 2x^3 - 8x - 2$. Then $f'(x) = 5x^4 - 6x^2 - 8$, and the quadratic formula can be used to show that the solutions of $f'(x) = 0$ are $x^2 = 2, -4/5$, yielding two real roots. Then $f(x)$ has one relative maximum and one relative minimum, and since the values of $f(x)$ change sign between -2 and -1, between -1 and 0, and between 2 and 3, it must have precisely three real roots.

By Theorem 8.3.10 (the fundamental theorem of algebra) there exists a splitting field F for $f(x)$ with $F \subseteq \mathbf{C}$. The polynomial $f(x)$ is irreducible by Eisenstein's criterion, and so adjoining a root of $f(x)$ gives an extension of degree 5. By the fundamental theorem of Galois theory, the Galois group of $f(x)$ over $\mathbf{Q}$ must contain a subgroup of index 5, so since its order is divisible by 5, it follows from Cauchy's theorem that it must contain an element of order 5. By Proposition 8.1.4, every element of the Galois group of $f(x)$ gives a permutation of the roots, and so the Galois group is easily seen to be isomorphic to a subgroup of S_5. This subgroup must contain an element of order 5, and it must also contain the transposition that corresponds to the element of the Galois group defined by complex conjugation. Therefore, by the previous lemma, the Galois group must be isomorphic to S_5. Applying Theorem 8.4.6 completes the proof, since S_5 is not a solvable group. $\square$

EXERCISES: SECTION 8.4

1. Show that $2x^5 - 10x + 5$ is irreducible over $\mathbf{Q}$ and is not solvable by radicals.

2. Find the primitive 8th roots of unity in $\mathbf{C}$, and show that they are the roots of the polynomial $x^4 + 1$.

   Hint: See Section A.5 of the appendix.

3. Find the Galois group of $x^5 - 1$ over $\mathbf{Q}$.

4. Find the Galois group of $x^9 - 1$ over $\mathbf{Q}$.

5. Let $\zeta \in \mathbf{C}$ be a primitive 9th root of unity. Show that the roots of the polynomial $x^3 - 3x + 1$ can be expressed as $\zeta + \zeta^8$, $\zeta^2 + \zeta^7$, and $\zeta^4 + \zeta^5$.

   Hint: Use the identity $\zeta^9 = 1$ and the fact that ζ^3 is a primitive third root of unity.

6. The determinant given below is called a **Vandermonde determinant**. Show that

$$\begin{vmatrix} 1 & \zeta_1 & \zeta_1^2 & \cdots & \zeta_1^{n-1} \\ 1 & \zeta_2 & \zeta_2^2 & \cdots & \zeta_2^{n-1} \\ \vdots & \vdots & \vdots & & \vdots \\ 1 & \zeta_n & \zeta_n^2 & \cdots & \zeta_n^{n-1} \end{vmatrix} = \prod_{1 \leq i < j \leq n} (\zeta_j - \zeta_i) .$$

Hint: Use induction. To make the inductive step, start from the right and subtract from each column ζ_n times the column to its left.

7. Let H be a subgroup of S_p, where p is prime. Show that if H contains a transposition and a cycle of length p, then $H = S_p$.

8. Prove that if $f(x) \in \mathbf{Q}[x]$ is irreducible of prime degree p and has exactly two non-real roots in $\mathbf{C}$, then the Galois group of $f(x)$ over $\mathbf{Q}$ is S_p.

   Hint: Show that complex conjugation gives a transposition in the Galois group, and apply Exercise 7.

8.5 Cyclotomic Polynomials

The complex roots of the polynomial $x^n - 1$ are the nth roots of unity. If we let α be the complex number $\alpha = \cos\theta + i\sin\theta$, where $\theta = 2\pi/n$, then $1, \alpha, \alpha^2, \ldots, \alpha^{n-1}$ are each roots of $x^n - 1$, and since they are distinct they must constitute the set of all nth roots of unity. Thus we have

$$x^n - 1 = \prod_{k=0}^{n-1}(x - \alpha^k) .$$

It is clear that the set of nth roots of unity is a cyclic subgroup of $\mathbf{C}^\times$ of order n. Thus there are $\varphi(n)$ generators of the group, which are the **primitive** nth roots of unity. If $d \mid n$, then any element of order d generates a subgroup of order d, which has $\varphi(d)$ generators. By Proposition 3.5.3 there is a one-to-one correspondence between subgroups and positive divisors of n, so there are precisely $\varphi(d)$ elements of order d.

If p is prime, then every nontrivial pth root of unity is primitive, and Corollary 4.4.7 shows that each pth root of unity is a root of the irreducible polynomial $x^{p-1} + x^{p-2} + \ldots + x + 1$, which is a factor of $x^p - 1$. The situation is more complicated when n is not prime. For example,

$$x^4 - 1 = (x - 1)(x + 1)(x^2 + 1) ,$$

and the primitive 4th roots of unity are the roots of $x^2 + 1$, consisting of $\pm i$. We also have

$$x^6 - 1 = (x - 1)(x + 1)(x^2 + x + 1)(x^2 - x + 1) .$$

The primitive cube roots of unity are roots of $x^2 + x + 1$, while the primitive 6th roots of unity are roots of $x^2 - x + 1$.

8.5.1 Definition. *Let n be a positive integer, and let α be the complex number $\alpha = \cos\theta + i\sin\theta$, where $\theta = 2\pi/n$. The polynomial*

$$\Phi_n(x) = \prod_{(k,n)=1,\, 1 \le k < n}(x - \alpha^k)$$

*is called the **nth cyclotomic polynomial**.*

8.5.2 Proposition. *Let n be a positive integer, and let* $\Phi_n(x)$ *be the nth cyclotomic polynomial. The following conditions hold:*

(a) $\deg(\Phi_n(x)) = \varphi(n);$

(b) $x^n - 1 = \prod_{d|n} \Phi_d(x);$

(c) $\Phi_n(x)$ *is monic, with integer coefficients.*

Proof. Let C_n denote the group of nth roots of unity, generated by $\alpha = \cos\theta + i \sin\theta$, where $\theta = 2\pi/n$. Since C_n is cyclic of order n, it is isomorphic to $\mathbf{Z}_n$.

(a) In the definition of Φ_n, we have $\varphi(n)$ linear factors.

(b) If $d|n$, then C_d is a subgroup of C_n, and the primitive dth roots of unity are the elements of C_n of order d. Grouping the linear factors $x - \alpha^k$ of $x^n - 1$ according to the order of α^k gives the required factorization of $x^n - 1$.

(c) We give a proof using induction on the integer n. To begin the induction, we have $\Phi_1(x) = x - 1$ and $\Phi_2(x) = x + 1$. By part (b) we have $x^n - 1 = \Phi_n(x)f(x)$, where $f(x)$ is a monic polynomial, and by the induction hypothesis, $f(x)$ has integer coefficients since it is a product of cyclotomic polynomials of degree less than n. Using the division algorithm over $\mathbf{Q}$, we can write $x^n - 1 = q(x)f(x)$, for some quotient $q(x)$. It is clear from the division algorithm that $q(x)$ has integer coefficients since $f(x)$ is monic, and so uniqueness implies that $\Phi_n(x) = q(x)$, and thus $\Phi_n(x)$ has integer coefficients. $\square$

The proof of Proposition 8.5.2 shows how $\Phi_n(x)$ can be computed inductively. Theorem 8.5.3 will prove that the answer is irreducible over $\mathbf{Q}$. As an example, we compute Φ_{12}.

Example 8.5.1.

To compute Φ_{12} we need to know Φ_d, for the divisors d of 12. We certainly already know that $\Phi_1(x) = x - 1$, $\Phi_2(x) = x + 1$, $\Phi_3(x) = x^2 + x + 1$, and $\Phi_4(x) = x^2 + 1$. To find $\Phi_6(x) = x^2 - x + 1$, we have

$$\Phi_6(x) = \frac{x^6 - 1}{(x - 1)(x + 1)(x^2 + x + 1)}$$

$$= \frac{(x^3 + 1)(x^3 - 1)}{(x + 1)(x^3 - 1)}$$

$$= x^2 - x + 1.$$

Then

$$\Phi_{12}(x) = \frac{x^{12} - 1}{(x - 1)(x + 1)(x^2 + x + 1)(x^2 + 1)(x^2 - x + 1)}$$

$$= \frac{(x^6 + 1)(x^6 - 1)}{(x^2 + 1)(x^6 - 1)}$$

$$= x^4 - x^2 + 1.$$

8.5.3 Theorem. *The nth cyclotomic polynomial* $\Phi_n(x)$ *is irreducible over* **Q**, *for every positive integer n.*

Proof. We give a proof by contradiction. Assume that $\Phi_n(x)$ is not irreducible over **Q**. Then we can factor $\Phi_n(x)$ into a product of irreducible polynomials, and by Theorem 4.4.5 we can assume that they each have integer coefficients. Let $f(x)$ be one of the irreducible factors of $\Phi_n(x)$, and let β be any root of $f(x)$. Note that $f(x)$ must be monic, and is the minimal polynomial of β over **Q**.

Now let p be any prime such that $p \nmid n$. Since $(p, n) = 1$, it follows from the definition of $\Phi_n(x)$ that β^p is a root of $\Phi_n(x)$. Suppose that β^p is not a root of $f(x)$. Then β^p must be a root of a different irreducible factor of $\Phi_n(x)$, say $g(x)$. We thus have $x^n - 1 = f(x)g(x)h(x)$, for some polynomial $h(x)$ with integer coefficients.

For later reference we note that since $g(x)$ has β^p as a root, the polynomial $g(x^p)$ must have β as a root. Since $f(x)$ is the minimal polynomial of β over **Q**, this implies that $g(x^p) = f(x)k(x)$, for some polynomial $k(x)$ with integer coefficients.

Let π be the function which assigns to each polynomial over **Z** the polynomial in $\mathbf{Z}_p[x]$ obtained by reducing each coefficient modulo p. For polynomials $s(x), t(x) \in \mathbf{Z}_p[x]$, we have $(s(x) + t(x))^p = s(x)^p + t(x)^p$, since the binomial coefficient $\binom{p}{i}$ has p as a factor unless $i = 0$ or $i = p$. Furthermore, if $a \in \mathbf{Z}_p$, then $a^p = a$ by Euler's theorem (see Example 3.2.12). Thus if $s(x) = a_m x^m + \ldots + a_1 x + a_0$, then

$$\begin{aligned} s(x)^p &= (a_m x^m + \ldots + a_1 x + a_0)^p = a_m^p x^{mp} + \ldots + a_1^p x^p + a_0^p \\ &= a_m (x^p)^m + \ldots + a_1 x^p + a_0 = s(x^p) \, . \end{aligned}$$

Applying this result to $\pi(g(x))$, we have $(\pi(g(x)))^p = \pi(g(x^p)) = \pi(f(x)k(x)) = \pi(f(x))\pi(k(x))$. (We have used the fact that π preserves products.) In $\mathbf{Z}_p[x]$, this shows that $\pi(g(x))$ and $\pi(f(x))$ must have an irreducible factor in common, so $x^n - 1$ has a repeated factor in $\mathbf{Z}_p[x]$. As in the proof of Proposition 8.2.3, this implies that $x^n - 1$ and its formal derivative nx^{n-1} must have an irreducible factor in common. This is an obvious contradiction, since $p \nmid n$ implies that the only possible irreducible factor is x.

We have obtained a contradiction to our initial supposition that β^p is not a root of $f(x)$. If k is any exponent with $1 \le k < n$ and $(k, n) = 1$, we can write $k = p_1 \cdots p_t$ where $p_1, \ldots, p_t$ are primes that are not divisors of n. Then an induction argument on t shows that β^k must be a root of $f(x)$. We conclude that $\Phi_n(x) = f(x)$, and therefore $\Phi_n(x)$ is irreducible over **Q**. $\square$

8.5.4 Theorem. *For every positive integer n, the Galois group of the nth cyclotomic polynomial* $\Phi_n(x)$ *over* **Q** *is isomorphic to* $\mathbf{Z}_n^\times$.

Proof. If α is a primitive nth root of unity, then the splitting field of $\Phi_n(x)$ over $\mathbf{Q}$ is $\mathbf{Q}(\alpha)$, and it follows from Theorem 8.5.3 that $[\mathbf{Q}(\alpha) : \mathbf{Q}] = \varphi(n)$. Thus if G is the Galois group of $\Phi_n(x)$ over $\mathbf{Q}$, it follows from Theorem 8.1.6 that $|G| = \varphi(n)$ since $\Phi_n(x)$ has no repeated roots over $\mathbf{Q}$. The proof of Proposition 8.4.2 shows that G is isomorphic to a subgroup of $\mathbf{Z}_n^\times$, and so $|G| = \varphi(n)$ implies that $G \cong \mathbf{Z}_n^\times$. $\square$

8.5.5 Corollary. *The Galois group of $\Phi_n(x)$ over $\mathbf{Q}$ is cyclic if and only if n is of the form 2, 4, p^k, or $2p^k$ for an odd prime p.*

Proof. This follows immediately from Theorem 8.5.4 and Corollary 7.5.13, which states the conditions under which $\mathbf{Z}_n^\times$ is cyclic. $\square$

Example 8.5.2 (Constructible polygons).

If a regular n-gon is constructible, then Corollary 6.3.7 can be used to show that a primitive nth root of unity lies in an extension F with $[F : \mathbf{Q}] = 2^k$, for some $k \geq 1$. It follows from Proposition 8.5.2 that $\varphi(n)$ must be a power of 2.

If p is an odd prime, then $\varphi(p^\alpha) = p^{\alpha-1}(p-1)$ is a power of 2 if and only if $\alpha = 1$ and $p - 1$ is a power of 2. Such a prime is called a **Fermat** prime, and must have the form $p = 2^k + 1$, where k is a power of 2 (see Exercise 23 of Section 1.2). The only known examples are 3, 5, $17 = 2^4 + 1$, $257 = 2^8 + 1$, and $65537 = 2^{16} + 1$. The next possibility, $2^{32} + 1$, is divisible by 641.

The complete answer to the constructibility question is that a regular n-gon is constructible if and only if $n = 2^k p_2 \cdots p_m$, where $k \geq 0$, and the factors p_i are distinct Fermat primes. (See Section 4.11 of Jacobson's *Basic Algebra I* for the "if" part of the proof.) $\square$

A set that satisfies all the axioms of a field except for commutativity of multiplication is called a **division ring** or **skew field**. We can use cyclotomic polynomials to give a proof of the following famous theorem of Joseph H. M. Wedderburn (1882–1948).

8.5.6 Theorem (Wedderburn). *Any finite division ring is a field.*

Proof. Let D be finite division ring, and let

$$F = \{x \in D \mid xd = dx \text{ for all } d \in D\},$$

the *center* of D. The verification that F is a field is left as an exercise. It follows from Proposition 6.5.1 that $F = \mathrm{GF}(p^m)$, for some prime number p and some positive integer m. Thus D is a vector space over F, and if it has dimension is n

over F, then D must have q^n elements, where $q = p^m$. To show that $D = F$, which will complete the proof, we only need to show that $n = 1$.

We now consider the class equation of the finite group $D^\times$. (See Definition 7.2.6.) Since the center of $D^\times$ is $F^\times$, we have

$$|D^\times| = |F^\times| + \sum_a [D^\times : C(a)] ,$$

where the sum ranges over one element a from each nontrivial conjugacy class, and $C(a) = \{x \in D^\times \mid xa = ax\}$ is the centralizer of a in $D^\times$. Thus

$$q^n - 1 = (q - 1) + \sum_a [D^\times : C(a)] .$$

For each $a \in D$, let $D_a = \{x \in D \mid xa = ax\}$. It is easy to check that D_a is a division ring, so it is a vector space over F, and must have q^k elements, for some positive integer k. Furthermore, it is clear that $C(a)$ is just the multiplicative group $D_a^\times$ of D_a. It follows from Lagrange's theorem that $q^k - 1$ is a divisor of $q^n - 1$, and Exercise 5 shows that k must be a divisor of n. We have now shown that for each $a \in D^\times$ we have

$$[D^\times : C(a)] = \frac{q^n - 1}{q^k - 1}$$

for some positive integer k such that $k \mid n$. Since $k \mid n$, Proposition 8.5.2 implies that $\Phi_n(x)$ is a divisor of $(x^n - 1)/(x^k - 1)$, and so $\Phi_n(q)$ must be a divisor of $(q^n - 1)/(q^k - 1)$. It follows from the class equation that $\Phi_n(q)$ is a divisor of $q - 1$.

To obtain the desired contradiction, substitute $x = q$ in the equation

$$\Phi_n(x) = \prod_{(j,n)=1, \, 1 \leq j < n} (x - \alpha^j)$$

and consider the magnitude of the resulting complex numbers. We have $|q - \alpha^j| \geq |q| - |\alpha^j| = q - 1$, and so

$$|\Phi_n(q)| = \prod_{(j,n)=1, \, 1 \leq j < n} |(q - \alpha^j)| \geq (q - 1)^{\varphi(n)} .$$

We conclude that $n = 1$, and thus $D = F$, showing that D is a field. $\square$

EXERCISES: SECTION 8.5

1. Compute the following cyclotomic polynomials.

   †(a) Φ_8

   (b) Φ_9

   †(c) Φ_{15} (See the comments after Corollary 4.4.7.)

   (d) Φ_{20}

2. Prove that if n is a power of 2, say $n = 2^k$ with $k > 1$, then $\Phi_n(x) = x^m + 1$, where $m = 2^{k-1}$.

3. Let n be an integer of the form $n = p^k$, where $k \geq 1$ and p is any odd prime. Prove that $\Phi_n(x) = \Phi_p(x^m) = \sum_{i=0}^{p-1} x^{mi}$, where $m = p^{k-1}$. (For example, $\Phi_{27} = x^{18} + x^9 + 1$.)

4. Let n be an integer of the form $n = 2p^k$, where $k \geq 1$ and p is any odd prime. Prove that $\Phi_n(x) = \Phi_p(-x^m) = \sum_{i=0}^{p-1} (-1)^i x^{mi}$, where $m = p^{k-1}$. (For example, $\Phi_{54} = x^{18} - x^9 + 1$.)

   Hint: First prove the result that if $q > 1$ is odd, then $\Phi_{2q}(x) = \Phi_q(-x)$.

5. Let $a \in \mathbf{Z}$ with $a > 1$, and let m, n be positive integers. Prove that if $a^m - 1$ is a divisor of $a^n - 1$, then m is a divisor of n.

6. This exercise extends Proposition 5.1.4 to arbitrary rings, which must satisfy all of the axioms of Definition 5.1.2 with the possible exception of the commutative law for multiplication. Let S be a ring, and let R be a nonempty subset of S. Then R is a subring of S (use Definition 5.1.3) if and only if for all $x, y \in R$, the elements $x - y$ and xy belong to R.

7. Let D be division ring, and let $F = \{x \in D \mid xd = dx \text{ for all } d \in D\}$. Prove that F is a field.

8. Let D be division ring, and let $a \in D$. Prove that $D_a = \{x \in D \mid xa = ax\}$ is a division ring.

9. Show that $\Phi_m(x) = \prod_{n \mid m} (x^n - 1)^{\mu(m/n)}$, where μ is the Möbius function defined in Section 6.6.

10. Verify the following identities for the cyclotomic polynomials $\Phi_n(x)$.

    (a) If p is a prime number, with $p \nmid m$, then $\Phi_{mp^k}(x) = \Phi_{mp}(x^{p^{k-1}})$.

    (b) If p is a prime number, with $p \nmid m$, then $\Phi_{pm}(x)\Phi_m(x) = \Phi_m(x^p)$.

    (c) If $n \geq 2$, then $\Phi_n(x) = \prod_{d \mid n} (1 - x^{n/d})^{\mu(d)}$.

    (d) If $n \geq 3$, and n is odd, then $\Phi_{2n}(x) = \Phi_n(-x)$.

8.6 Computing Galois Groups

In this section we will investigate some techniques that help in actually computing the Galois group of a polynomial. We warn the reader in advance that some of these techniques are based on theoretical results that are beyond the scope of this book.

A reasonable question to ask is whether there are any restrictions on the groups that can occur as Galois groups. It can be shown that for any finite group G there exist a field K and an extension field F with $\text{Gal}(F/K) = G$. On the other hand, whether or not there exists a polynomial in $\mathbf{Q}[x]$ whose Galois group over $\mathbf{Q}$ is G is still an open question.

If we are given a polynomial $f(x)$ of degree n that is irreducible over $\mathbf{Q}$, then we can find some restrictions on the possible candidates for the Galois group of $f(x)$ over $\mathbf{Q}$. To do this, we need the definition of a transitive subgroup of S_n (which also appears in Exercise 14 of Section 7.3).

8.6.1 Definition. *Let G be a group acting on a set S. We say that G acts **transitively** on S if for each pair of elements $x, y \in S$ there exist an element $g \in G$ such that $y = gx$.*

*If G is a subgroup of the symmetric group S_n, then G is called a **transitive** group if it acts transitively on the set $\{1, 2, \ldots, n\}$.*

8.6.2 Proposition. *Let $f(x)$ be a separable polynomial over the field K, with $f(x) = p_1(x)p_2(x) \cdots p_k(x)$ its factorization in $K[x]$ as a product of distinct irreducible polynomials. If F is the splitting field of $f(x)$ over K, then $f(x)$ is irreducible over K if and only if $\mathrm{Gal}(F/K)$ acts transitively on the roots of $f(x)$.*

Proof. Assume that $f(x)$ is irreducible over K. For any roots r_i, r_j of $f(x)$, since $f(x)$ is irreducible there exists an isomorphism from $K(r_i)$ onto $K(r_j)$ that maps r_i to r_j. This can be extended to an automorphism of the splitting field F, yielding an element of $\mathrm{Gal}(F/K)$ which maps r_i to r_j.

Conversely, suppose that $\mathrm{Gal}(F/K)$ acts transitively. Let $p_1(x)$ be the irreducible factor of $f(x)$ which has r_1 as a root. Then since $\mathrm{Gal}(F/K)$ acts transitively, and any of its elements take r_1 to another root of $p_1(x)$, we see that every root of $f(x)$ is a root of $p_1(x)$. Since $f(x) = p_1(x)p_2(x) \cdots p_k(x)$ has distinct factors, it follows that $f(x) = p_1(x)$, and hence $f(x)$ is irreducible over K. $\square$

We can now give a list of the possible Galois groups of equations of small degree. Let $f(x)$ be a separable polynomial over the field K. If $f(x)$ is irreducible of degree 3, then its Galois group over K must be

$$S_3 \text{ or } \mathbf{Z}_3 \ .$$

Using Exercise 5, if $f(x)$ is irreducible of degree 4, then its Galois group over K must be

$$S_4, \quad A_4, \quad D_4, \quad \mathbf{Z}_4, \text{ or } \mathbf{Z}_2 \times \mathbf{Z}_2 \ .$$

Using Exercise 8, if $f(x)$ is irreducible of degree 5, then its Galois group over K must be

$$S_5, \quad A_5, \quad F_{20} \ D_5, \text{ or } \mathbf{Z}_5 \ .$$

Here F_{20} is the Frobenius group of order 20, studied in Exercises 12 and 13 of Section 7.1 and Exercise 14 of Section 7.2. We can identify $\mathbf{Z}_5$ and D_5 with subgroups of F_{20}, with $\mathbf{Z}_5 \subset D_5 \subset F_{20}$. These are the solvable groups on the list, so a polynomial $f(x)$ of degree 5 is solvable by radicals if and only if its Galois group is isomorphic to a subgroup of F_{20}.

Example 8.6.1 (Galois group of $x^4 - 2$ over Q).

In Example 8.3.3 we computed the Galois group of $x^4 - 2$ over **Q**. With our list of transitive subgroups of S_4 in hand, as soon as we know that the order of the Galois group is 8 we can be certain that it is isomorphic to D_4. □

Example 8.6.2 (Galois group of $x^5 - 2$ over Q).

To compute the Galois group of $x^5 - 2$ over **Q**, we first need to find the splitting field of the polynomial. The polynomial is irreducible by Eisenstein's criterion, and it has 5 distinct roots, given by $\sqrt[5]{2}$, $\gamma \sqrt[5]{2}$, $\gamma^2 \sqrt[5]{2}$, $\gamma^3 \sqrt[5]{2}$, and $\gamma^4 \sqrt[5]{2}$, where γ is a primitive 5th root of unity. We have $[\mathbf{Q}(\sqrt[5]{2}) : \mathbf{Q}] = 5$, and since the minimal polynomial of γ over **Q** is $x^4 + x^3 + x^2 + x + 1$, we have $[\mathbf{Q}(\gamma) : \mathbf{Q}] = 4$. The splitting field for $x^5 - 2$ over **Q** is $\mathbf{Q}(\sqrt[5]{2}, \gamma)$, so its degree over **Q** is divisible by 4 and 5, which shows that $[\mathbf{Q}(\sqrt[5]{2}, \gamma) : \mathbf{Q}] = 20$. From our list of transitive subgroups of S_5, it follows that the Galois group over **Q** is F_{20}. □

If $f(x)$ is irreducible of prime degree p over **Q**, then the next three results give some information about the case in which $f(x)$ is solvable by radicals.

8.6.3 Lemma. *Let p be a prime number, and let G be a transitive subgroup of S_p. Then any nontrivial normal subgroup of G is also transitive.*

Proof. Assume that S_p acts on $P = \{1, \ldots, p\}$ in the usual way, that G is a transitive subgroup of S_p, and that N is a nontrivial normal subgroup of G. Since N is nontrivial, $|Nx| > 1$ for some orbit Nx with $x \in P$. For $x, y \in P$, since G is transitive there exists $\tau \in G$ with $\tau(x) = y$. For any $\sigma \in N$, define $f(\sigma x) = \tau\sigma(x) = \tau\sigma\tau^{-1}(y)$. Since N is normal in G, we have $\tau\sigma\tau^{-1} \in N$, and so we have defined a function $f : Nx \to Ny$ from the orbit of x into the orbit of y. This function has an inverse defined by $g(\sigma y) = \tau^{-1}\sigma(y)$, and so Nx and Ny have the same number of elements. Since P is the union of the orbits under N, it follows that $|Nx|$ is a divisor of p, and since $|Nx| > 1$ we must have $|Nx| = p$. Thus $Nx = P$, and so N is also a transitive subgroup of S_p. □

8.6.4 Lemma. *Let p be a prime number, and let G be a solvable, transitive subgroup of S_p. Then G contains a cycle of length p.*

Proof. Since G is solvable, by Proposition 7.6.2 it has a composition series $G = N_0 \supset N_1 \supset \cdots \supset N_{k-1} \supset N_k = \{(1)\}$ in which each factor group is cyclic of prime order. By Lemma 8.6.3, each subgroup in the series is a transitive subgroup, and so N_{k-1} must be cyclic of order p. $\square$

8.6.5 Proposition. *Let p be a prime number, and let G be a solvable, transitive subgroup of S_p. Then G is a subgroup of the normalizer in S_p of a cyclic subgroup of order p.*

Proof. Let $G = N_0 \supset N_1 \supset \cdots \supset N_{k-1} \supset N_k = \{(1)\}$ be a composition series for G, and let N be the normalizer of N_{k-1} in S_p. By Lemma 8.6.4, N_{k-1} is cyclic of order p, and since it is normal in N_{k-2}, it follows that N_{k-2} is contained in N.

We now give a proof by induction, in which we assume that N_i is contained in N and show that N_{i-1} is contained in N. Since N_{k-1} is normal in N_i, it follows that N_{k-1} is the only Sylow p-subgroup of N_i. Because the factor group N_{i-1}/N_i is cyclic of prime order, say $[N_{i-1} : N_i] = q$, it has no proper nontrivial subgroups, and therefore there are no proper subgroups of N_{i-1} that properly contain N_i. Thus the normalizer of N_{k-1} in N_{i-1} is either N_i or N_{i-1}. In the first case we then have q conjugates of N_{k-1} in N_{i-1}, and so the number of Sylow p-subgroups of N_{i-1} is q. This violates the Sylow theorems, since $q < p$ implies that $q \not\equiv 1 \pmod{p}$. We conclude that the normalizer of N_{k-1} in N_{i-1} is N_{i-1}, showing that N_{i-1} is contained in N. $\square$

Suppose that C_p is a cyclic subgroup of S_p of order p, and that $N(C_p)$ is the normalizer of C_p in S_p. It is easy to determine the order of $N(C_p)$, since there are $(p-1)!$ cycles of length p in S_p, and these combine to give $(p-2)!$ cyclic subgroups of order p. These subgroups are conjugate in S_p, since all p-cycles are conjugate, and so the index of $N(C_p)$ in S_p is $(p-2)!$. It follows that $|N(C_p)| = p(p-1)$.

Exercise 9 provides another description of $N(C_p)$. It shows that any such normalizer must be isomorphic to the group H_p constructed as follows (a generalization of F_{20}). Let H_p be the subgroup of $GL_2(\mathbf{Z}_p)$ consisting of all 2×2 matrices of the form $\begin{bmatrix} m & b \\ 0 & 1 \end{bmatrix}$ such that $m \in \mathbf{Z}_p^\times$ and $b \in \mathbf{Z}_p$. This group has order $p(p-1)$, and contains a normal cyclic subgroup of order p, given by all matrices of the form $\begin{bmatrix} 1 & b \\ 0 & 1 \end{bmatrix}$ such that $b \in \mathbf{Z}_p$.

Example 8.6.3 (Galois group of $x^p - 2$ over Q).

In this example we will show that for any prime number p, the Galois group of $x^p - 2$ over $\mathbf{Q}$ is H_p. The polynomial is irreducible over $\mathbf{Q}$ by Eisenstein's criterion, so $[\mathbf{Q}(\sqrt[p]{2}) : \mathbf{Q}] = p$. To obtain a splitting field for $x^p - 2$, in addition to adjoining $\sqrt[p]{2}$ we must adjoin a primitive pth root of unity, say ζ.

Since the minimal polynomial of ζ over $\mathbf{Q}$ is $x^{p-1} + x^{p-2} + \ldots + x + 1$, we have $[\mathbf{Q}(\zeta) : \mathbf{Q}] = p - 1$, and therefore $[\mathbf{Q}(\sqrt[p]{2}, \zeta) : \mathbf{Q}] = p(p - 1)$. The polynomial is solvable by radicals, so by Proposition 8.6.5 its Galois group is isomorphic to a subgroup of H_p. The order of the Galois group is order $p(p - 1)$, so the two groups must coincide. $\square$

Let $f(x)$ be a polynomial of degree n over the field K, and assume that $f(x)$ has roots $r_1, r_2, \ldots, r_n$ in its splitting field F. The element Δ of F defined by

$$\Delta = \prod_{1 \le i < j \le n} (r_i - r_j)^2$$

is called the **discriminant** of $f(x)$.

For example, the discriminant of the polynomial $ax^2 + bx + c$ is the familiar expression $b^2 - 4ac$. For a cubic polynomial $ax^3 + bx^2 + cx + d$, we first assume that $a = 1$, and then the substitution $x = y - b/3$ reduces the polynomial to $y^3 + py + q$, where $p = c - b^2/3$ and $q = d - bc/3 + 2b^3/27$. Because this substitution is linear, it does not change the discriminant of the polynomial. In Section A.6 of the appendix, this discriminant is found to be $\Delta = -4p^3 - 27q^2$.

It can be shown that the discriminant of any polynomial $f(x)$ can be expressed as a polynomial in the coefficients of $f(x)$, with integer coefficients. This requires use of elementary symmetric functions, and lies beyond the scope of what we have chosen to cover in the book.

We have the following properties of the discriminant:

(i) $\Delta \ne 0$ if and only if $f(x)$ has distinct roots;
(ii) $\Delta \in K$;
(iii) if $\Delta \ne 0$, then a permutation $\sigma \in S_n$ is even if and only if it leaves the sign of $\prod_{1 \le i < j \le n} (r_i - r_j)$ unchanged.

The first statement is obvious from the definition of Δ. If $\Delta = 0$, then it certainly belongs to K. If $\Delta \ne 0$, then $f(x)$ has distinct roots, and so the fixed field of its Galois group is K (by Proposition 8.3.3). Any permutation of the roots of $f(x)$ leaves Δ unchanged, and so Δ must belong to K. This verifies part (ii). To prove part (iii), we note that any permutation $\sigma \in S_n$ acts on $\prod_{1 \le i < j \le n} (r_i - r_j)$ by permuting the subscripts, and so (iii) follows directly from Theorem 3.6.6.

8.6.6 Proposition. *Let $f(x)$ be a separable polynomial over the field K, with discriminant Δ, and let F be its splitting field over K. Then every permutation in $\mathrm{Gal}(F/K)$ is even if and only if Δ is the square of some element in K.*

Proof. If every permutation in $\mathrm{Gal}(F/K)$ is even, then every permutation leaves $\prod_{1 \le i < j \le n} (r_i - r_j)$ fixed, so both this element and its square Δ belong to K.

Conversely, if $\Delta = a^2$ for some $a \in K$, then every permutation in $\mathrm{Gal}(F/K)$ must leave $\prod_{1 \le i < j \le n} (r_i - r_j) = \pm a$ fixed, and hence must be even. $\square$

Example 8.6.4 (Galois group of $x^3 - 2$ over Q).

In Example 8.3.2 we showed that the Galois group of the polynomial $x^3 - 2$ over **Q** is S_3. With the new knowledge in this section, we know at the outset that the Galois group is either S_3 or A_3. The discriminant of the polynomial is -216, which is not a square, so the Galois group is not contained in A_3, and therefore must be S_3.   □

Example 8.6.5 (Galois group of $x^3 - 3x + 1$ over Q).

In this example we consider the polynomial $x^3 - 3x + 1$ over **Q**. This is irreducible over **Q**, since the substitution $x - 1$ gives $x^3 - 3x^2 + 3$, which satisfies Eisenstein's criterion. The next step is to compute the discriminant, which is $-4(-3)^3 - 27(1)^2 = 81$, and is thus a square in **Q**. By Proposition 8.6.6 the Galois group is contained in A_3, so it must be equal to A_3.   □

We now restrict our attention to polynomials with rational coefficients. The next lemma shows that in computing Galois groups it is enough to consider polynomials with integer coefficients. Then a powerful technique is to reduce the integer coefficients modulo a prime and consider the Galois group of the reduced equation over the field GF(p).

8.6.7 Lemma. *Let $f(x) = x^n + a_{n-1}x^{n-1} + \ldots + a_1 x + a_0 \in \mathbf{Q}[x]$, and assume that $a_i = b_i/d$ for $d, b_0, b_1, \ldots, b_{n-1} \in \mathbf{Z}$. Then $d^n f(x/d)$ is monic with integer coefficients, and has the same splitting field over $\mathbf{Q}$ as $f(x)$.*

Proof. Exercise.   □

If p is a prime number, we have the natural mapping $\pi : \mathbf{Z}[x] \to \mathbf{Z}_p[x]$ which reduces each coefficient modulo p. We will use the notation $\pi(f(x)) = f_p(x)$.

The next theorem is due to Dedekind. For its proof we refer the reader to Section 8.10 of *Algebra* by van der Waerden or Section 4.16 of *Basic Algebra I* by Jacobson.

Theorem (Dedekind). *Let $f(x)$ be a monic polynomial of degree n, with integer coefficients and Galois group G over $\mathbf{Q}$, and let p be a prime such that $f_p(x)$ has distinct roots. If $f_p(x)$ factors in $\mathbf{Z}_p[x]$ as a product of irreducible factors of degrees $n_1, n_2, \ldots, n_k$, then G contains a permutation with the cycle decomposition*

$$(1, 2, \ldots, n_1)(n_1 + 1, n_1 + 2, \ldots, n_1 + n_2) \cdots (n - n_k + 1, \ldots, n) ,$$

relative to a suitable ordering of the roots.

For the cases $n = 4$ and $n = 5$, Tables 8.6.1 and 8.6.2 list the shapes of the elements of the various transitive subgroups of S_4 and S_5.

Table 8.6.1: Number of Elements of Various Shapes in S_4

	(1)	(a,b)	$(a,b)(c,d)$	(a,b,c)	(a,b,c,d)
S_4	1	6	3	8	6
A_4	1		3	8	
D_4	1	2	3		2
Z_4	1		1		2
V	1		3		

Table 8.6.2: Number of Elements of Various Shapes in S_5

	(1)	(a,b)	$(a,b)(c,d)$	(a,b,c)	(a,b,c,d)	$(a,b,c)(d,e)$	(a,b,c,d,e)
S_5	1	10	15	20	30	20	24
A_5	1		15	20			24
F_{20}	1		5		10		4
D_5	1		5				4
Z_5	1						4

In the proof of Theorem 8.4.8, to exhibit a polynomial of degree 5 that is not solvable by radicals, we showed that a polynomial irreducible of degree 5 over $\mathbf{Q}$ with precisely three real roots must have a Galois group equal to S_5. The next example shows how to use techniques of this section to find polynomials of degree 5 over $\mathbf{Q}$ that are not solvable by radicals.

Example 8.6.6.

As seen in Table 8.6.2, no proper transitive subgroup of S_5 contains a cycle of the form $(a, b, c)(d, e)$. Thus to construct a polynomial with Galois group S_5 it suffices to guarantee that modulo some prime number the polynomial has cubic and quadratic factors.

For example, we can choose the modulus 2 and the irreducible factors $x^3 + x^2 + 1$ and $x^2 + x + 1$. Thus any irreducible polynomial with integer coefficients that reduces modulo 2 to the polynomial $x^5 + x + 1 = (x^3 + x^2 + 1)(x^2 + x + 1)$ will have a Galois group equal to S_5. Thus in the polynomial $f(x) = a_5x^5 + a_4x^4 + a_3x^3 + a_2x^2 + a_1x + a_0$ we need to require that $f(x)$ is irreducible, that the coefficients a_4, a_3, and a_2 are even, while a_5, a_1, and a_0 are odd. For instance, we can choose $f(x) = x^5 + 3x + 3$, which is irreducible over $\mathbf{Q}$ by Eisenstein's criterion. Note that this polynomial has only one real root, since the fact that its derivative is positive everywhere shows it to be an increasing function. $\square$

In the next example we return to the polynomial $x^5 - 2x^3 - 8x - 2$, which was used in Theorem 8.4.8. For this particular polynomial, it would be rather difficult to do the calculations by hand, and this points out the usefulness of programs that can do symbolic computations.

Example 8.6.7.

For the polynomial $x^5 - 2x^3 - 8x - 2$, the following factorizations were found on the computer. They represent the smallest moduli that yield significant information.

Reducing modulo 7, we have the factorization $x^5 - 2x^3 - 8x - 2 = (x^4 - x^3 - x^2 + x - 2)(x + 1)$. After checking that the degree 4 factor has no roots, it is not difficult to show that it cannot be factored into a product of two monic polynomials of degree 2. This is left as an exercise. Reducing modulo 7 shows that the Galois group must contain a 4-cycle, so it must be either S_5 or F_{20}.

Reducing modulo 37, we have the factorization $x^5 - 2x^3 - 8x - 2 = (x^3 - 12x^2 - 11x + 7)(x^2 + 12x + 5)$. It is left as an exercise to show that the cubic factor is irreducible since it has no roots in GF(37). Thus the Galois group must contain a 3-cycle.

This establishes that the Galois group of $x^5 - 2x^3 - 8x - 2$ over $\mathbf{Q}$ is S_5, since from Table 8.6.2 the only transitive subgroup that contains a 4-cycle and a 3-cycle is S_4. □

The previous example shows that the techniques we have discussed are really only practical if you have access to a computer program that is capable of doing symbolic algebra. Even in that case, some additional techniques are likely to be necessary. In Section A.6 of the appendix, the solution by radicals of an equation of degree 4 uses a resolvent equation of degree 3. After a standard substitution the general equation of degree 4 reduces to $y^4 + py^2 + qy + r = 0$. The resolvent equation is $z^3 - pz^2 - 4rz + (4pr - q^2) = 0$, and any real root of this equation leads to a solution of $y^4 + py^2 + qy + r = 0$.

When developing computational techniques for finding Galois groups, it is necessary to consider resolvent equations, in addition to the above techniques that we have mentioned. In higher degrees this becomes rather complicated. For example, a polynomial equation of degree 5 leads to a resolvent equation of degree 6. In any case, factorizations of the resolvent equation yield some information about the original group, because they correspond to shapes of elements in appropriate homomorphic images of the Galois group.

EXERCISES: SECTION 8.6

1. Prove Lemma 8.6.7.

2. Let $f(x)$ be a cubic polynomial that is irreducible over the field K. Prove that if the discriminant of $f(x)$ is a square of some element of K, then its Galois group is cyclic of order 3. Prove that if this is not the case, then the Galois group of $f(x)$ is the symmetric group on 3 elements.

3.†Compute the Galois group of the polynomial $x^5 - x - 1$ over $\mathbf{Q}$, by reducing the coefficients modulo 2 and then modulo 3.

4. Use techniques of this section to find the Galois group of $x^4 + 2x^2 + x + 3$ over $\mathbf{Q}$.

5. Show that the following is a complete list of the transitive subgroups of S_4: (i) S_4; (ii) A_4; (iii) the Sylow 2-subgroups (isomorphic to D_4); (iv) the cyclic subgroups of order 4; and (v) the subgroup $V = \{(1),\ (1, 2)(3, 4),\ (1, 3)(2, 4),\ (1, 4)(2, 3)\}$.

6. Let $p(x) = 2x^5 - 10x + 5$ be the polynomial in Exercise 1 of Section 8.4.
   (a) Check that $p(x) = 2(x^4 + 2x^3 + 4x^2 + x + 4)(x + 5)$ modulo 7.
   (b) Check that $p(x) = 2(x^3 + x^2 + 2x + 3)(x + 3)(x + 7)$ modulo 11.
   (c) Use the techniques of this section to find the Galois group over $\mathbf{Q}$ of $p(x)$.

7. Show that if G is a transitive subgroup of S_n that contains an $(n-1)$-cycle and a transposition, then $G = S_n$.

8. Show that the following is a complete list of the transitive subgroups of S_5: (i) S_5; (ii) A_5; (iii) any cyclic subgroup of order 5; (iv) the normalizer in A_5 of any cyclic subgroup of order 5 (isomorphic to the dihedral group D_5); (v) the normalizer in S_5 of any cyclic subgroup of order 5 (isomorphic to the Frobenius group F_{20}).

9. Let H_p be the subgroup of $GL_2(\mathbf{Z}_p)$ consisting of all 2×2 matrices of the form $\begin{bmatrix} m & b \\ 0 & 1 \end{bmatrix}$ such that $m \in \mathbf{Z}_p^{\times}$ and $b \in \mathbf{Z}_p$. This group is known as the **holomorph** of $\mathbf{Z}_p$. Show that the normalizer in S_p of any cyclic subgroup of order p is isomorphic to H_p.

10. Show that $x^4 - x^3 - x^2 + x - 2$ is irreducible over GF(7).

11. Show that $x^3 - 12x^2 - 11x + 7$ is irreducible over GF(37).

12. Show that over GF(31) the polynomial $x^5 - 2x^3 - 8x - 2$ has $x^3 + 15x^2 + 4x - 1$ and $(x + 8)$ as irreducible factors.

Chapter 9

UNIQUE FACTORIZATION

In the notes at the end of Chapter 1, we discussed "Fermat's last theorem," which states that the equation $x^n + y^n = z^n$ has no solution in the set of positive integers, when $n > 2$. Attempts to prove this theorem led to the development of some very rich areas of mathematics. In this chapter we will discuss unique factorization in commutative rings, and show how this theory provides a proof of Fermat's last theorem in certain cases.

Ernst Kummer (1810–1891) became interested in higher reciprocity laws. (See the discussion of quadratic reciprocity in Section 6.7.) In his research he found it necessary to investigate unique factorization in subrings of the field $\mathbf{C}$ of complex numbers. When he found that elements do not always have a unique factorization into products of primes, he considered sets of numbers that he called "ideal" numbers. Then the "greatest common divisor" of two elements could be represented by the sum of the "ideal numbers" which they determine. This is a generalization of the way in which we defined the greatest common divisor of two integers, and of two polynomials.

In 1847, Gabriel Lamé (1795–1870) announced to the Paris Academy that he had proved Fermat's last theorem. His proof was based on factoring $x^n + y^n$ into a product of linear factors. For example, since the complex number i belongs to $\mathbf{Z}[i] = \{m + ni \mid m, n \in \mathbf{Z}\}$, in this ring we have the factorization

$$x^2 + y^2 = (x + iy)(x - iy) .$$

It is easy to reduce Fermat's last theorem to the cases $n = 4$ and $n = p$, where p is an odd prime. To obtain a factorization in the general case, it is enough to work in a subring which includes a primitive pth root of unity, say α. Then over this subring we have the factorization

$$x^p + y^p = \prod_{i=1}^{p}(x + \alpha^i y) .$$

The proof given by Lamé required unique factorization of the numbers he needed to use in obtaining the factorization of $x^p + y^p$.

The announcement by Lamé created quite a stir, with divided opinions as to whether or not his proof was correct. In fact, Kummer had published his results in 1844, and it was soon realized that this approach could not handle every case of Fermat's last theorem. However, it does work for those values of n for which a unique factorization theorem holds. In Section 9.3 we show how such an argument can be given for the case $n = 3$.

The modern definition of an ideal was given later by Richard Dedekind (1831–1916), in work which appeared in 1871. He proved that in certain subrings of **C** every nonzero ideal can be expressed uniquely as a product of prime ideals.

9.1   Principal Ideal Domains

In our study of the ring of integers **Z** and the ring $F[x]$ of polynomials over a field F, we began with a proof of the division algorithm. Then we were able to define the notion of a greatest common divisor, and finally obtained results on unique factorization.

In this section we will show that the proofs that **Z** and $F[x]$ are principal ideal domains are special cases of a general argument that shows that any integral domain which satisfies a division algorithm must be a principal ideal domain. We then prove that a unique factorization theorem holds in any principal ideal domain.

In stating a division algorithm, we need some notion of the "size" of an element. In **Z**, we used the absolute value of a number, and in $F[x]$ we used the degree of a polynomial. We now need to determine the crucial properties that will enable us to prove that every ideal is principal. First, the size of an element should be a positive integer. For the absolute value of integers we have $|mn| = |m||n|$, while for the degree of nonzero polynomials we have $\deg(f(x)g(x)) = \deg(f(x)) + \deg(g(x))$. To find a common property, we can at least write $|mn| \geq |n|$ and $\deg(f(x)g(x)) \geq \deg(g(x))$. In general, we will use the term **norm** or **degree**, and write $\delta(r)$ for the norm of an element r of a ring. Then we will need to require that $\delta(rs) \geq \delta(s)$.

In several examples our norm will be based on the length of a complex number. Recall that for any complex numbers z and w we have

$$|wz| = (wz\overline{wz})^{1/2} = (w\overline{w}z\overline{z})^{1/2} = (w\overline{w})^{1/2}(z\overline{z})^{1/2} = |w|\,|z|\,.$$

Example 9.1.1 (Norm for Z[i]).

In the ring **Z**[i] of Gaussian integers, each element has the form $m + ni$, for $m, n \in$ **Z**. We can define a norm by letting $\delta(m + ni) = m^2 + n^2$. Then $\delta(a)$ is always a positive integer, provided $a = m + ni \neq 0$. Furthermore, this norm is based on the usual norm for complex numbers. Thus for nonzero elements $a = m + ni$ and $b = p + qi$ in $\mathbf{Z}[i]$ we have

$$\delta(ab) = |ab|^2 = |a|^2|b|^2 \geq |b|^2 = \delta(b) ,$$

since $|a|^2 \geq 1$. $\quad\square$

9.1.1 Definition. *An integral domain D is called a **Euclidean domain** if there exists a function δ from the nonzero elements of D to the nonnegative integers such that*

(i) *$\delta(ab) \geq \delta(b)$ for all nonzero $a, b \in D$, and*

(ii) *for any nonzero elements $a, b \in D$ there exist $q, r \in D$ such that $a = bq + r$, where either $r = 0$ or $\delta(r) < \delta(b)$.*

Example 9.1.2 (Units of a Euclidean domain).

We will show that if D is a Euclidean domain, then an element $a \in D$ is a unit of D if and only if $\delta(a) = \delta(1)$. First, we observe that for any nonzero $x \in D$ we have $\delta(1) \leq \delta(1 \cdot x) = \delta(x)$. If a is a unit of D, with $ab = 1$, then $\delta(a) \leq \delta(ab) = \delta(1)$. Conversely, if $\delta(a) = \delta(1)$, let us write $1 = aq + r$, with $r = 0$ or $\delta(r) < \delta(a)$. The second condition is impossible, since $\delta(r) \geq \delta(1) = \delta(a)$. Hence $r = 0$ and a is a unit. Thus a is a unit if and only if $\delta(a) = \delta(1)$. $\quad\square$

Example 9.1.3 ($\mathbf{Z}[i]$ is a Euclidean domain).

For each nonzero element $a = m + ni$ of $\mathbf{Z}[i]$, we define $\delta(a) = m^2 + n^2$. We have already shown that if $b = s + ti$ is nonzero, then $\delta(ab) \geq \delta(b)$. For nonzero elements a and b, we have $a/b \in \mathbf{C}$. Let $a/b = x + yi$, where $x, y \in \mathbf{R}$. We can choose integers u and v such that $|x - u| \leq \frac{1}{2}$ and $|y - v| \leq \frac{1}{2}$. Set $q = u + vi$, and let $r = a - bq$. If $r \neq 0$, then

$$r = a - bq = b\left(\frac{a}{b} - q\right) ,$$

and so

$$
\begin{aligned}
\delta(r) &= |b((x + yi) - (u + vi))|^2 = |b|^2|(x - u) + (y - v)i|^2 \\
&= |b|^2((x - u)^2 + (y - v)^2) \leq |b|^2\left(\frac{1}{4} + \frac{1}{4}\right) .
\end{aligned}
$$

Thus $\delta(r) \leq \delta(b)/2$, and so we have verified the division algorithm in $\mathbf{Z}[i]$, making it a Euclidean domain. $\quad\square$

Example 9.1.4 (Units of Z[i]).

Example 9.1.2 allows us to find the units of $\mathbf{Z}[i]$. Since $m + ni$ is a unit if and only if $\delta(m + ni) = 1$, the only units are ± 1 and $\pm i$. Because i has order 4, the group of units $\mathbf{Z}[i]^\times$ is isomorphic to $\mathbf{Z}_4$. □

9.1.2 Theorem. *Any Euclidean domain is a principal ideal domain.*

Proof. Let D be a Euclidean domain, and let I be any nonzero ideal of D. By the well-ordering principle there is an element $d \in I$ whose norm $\delta(d)$ is minimal in the set $\{\delta(x) \mid x \in I$ and $x \neq 0\}$. We claim that $I = dD$. Since $d \in I$, it is clear that $dD \subseteq I$, and so we only need to show that $I \subseteq dD$. Given any $a \in I$ we can write $a = dq + r$, where $r = 0$ or $\delta(r) < \delta(d)$. We note that $r = a - dq \in I$ since $a, d \in I$. But then the remainder r cannot be nonzero, since this would give $\delta(r) < \delta(d)$, contradicting the way in which d was chosen. We conclude that $a = dq \in dD$ and thus $I = dD$. □

In any commutative ring R it is possible to develop a general theory of divisibility. For $a, b \in R$, we say that b is a **divisor** of a if $a = bq$ for some $q \in R$. We also say that a is a **multiple** of b, and we will use the standard notation $b|a$ when b is a divisor of a. Divisibility can also be expressed in terms of the principal ideals generated by a and b. Since aR is the smallest ideal that contains a, we have $aR \subseteq bR$ if and only if $a \in bR$, and this occurs if and only if $a = bq$ for some $q \in R$. Thus we have shown that $aR \subseteq bR$ if and only if $b|a$.

Let a and b be elements of a commutative ring R. Then a is called an **associate** of b if $a = bu$ for some unit $u \in R$. Of course, if $a = bu$, then $b = au^{-1}$, and so b is also an associate of a. In the ring $\mathbf{Z}$, two integers are associates if they are equal or differ in sign. In the ring $F[x]$, since the only units are the nonzero constant polynomials, two polynomials are associates if and only if one is a constant multiple of the other. Part (d) of the following proposition shows that for nonzero elements a and b of any integral domain, $aR = bR$ if and only if a and b are associates.

9.1.3 Proposition. *Let R be a commutative ring with identity, with $a, b, c \in R$.*

(a) *If $c|b$ and $b|a$, then $c|a$.*

(b) *If $c|a$, then $c|ab$.*

(c) *If $c|a$ and $c|b$, then $c|(ax + by)$, for any $x, y \in R$.*

(d) *If R is an integral domain, a is nonzero, and both $b|a$ and $a|b$ hold, then a and b are associates.*

Proof. (a) If $b = cq_1$ and $a = bq_2$, then $a = c(q_1 q_2)$.

(b) If $a = cq$, then $ab = c(qb)$.

(c) If $a = cq_1$ and $b = cq_2$, then $ax + by = c(q_1 x + q_2 y)$.

(d) If $a = bq_1$ and $b = aq_2$, then substituting for b in the first equation yields $a = aq_2q_1$. If R is an integral domain, then we can cancel a to obtain $1 = q_2q_1$. Thus q_1 and q_2 are units, showing that a and b are associates. $\square$

If I and J are ideals of the commutative ring R, then their **sum** is defined to be

$$I + J = \{x \in R \mid x = a + b \text{ for some } a \in I, \ b \in J\}.$$

It is easy to check that $I + J$ is an ideal of R.

9.1.4 Definition. *Let $a_1, \ldots, a_n$ be elements of a commutative ring R. A nonzero element d of R is called a **greatest common divisor** of $a_1, \ldots, a_n$ if*

(i) *$d \mid a_i$ for $1 \leq i \leq n$, and*

(ii) *if $c \mid a_i$ for $1 \leq i \leq n$, for $c \in R$, then $c \mid d$.*

If d and d' are greatest common divisors of a and b, then $d \mid d'$ and $d' \mid d$, that is, d and d' are associates. Conversely, any associate of a greatest common divisor of a and b is also a greatest common divisor of a and b. In the rings $\mathbf{Z}$ and $F[x]$, where F is a field, we required the greatest common divisors to be nonnegative integers and monic polynomials, respectively. Thus we achieved uniqueness by excluding all but one member of each family of associates. There is no natural way of imposing uniqueness in arbitrary Euclidean domains.

9.1.5 Lemma *Let R be a commutative ring, and let $a, b, d \in R$, with d nonzero. If $aR + bR = dR$, then d is a greatest common divisor of a and b.*

Proof. If $aR + bR = dR$, then $a, b \in dR$, so $d \mid a$ and $d \mid b$. If $c \mid a$ and $c \mid b$, then $a, b \in cR$, and so $aR + bR \subseteq cR$, since $aR + bR$ is the smallest ideal that contains both a and b. Hence $dR = aR + bR \subseteq cR$, and so $c \mid d$. $\square$

9.1.6 Proposition. *Let D be a principal ideal domain. If a and b are nonzero elements of D, then D contains a greatest common divisor of a and b, of the form $as + bt$ for $s, t \in D$. Furthermore, any two greatest common divisors of a and b are associates.*

Proof. Given nonzero elements a and b, since D is a principal ideal domain we can choose a nonzero generator d for the ideal $aR + bR$. By the preceding lemma, d is a greatest common divisor of a and b, and has the form $as + bt$ for some $s, t \in R$. If d' is another greatest common divisor, then $d' \mid d$ and $d \mid d'$, and so d and d' are associates. $\square$

We will now begin our development of a theory of unique factorization for principal ideal domains. We first need to define an analog of prime numbers in $\mathbf{Z}$ and irreducible polynomials in $F[x]$. We also need to take into account the fact that factorizations are unique only up to units.

9.1.7 Definition. *Let R be a commutative ring. A nonzero element p of R is said to be **irreducible** if*

   (i) *p is not a unit of R, and*

   (ii) *if $p = ab$ for $a, b \in R$, then either a or b is a unit of R.*

The first result on irreducible elements generalizes a familiar property of integers and polynomials.

9.1.8 Proposition. *Let p be an irreducible element of the principal ideal domain D. If $a, b \in D$ and $p|ab$, then either $p|a$ or $p|b$.*

Proof. Assume that $p|ab$. If $p \nmid a$, then the only common divisors of p and a are units, so 1 is a greatest common divisor of p and a. Thus there exist $q_1, q_2 \in D$ such that $1 = pq_1 + aq_2$, and multiplying by b gives $b = p(bq_1) + (ab)q_2$. Finally, since $p|ab$ we must have $p|b$. $\square$

9.1.9 Proposition. *Let D be a principal ideal domain, and let p be a nonzero element of D. Then p is irreducible in D if and only if pD is a prime ideal of D.*

Proof. Assume first that p is irreducible in D. If $ab \in pD$ for $a, b \in D$, then Proposition 9.1.8 shows that either $a \in pD$ or $b \in pD$, and thus pD is a prime ideal.

Conversely, assume that pD is a prime ideal of D. If p is not irreducible, then $p = ab$ for some nonunits a, b of D. By assumption, $ab \in pD$ implies $a \in pD$ or $b \in pD$, so either $p \mid a$ or $p \mid b$. If $p \mid a$, then $a = pc$ for some c, and so $p = pcb$. In a domain the cancellation law holds, so we have $1 = cb$, contradicting the assumption that b is not a unit. If $p \mid b$, we obtain a similar contradiction. Since pD is a proper ideal, p is not a unit, and thus p is irreducible. $\square$

9.1.10 Definition. *Let D be an integral domain. Then D is called a **unique factorization domain** if*

   (i) *each nonzero element a of D that is not a unit can be expressed as a product of irreducible elements of D, and*

   (ii) *in any two such factorizations $a = p_1 p_2 \cdots p_n = q_1 q_2 \cdots q_m$ the integers n and m are equal and it is possible to rearrange the factors so that q_i is an associate of p_i, for $1 \leq i \leq n$.*

9.1.11 Lemma. *Let D be a principal ideal domain. In any collection of ideals $I_1 \subseteq I_2 \subseteq I_3 \subseteq \ldots$, there is a subscript m such that $I_n = I_m$ for all $n > m$.*

Proof. Let $I_1 \subseteq I_2 \subseteq I_3 \subseteq \ldots$ be any ascending chain of ideals, and let $I = \cup_{n=1}^{\infty} I_n$. If $x, y \in I$, then $x \in I_j$ and $y \in I_k$ for some j, k. We can assume that $j \geq k$, and then $x, y \in I_j$, so $x + y \in I_j \subseteq I$. If $r \in D$, then $rx \in I_j \subseteq I$. Thus I is an ideal of D, and so $I = aD$ for some $a \in D$. But then $a \in I_m$ for some m, and so it follows that $I = aD \subseteq I_m$, and thus $I_n = I = I_m$ for all $n > m$. $\square$

9.1.12 Theorem. *Any principal ideal domain is a unique factorization domain.*

Proof. Let D be a principal ideal domain, and let d be a nonzero element of D that is not a unit. Suppose that d cannot be written as a product of irreducible elements. Then d is not irreducible, and so $d = a_1 b_1$, where neither a_1 nor b_1 is a unit and either a_1 or b_1 cannot be written as a product of irreducible elements. Assume that a_1 cannot be written as a product of irreducible elements. Now b_1 is not a unit, and so we have $dD \subset a_1 D$. We can continue this argument to obtain a factor a_2 of a_1 that cannot be written as a product of irreducible elements and such that $a_1 D \subset a_2 D$. Thus the assumption that d cannot be written as a product of irreducible elements allows us to construct a strictly ascending chain of ideals

$$dD \subset a_1 D \subset a_2 D \subset a_3 D \subset \ldots \, .$$

According to Lemma 9.1.11, this contradicts the fact that D is a principal ideal domain.

Now suppose that d can be written in two ways as a product of irreducible elements, say $d = p_1 p_2 \cdots p_n = q_1 q_2 \cdots q_m$, where $n \leq m$. We will proceed by induction on n. Since p_1 is irreducible and a divisor of $q_1 q_2 \cdots q_m$, it follows that p_1 is a divisor of q_i, for some i, and we can assume that $i = 1$. Since both p_1 and q_1 are irreducible and $p_1 | q_1$, it follows that they are associates. Then we can cancel to obtain $p_2 \cdots p_n = u q_2 \cdots q_m$, where u is a unit, and so by induction we have $n - 1 = m - 1$, and we can rearrange the elements q_i so that q_i and p_i are associates. $\square$

EXERCISES: SECTION 9.1

1. An element p of an integral domain D is called **prime** if p is not a unit of D and $p \mid ab$ implies $p \mid a$ or $p \mid b$, for all $a, b \in D$. Prove that in a principal ideal domain a nonzero element is prime if and only if it is irreducible.

2. Let D be a Euclidean domain with norm δ. Prove that if $a, b \in D$ such that $b | a$ but $a \nmid b$, then $\delta(b) < \delta(a)$.

prime 3. Let D be a Euclidean domain, with norm δ, and assume that $\delta(x)\,\delta(y) = \delta(xy)$ for all $x, y \in D$. Prove that $\delta(1) = 1$, and that if $\delta(a)$ is a prime number, then a is irreducible in D.

4. Prove that in any unique factorization domain, any two nonzero elements have a greatest common divisor.

5. Let D be a unique factorization domain. Formulate a definition of the least common multiple of two elements, and show that if a, b are nonzero elements of D, then ab is an associate of the product of the greatest common divisor and the least common multiple of a and b.

gcd 6. Let R be a subring of the commutative ring S. Show that if d is a greatest common divisor of a and b in R with $d = ra + tb$ for some $r, t \in R$, then d is a greatest common divisor of a and b in S.

7. Let $S = F[x, y, z]$, where F is a field, and let $R = \{f(xy, xz) \mid f(u, v) \in F[u, v]\}$. Show that for the rings $R \subseteq S$ the conclusion of Exercise 6 is false without the assumption that $d = ra + tb$ for some $r, t \in R$.

8. Let R be the ring $\mathbf{Z}[x]/\langle 5x \rangle$.

   (a) Show that R is not an integral domain.

   (b) Show that the elements $x + \langle 5x \rangle$ and $2x + \langle 5x \rangle$ of R are each divisors of the other, but are not associates.

9. Show that $\mathbf{Z}[x]$ is not a principal ideal domain.

10. Prove that in a principal ideal domain every proper ideal is contained in a maximal ideal.

11. Let a and b be integers, with $n = a^2 + b^2$.

   (a) Show that if $(a, b) = 1$, then $\mathbf{Z}[i]/\langle a + bi \rangle \cong \mathbf{Z}_n$.

   (b) Show that if $(a, b) \neq 1$, then $\mathbf{Z}[i]/\langle a + bi \rangle$ is not isomorphic to $\mathbf{Z}_m$, for any positive integer m.

(unit) 12. Let D be an integral domain for which there exists a function ∂ from the nonzero elements of D to the nonnegative integers such that for any nonzero elements $a, b \in D$ there exist $q, r \in D$ such that $a = bq + r$, where either $r = 0$ or $\partial(r) < \partial(b)$.

   (a) Show that if $\partial(ub) = \partial(b)$ for all $b \in D$ and all units $u \in D$, then $\partial(ab) \geq \partial(b)$ for all nonzero $a, b \in D$.

   (b) Show that D is a Euclidean domain.

   Hint: For each nonzero element $a \in D$, let $\delta(a) = \min_{u \in D^\times}\{\partial(ua)\}$.

13. Let S be a subring of $\mathbf{R}$ for which there exists a unit $u \in S^\times$ with the property that $u > 1$, but there are no units between 1 and u. Prove that $S^\times \cong \mathbf{Z}_2 \times \mathbf{Z}$.

14. Let $S = \mathbf{Z}[\sqrt{2}]$. Show that $S^\times \cong \mathbf{Z}_2 \times \mathbf{Z}$. (See Exercise 4 of Section 5.1.)

9.2 Unique Factorization Domains

If F is a field, then we know that the polynomial ring $F[x]$ is a Euclidean domain, and therefore a unique factorization domain. If we relax the requirements so that the coefficients no longer need to form a field, but just a ring, then we may lose the division algorithm, and so the polynomial ring may not be a principal ideal domain. In fact, as shown by Exercise 9 of Section 9.1, the ring $\mathbf{Z}[x]$ of polynomials with integer coefficients is not a principal ideal domain. In this section we will show that all is not lost, since if the coefficients come from a unique factorization domain D, then the ring of polynomials $D[x]$ still has unique factorization.

Even though D is a principal ideal domain, the polynomial ring $D[x]$ need not have the same property. In fact, $D[x]$ is only a principal ideal domain if D is a field. But a weaker form of this property does hold, in the sense that every ideal of $D[x]$ can be generated by a finite number of elements. In honor of Emmy Noether, a commutative ring R is said to be **Noetherian** if each ideal of R has a finite set of generators. That is, any ideal I of R has a set of generators $x_1, x_2, \ldots, x_n \in I$, such that for any $x \in I$ there exist $a_1, a_2, \ldots, a_n \in R$ with $x = a_1 x_1 + a_2 x_2 + \ldots + a_n x_n$. In the exercises, we outline a proof of the theorem that if R is a Noetherian ring, then so is $R[x]$.

We begin the section with a generalization of Euclid's lemma (Lemma 1.2.5).

9.2.1 Lemma. *Let D be a unique factorization domain, and let p be an irreducible element of D. If $a, b \in D$ and $p|ab$, then $p|a$ or $p|b$.*

Proof. Assume that p is irreducible and that $p|ab$. Then $ab = pc$ for some $c \in D$, and we can assume that $ab \neq 0$. Let $a = q_1 q_2 \cdots q_n$, $b = r_1 r_2 \cdots r_m$, and $c = p_1 p_2 \cdots p_k$ be the factorizations of a, b, and c into products of irreducible elements. Then ab has the two factorizations $ab = q_1 q_2 \cdots q_n r_1 r_2 \cdots r_m = p p_1 p_2 \cdots p_k$, and it follows that from the definition of a unique factorization domain that p is an associate of q_i or r_j, for some i or j. Thus $p|a$ or $p|b$. $\square$

9.2.2 Proposition. *Let D be a unique factorization domain. Any finite set of nonzero elements of D has a greatest common divisor in D.*

Proof. Let D be a unique factorization domain, and let $a_1, \ldots, a_n$ be a set of nonzero elements of D. If one of these elements is a unit, then it is clear that it is a greatest common divisor of the set. If not, for each i we consider a factorization of a_i into a product of irreducible elements, and let $p_1, \ldots, p_m$ be the collection of all irreducible elements that occur in these factorizations. By choosing one irreducible element among each set of associates p_j's, we can now write a_i in the form $a_i = u_i p_1^{\alpha_{1i}} p_2^{\alpha_{2i}} \cdots p_k^{\alpha_{ki}}$, where u_i is a unit, and the elements $p_1, \ldots, p_k$ are irreducible and not associates. We allow the exponents α_{ji} to be nonnegative integers, so that we can use a common set of irreducible factors. For each j with $1 \leq j \leq k$, let $\mu_j = \min\{\alpha_{j1}, \ldots, \alpha_{jn}\}$, and let $d = p_1^{\mu_1} p_2^{\mu_2} \cdots p_k^{\mu_k}$. Clearly $d|a_i$

for $1 \leq i \leq n$. If $c|a_i$ for $1 \leq i \leq n$, then $c = wp_1^{\gamma_1} p_2^{\gamma_2} \cdots p_k^{\gamma_k}$, where w is a unit, and $0 \leq \gamma_j \leq \alpha_{ji}$ for $1 \leq i \leq n$. Hence $\gamma_j \leq \mu_j$ for $1 \leq j \leq k$, and so $c|d$. We conclude that d is a greatest common divisor of $a_1, \ldots, a_n$.   □

9.2.3 Definition. *Let D be a unique factorization domain. We say that the elements $d_1, d_2, \ldots, d_n \in D$ are* **relatively prime in D** *if there is no irreducible element $p \in D$ such $p \mid a_i$ for each $i = 1, \ldots, n$.*

9.2.4 Definition. *Let D be a unique factorization domain. A nonconstant polynomial $f(x) = a_n x^n + a_{n-1} x^{n-1} + \ldots + a_1 x + a_0$ in $D[x]$ is called* **primitive** *if its coefficients are relatively prime in D.*

Note that any nonconstant factor of a primitive polynomial is primitive.

9.2.5 Theorem. *Let D be a unique factorization domain. The product of two primitive polynomials in $D[x]$ is primitive in $D[x]$.*

Proof. Let p be any irreducible element, and let $f(x) = g(x)h(x)$, where $g(x)$ and $h(x)$ are primitive polynomials in $D[x]$ and $f(x) = a_m x^m + \ldots + a_1 x + a_0$, $g(x) = b_n x^n + \ldots + b_1 x + b_0$, and $h(x) = c_k x^k + \ldots + c_1 x + c_0$. If b_s and c_t are the coefficients of $g(x)$ and $h(x)$ of least index not divisible by p, then a_{s+t} is the coefficient of $f(x)$ of least index not divisible by p. This is proved by observing that the coefficient a_{s+t} of $f(x)$ is equal to

$$a_{s+t} = b_0 c_{s+t} + b_1 c_{s+t-1} + \ldots + b_{s-1} c_{t+1} + b_s c_t + b_{s+1} c_{t-1} + \ldots + b_{s+t} c_0 .$$

We know that $p \nmid b_s c_t$, since p is irreducible and $p \nmid b_s$ and $p \nmid c_t$. Because each of the coefficients $b_0, b_1, \ldots, b_{s-1}$ and $c_{t-1}, \ldots, c_0$ is divisible by p by assumption, each term in the above sum is divisible by p, except for $b_s c_t$. Therefore a_{s+t} is not divisible by p. It is clear that in any coefficient of $f(x)$ of lower index, each term in the sum is divisible by p, and thus a_{s+t} is the coefficient of least index not divisible by p.

Since $g(x)$ and $h(x)$ are primitive, each has a coefficient not divisible by p, so it follows that $f(x)$ has at least one coefficient not divisible by p. Since this is true for every irreducible element p of D, we conclude that $f(x)$ is primitive.   □

Let D be a unique factorization domain. For any polynomial $f(x) \in D[x]$ of positive degree, we can write $f(x) = df^*(x)$, where $f^*(x)$ is primitive, and d is a greatest common divisor of the nonzero coefficients of $f(x)$. To see this, we only need to observe that if $f(x) = a_n x^n + a_{n-1} x^{n-1} + \ldots + a_1 x + a_0 \in D[x]$, then by Proposition 9.2.2 the nonzero coefficients of $f(x)$ have a greatest common divisor $d \in D$. It is clear that factoring out d leaves a primitive polynomial. In Lemma 9.2.6 we will obtain a stronger result, stated for polynomials over the field of quotients of D. Recall that by Theorem 5.4.4, an integral domain D has a field of quotients Q in which each element has the form a/b, where $a, b \in D$ and $b \neq 0$.

9.2.6 Lemma. *Let D be a unique factorization domain.*

(a) *Let $a, b, c, d \in D$, and suppose that ad and bc are associates. If a and b are relatively prime, and c and d are relatively prime, then a and c are associates, and b and d are associates.*

(b) *Let Q be the quotient field of D, and let $f(x) \in Q[x]$. Then $f(x)$ can be written in the form $f(x) = (a/b)f^*(x)$, where $f^*(x)$ is a primitive element of $D[x]$, $a, b \in D$, and a and b are relatively prime.*

This expression is unique up to associates. That is, if $(a/b)f^*(x) = (c/d)g^*(x)$, where $g^*(x)$ is primitive and c and d are relatively prime in D, then a and c are associates, b and d are associates, and $f^*(x)$ and $g^*(x)$ are associates.

Proof. (a) Suppose that $ad = ubc$ for some unit $u \in D$, where a and b are relatively prime. Let p^α be a factor of a, where p is irreducible in D. Since a and b are relatively prime, they have no irreducible factors in common, and so Lemma 9.2.1 implies that p^α is a factor of c. Similarly, since c and d are relatively prime, each term in a factorization of c is also a factor of a. It follows that a and c are associates. It can be shown in the same way that b and d are associates.

(b) Let Q denote the quotient field of D, described as elements of the form a/b, where $a, b \in D$ and $b \neq 0$. The required factorization of $f(x)$ can be found as follows: let

$$f(x) = (a_n/b_n)x^n + \ldots + (a_1/b_1)x + (a_0/b_0) ,$$

where $a_i, b_i \in D$ for $0 \leq i \leq n$. We can find a common nonzero multiple of the denominators $b_0, b_1, \ldots, b_n$, say t, and then $f(x) = (1/t) \cdot tf(x)$, and $tf(x)$ has coefficients in D. Next we can write $tf(x) = sf^*(x)$, where $f^*(x)$ is primitive in $D[x]$. Finally, we can factor out common irreducible divisors of s and t to reduce the fraction s/t to a fraction a/b in which a and b are relatively prime. This finally gives us the required form $f(x) = (a/b)f^*(x)$.

To show uniqueness, let $(a/b)f^*(x) = (c/d)g^*(x)$, where $g^*(x)$ is primitive and c and d are relatively prime. Then $adf^*(x) = bcg^*(x)$, so the irreducible factors of ad and bc must be the same. Since a and b are relatively prime, and c and d are relatively prime, part (a) shows that a and c are associates and that b and d are associates. Thus $f^*(x)$ and $g^*(x)$ are associates. $\square$

9.2.7 Proposition. *Let D be a unique factorization domain, let Q be the quotient field of D, and let $f(x)$ be a primitive polynomial in $D[x]$. Then $f(x)$ is irreducible in $D[x]$ if and only if $f(x)$ is irreducible in $Q[x]$.*

Proof. If $f(x)$ is irreducible in $Q[x]$, then any factorization of $f(x)$ in $D[x]$ already takes place in $Q[x]$, and so $f(x)$ cannot have a proper factorization into polynomials of lower degree. Since $f(x)$ is primitive in $D[x]$, it cannot have a factor of degree 0 that is not a unit.

On the other hand, suppose that $f(x)$ is irreducible in $D[x]$, but has a proper factorization in $Q[x]$. Then since $f(x)$ is primitive, the only factorization into nonunits of $Q[x]$ must have the form $f(x) = g(x)h(x)$, where $g(x)$ and $h(x)$ belong to $Q[x]$ and both have lower degree than $f(x)$. We can then write $g(x) = (a/b)g^*(x)$ and $h(x) = (c/d)h^*(x)$, where $g^*(x)$ and $h^*(x)$ are primitive polynomials in $D[x]$.

The next step is to factor out common irreducible divisors of ac and bd, to obtain s/t, where $s, t \in D$ and s and t are relatively prime. Now $tf(x) = sg^*(x)h^*(x) \in D[x]$, and if p is any irreducible factor of t, then p is a divisor of every coefficient of $sg^*(x)h^*(x)$. Since p is not a divisor of s, by our choice of s and t, it follows that p is a divisor of every coefficient of $g^*(x)h^*(x)$. By Theorem 9.2.5, the product of primitive polynomials in $D[x]$ is again primitive, and so this cannot occur.

We conclude that t is a unit of D, and so the factorization $f(x) = g(x)h(x)$ in $Q[x]$ actually gives rise to a factorization $f(x) = tg^*(x)h^*(x)$ in $D[x]$. This construction does not change the degrees of the polynomials involved in the factorization, and so it contradicts the assumption that $f(x)$ is irreducible in $D[x]$.   □

9.2.8 Theorem. *If D is a unique factorization domain, then so is the ring $D[x]$ of polynomials with coefficients in D.*

Proof. We first show that each nonzero element of $D[x]$ that is not a unit can be written as a product of irreducible elements. Let $f(x)$ be a nonzero element of $D[x]$, and assume that $f(x)$ is not a unit. We will use induction on the degree of $f(x)$.

As the first step in the proof we will write $f(x) = df^*(x)$, where $d \in D$ and $f^*(x)$ is primitive. If $f(x)$ has degree 0, then d is not a unit. We can use the fact that D is a unique factorization domain to express d as a product of irreducible elements of D, and these elements are still irreducible in $D[x]$. Now assume that $f(x)$ has degree n, and that the result holds for all polynomials of degree less than n. If $f^*(x)$ is not irreducible, then it can be expressed as a product of polynomials of lower degree, which must each be primitive, and the induction hypothesis can be applied to each factor. If d is a unit we are done, and if not we can factor d in D. The combined factorizations give the required factorization of $df^*(x)$.

Now we must show that the factorization is unique, up to associates. If $f(x)$ is primitive, then it can be factored into a product of irreducible primitive polynomials, and these are irreducible over $Q[x]$. Since $Q[x]$ is a unique factorization domain, in any two factorizations the irreducible factors of one factorization will be associates in $Q[x]$ of the irreducible factors of the other. Lemma 9.2.6 implies that they will be associates in $D[x]$. If $f(x)$ is not primitive, we can write $f(x) = df^*(x)$, where $f^*(x)$ is primitive. If we also have $f(x) = cg^*(x)$, then by Lemma 9.2.6, c and d must be associates, and $f^*(x)$ and $g^*(x)$ must be associates. The previous remarks about primitive polynomials together with the fact that D is a unique factorization domain finish the proof.   □

9.2.9 Corollary. *For any field F, the ring of polynomials $F[x_1, x_2, \ldots, x_n]$ in n indeterminates is a unique factorization domain.*

Proof. Since $F[x_1]$ is a unique factorization domain, so is the ring $(F[x_1])[x_2]$, and this is identical to the ring $F[x_1, x_2]$. This argument can be extended by induction to the ring of polynomials in any finite number of indeterminates. $\square$

We end this section with two examples of integral domains which do not enjoy the unique factorization property.

 Example 9.2.1 ($Z[\sqrt{-5}]$ is not a unique factorization domain).

It is easily checked that $D = \{m + n\sqrt{-5} \mid m, n \in Z\}$ is a subdomain of C. For $a = m + n\sqrt{-5} \in D$, define $\delta(a) = a\bar{a} = m^2 + 5n^2$. We first note that if a is a unit in D, then $\delta(a)\delta(a^{-1}) = \delta(aa^{-1}) = 1$, and so we must have $\delta(a) = 1$. The only possibility is that $m = \pm 1$ and $n = 0$, and so the units of D are ± 1.

Multiplying $1 + 2\sqrt{-5}$ by its conjugate produces the integer 21, which of course can also be factored in Z. Thus we have two factorizations of 21:

$$21 = 3 \cdot 7 \quad \text{and} \quad 21 = (1 + 2\sqrt{-5})(1 - 2\sqrt{-5}) .$$

If we can show that $3, 7, 1 + 2\sqrt{-5}$, and $1 - 2\sqrt{-5}$ are each irreducible in D, then we will have shown that D is not a unique factorization domain.

It is easy to check that there are no integer solutions to the equations $m^2 + 5n^2 = 3$ and $m^2 + 5n^2 = 7$, and hence D has no elements with norm 3 or 7. If $3 = ab$, where $a, b \in D$ are not units, then $\delta(a)\delta(b) = \delta(3) = 9$, and so $\delta(a) > 1$ and $\delta(b) > 1$ together imply that $\delta(a) = 3$ and $\delta(b) = 3$, a contradiction. A similar argument shows that 7 is irreducible in D. If $1 + 2\sqrt{-5} = ab$, where $a, b \in D$ are not units, then $\delta(a)\delta(b) = \delta(1 + 2\sqrt{-5}) = 21$. This implies $\delta(a) = 3$ or $\delta(a) = 7$, a contradiction. The same argument implies that $1 - 2\sqrt{-5}$ is irreducible in D, completing the proof that D is not a unique factorization domain. $\square$

Example 9.2.2.

Let F be a field and let

$$
\begin{aligned}
D &= F[x^2, x^3] = \{f(x) \in F[x] \mid f'(0) = 0\} \\
&= \{f(x) \in F[x] \mid f(x) = a_0 + a_2 x^2 + a_3 x^3 + \ldots + a_n x^n\} .
\end{aligned}
$$

It is easy to verify that D is a subring of $F[x]$, and that x^2 and x^3 are irreducible elements of D, since D contains no polynomials of degree 1. The domain D is not a unique factorization domain since $(x^2)^3 = (x^3)^2$.

It is interesting to note that the quotient field $Q(D)$ of D is isomorphic to the quotient field $Q(F[x]) \cong F(x)$ of $F[x]$. To see this, set $y = x^3/x^2$

in $Q(D)$. Then $y^2 = (x^3 \cdot x^3)/(x^2 \cdot x^2) = (x^2)^3/(x^2)^2 = x^2$, and $y^3 = (x^3)^3/(x^2)^3 = x^3$. Thus $D \subseteq F[y] \subseteq Q(D)$, so $Q(D) = Q(F[y]) \cong F(y)$. Then $Q(D) \cong Q(F[x])$ since $F[y] \cong F[x]$. $\square$

EXERCISES: SECTION 9.2

1. Factor these polynomials as elements of $(\mathbf{Z}[x])[y]$, and then find their greatest common divisor:

†(a) $x^3y^2 - 2xy^2 - 2y - x$ and $x^2y^2 + 3xy + 2$;

(b) $x^3y^2 + x^4y + 2x^2y^2 + 3x^3y + x^4 + xy^2 + 2y^2 + xy$ and $x^4y^2 + x^5y - 2x^2y^2 - 3x^3y + x^2y + x^3 - 3y^2 + y$.

2. Let D be the subring of $\mathbf{C}$ defined by $D = \mathbf{Z}[2i] = \{m + 2ni \mid m, n \in \mathbf{Z}\}$.

(a) Prove that the quotient field of D is $Q(D) = \mathbf{Q}(i)$.

(b) Prove that the polynomial $x^2 + 1$ has a proper factorization over $Q(D)$, but not over D.

(c) Prove that D is not a unique factorization domain.

3. Let D be a principal ideal domain, and let a, b be nonzero elements of D. Show that a and b are relatively prime if and only if $aD + bD = D$.

4. Let D be an integral domain. Prove that if $D[x]$ is a principal ideal domain, then D must be a field.

5. Let D be an integral domain. Show that D is a unique factorization domain if and only if every nonzero element of D that is not a unit can be written as a product of prime elements. (See Exercise 1 of Section 9.1 for the definition of a prime element.)

6. (Chinese remainder theorem) Let R be a commutative ring.

(a) Prove that if I, J are proper ideals of R with $I + J = R$, then for any $a, b \in R$ there exists $x \in R$ such that $x + I = a + I$ and $x + J = b + J$.

(b) Show that if I, J are ideals of R with the property that for all $a, b \in R$ there exists $x \in R$ such that $x + I = a + I$ and $x + J = b + J$, then $I + J = R$.

(c) Give an example of a unique factorization domain R with relatively prime elements r and s and elements $a, b \in R$ such that there is no $x \in R$ with $x + \langle r \rangle = a + \langle r \rangle$ and $x + \langle s \rangle = a + \langle s \rangle$.

7. Show that the results of parts (a) and (b) of Exercise 6 remain true for subgroups I, J of an abelian group A.

8. A commutative ring R is said to be a *Noetherian* ring if every ideal of R has a finite set of generators. Prove that if R is a commutative ring, then R is Noetherian if and only if for any ascending chain of ideals $I_1 \subseteq I_2 \subseteq \cdots$ there exists a positive integer n such that $I_k = I_n$ for all $k > n$.

9. Let R be a Noetherian ring. This exercise provides an outline of the steps in a proof of the Hilbert basis theorem, which states that the polynomial ring $R[x]$ is a Noetherian ring.

(a) Let I be any ideal of $R[x]$, and let I_k be the set of all $r \in R$ such that $r = 0$ or r occurs as the leading coefficient of a polynomial of degree k in I. Prove that I_k is an ideal of R.

(b) For the ideals I_k in part (a), prove that there exists an integer n such that $I_n = I_{n+1} = \cdots$.

(c) By assumption, each left ideal I_k is finitely generated (for $k \leq n$), and we can assume that it has $m(k)$ generators. Each generator of I_k is the leading coefficient of a polynomial of degree k, so we let $\{p_{jk}(x)\}_{j=1}^{m(k)}$ be the corresponding polynomials. Prove that $\mathcal{B} = \cup_{k=1}^{n} \{p_{jk}(x)\}_{j=1}^{m(k)}$ is a set of generators for I.

Hint: If not, then among the polynomials that cannot be expressed as linear combinations of polynomials in $\mathcal{B}$ there exists one of minimal degree.

10. Let F be a field, and let $f_1, f_2, \ldots, f_n \in F[x]$. Let $y_1, y_2, \ldots, y_n$ be indeterminates. Define the mapping $\phi : F[y_1, y_2, \ldots, y_n] \to F[x]$ by $\phi(a) = a$ for all $a \in F$ and $\phi(y_j) = f_j$ for $j = 1, \ldots, n$. Then ϕ is a homomorphism, and we denote the subring $\phi(F[y_1, y_2, \ldots, y_n])$ of $F[x]$ by $F[f_1, f_2, \ldots, f_n]$.

If R is any subring of $F[x]$ that contains F, prove that there exists a finite set of polynomials $f_1, f_2, \ldots, f_n \in F[x]$ such that $R = F[f_1, f_2, \ldots, f_n]$.

9.3   Some Diophantine Equations

In this section we give two number theoretic applications of our results on unique factorization. Using the arithmetic of the Gaussian integers, we will be able to prove that every prime in $\mathbf{Z}$ of the form $4k + 1$ can be written uniquely as the sum of two squares. As a second application, we introduce another subring of the complex numbers, based on a cube root of unity, and use its arithmetic to prove that $x^3 + y^3 = z^3$ has no solution in the set of positive integers. In both cases we need to first determine the irreducible elements of the relevant unique factorization domain. Polynomial equations such as $x^n + y^n = z^n$ for which solutions are sought in $\mathbf{Z}$ are usually referred to as *Diophantine* equations after the Greek mathematician Diophantus (c. 250 A.D.), who first studied such equations.

9.3.1 Lemma. *The following conditions are equivalent for a prime p of $\mathbf{Z}$:*

(1) *p is an irreducible element in $\mathbf{Z}[i]$;*

(2) *$x^2 + 1$ is an irreducible polynomial in $\mathbf{Z}_p[x]$;*

(3) *$p \equiv 3 \pmod 4$.*

Proof. (1) if and only if (2): In any principal ideal domain, a nonzero element is irreducible if and only if the ideal it generates is a prime ideal (Proposition 9.1.9), and then since nonzero prime ideals are maximal (Theorem 5.3.10), this is equivalent to the statement that the factor ring it determines is a field. The ideal of $\mathbf{Z}[i]$ generated by p is $p\mathbf{Z}[i] = \{m + ni \mid p|m \text{ and } p|n\}$. If we can show that $\mathbf{Z}[i]/p\mathbf{Z}[i]$ is isomorphic to $\mathbf{Z}_p[x]/\langle x^2 + 1\rangle$, then we can conclude that $\mathbf{Z}[i]/p\mathbf{Z}[i]$ is a field if and only if $\mathbf{Z}_p[x]/\langle x^2 + 1\rangle$ is a field, and hence that p is irreducible in $\mathbf{Z}[i]$ if and only if $x^2 + 1$ is irreducible in $\mathbf{Z}_p[x]$.

Define $\phi : \mathbf{Z}[i] \to \mathbf{Z}_p[x]/\langle x^2 + 1\rangle$ by $\phi(m + ni) = [m]_p + [n]_p x + \langle x^2 + 1\rangle$, for all elements $m + ni \in \mathbf{Z}[i]$. Since $x^2 + 1$ has degree 2, each equivalence class of elements of the ring $\mathbf{Z}_p[x]/\langle x^2 + 1\rangle$ contains a unique element of the form $[m]_p + [n]_p x$, and so ϕ maps $\mathbf{Z}[i]$ onto $\mathbf{Z}_p[x]/\langle x^2 + 1\rangle$. It is clear that ϕ is additive, and the proof that ϕ respects multiplication follows from

$$i^2 = -1 \quad \text{and} \quad x^2 \equiv -1 \ (\text{mod} \ (x^2 + 1)) \ .$$

It is easy to see that $\ker(\phi) = \{m + ni \mid p|m \text{ and } p|n\}$, and this set is precisely the ideal $p\mathbf{Z}[i]$. The fundamental homomorphism theorem (Theorem 5.2.6) implies that $\mathbf{Z}[i]/\ker(\phi)$ is isomorphic to $\mathbf{Z}_p[x]/\langle x^2 + 1\rangle$.

(2) if and only if (3): The polynomial $x^2 + 1$ is irreducible in $\mathbf{Z}_p[x]$ if and only if it has no roots in $\mathbf{Z}_p$. We will show that $x^2 + 1$ has a root if and only if either $p = 2$ or $p \equiv 1 \ (\text{mod} \ 4)$, and thus $x^2 + 1$ is irreducible if and only if $p \equiv 3 \ (\text{mod} \ 4)$.

If $p = 2$, then $[1]$ is a root, and so we concentrate on the case when p is odd. In that case neither $[1]$ nor $[-1]$ is a root of $x^2 + 1$. Since we have the factorization $x^4 - 1 = (x^2 + 1)(x^2 - 1)$, to find a root of $x^2 + 1$ it suffices to find a root of $x^4 - 1$ that is not a root of $x^2 - 1$. Thus we can find a root of $x^2 + 1$ in $\mathbf{Z}_p$ if and only if there is an element of order 4 in the multiplicative group $\mathbf{Z}_p^\times$. By Theorem 6.5.10, the multiplicative group of nonzero elements of any finite field is cyclic, and so this occurs if and only if the order $p - 1$ of $\mathbf{Z}_p^\times$ is divisible by 4, or, equivalently, if and only if $p \equiv 1 \ (\text{mod} \ 4)$. $\square$

We note a useful fact that is verified in the proof of the next proposition. If D is a Euclidean domain with a norm δ that is multiplicative (that is, $\delta(ab) = \delta(a)\,\delta(b)$ for all $a, b \in D$), then any element $p \in D$ such that $\delta(p)$ is a prime in $\mathbf{Z}$ must be an irreducible element of D.

9.3.2 Theorem. *Let a be a nonzero element of $\mathbf{Z}[i]$. Then a is irreducible in $\mathbf{Z}[i]$ if and only if one of the following conditions holds:*

(i) *$a = \pm p$ or $a = \pm pi$, where p is a prime of $\mathbf{Z}$ such that $p \equiv 3 \ (\text{mod} \ 4)$, or*

(ii) *$\delta(a)$ is prime in $\mathbf{Z}$.*

Proof. If condition (i) holds, then it follows from Lemma 9.3.1 that a is irreducible. If condition (ii) holds and $a = bc$ in $\mathbf{Z}[i]$, then $\delta(b)\,\delta(c) = \delta(bc) = \delta(a)$, and so either $\delta(b) = 1$ or $\delta(c) = 1$, since $\delta(a)$ is prime in $\mathbf{Z}$. Thus either b or c is a unit, and so a is irreducible.

Conversely, assume that $a = m + ni$ is irreducible. Then the conjugate $\overline{a} = m - ni$ of a is also irreducible, since $\overline{a} = bc$ implies $a = \overline{b}\overline{c}$. We first consider the case in which $\overline{a}$ is an associate of a. Since the units of $\mathbf{Z}[i]$ are ± 1 and $\pm i$, one of the following equalities must hold.

$$m + ni = m - ni \qquad\qquad m + ni = -m + ni$$

$$m + ni = n + mi \qquad\qquad m + ni = -n - mi$$

The first two equations imply either $n = 0$ or $m = 0$, and so it follows from Lemma 9.3.1 that condition (i) holds. The second two equations imply that a has $1 + i$ as a factor, and so since a is irreducible we must have $\delta(a) = 2$, and thus condition (ii) holds.

We next consider the case in which $\overline{a}$ is not an associate of a, which implies that a and $\overline{a}$ are relatively prime. Now suppose that $\delta(a)$ is a composite number, say $\delta(a) = uv$, where u, v are positive integers. Then in $\mathbf{Z}[i]$ we have $a|u$ or $a|v$, say $u = aq$. Since u is real, $u = \overline{u} = \overline{aq}$, showing that $\overline{a}|u$. But then our observation that a and $\overline{a}$ are relatively prime implies that $\delta(a) = a\overline{a}$ divides u, so we must have $u = \delta(a)$ and $v = 1$. This shows that $\delta(a)$ is prime in $\mathbf{Z}$, and so condition (ii) holds. $\square$

9.3.3 Theorem. *Every prime in $\mathbf{Z}$ of the form $4k + 1$ can be written as the sum of two squares, in an essentially unique manner.*

Proof. Assume that p is a prime in $\mathbf{Z}$ such that $p \equiv 1 \pmod 4$. Then p is not irreducible in $\mathbf{Z}[i]$, by Lemma 9.3.1, and so in $\mathbf{Z}[i]$ we have $p = ac$, where $a = m + ni$ is an irreducible factor. Thus $p^2 = \delta(p) = \delta(a)\,\delta(c)$, so $\delta(a) = \delta(c) = p$ since neither a nor c is a unit, and then since $a = m + ni$ we have $p = m^2 + n^2$.

The above solution is essentially unique. If $p = m^2 + n^2$ is prime in $\mathbf{Z}$ and can also be written as $p = u^2 + v^2$ for integers u, v, then in $\mathbf{Z}[i]$ we would have $p = (u + vi)(u - vi)$, showing that $u + vi$ must differ from $m + ni$ by a unit of $\mathbf{Z}[i]$. Since the only units are ± 1 and $\pm i$, either $u = \pm m$ and $v = \pm n$ or $u = \pm n$ and $v = \pm m$. $\square$

We are now ready to begin a proof due to Gauss of Fermat's last theorem for the exponent 3. The case for the exponent 4 is easier than that for the exponent 3, and we suggest that the reader should try Exercises 5 and 6 at the end of the section. We want to show that

$$x^3 + y^3 = z^3$$

has no solution in the set of positive integers. If we allow negative integers as solutions, and call any solution with one of x, y, z equal to zero a trivial solution, then we can restate the problem as follows: $x^3 + y^3 = z^3$ has no nontrivial solution in $\mathbf{Z}$. This has the advantage of allowing us to rewrite the equation as $x^3 + y^3 + (-z)^3 = 0$, which provides more symmetry in the problem.

As a polynomial over $\mathbf{Z}$, we have the factorization

$$x^3 + y^3 = (x + y)(x^2 - xy + y^2) \, .$$

To factor $x^2 - xy + y^2$ completely, we can use the quadratic formula to solve $x^2 + (-y)x + y^2 = 0$, obtaining

$$x = \frac{y \pm \sqrt{y^2 - 4y^2}}{2} = y \left(\frac{1 \pm \sqrt{3}i}{2} \right) \, .$$

This yields $x = -\omega y$ or $x = -\overline{\omega} y$, for the cube root of unity $\omega = (-1 + \sqrt{3}i)/2$. As a polynomial over the field $\mathbf{C}$, we have

$$x^3 + y^3 = (x + y)(x + \omega y)(x + \overline{\omega} y) = (x + y)(x + \omega y)(x + \omega^2 y) \, .$$

If x and y are integers, then we can obtain this factorization in any subring of $\mathbf{C}$ that contains $\mathbf{Z}$ and ω. We introduce the appropriate subring in the next example.

Example 9.3.1 ($\mathbf{Z}[\omega]$).

Let $\omega = (-1 + \sqrt{3}i)/2$. Since ω is a cube root of unity, it is a root of the polynomial $x^3 - 1$, and more specifically, of the factor $x^2 + x + 1$. Thus $\omega^2 = -1 - \omega$. The other root of $x^2 + x + 1$ is $\overline{\omega} = \omega^2$, so we have $\omega + \overline{\omega} = -1$ and $\omega\overline{\omega} = \omega\omega^2 = 1$.

If $a, b, c, d \in \mathbf{Z}$, then in the field of complex numbers we have

$$
\begin{aligned}
(a + b\omega)(c + d\omega) &= ac + (ad + bc)\omega + bd\omega^2 \\
&= (ac - bd) + (ad + bc - bd)\omega \, .
\end{aligned}
$$

This shows that the set $\{a + b\omega \mid a, b \in \mathbf{Z}\}$ is closed under multiplication, and so it is easily seen to be a subring of $\mathbf{C}$, which we will denote by $\mathbf{Z}[\omega]$. We can also write $a + b\omega = (a - b/2) + (b\sqrt{3}/2)i$, and so the uniqueness of representation for complex numbers shows that each element of $\mathbf{Z}[\omega]$ has a unique representation in the form $a + b\omega$.

We note that just as with $\mathbf{Z}[i]$, there is a natural norm on $\mathbf{Z}[\omega]$ based on the length $|a + b\omega|$ of an element simply considered as a complex number. In order to obtain an integer valued norm we must use the square of the length. $\square$

Any integer solution to $x^3 + y^3 = z^3$ will remain a solution in the ring $\mathbf{Z}[\omega]$, and so we only need to find the solutions in $\mathbf{Z}[\omega]$. The complete factorization of $x^3 + y^3$ in $\mathbf{Z}[\omega]$ is the reason that it is useful to translate the problem in $\mathbf{Z}$ into a problem in $\mathbf{Z}[\omega]$. We will need the fact that $\mathbf{Z}[\omega]$ is a unique factorization domain. The proof that the division algorithm holds is similar to the proof for the Gaussian integers.

9.3.4 Proposition. *For $\omega = (-1 + \sqrt{3}i)/2$, the ring $\mathbf{Z}[\omega]$ is a Euclidean domain.*

Proof. Since $\mathbf{Z}[\omega] = \{m + n\omega \mid m, n \in \mathbf{Z}\}$, for any element $a = m + n\omega \in \mathbf{Z}[\omega]$ we let

$$
\begin{aligned}
\delta(a) &= |a|^2 = (m + n\omega)(m + n\overline{\omega}) \\
&= m^2 + mn(\omega + \overline{\omega}) + n^2 \omega \overline{\omega} \\
&= m^2 + n^2 - mn \ .
\end{aligned}
$$

For nonzero elements $a, b \in \mathbf{Z}[\omega]$, we have $\delta(ab) = |ab|^2 \geq |b|^2 = \delta(b)$.

For nonzero elements a and b of $\mathbf{Z}[\omega]$, we consider the quotient a/b (as a complex number). Then for $a = m + n\omega$ and $b = s + t\omega$ we have

$$
\begin{aligned}
\frac{a}{b} &= \frac{(m + n\omega)(s + t\overline{\omega})}{(s + t\omega)(s + t\overline{\omega})} = \frac{(ms + ns\omega + mt\overline{\omega} + nt\omega\overline{\omega})}{s^2 + t^2 - st} \\
&= \left(\frac{ms + nt - mt}{s^2 + t^2 - st} \right) + \left(\frac{ns - mt}{s^2 + t^2 - st} \right) \omega \ .
\end{aligned}
$$

We can choose the integers u and v, respectively, that are closest to

$$
c = (ms + nt - mt)/(s^2 + t^2 - st) \quad \text{and} \quad d = (ns - mt)/(s^2 + t^2 - st) \ .
$$

Then $|c - u| \leq \frac{1}{2}$ and $|d - v| \leq \frac{1}{2}$. Let $q = u + v\omega$, and let $r = a - bq$. If $r \neq 0$, then

$$
r = a - bq = b \left(\frac{a}{b} - q \right) ,
$$

and so we have

$$
\begin{aligned}
\delta(r) &= |b((c - u) + (d - v)\omega)|^2 \\
&= |b|^2 |(c - u) + (d - v)\omega|^2 \\
&= |b|^2 \left((c - u)^2 + (d - v)^2 - (c - u)(d - v) \right) \\
&\leq |b|^2 \left(\frac{1}{4} + \frac{1}{4} + \frac{1}{4} \right) .
\end{aligned}
$$

Thus $\delta(r) \leq \frac{3}{4} \delta(b) < \delta(b)$, completing the proof. $\square$

Example 9.3.2 (Units of Z[ω]).

> Since $\omega\omega^2 = 1$ and $\omega^2 = -1 - \omega$, it is clear that ± 1, $\pm\omega$, and $\pm(1 + \omega)$
> are units of $\mathbf{Z}[\omega]$. These can also be listed as ± 1, $\pm\omega$, and $\pm\overline{\omega}$. On the other
> hand, if $a = m + n\omega$ is a unit of $\mathbf{Z}[\omega]$, then we must have $\delta(a) = 1$, and so
> $m^2 - nm + (n^2 - 1) = 0$. Using the quadratic formula shows that the only
> integer solution with both $m \neq 0$ and $n \neq 0$ requires a positive discriminant,
> and so we must have $n^2 - 4(n^2 - 1) > 0$, or $3n^2 < 4$. This solution yields
> $n = m = 1$ or $n = m = -1$, and so the six units listed above are the only
> ones. It follows immediately that $\mathbf{Z}[\omega]^{\times} \cong \mathbf{Z}_6$. $\square$

9.3.5 Lemma. *Let* $\omega = (-1 + \sqrt{3}i)/2$. *The following conditions are equivalent
for a prime* p *of* $\mathbf{Z}$:

(1) *p is an irreducible element in* $\mathbf{Z}[\omega]$;

(2) *$x^2 + x + 1$ is an irreducible polynomial in* $\mathbf{Z}_p[x]$;

(3) *$p \equiv 2 \pmod{3}$.*

Proof. (1) if and only if (2):   We define $\phi : \mathbf{Z}[\omega] \to \mathbf{Z}_p[x]/\langle x^2 + x + 1 \rangle$ by
$\phi(m + n\omega) = [m]_p + [n]_p x + \langle x^2 + x + 1 \rangle$, for all elements $m + n\omega \in \mathbf{Z}[\omega]$.
Since $x^2 + x + 1$ has degree 2, each equivalence class of elements of the ring
$\mathbf{Z}_p[x]/\langle x^2 + x + 1 \rangle$ contains a unique element of the form $[m]_p + [n]_p x$, and so
ϕ maps $\mathbf{Z}[\omega]$ onto $\mathbf{Z}_p[x]/\langle x^2 + x + 1 \rangle$. The fact that ϕ is a ring homomorphism
follows from

$$\omega^2 = -1 - \omega \quad \text{and} \quad x^2 \equiv -1 - x \pmod{(x^2 + x + 1)} .$$

It can be checked that

$$\ker(\phi) = \{m + n\omega \mid p|m \text{ and } p|n\} = p\mathbf{Z}[\omega] ,$$

so $\mathbf{Z}[\omega]/\ker(\phi)$ is isomorphic to $\mathbf{Z}_p[x]/\langle x^2 + x + 1 \rangle$. Thus p is irreducible in $\mathbf{Z}[\omega]$
if and only if the factor ring $\mathbf{Z}_p[x]/\langle x^2 + x + 1 \rangle$ is a field, and this occurs if and
only if $x^2 + x + 1$ is irreducible over $\mathbf{Z}_p$.

(2) if and only if (3): Note first that $x^2 + x + 1$ is irreducible over $\mathbf{Z}_2$, but has
the root [1] over $\mathbf{Z}_3$. We will show that if $p > 3$, then $x^2 + x + 1$ has a root in $\mathbf{Z}_p$
if and only if $p \equiv 1 \pmod 3$.

We use the factorization $x^3 - 1 = (x - 1)(x^2 + x + 1)$. If $p > 3$, then [1]
is not a root of $x^2 + x + 1$, and so $\mathbf{Z}_p$ contains a root of $x^2 + x + 1$ if and only
if it contains a root of $x^3 - 1$ different from [1]. Thus the roots of $x^2 + x + 1$
correspond to elements of order 3 in the multiplicative group $\mathbf{Z}_p^{\times}$, and such elements
exist if and only if $3 \mid (p - 1)$. Therefore $x^2 + x + 1$ is irreducible if and only if
$p \equiv 2 \pmod 3$. $\square$

In the statement of condition (3) of Lemma 9.3.5, note that $p \equiv 2 \pmod 3$ if and only if $p \equiv 2 \pmod 6$ or $p \equiv 5 \pmod 6$. If $p \equiv 2 \pmod 6$, then p must be even, and hence $p = 2$. Thus the condition may be expressed by stating that either $p = 2$ or $p \equiv 5 \pmod 6$.

9.3.6 Theorem. *Let $\omega = (-1 + \sqrt{3}i)/2$, and let a be a nonzero element of $\mathbf{Z}[\omega]$. Then a is an irreducible element of $\mathbf{Z}[\omega]$ if and only if one of the following conditions holds:*

(i) $a = \pm p$ or $a = \pm \omega p$ or $a = \pm \omega^2 p$, where p is a prime of $\mathbf{Z}$ such that $p \equiv 2 \pmod 3$, or

(ii) $\delta(a)$ is prime in $\mathbf{Z}$.

Proof. If a satisfies condition (i), then a is an associate of an irreducible element of the type characterized in Lemma 9.3.5, since the units of $\mathbf{Z}[\omega]$ are ± 1, $\pm \omega$, and $\pm \overline{\omega}$, where $\overline{\omega} = \omega^2 = -1 - \omega$. If a satisfies condition (ii), then $\delta(a)$ is a prime in $\mathbf{Z}$, and it follows that a is irreducible.

Conversely, assume that $a = m + n\omega$ is irreducible. Then the conjugate of a, given by $\overline{a} = m + n\overline{\omega} = (m - n) - n\omega$ is also irreducible, and we first consider the case in which $\overline{a}$ is an associate of a. We have six cases, namely when $\overline{a} = a, a\omega, a\overline{\omega}, -a, -a\omega, -a\overline{\omega}$. The first three cases lead to the respective equations $n = 0$, $m = 0$, and $m = n$, which imply that a must satisfy condition (i). The last three cases lead to the respective equations $2m = n$, $m = 2n$, and $m = -n$. These equations imply, respectively, that a must have the irreducible factor $1 + 2\omega$, $2 + \omega$, or $1 - \omega$. Since a itself is irreducible, in each case we must have $\delta(a) = 3$.

In the second case, if $\overline{a}$ and a are not associates, then they must be relatively prime. Now suppose that $\delta(a)$ is a composite number, say $\delta(a) = uv$, where u, v are positive integers. In $\mathbf{Z}[\omega]$ we also have the factorization $\delta(a) = a\overline{a}$, so either u or v must be divisible by a, say $u = aq$. Since u is real, $u = \overline{u} = \overline{aq}$, showing that $\overline{a}|u$. The fact that a and $\overline{a}$ are relatively prime implies that $a\overline{a} \mid u$, so we must have $u = \delta(a)$ and $v = 1$. This shows that $\delta(a)$ is a prime in $\mathbf{Z}$. $\square$

9.3.7 Lemma. *Let $\omega = (-1 + \sqrt{3}i)/2$.*

(a) *The ring $\mathbf{Z}[\omega]/\langle 1 - \omega \rangle$ is isomorphic to $\mathbf{Z}_3$.*

(b) *If $[x] = [1]$ in $\mathbf{Z}[\omega]/\langle 1 - \omega \rangle$, then $x^3 - 1$ is divisible by $(1 - \omega)^4$ in $\mathbf{Z}[\omega]$. If $[x] = [-1]$ in $\mathbf{Z}[\omega]/\langle 1 - \omega \rangle$, then $x^3 + 1$ is divisible by $(1 - \omega)^4$ in $\mathbf{Z}[\omega]$.*

(c) *In any solution of $x^3 + y^3 = z^3$ in $\mathbf{Z}[\omega]$, one of x^3, y^3, z^3 must be divisible by $(1 - \omega)^4$.*

Proof. (a) We have $\delta(1 - \omega) = (1 - \omega)(1 - \omega^2) = 3$, so $1 - \omega$ is irreducible in $\mathbf{Z}[\omega]$ by Proposition 9.3.6. This also gives us the factorization of 3 in $\mathbf{Z}[\omega]$, since $1 + \omega = -\omega^2$ is a unit and $3 = (1 - \omega)^2(1 + \omega)$. The factor ring $\mathbf{Z}[\omega]/\langle 1 - \omega \rangle$ is a field, which we will denote by F, with elements $[m + n\omega]$.

To show that F is isomorphic to $\mathbf{Z}_3$, define $\phi : \mathbf{Z} \to F$ by $\phi(m) = [m]$, for all $m \in \mathbf{Z}$. This is a ring homomorphism, and using the defining equation $[\omega] = [1]$ we see that ϕ is onto since $[m + n\omega] = [m + n] = \phi(m + n)$. Since 3 is divisible by $1 - \omega$, we must have $\ker(\phi) = 3\mathbf{Z}$. It follows from the fundamental homomorphism theorem (Theorem 5.2.6) that F is isomorphic to $\mathbf{Z}/3\mathbf{Z} = \mathbf{Z}_3$.

(b) If $[x] = [1]$ in F, then $x - 1$ is divisible by $1 - \omega$, say $x - 1 = (1 - \omega)q$ for some $q \in \mathbf{Z}[\omega]$. In general, we have

$$x^3 + y^3 = (x + y)(x + \omega y)(x + \omega^2 y) \,,$$

so if we set $y = -1$ we obtain

$$
\begin{aligned}
x^3 - 1 &= (x - 1)(x - \omega)(x - \omega^2) \\
&= (x - 1)(x - 1 + 1 - \omega)(x - 1 + 1 - \omega^2) \\
&= (1 - \omega)q(1 - \omega)(q + 1)(1 - \omega)(q + 1 + \omega) \\
&= (1 - \omega)^3 q(q + 1)(q + 1 + \omega) \,.
\end{aligned}
$$

In F we have $[q + 1 + \omega] = [q - 1]$, and since F is isomorphic to $\mathbf{Z}_3$ by part (a), one of the elements $[q]$, $[q + 1]$, $[q - 1]$ must be zero. Thus one of the factors q, $q + 1$, $q + 1 + \omega$ must be divisible by $1 - \omega$. This shows that $x^3 - 1$ is divisible by $(1 - \omega)^4$.

If $[x] = [-1]$, then $[-x] = [1]$, and so $x^3 + 1 = -((-x)^3 - 1)$ is divisible by $(1 - \omega)^4$.

(c) With only a small number of possibilities to check, it is easy to find all solutions of $[x]^3 + [y]^3 = [z]^3$ in F. The only solutions in which $[x]$, $[y]$, and $[z]$ are all nonzero are $[x] = [1]$, $[y] = [1]$, $[z] = [-1]$ and $[x] = [-1]$, $[y] = [-1]$, $[z] = [1]$. We will show that neither of these solutions arises from a solution in $\mathbf{Z}[\omega]$. Thus in any solution that comes from a solution in $\mathbf{Z}[\omega]$, either $[x] = [0]$ or $[y] = [0]$ or $z = [0]$, showing that one of x, y, or z is divisible by $1 - \omega$.

Let x, y, z be a solution of $x^3 + y^3 = z^3$ in $\mathbf{Z}[\omega]$, such that $[x] = [1]$, $[y] = [1]$, $[z] = [-1]$ in F. Then by part (b),

$$(x^3 - 1) + (y^3 - 1) + ((-z)^3 - 1) = -3$$

is divisible by $(1 - \omega)^4$. Because we have unique factorization in $\mathbf{Z}[\omega]$, this contradicts the factorization $3 = (1 + \omega)(1 - \omega)^2$. A similar argument eliminates the case $[x] = [-1]$, $[y] = [-1]$, $[z] = [1]$.

Again, let x, y, z be a solution of $x^3 + y^3 = z^3$ in $\mathbf{Z}[\omega]$. Since a divisor of two of x, y, z is also a divisor of the third, we will assume that x, y, z are relatively prime in pairs. We have shown that $[x] = [0]$, $[y] = [0]$, or $[z] = [0]$, so by changing variables in the equation we can assume without loss of generality that $[z] = [0]$.

Thus in F we have $[x]^3 + [y]^3 = [0]$, and the only possible solutions are $[x] = [1]$, $[y] = [-1]$ and $[x] = [-1]$, $[y] = [1]$. It follows from by part (b) and the equation

$$x^3 + y^3 = (x^3 - 1) + (y^3 + 1) = (x^3 + 1) + (y^3 - 1)$$

that $z^3 = x^3 + y^3$ is divisible by $(1 - \omega)^4$ in either case. $\square$

9.3.8 Theorem. *The equation $x^3 + y^3 = z^3$ has no solution in the set of positive integers.*

Proof. We actually prove the stronger result that $x^3 + y^3 = z^3$ has no nontrivial solution in $\mathbf{Z}[\omega]$.

Assume that the nonzero elements x, y, z of $\mathbf{Z}[\omega]$ represent a solution to the equation $x^3 + y^3 = z^3$. Any irreducible element of $\mathbf{Z}[\omega]$ that is a divisor of two of x, y, and z must also be a divisor of the third, and so we can assume that $\gcd(x, y) = \gcd(x, z) = \gcd(y, z) = 1$ in $\mathbf{Z}[\omega]$.

Recall that in $\mathbf{Z}[\omega]$ we have $\omega\omega^2 = 1$ and $1 + \omega + \omega^2 = 0$, and then $\overline{\omega} = \omega^2$ implies further that $\omega\overline{\omega} = 1$ and $\omega + \overline{\omega} = -1$. By Lemma 9.3.7 (c), and renaming variables if necessary, we can assume without loss of generality that the element z^3 must be divisible by $(1 - \omega)^4$.

Among all solutions in $\mathbf{Z}[\omega]$ of the above type, where (i) x, y, z are pairwise relatively prime and (ii) z^3 is divisible by $(1 - \omega)^4$, we can choose one in which the exponent of $1 - \omega$ in the factorization of z is minimal. (This depends on having unique factorization in $\mathbf{Z}[\omega]$.) Thus there is a positive integer $k \geq 2$ such that $(1 - \omega)^k$ is a divisor of z but $(1 - \omega)^{k+1}$ is not, while in any other solution that satisfies (i) and (ii) we know that $(1 - \omega)^k$ is a divisor of z. We will arrive at a contradiction by showing that we can always find another solution x_*, y_*, z_* in which z_* is divisible by $1 - \omega$, but not $(1 - \omega)^k$.

Consider the factorization

$$z^3 = x^3 + y^3 = (x + y)(x + \omega y)(x + \omega^2 y).$$

Because z is assumed to be divisible by $1 - \omega$, one of the factors $x + y$, $x + \omega y$, $x + \omega^2 y$ must be divisible by $1 - \omega$. Since $[\omega] = [1]$ in $\mathbf{Z}[\omega]/\langle 1 - \omega \rangle$, we have $[x + \omega y] = [x + \omega^2 y] = [x + y]$, and this shows that each of the factors must be divisible by $1 - \omega$.

To find the new solution that we seek, let

$$x_0 = \frac{x + y}{1 - \omega}, \qquad y_0 = \frac{\omega(x + \omega y)}{1 - \omega}, \qquad z_0 = \frac{\omega^2(x + \omega^2 y)}{1 - \omega}.$$

We note that

$$x_0 + y_0 + z_0 = \frac{(1 + \omega + \omega^2)(x + y)}{1 - \omega} = 0.$$

The following equations, which can easily be verified by direct computation, show that x_0, y_0, z_0 are pairwise relatively prime, since x and y are relatively prime.

$$x = -\omega x_0 + \omega^2 y_0 = x_0 - \omega^2 z_0 = -y_0 + \omega z_0$$
$$y = x_0 - \omega^2 y_0 = -\omega x_0 + \omega^2 z_0 = \omega y_0 - z_0$$

By our choice of x_0, y_0, and z_0 we have

$$z^3 = (1 - \omega)^3 x_0 y_0 z_0 .$$

The product

$$x_0 y_0 z_0 = (z/(1 - \omega))^3$$

is a cube, so the fact that x_0, y_0, z_0 are pairwise relatively prime implies that there exist elements x_1, y_1, z_1 and units a, b, c with $x_0 = a x_1^3$, $y_0 = b y_1^3$, and $z_0 = c z_1^3$. (This again depends on having unique factorization in $\mathbf{Z}[\omega]$.) Furthermore, one of x_0, y_0, z_0 must be divisible by $1 - \omega$ because z^3 is divisible by $(1 - \omega)^4$. Hence one of x_1, y_1, z_1 must be divisible by $1 - \omega$. Since x_0, y_0, z_0 are pairwise relatively prime, we have that x_1, y_1, z_1 are pairwise relatively prime. Because z is not divisible by $(1 - \omega)^{k+1}$, whichever of x_1, y_1, z_1 is divisible by $1 - \omega$ is not divisible by $(1 - \omega)^k$. Since $x_0 + y_0 + z_0 = 0$, we have

$$a x_1^3 + b y_1^3 + c z_1^3 = 0 .$$

We also note that

$$abc = \frac{z^3}{(x_1^3 y_1^3 z_1^3)(1 - \omega)^3}$$

is a unit and a cube, and so $abc = \pm 1$.

We now have a solution to the equation

$$au^3 + bv^3 + cw^3 = 0 ,$$

where a, b, c are units, abc is a cube, and the solution has the property that exactly one of u, v, w is divisible by $1 - \omega$, and that one is not divisible by $(1 - \omega)^k$. We will next show that a, b, c differ from each other only by a sign. Without loss of generality we may assume that $1 - \omega$ divides w, and that u and v are not divisible by $1 - \omega$ in $\mathbf{Z}[\omega]$. Thus in $\mathbf{Z}[\omega]/\langle 1 - \omega \rangle \cong \mathbf{Z}_3$ we have $[u] = [\pm 1]$ and $[v] = [\pm 1]$. Since

$$au^3 + bv^3 + cw^3 = 0$$

and since $(1 - \omega)^3$ divides w^3, we see that

$$\pm a \pm b = a(u^3 \pm 1) + b(v^3 \pm 1) + cw^3$$

is divisible by $(1 - \omega)^3$ for some choice of signs. Since a is a unit, $\pm a \pm b = \pm a(1 \pm b/a)$ and since b/a is a unit, $b/a = \pm 1$ or $\pm \omega$ or $\pm \omega^2$. The only choice of b/a for which $1 \pm b/a$ is divisible by $(1 - \omega)^3$ is $1 \pm b/a = 0$. Thus $a = \pm b$. Since $abc = \pm 1$, we have $a^2 c = \pm 1$. Since $a^3 = \pm 1$, we have $c = \pm a$. Thus we may cancel a to obtain

$$u^3 + (\pm v)^3 + (\pm w)^3 = 0 ,$$

where w is divisible by $1 - \omega$ but not by $(1 - \omega)^k$. Letting $x_* = \pm u$, $y_* = \pm v$, and $z_* = \pm w$, the appropriate choice of signs yields the desired contradiction. We have

$$x_*^3 + y_*^3 = z_*^3 ,$$

where z_* is divisible by $1 - \omega$ but not by $(1 - \omega)^k$, completing the proof. $\square$

EXERCISES: SECTION 9.3

1.†For each of the prime numbers less than 20, give the factorization into a product of irreducible elements in $\mathbf{Z}[i]$.

2. For each of the prime numbers less than 20, give the factorization into a product of irreducible elements in $\mathbf{Z}[\omega]$.

3. The element $5 - i$ is not irreducible in $\mathbf{Z}[i]$ (Exercise 22 of Chapter 5 shows that the ideal $\langle 5 - i \rangle$ is not a prime ideal of $\mathbf{Z}[i]$). Show that $5 - i$ can be factored as $5 - i = (1 - i)(3 + 2i)$, and that $1 - i$ and $3 + 2i$ are irreducible in $\mathbf{Z}[i]$.

4. Let n be a positive integer. Show that $x^2 + 1 \equiv 0 \pmod{n}$ has a solution if and only if $n = a^2 + b^2$, where $(a, b) = 1$.

   Hint: Use Exercise 11 of Section 9.1.

5. This exercise outlines a proof that the equation $x^4 + y^4 = z^2$ has no solution in $\mathbf{Z}^+$.

   (a) Suppose that there is a positive triple x, y, z such that $x^4 + y^4 = z^2$. Show that we may assume that $(x, y) = 1$, and that $(x^2, y^2, z) = 1$.

   (b) Show that there exists a least positive integer z such that $x^4 + y^4 = z^2$, with $(x, y) = 1$, $x > 0$, and $y > 0$.

   (c) Show that $x \not\equiv y \pmod 2$.

   (d) Without loss of generality, suppose that x is even and y is odd. Show that there exist positive integers $r < s$, $(r, s) = 1$, $r \not\equiv s \pmod 2$ such that $x^2 = 2sr$, $y^2 = s^2 - r^2$, and $z = s^2 + r^2$.

   (e) Show that r is even, and s is odd.

   (f) Say that $r = 2t$. Show that $(t, s) = 1$, and that both t and s are squares.

   (g) Show that there exist integers m, n such that $0 < m < n$, $(m, n) = 1$, and $t = mn$, $y = n^2 - m^2$, and $s = n^2 + m^2$.

(h) Show that both m and n are squares.

(i) Say $m = a^2$ and $n = b^2$. Show that there exists $k \in \mathbf{Z}$ such that $a^4 + b^4 = k^2$, and obtain a contradiction to the choice of z in part (b) of this exercise.

6. (a) Show that the equation $x^4 + y^4 = z^4$ has no integer solution with $xyz \neq 0$.

(b) In order to prove Fermat's last theorem, show this it *suffices* to prove that for any odd prime p, the equation $x^p + y^p = z^p$ has no integer solution with $xyz \neq 0$.

APPENDIX

A.1 Sets

We have assumed that the reader is familiar with the basic language of set theory, allowing us to begin our book with elementary number theory. This section of the appendix is designed to serve two purposes: to establish our notation, and to provide a quick review of some basic facts.

If S is a set (or collection) of elements denoted by a, b, c, etc., then we indicate that a belongs to S by writing $a \in S$. If $a \in S$, we say that a is an **element** of S, that a is a **member** of S, or simply that a is **in** S. If a is not an element of S, we write $a \notin S$. When a set A consists entirely of elements of S, it is said to be a **subset** of S, denoted by $A \subseteq S$. More formally, $A \subseteq S$ if and only if $a \in S$ for all $a \in A$.

Two sets A and B are said to be equal if they contain precisely the same elements. Thus to show that $A = B$, it is necessary to show that each element of A is also an element of B and that each element of B is also an element of A. In practice, equality is often proved by showing that $A \subseteq B$ and $B \subseteq A$, since different arguments may apply in the two different situations. If $A \subseteq B$ but $A \neq B$, then we say that A is a **proper** subset of B, and we will use the notation $A \subset B$. The meaning of the notations $B \supseteq A$ and $B \supset A$ should be obvious.

Let A and B be subsets of a given set S. There are several useful ways of constructing new subsets from A and B. The **intersection** of A and B is the set of all elements which belong to both A and B, and is denoted by $A \cap B$. In symbols we would write

$$A \cap B = \{x \in S \mid x \in A \text{ and } x \in B\}.$$

This notation $\{ \mid \}$ requires some explanation. The braces $\{ \}$ are used to denote a set, and the vertical bar (occasionally replaced by a colon) is read "such that." Thus

$$\{x \in S \mid x \in A \text{ and } x \in B\}$$

is read "the set of all x in S such that x is an element of A and x is an element of B." The intersection of two sets may very well not contain any elements, and this points up the necessity of considering the **empty set**, or **null set**, which we denote by $\emptyset$.

We also define the **union** of sets A and B in the following way:

$$A \cup B = \{x \in S \mid x \in A \text{ or } x \in B\}.$$

The word "or" is used in the way generally accepted by mathematicians; that is, we use it in an inclusive rather than an exclusive way, so that $A \cup B$ contains all elements either in A or in B or in both A and B.

We need notations for the intersection and union of a family of subsets. If $\{A_\lambda\}_{\lambda \in \Lambda}$ is a collection of subsets of S indexed by the set Λ, then we write

$$\bigcap\nolimits_{\lambda \in \Lambda} A_\lambda = \{x \in S \mid x \in A_\lambda \text{ for each } \lambda \in \Lambda\}$$

and

$$\bigcup\nolimits_{\lambda \in \Lambda} A_\lambda = \{x \in S \mid x \in A_\lambda \text{ for some } \lambda \in \Lambda\}$$

for the intersection and union, respectively, of the collection. For example, let $S = \mathbf{R}^2$ and let ℓ_a be the vertical line through $(a, 0)$. We can express the fact that the plane is the union of these lines by letting $\Lambda = \mathbf{R}$ and writing $\mathbf{R}^2 = \bigcup_{a \in \mathbf{R}} \ell_a$.

We also define the **difference** of the two sets as follows:

$$A - B = \{x \in A \mid x \notin B\}.$$

Thus $A - B$ is the set obtained by taking from A all elements which belong to B.

If S is the set we are working with, and A is a subset of S, then we call $S - A$ the **complement** of A in S, and denote it by $\overline{A}$. The two important identities given below are known as DeMorgan's laws:

$$\overline{A \cap B} = \overline{A} \cup \overline{B}, \qquad \overline{A \cup B} = \overline{A} \cap \overline{B}.$$

Example A.1.1.

Let A and B be sets. We will show that $A \subseteq B$ if and only if $A \cup B = B$.

First assume that $A \subseteq B$. To show that $A \cup B = B$ we must show that $A \cup B \subseteq B$ and that $B \subseteq A \cup B$. For this purpose, let $x \in A \cup B$. Thus $x \in A$ or $x \in B$. Since $A \subseteq B$, in either case $x \in B$, and thus $A \cup B \subseteq B$. On the other hand, it is always true that B is contained in $A \cup B$.

Conversely, assume that $A \cup B = B$, and let $x \in A$. Since x is in A, it is in $A \cup B$ and hence in B. This shows that $A \subseteq B$, completing the proof. $\square$

In several places in the book we need to work with ordered pairs of elements. If $a \in A$ and $b \in B$, the notion of an **ordered pair** distinguishes between the pairs (a, b) and (b, a). These are different from the set $\{a, b\}$, which is equal as a set to $\{b, a\}$. Since ordered pairs are familiar from calculus and linear algebra, we will not go into more detail.

Let A and B be any sets. The **Cartesian product** of A and B is formed from all ordered pairs whose first element is in A and whose second element is in B. Formally, we define the Cartesian product of A and B as

$$A \times B = \{(a, b) \mid a \in A \text{ and } b \in B\}.$$

In $A \times B$, ordered pairs (a_1, b_1) and (a_2, b_2) are equal if and only if $a_1 = a_2$ and $b_1 = b_2$.

We can extend the definition of the Cartesian product to n sets by considering n-tuples in which the ith entry belongs to the ith set. The n-dimensional vector space $\mathbf{R}^n$ is just the Cartesian product of $\mathbf{R}$ with itself n times, together with the algebraic structure that defines addition of vectors and scalar multiplication.

Example A.1.2.

For example, if $A = \{1, 2, 3\}$ and $B = \{u, v\}$, then the Cartesian product of A and B has a total of six distinct elements:

$$A \times B = \{(1, u), (1, v), (2, u), (2, v), (3, u), (3, v)\}.$$

The Cartesian product $B \times A$ is quite different:

$$B \times A = \{(u, 1), (u, 2), (u, 3), (v, 1), (v, 2), (v, 3)\} . \quad \square$$

We have listed below some of the important facts about sets. They will provide good exercises for the reader who needs some review.

EXERCISES: SECTION A.1

Let A, B, C be subsets of a given set S. Prove the following statements.

1. If $A \subseteq B$ and $B \subseteq C$, then $A \subseteq C$.

2. $A \cap B \subseteq A$ and $A \cap B \subseteq B$.

3. $A \subseteq A \cup B$ and $B \subseteq A \cup B$.

4. If $A \subseteq B$, then $A \cup C \subseteq B \cup C$.

5. If $A \subseteq B$, then $A \cap C \subseteq B \cap C$.

6. $A \subseteq B$ if and only if $A \cap B = A$.

7. $A \cup B = (A \cap B) \cup (A - B) \cup (B - A)$.

8. $A \cup (B \cap C) = (A \cup B) \cap (A \cup C)$.

9. $A \cap (B \cup C) = (A \cap B) \cup (A \cap C)$.

10. $(A - B) \cup (B - A) = (A \cup B) - (A \cap B)$.

11. $(A \cup B) \times C = (A \times C) \cup (B \times C)$.

A.2 Construction of the Number Systems

The purpose of this section is to provide an outline of the logical development of our number systems—the natural numbers, integers, rational numbers, real numbers, and complex numbers. At best, we hope to whet the reader's appetite. We will only rarely attempt to give proofs for our statements. However, elsewhere in the text we study general constructions which include as special cases the construction of the rational numbers from the integers and the construction of the complex numbers from the real numbers.

In Chapter 1 we take a naive approach in working with the set of integers. We have assumed that the reader is willing to accept the familiar properties of the operations of addition and multiplication. However, it is possible to derive these properties from a very short list of postulates. They are called the "Peano postulates," formulated about the turn of the last century by Giuseppe Peano (1858–1932). A similar set of axioms was stated by Richard Dedekind at about the same time. These axioms provide a description of the natural numbers (nonnegative integers $0, 1, 2, \ldots$), denoted by $\mathbf{N}$.

We have chosen to take the language and concepts of set theory as the starting point of the development of the number systems. This means that the Peano postulates must be stated in set theoretic terms alone. Intuitively, to describe the natural numbers we begin with 0 and then list successive numbers. The process that extends the set from one natural number to the next can be described as a function, which we denote by S in the postulates. We have in mind the formula $S(m) = m + 1$, although the formula does not yet make sense since $+$ has not been defined. The third postulate is a statement of the principle of mathematical induction (see Section A.4).

A.2.1 Axiom (Peano postulates). *The system $\mathbf{N}$ of natural numbers is a set $\mathbf{N}$ with a distinguished element 0 and a function S from $\mathbf{N}$ into $\mathbf{N}$ which satisfies*
  (i) $S(n) \neq 0$ *for all members n of $\mathbf{N}$,*
  (ii) $S(n_1) \neq S(n_2)$ *for all members $n_1 \neq n_2$ of $\mathbf{N}$, and*
  (iii) *any subset $\mathbf{N}'$ of $\mathbf{N}$ which contains 0 and which contains $S(n)$ for all n in $\mathbf{N}'$ must be equal to $\mathbf{N}$.*

The function S utilized in the Peano postulates is called the **successor function**. We will use it below to define addition and multiplication of natural numbers. Note that the assumption that S is a function means that it is possible to define the composition of S with itself n times, which we denote by S^n. We define S^0 to be the identity function.

A.2.2 Definition. *With the notation of the Peano postulates, let $m, n \in \mathbf{N}$. We define operations of addition and multiplication on $\mathbf{N}$ as follows:*
$$m + n = S^n(m) \quad and \quad m \cdot n = (S^m)^n(0).$$
We define $m \geq n$ if the equation $m = n + x$ has a solution $x \in \mathbf{N}$.

It is possible to derive the basic arithmetic and order properties of the natural numbers from the Peano postulates, but that is beyond the scope of what we have set out to do. After defining the integers **Z** in terms of **N**, it is then possible to extend properties of **N** to **Z**. This indication of how the properties which are listed in Section A.3 can be proved is as much detail as we can provide without digressing.

In Section 1.1 we take the well-ordering principle to be an axiom. Here we show that it is a direct consequence of the Peano postulates. In Section A.4 we show that the well-ordering principle implies the principles of mathematical induction, and so the well-ordering principle is logically equivalent to induction. The last sentence of the proof of the following theorem depends on the nontrivial fact that $\{m \in \mathbf{N} \mid n < m < S(n)\}$ is empty.

A.2.3 Theorem (Well-Ordering Principle). *Any nonempty set of natural numbers contains a smallest element.*

Proof. (Outline) Let T be a nonempty subset of **N** and let L be the set of natural numbers x such that $x \leq t$ for all $t \in T$. We cannot have $L = \mathbf{N}$ since there is some natural number t in T, and then $t + 1 = S(t)$ is not in L. (We are making use of the function S from the Peano postulates.) This means that L cannot satisfy the assumptions of postulate (iii), and since we certainly have $0 \in L$, there must be some n in L with $S(n) \notin L$. Thus we have $n \leq t$ for all $t \in T$, and to finish the proof we only need to show that $n \in T$. If this were not the case, then in fact $n < t$ for all $t \in T$, and therefore $S(n) \leq t$ for all $t \in T$, a contradiction. □

The next step is to use natural numbers to define the set of integers. We can do this by considering ordered pairs of natural numbers. We know that any negative integer can be expressed (in many ways) as a difference of natural numbers. To avoid the use of subtraction, which is as yet undefined, we consider the set $\mathbf{N} \times \mathbf{N}$, where an ordered pair (a, b) in $\mathbf{N} \times \mathbf{N}$ is thought of as representing $a - b$.

Just as with fractions, there are many ways in which a particular integer can be written as the difference of two natural numbers. For example, $(0, 2)$, $(1, 3)$, $(2, 4)$, etc., all represent what we know should be -2. We need a notion of equivalence of ordered pairs, and since we know that we should have $a - b = c - d$ if and only if $a + d = c + b$, we can avoid the use of subtraction in the definition. The formulas given below for addition and multiplication are motivated by the fact that we know that we should get $(a - b) + (c - d) = (a + c) - (b + d)$ and $(a - b)(c - d) = (ac + bd) - (ad + bc)$. If we were going to prove all of our assertions, we would have to show that the definitions of addition and multiplication of integers do not depend on the particular ordered pairs of natural numbers which we choose to represent them.

A.2.4 Definition. *The set of **integers**, denoted by* $\mathbf{Z}$, *is defined via the set* $\mathbf{N} \times \mathbf{N}$, *where we specify that ordered pairs* (a, b) *and* (c, d) *are equivalent if and only if* $a + d = b + c$.

We define addition and multiplication of ordered pairs as follows:

$$(a, b) + (c, d) = (a + c, b + d) \quad \text{and} \quad (a, b)(c, d) = (ac + bd, ad + bc) \,.$$

It is possible to verify all of the properties of $\mathbf{Z}$ that are listed in Section A.3, using the above definitions and the properties of $\mathbf{N}$. Furthermore, the set $\mathbf{N}$ can be identified with the set of ordered pairs (a, b) such that $a \geq b$, and so we can view $\mathbf{N}$ as the set of nonnegative integers. The well-ordering principle can easily be extended to the statement that any set of integers that is bounded below must contain a smallest element.

The next step is to construct the set of rational numbers $\mathbf{Q}$ from the set of integers. This is a special case of a general construction given in Section 5.4, where detailed proofs are provided.

A.2.5 Definition. *The set of **rational numbers**, denoted by* $\mathbf{Q}$, *is defined via the set of ordered pairs* (m, n) *such that* $m, n \in \mathbf{Z}$ *and* $n > 0$, *where we agree that* (a, b) *is equivalent to* (c, d) *if and only if* $ad = bc$. *We define addition and multiplication as follows:*

$$(a, b) + (c, d) = (ad + bc, bd) \quad \text{and} \quad (a, b)(c, d) = (ac, bd) \,.$$

It is more difficult to describe the construction of the set of real numbers from the set of rational numbers. The Greeks used a completely geometric approach to real numbers, and initially considered numbers to be simply the ratios of lengths of line segments. However, the length of a diagonal of a square with sides of length 1 cannot be expressed as the ratio of two integer lengths, since $\sqrt{2}$ is not a rational number. This makes it necessary to introduce irrational numbers.

A sequence $\{a_n\}_{n=1}^{\infty}$ of rational numbers is said to be a **Cauchy sequence** if for each $\epsilon > 0$ there exists N such that $|a_n - a_m| < \epsilon$ for all $n, m > N$. It is then possible to define the set of **real numbers** $\mathbf{R}$ as the set of all Cauchy sequences of rational numbers, where such sequences are considered to be equivalent if the limit of the difference of the sequences is 0. To verify all of the properties of the real numbers is then quite an involved process.

We note only a few of the properties of real numbers: $\mathbf{R}$ is a field (see Section 4.1 for the definition and properties of a field) ordered by $\leq$. The set $\mathbf{Q}$ is dense in $\mathbf{R}$, in the sense that between any two distinct real numbers there is a rational number. Any set of real numbers that has a lower bound has a greatest lower bound, and any set that has an upper bound has a least upper bound. The **Archimedean property** holds; i.e., for any two positive real numbers a, b there exists an integer n such that $na > b$.

Finally, the set **C** of **complex numbers** is described in Section A.5. One method of construction is to use Kronecker's theorem, Theorem 4.3.8. An alternative is to consider ordered pairs of real numbers. Then addition and multiplication are defined as follows:

$$(a, b) + (c, d) = (a + c, b + d) \quad \text{and} \quad (a, b) \cdot (c, d) = (ac - bd, ad + bc).$$

The ordered pair (a, b) is usually written $a + bi$, where $i^2 = -1$.

Detailed proofs of the assertions in this section can be found in various text books such as those by Landau and by Cohen and Ehrlich. The construction of the real numbers from the rationals is usually viewed as a part of analysis rather than algebra.

A.3 Basic Properties of the Integers

We assume that the reader is familiar with the arithmetic and order properties of the integers, and indeed, we have freely used these properties throughout the book. In the interest of completeness we now explicitly list these properties, as well as their names.

A.3.1 (Properties of Addition).

    (a) Closure: *Given any two integers a and b, there is a unique integer $a + b$.*

    (b) Associativity: *Given integers a, b, c, we have $(a + b) + c = a + (b + c)$.*

    (c) Commutativity: *Given integers a, b, we have $a + b = b + a$.*

    (d) Zero element: *There exists a unique integer 0 such that $a + 0 = a$ for any integer a.*

    (e) Inverses: *Given an integer a, there exists a unique integer, denoted by $-a$, such that $a + (-a) = 0$.*

A.3.2 (Properties of Multiplication).

    (a) Closure: *Given any two integers a and b, there is a unique integer $a \cdot b = ab$.*

    (b) Associativity: *Given integers a, b, c, we have $(a \cdot b) \cdot c = a \cdot (b \cdot c)$.*

    (c) Commutativity: *Given integers a, b, we have $a \cdot b = b \cdot a$.*

    (d) Identity element: *There exists a unique integer $1 (\neq 0)$ such that $a \cdot 1 = a$ for any integer a.*

A.3.3 (Joint Property of Addition and Multiplication).

    Distributivity: *Given integers a, b, c, we have $a(b + c) = ab + ac$.*

A.3.4 (Properties of Order). *There exists a subset $\mathbf{Z}^+ \subset \mathbf{Z}$, called the* set of positive integers, *which satisfies the following properties:*

    (a) Closure under addition: *If $a, b \in \mathbf{Z}^+$, then $a + b \in \mathbf{Z}^+$.*

    (b) Closure under multiplication: *If $a, b \in \mathbf{Z}^+$, then $ab \in \mathbf{Z}^+$.*

    (c) Trichotomy: *Given $a \in \mathbf{Z}$, exactly one of the following holds:*

       (i) $a \in \mathbf{Z}^+$,   (ii) $a = 0$,   (iii) $-a \in \mathbf{Z}^+$.

A number of these properties are redundant. Our purpose is to provide a working knowledge of the system of integers, and so we have not given the most economical list of properties. Rather than investigating the foundations of the number systems, we will be content with simply making the following statement: Together with the well-ordering principle, the above list of thirteen properties completely characterizes the set of integers.

The following proposition lists some of the usual arithmetic properties of the set of integers. These properties hold in a more general setting, which is studied in Chapter 5. We will use the notation $a - b$ for $a + (-b)$.

A.3.5 Proposition. *Let $a, b, c \in \mathbf{Z}$.*
    (a) *If $a + b = a + c$, then $b = c$.*
    (b) $-(-a) = a$.
    (c) $a \cdot 0 = 0$.
    (d) $(-a)(-b) = ab$.

We introduce the usual order symbols as follows. We say that a is greater than b, denoted by $a > b$, if $a - b \in \mathbf{Z}^+$. For $a > b$ we also write $b < a$ (read "b is less than a"), and $a \geq b$ (read "a is greater than or equal to b") denotes that $a = b$ or $a > b$. Finally, the absolute value of a, denoted by $|a|$, is equal to a if $a \in \mathbf{Z}^+$ or $a = 0$ and is equal to $-a$ if $-a \in \mathbf{Z}^+$. The proof of the next proposition is left as an exercise.

A.3.6 Proposition. *Let $a, b, c \in \mathbf{Z}$.*
    (a) *If $a > 0$, then $a \geq 1$.*
    (b) *If $a > b$ and $b > c$, then $a > c$.*
    (c) *If $a > b$, then $a + c > b + c$.*
    (d) $a < 0$ *if and only if $-a \in \mathbf{Z}^+$.*
    (e) *If $a > b$ and $c > 0$, then $ac > bc$.*
    (f) *If $a > b$ and $c < 0$, then $ac < bc$.*
    (g) $|a| \geq 0$, *and $|a| = 0$ if and only if $a = 0$.*
    (h) *If $a > 0$, then $|b| \leq a$ if and only if $-a \leq b \leq a$.*
    (k) $|ab| = |a||b|$.
    (m) $|a + b| \leq |a| + |b|$.
    (n) *If $ab = ac$ and $a \neq 0$, then $b = c$.*

A.4   Induction

If one develops the natural numbers from the Peano postulates, then mathematical induction is taken to be one of the postulates. On the other hand, if one uses the list of properties given in Section A.3 as a starting point, then the well-ordering principle is usually chosen as an axiom. We begin by showing that mathematical induction can be deduced from the well-ordering principle. We will let $\mathbf{Z}^+$ denote the set $\{1, 2, 3, \ldots\}$ of positive integers.

A.4.1 Theorem. *The well-ordering principle implies that if* $S \subseteq \mathbf{Z}^+$ *and*
  (i) $1 \in S$
  (ii) $n + 1 \in S$ *whenever* $n \in S$,
then $S = \mathbf{Z}^+$.

Proof. Suppose that $S \neq \mathbf{Z}^+$. Then the set $T = \{n \in \mathbf{Z}^+ \mid n \notin S\}$ is not empty and by the well-ordering principle has a least element k. Since $1 \in S$, $k \neq 1$, and so $k - 1 \in \mathbf{Z}^+$. Since $k - 1 < k$, we have $k - 1 \in S$. But by (ii), then we have $k = (k - 1) + 1 \in S$, a contradiction. Thus $T = \emptyset$ and $S = \mathbf{Z}^+$. $\square$

This theorem is applied in the principle of mathematical induction, which is of paramount importance. The principle of mathematical induction applies to statements which involve an arbitrary positive integer n. Examples of such statements are:

1. $1 + 2 + \ldots + n = n(n + 1)/2$.

2. $3 | (10^n - 1)$.

3. Let $a_1, \ldots, a_n$ be positive real numbers. Then $\sqrt[n]{a_1 a_2 \cdots a_n} \leq \dfrac{1}{n} \sum_{i=1}^{n} a_i$.

4. $n^2 - n + 41$ is a prime number.

Observe that each statement depends on the positive integer n and becomes either true or false when some value is substituted for n. The last example is true when $n = 2$, but false when $n = 41$.

Suppose that P_n is a statement depending on the positive integer n. If P_n becomes true for each choice of n, then the principle of mathematical induction frequently allows us to establish this fact. Let us state the principle, prove that it holds, and then apply it to some examples. Note that we could begin numbering with any integer, say $P_0, P_1, \ldots$ or even $P_{-297}, P_{-296}, \ldots$.

A.4.2 Theorem (Principle of Mathematical Induction). *Let* $P_1, P_2, \ldots$ *be a sequence of propositions. Suppose that*
  (i) P_1 *is true, and*
  (ii) *if* P_k *is true, then* P_{k+1} *is true for all positive integers* k.
Then P_n *is true for all positive integers* n.

Proof. Let $S = \{n \in \mathbf{Z}^+ \mid P_n \text{ is true}\}$. By (i), $1 \in S$; and (ii), if $n \in S$, then $n + 1 \in S$. Apply Theorem A.4.1 to get $S = \mathbf{Z}^+$. Thus P_n is true for all positive integers. $\square$

Example A.4.1.

To establish that $1 + 2 + \ldots + n = n(n+1)/2$, let P_n be the statement $1 + 2 + \ldots + n = n(n+1)/2$. Then $1 = 1(1+1)/2$, so P_1 is true. The next step is to show that P_k implies P_{k+1}. Assume that P_k is true, so that we have $1 + 2 + \ldots + k = k(k+1)/2$. Add $k + 1$ to both sides of this equation to get

$$1 + 2 + \ldots + k + (k+1) = \frac{k(k+1)}{2} + (k+1) = \frac{k(k+1) + 2(k+1)}{2}$$

$$= \frac{(k+1)(k+2)}{2} = \frac{(k+1)[(k+1)+1]}{2}.$$

Thus P_{k+1} is true, and so by induction P_n holds for all $n \in \mathbf{Z}^+$.   □

This is a good point at which to emphasize that when we are using the principle of mathematical induction, we must establish the truth of P_1. However, when we establish the truth of P_{k+1}, we get to assume the truth of P_k without having to prove anything about the truth of P_k.

Example A.4.2.

To prove that $3|(10^n - 1)$ for all positive integers n, let P_n be the statement $3|(10^n - 1)$. Now P_1 says $3|(10^1 - 1)$ or $3|9$, which is true. Assume that P_k is true, that is, that $3|(10^k - 1)$. Then since

$$10^{k+1} - 1 = 10 \cdot 10^k - 1 = 10 \cdot 10^k - 10 + 10 - 1 = 10 \cdot (10^k - 1) + 9$$

we have that $3|(10^{k+1} - 1)$ since $3|(10^k - 1)$ and $3|9$. Hence P_{k+1} is true. By the principle of mathematical induction, P_n holds for all positive integers n.   □

A second form of mathematical induction is more useful for some purposes.

A.4.3 Theorem (Second Principle of Mathematical Induction).

Let $P_1, P_2, \ldots$ be a sequence of propositions. Suppose that
  (i) *P_1 is true, and*
  (ii) *if P_m is true for all $m \leq k$, then P_{k+1} is true for all $k \in \mathbf{Z}^+$.*
Then P_n is true for all positive integers n.

Proof. Let $S = \{n \in \mathbf{Z}^+ \mid P_n \text{ is true }\}$. If $S \neq \mathbf{Z}^+$, then the set $T = \{n \in \mathbf{Z}^+ \mid P_n \text{ is false}\}$ is nonempty. By the well-ordering principle, T has a least element k. Since $1 \notin T$, $k \neq 1$. Thus for all $m < k$ we have $m \in S$; that is, P_m is true for all $m \leq k - 1$. By hypothesis P_k is true and so $k \notin T$, a contradiction. Thus P_n is true for all natural numbers n.   □

Example A.4.3 (Fibonacci numbers).

Define a sequence of natural numbers as follows: Let $F_1 = F_2 = 1$, and $F_n = F_{n-1} + F_{n-2}$ for $n \geq 3$. Thus $F_3 = 2$, $F_4 = 3$, $F_5 = 5$, $F_6 = 8$, etc. The sequence $F_1, F_2, \ldots$ is called the *Fibonacci sequence*. We will show that $F_n < (7/4)^n$ for all positive integers n.

Let P_n be the statement $F_n < (7/4)^n$. Then P_1 says $F_1 = 1 < (7/4)^1$, which is true. Assuming P_m for all $m \leq k$, we have that

$$F_{k+1} = F_k + F_{k-1} < \left(\frac{7}{4}\right)^k + \left(\frac{7}{4}\right)^{k-1}$$

$$= \left(\frac{7}{4}\right)^{k-1}\left(\frac{7}{4} + 1\right) = \left(\frac{7}{4}\right)^{k-1}\left(\frac{11}{4}\right)$$

$$< \left(\frac{7}{4}\right)^{k-1}\left(\frac{49}{16}\right) = \left(\frac{7}{4}\right)^{k+1},$$

and so P_{k+1} holds, unless $k = 1$, in which case the formula $F_{k+1} = F_k + F_{k-1}$ does not make sense since F_0 is not defined. For this reason we must directly verify that P_2 holds. Since $F_2 = 1 < (7/4)^2$, we have that if P_m is true for all $m \leq k$, then P_{k+1} is true. Thus the fact that $F_n < (7/4)^n$ for all positive integers n follows from the second principle of mathematical induction. $\square$

EXERCISES: SECTION A.4

Use the principle of mathematical induction to establish each of the following, where n is any positive integer:

1. $n < 2^n$

2. $1^2 + 2^2 + \ldots + n^2 = n(n+1)(2n+1)/6$

3. $1^3 + 2^3 + \ldots + n^3 = n^2(n+1)^2/4$

4. $2 + 2^2 + \ldots + 2^n = 2^{n+1} - 2$

5. $x + 4x + 7x + \ldots + (3n-2)x = n(3n-1)x/2$

6. $10^{n+1} + 10^n + 1$ is divisible by 3.

7. $10^{n+1} + 3 \times 10^n + 5$ is divisible by 9.

8. $4 \times 10^{2n} + 9 \times 10^{2n-1} + 5$ is divisible by 99.

9. Let $a_1, \ldots, a_n$ be positive real numbers, $G_n = \sqrt[n]{a_1 a_2 \cdots a_n}$, and $A_n = \frac{1}{n} \sum_{i=1}^n a_i$. Then G_n is called the **geometric mean** and A_n is called the **arithmetic mean**. We wish to show that $G_n \leq A_n$.

(i) Show that $G_2 \leq A_2$.

(ii) Show that $G_{2^n} \leq A_{2^n}$ by using induction on n.

(iii) Show that $G_n \leq A_n$.

Hint: Let m be such that $2^m \geq n$, and set $a_{n+1} = a_{n+2} = \ldots = a_{2^m} = A_n$ and apply part (ii).

10. Let a and b be real numbers. Prove the **binomial theorem**, which states that

$$(a + b)^n = \sum_{i=0}^n \binom{n}{i} a^i b^{n-i} \qquad \text{where} \qquad \binom{n}{i} = \frac{n!}{i!(n-i)!}$$

and $n! = n(n-1) \cdots 2 \cdot 1$ for $n \geq 1$ and $0! = 1$.

Hint: $\binom{m+1}{k} = \binom{m}{k} + \binom{m}{k-1}$.

11. Find a formula for the derivative of the product of n functions, and give a detailed proof by induction (assuming the product rule for the derivative of two functions).

12. Find a formula for the nth derivative of the product of two functions, and give a detailed proof by induction.

A.5 Complex Numbers

The equation $x^2 + 1 = 0$ has no real root since for any real number x we have $x^2 + 1 \geq 1$. The purpose of this section is to construct a set of numbers that extends the set of real numbers and includes a root of this equation. If we had a set of numbers that contained the set of real numbers, was closed under addition, subtraction, multiplication, and division, and contained a root i of the equation $x^2 + 1 = 0$, then it would have to include all numbers of the form $a + bi$ where a and b are real numbers. The addition and multiplication would be given by

$$(a + bi) + (c + di) = (a + c) + (b + d)i ,$$

$$(a + bi)(c + di) = ac + (bc + ad)i + bdi^2 = (ac - bd) + (ad + bc)i .$$

Here we have used the fact that $i^2 = -1$ since $i^2 + 1 = 0$.

Our construction for the desired set of numbers is to invent a symbol i for which $i^2 = -1$, and then to consider all pairs of real numbers a and b, in the form $a + bi$. In Chapter 4 we show how this construction can be done formally, by working with congruence classes of polynomials. At the end of this section, we also indicate how 2×2 matrices can be used to construct a set in which the equation $x^2 + 1 = 0$ has a solution. At this point, we simply ask the reader to accept the "invention" of the symbol i at an informal, intuitive level.

A.5.1 Definition. *The set* $\mathbf{C} = \{a + bi \mid a, b \in \mathbf{R} \text{ and } i^2 = -1\}$ *is called the set of complex numbers. Addition and multiplication of complex numbers are defined as follows:*

$$(a + bi) + (c + di) = (a + c) + (b + d)i ,$$

$$(a + bi)(c + di) = (ac - bd) + (ad + bc)i .$$

Note that $a + bi = c + di$ if and only if $a = c$ and $b = d$. If $c + di$ is nonzero, that is, if $c \neq 0$ or $d \neq 0$, then division by $c + di$ is possible:

$$\frac{a + bi}{c + di} = \frac{(a + bi)(c - di)}{(c + di)(c - di)} = \frac{ac + bd}{c^2 + d^2} + \frac{bc - ad}{c^2 + d^2} i .$$

A useful model for the set of complex numbers is a geometric model in which the number $a + bi$ is viewed as the ordered pair (a, b) in the plane. (See Figure A.5.1.) Note that i corresponds to the pair $(0, 1)$.

Figure A.5.1:

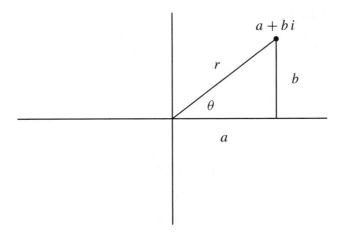

In polar coordinates, $a + bi$ is represented by (r, θ), where $r = \sqrt{a^2 + b^2}$ and $\cos \theta = a/r$, $\sin \theta = b/r$. The value r is called the **absolute value** of $a + bi$, and we write $|a + bi| = \sqrt{a^2 + b^2}$. This gives $a + bi = r(\cos \theta + i \sin \theta)$. In this form we can compute the product of two complex numbers as follows:

$$r(\cos \theta + i \sin \theta) \quad \cdot \quad t(\cos \phi + i \sin \phi)$$
$$= rt((\cos \theta \cos \phi - \sin \theta \sin \phi) + i(\sin \theta \cos \phi + \cos \theta \sin \phi))$$
$$= rt(\cos(\theta + \phi) + i \sin(\theta + \phi)) .$$

This simplification of the product comes from the trigonometric formulas for the cosine and sine of the sum of two angles. Thus to multiply two complex numbers represented in polar form we multiply their absolute values and add their angles. A repeated application of this formula to $\cos\theta + i\sin\theta$ gives the following theorem.

A.5.2 Theorem (DeMoivre). *For any positive integer* n,

$$(\cos\theta + i\sin\theta)^n = \cos(n\theta) + i\sin(n\theta) .$$

A.5.3 Corollary. *For any positive integer* n, *the equation* $z^n = 1$ *has* n *distinct roots in the set of complex numbers.*

Proof. For $k = 0, 1, \ldots, n-1$, the values $\cos\dfrac{2k\pi}{n} + i\sin\dfrac{2k\pi}{n}$ are distinct and

$$\left(\cos\frac{2k\pi}{n} + i\sin\frac{2k\pi}{n}\right)^n = \cos 2k\pi + i\sin 2k\pi = 1 . \quad \square$$

The complex roots of $z^n = 1$ are called the **nth roots of unity**. When plotted in the complex plane, they form the vertices of a regular polygon with n sides inscribed in a circle of radius 1 with center at the origin.

Example A.5.1 (Cube roots of unity).

The cube roots of unity are 1, $\cos\dfrac{2\pi}{3} + i\sin\dfrac{2\pi}{3}$, and $\cos\dfrac{4\pi}{3} + i\sin\dfrac{4\pi}{3}$, or equivalently, 1, $\omega = -\dfrac{1}{2} + \dfrac{\sqrt{3}}{2}i$, and $\omega^2 = -\dfrac{1}{2} - \dfrac{\sqrt{3}}{2}i$. Note that $\omega^2 + \omega + 1 = 0$, since ω is a root of $z^3 - 1 = (z-1)(z^2 + z + 1) = 0$. (See Figure A.5.2.) $\quad \square$

Figure A.5.2:

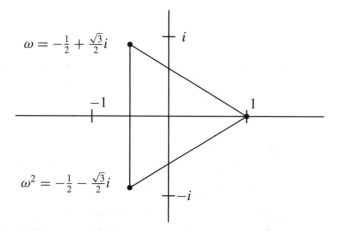

Example A.5.2 (Fourth roots of unity).

The fourth roots of unity are 1, i, $i^2 = -1$, and $i^3 = -i$. (See Figure A.5.3.) □

Figure A.5.3:

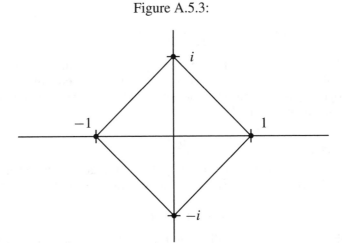

Example A.5.3 (Finding nth roots).

If $z^n = u$, then $(z\omega)^n = u$, where ω is any nth root of unity. Thus if all nth roots of unity are already known, it is easy to find the nth roots of any complex number, provided we know one of them. In general, the nth roots of $r(\cos\theta + i\sin\theta)$ are

$$r^{1/n}\left(\cos\frac{\theta + 2k\pi}{n} + i\sin\frac{\theta + 2k\pi}{n}\right), \quad \text{for} \quad 1 \le k \le n.$$

To find the square root of a complex number it may be helpful to use the formulas

$$\sin\frac{\theta}{2} = \pm\sqrt{\frac{1 - \cos\theta}{2}} \quad \text{and} \quad \cos\frac{\theta}{2} = \pm\sqrt{\frac{1 + \cos\theta}{2}}. \quad \square$$

We have noted that the powers of i are i, $i^2 = -1$, $i^3 = -i$, $i^4 = 1$. Since $i^4 = 1$, the powers repeat. For example, $i^5 = i^4 i = i$, $i^6 = i^4 i^2 = -1$, and so on. For any integer n, the power i^n depends on the remainder of n when divided by 4, since if $n = 4q + r$, then $i^n = i^{4q+r} = (i^4)^q i^r = i^r$. In particular, $i^{-1} = i^3 = -i$, and a similar computation can be given for any negative exponent.

We can use our knowledge of powers of i to give another way to express complex numbers. We need to recall the Taylor series expansions for e^x, $\sin x$, and $\cos x$:

$$e^x = 1 + x + \frac{x^2}{2!} + \frac{x^3}{3!} + \frac{x^4}{4!} + \cdots ,$$

$$\cos x = 1 - \frac{x^2}{2!} + \frac{x^4}{4!} - \frac{x^6}{6!} + \cdots ,$$

$$\sin x = x - \frac{x^3}{3!} + \frac{x^5}{5!} - \frac{x^7}{7!} + \cdots .$$

We extend the definition of e^x to complex values of x by using the same power series. For real values of θ, in $e^{i\theta}$ we can reduce the exponents of i modulo 4 and then group together those terms that contain i and those terms that do not contain i. This gives us the expression

$$e^{i\theta} = \cos\theta + i\,\sin\theta .$$

Thus complex numbers can be written in polar coordinates in the form $re^{i\theta}$. The product $re^{i\theta} \cdot te^{i\phi}$ is then equal to $rte^{i(\theta+\phi)}$. Substituting $\theta = \pi$ yields the remarkable formula

$$e^{\pi i} + 1 = 0 ,$$

which combines $0, 1, \pi, e, i$, equality, addition, multiplication, and exponentiation.

We will now construct a concrete model for the set of complex numbers, using 2×2 matrices over **R**. We can identify the set **R** of real numbers with the set of all scalar 2×2 real matrices. That is, the real number a corresponds to the matrix $\begin{bmatrix} a & 0 \\ 0 & a \end{bmatrix}$. As you should recall from your study of linear algebra, scalar matrices are added and multiplied just like real numbers. We can now ask for a root of the equation $x^2 + 1 = 0$ in the set of 2×2 matrices, where we can identify 1 with the scalar matrix $\begin{bmatrix} 1 & 0 \\ 0 & 1 \end{bmatrix}$. In fact, the matrix $\begin{bmatrix} 0 & 1 \\ -1 & 0 \end{bmatrix}$ is one such root, since

$$\begin{bmatrix} 0 & 1 \\ -1 & 0 \end{bmatrix}^2 + \begin{bmatrix} 1 & 0 \\ 0 & 1 \end{bmatrix} = \begin{bmatrix} -1 & 0 \\ 0 & -1 \end{bmatrix} + \begin{bmatrix} 1 & 0 \\ 0 & 1 \end{bmatrix} = \begin{bmatrix} 0 & 0 \\ 0 & 0 \end{bmatrix} .$$

If we consider all matrices of the form

$$a\begin{bmatrix} 1 & 0 \\ 0 & 1 \end{bmatrix} + b\begin{bmatrix} 0 & 1 \\ -1 & 0 \end{bmatrix} = \begin{bmatrix} a & b \\ -b & a \end{bmatrix}$$

we have a set that is closed under addition, subtraction, and multiplication, since

$$\begin{bmatrix} a & b \\ -b & a \end{bmatrix} \pm \begin{bmatrix} c & d \\ -d & c \end{bmatrix} = \begin{bmatrix} a \pm c & b \pm d \\ -(b \pm d) & a \pm c \end{bmatrix}$$

and

$$\begin{bmatrix} a & b \\ -b & a \end{bmatrix} \cdot \begin{bmatrix} c & d \\ -d & c \end{bmatrix} = \begin{bmatrix} ac - bd & ad + bc \\ -(ad + bc) & ac - bd \end{bmatrix}.$$

Finally, if $\begin{bmatrix} a & b \\ -b & a \end{bmatrix}$ is not the zero matrix, then it has an inverse

$$\frac{1}{a^2 + b^2} \begin{bmatrix} a & -b \\ b & a \end{bmatrix}.$$

The set of 2×2 matrices of the form $\begin{bmatrix} a & b \\ -b & a \end{bmatrix}$ has the properties we are looking for,

and the correspondence $a + bi \leftrightarrow \begin{bmatrix} a & b \\ -b & a \end{bmatrix}$ preserves addition and multiplication.
We have thus constructed a concrete model for the set of complex numbers, for those who were worried by our "invention" of the root i of $x^2 + 1 = 0$. The matrix representation for $\mathbf{C}$ is a convenient form in which to verify that the associative and distributive laws hold for the set of complex numbers, since these laws are known to hold for matrix multiplication. It is routine to check that these special 2×2 matrices satisfy the commutative law for multiplication.

EXERCISES: SECTION A.5

1. Compute each of the following:
   (a) $(\frac{1}{\sqrt{2}} + \frac{i}{\sqrt{2}})^6$
   (b) $(1 + i)^8$
   (c) $(\cos 20° + i \sin 20°)^9$

2. Find $(\cos \theta + i \sin \theta)^{-1}$.

3. (a) Find the 6th roots of unity.
   (b) Find the 8th roots of unity.

4. (a) Find the cube roots of $-8i$.
   (b) Find the cube roots of $-4\sqrt{2} + 4\sqrt{2}i$.
   (c) Find the cube roots of $2 + 2i$.
   (d) Find the fourth roots of $1 + i$.

5. Solve the equation $z^2 + z + (1 + i) = 0$.

6. Solve the equation $x^3 - 3x^2 - 6x - 20 = 0$, given that one root is $-1 + \sqrt{3}i$.

7. Use DeMoivre's theorem to find formulas for $\cos 3\theta$ (in terms of $\cos \theta$) and $\sin 3\theta$ (in terms of $\sin \theta$).

8. If η is an nth root of unity, it is called a **primitive** nth root of unity if it is not a root of $z^k - 1 = 0$ for any k such that $1 \leq k < n$.

(a) Show that $\cos(2\pi/n) + i \sin(2\pi/n)$ is a primitive nth root of unity.

(b) If $\eta = \cos(2\pi/n) + i \sin(2\pi/n)$, show that η^m is a primitive nth root of unity if and only if n and m are relatively prime.

A.6 Solution of Cubic and Quartic Equations

In this section we will discuss the solution by radicals of cubic and quartic equations with real coefficients.

A.6.1. Solution of the General Quadratic Equation

To solve the equation

$$ax^2 + bx + c = 0$$

with $a, b, c \in \mathbf{R}$ and a nonzero, we can divide through by a and make the substitution $x = y - b/2a$. (We have rearranged the usual approach of completing the square in order to parallel later work.) This gives

$$y^2 + p = 0, \quad \text{where} \quad -p = \frac{b^2 - 4ac}{4a^2},$$

and so

$$y = \frac{\pm\sqrt{b^2 - 4ac}}{2a}.$$

Thus the general solution is

$$x = \frac{-b \pm \sqrt{b^2 - 4ac}}{2a}.$$

A.6.2. Discriminant of a Quadratic Equation

For the real quadratic equation $ax^2 + bx + c = 0$, the **discriminant**

$$\Delta = b^2 - 4ac$$

determines whether the solutions are real numbers (when $\Delta \geq 0$) or imaginary numbers (when $\Delta < 0$). The equation has a multiple root if and only if $\Delta = 0$, and this occurs if and only if

$$ax^2 + bx + c = (mx + k)^2$$

for some $m, k \in \mathbf{R}$. If $a = 1$, then $\Delta = (x_1 - x_2)^2$, where x_1, x_2 are the solutions of the equation $x^2 + bx + c = 0$.

A.6.3. Discriminant of a Cubic Equation

In discussing the general cubic equation

$$ax^3 + bx^2 + cx + d = 0$$

with real coefficients, we will assume that $a = 1$. The discriminant Δ of the equation is then defined by

$$\Delta = (x_1 - x_2)^2(x_1 - x_3)^2(x_2 - x_3)^2$$

where x_1, x_2, x_3 are the roots of the equation, and it gives the following information: if the roots are all real, then $\Delta \geq 0$, and $\Delta = 0$ if and only if at least two of the roots coincide. If not all of the roots are real, say x_2 is imaginary, then one of the other roots must be its complex conjugate, say $x_3 = \overline{x}_2$, and the remaining root x_1 is real. Since x_1 is real,

$$(x_1 - x_2)(x_1 - \overline{x}_2) = (x_1 - x_2)\overline{(x_1 - x_2)}$$

is a real number, so

$$(x_1 - x_2)^2(x_1 - \overline{x}_2)^2 > 0 \,.$$

But $x_2 - x_3 = x_2 - \overline{x}_2$ is purely imaginary, so $(x_2 - x_3)^2 < 0$, and therefore $\Delta < 0$. Thus we have shown that $\Delta \geq 0$ when all roots are real, $\Delta < 0$ when two roots are imaginary, and $\Delta = 0$ when there is a multiple root.

If we make the substitution $x = y - b/3$, which is the first step in solving the general cubic equation in F.4, we obtain the reduced equation $y^3 + py + q = 0$. This reduced equation has the same discriminant, which can now be computed as $\Delta = -4p^3 - 27q^2$. This can be shown either by using the relations $y_1 + y_2 + y_3 = 0$, $y_1y_2 + y_1y_3 + y_2y_3 = p$, and $y_1y_2y_3 = -q$ for the roots y_1, y_2, y_3 of the reduced equation, or by a direct computation involving the roots (which will be determined later). See the exercises at the end of this section for hints in making this computation.

A.6.4. Solution of the General Cubic Equation

The first step in solving the equation

$$x^3 + bx^2 + cx + d = 0$$

is to substitute $x = y - b/3$, which eliminates the quadratic term. This gives the reduced cubic equation

$$y^3 + py + q = 0 \,,$$

where

$$p = c - \frac{b^2}{3}$$

and

$$q = d - \frac{bc}{3} + \frac{2b^3}{27} .$$

The roots of the original equation can easily be found from those of the reduced equation. The method we will use to solve the reduced cubic equation is essentially the same as that used in 1591 by François Viète (1540–1603). We let

$$\omega = -\frac{1}{2} + \frac{\sqrt{3}}{2} i$$

be a complex cube root of unity.

We next make the substitution $y = z - p/3z$, to obtain the equation

$$z^3 - \frac{p^3}{27z^3} + q = 0 .$$

Multiplying through by z^3 gives the **resolvent** equation

$$(z^3)^2 + q(z^3) - \left(\frac{p}{3}\right)^3 = 0 ,$$

which is a quadratic equation in z^3. Using the quadratic formula we obtain

$$z^3 = \frac{1}{2}\left(-q \pm \sqrt{q^2 + 4\left(\frac{p}{3}\right)^3}\right) = -\frac{q}{2} \pm \sqrt{\left(\frac{p}{3}\right)^3 + \left(\frac{q}{2}\right)^2} .$$

If z_1^3 and z_2^3 are the two solutions, then $z_1^3 z_2^3 = -(p/3)^3$. It is possible to choose the cube root in such a way that $z_1 z_2 = -p/3$, or $z_2 = -p/3z_1$, and then the other cube roots will be ωz_1, $\omega^2 z_1$, ωz_2, and $\omega^2 z_2$. Now substituting z_1 in the equation $y = z - p/3z$ to find the corresponding root y_1, we find that $y_1 = z_1 - p/3z_1 = z_1 + z_2$. Since $(\omega z_1)(\omega^2 z_2) = (\omega^2 z_1)(\omega z_2) = -p/3$, in the same way we have $y_2 = \omega z_1 + \omega^2 z_2$ and $y_3 = \omega^2 z_1 + \omega z_2$. Thus the roots of $z^6 + qz^3 - (p/3)^3 = 0$ give at most three distinct values when substituted into $y = z - p/3z$, and these are the roots of $y^3 + py + q = 0$.

Using the identities $\omega^3 = 1$ and $1 + \omega + \omega^2 = 0$ it is not difficult to check that

$$y_1 = z_1 + z_2 ,$$

$$y_2 = \omega z_1 + \omega^2 z_2 ,$$

and

$$y_3 = \omega^2 z_1 + \omega z_2$$

satisfy the relations

$$y_1 + y_2 + y_3 = 0 ,$$

$$y_1 y_2 + y_1 y_3 + y_2 y_3 = p \,,$$

and

$$y_1 y_2 y_3 = -q \,,$$

so they are the three roots of the reduced equation. We now give additional details for finding the roots in practice.

Finding the roots when $\Delta = 0$

If $-4p^3 - 27q^2 = 0$, then

$$\frac{p^3}{27} + \frac{q^2}{4} = 0 \,,$$

and so $z_1 = z_2$. Thus

$$y_1 = z_1 + z_2 = 2\sqrt[3]{\frac{q}{2}} = \sqrt[3]{-4q} \,.$$

Since $\omega + \omega^2 = -1$, we have $y_2 = -z_1 = \sqrt[3]{q}/\sqrt[3]{2}$, and similarly $y_3 = \sqrt[3]{q}/\sqrt[3]{2}$. Thus if $\Delta = 0$ we have roots $y = \sqrt[3]{-4q}$ and $y = \sqrt[3]{q}/\sqrt[3]{2}$, the latter with multiplicity 2.

Finding the roots when $\Delta < 0$

If $-4p^3 - 27q^2 < 0$, then

$$\frac{p^3}{27} + \frac{q^2}{4} > 0$$

and

$$\sqrt{\left(\frac{p}{3}\right)^3 + \left(\frac{q}{2}\right)^2}$$

is a real number. Thus we can take

$$z_1 = \sqrt[3]{\frac{-q}{2} + \sqrt{\left(\frac{p}{3}\right)^3 + \left(\frac{q}{2}\right)^2}} \,,$$

$$z_2 = \sqrt[3]{\frac{-q}{2} - \sqrt{\left(\frac{p}{3}\right)^3 + \left(\frac{q}{2}\right)^2}} \,,$$

and $y_1 = z_1 + z_2$. Then

$$y_2 = -\frac{1}{2}(z_1 + z_2) + \frac{1}{2}(z_1 - z_2)\sqrt{3}i$$

and

$$y_3 = -\frac{1}{2}(z_1 + z_2) - \frac{1}{2}(z_1 - z_2)\sqrt{3}i \ .$$

The real root y_1 leads to the factorization

$$y^3 + py + q = (y - y_1)(y^2 + y_1 y + (p + y_1^2)) \ ,$$

and using the quadratic formula gives the roots

$$\frac{-y_1}{2} \pm \frac{i}{2}\sqrt{3y_1^2 + 4p} \ .$$

Once it has been checked that y_1 is a root, this formula can be used to check y_2 and y_3.

Finding the roots when $\Delta > 0$

When $\Delta > 0$, all roots are real, but

$$\sqrt{\left(\frac{p}{3}\right)^3 + \left(\frac{q}{2}\right)^2}$$

is imaginary, so to compute the roots of the reduced equation we must first find the cube roots of two imaginary numbers. It can be proved that no formula involving only real radicals can be given.

In this case it is easier to use another method for finding roots, although we must admit that it does not yield a solution by radicals. Note that since $-4p^3 - 27q^2 > 0$, we have $-4p^3 > 27q^2$ and so p must be negative.

The trigonometric formula for the cosine of the sum of two angles can be used to show that $\cos 3\theta = 4\cos^3 \theta - 3\cos \theta$, so that $z = \cos \theta$ is a root of the equation $4z^3 - 3z = k$ if $\cos 3\theta = k$, that is, if $\theta = (\arccos k)/3$. In this case, $\cos(\theta + 2\pi/3)$ and $\cos(\theta + 4\pi/3)$ are also solutions, since $\cos 3(\theta + 2n\pi/3) = k$. This gives a method for solving the equation $4z^3 - 3z = k$, if $|k| \le 1$.

In the reduced equation $y^3 + py + q = 0$, we make the substitution

$$y = 2\sqrt{\frac{-p}{3}}\,z \ .$$

(Remember that $-p > 0$ since $\Delta > 0$.) We obtain the equation

$$4\left(2\sqrt{-\left(\frac{p}{3}\right)^3}\right)z^3 - 3\left(2\left(-\frac{p}{3}\right)\sqrt{-\frac{p}{3}}\right)z + q = 0 \ ,$$

or

$$4z^3 - 3z = k \ ,$$

where

$$k = \frac{-q}{2}\sqrt{-\left(\frac{3}{p}\right)^3} = \frac{-q}{2}\left(\frac{3}{-p}\right)^{3/2}.$$

Since $\Delta > 0$, we showed above that $27q^2 < -4p^3$, so $(27q^2)/(-4p^3) = k^2 < 1$, and thus $z_1 = \cos\theta$ is a root, where

$$\theta = \frac{1}{3}\arccos\left(\frac{-q}{2}\left(\frac{3}{-p}\right)^{3/2}\right).$$

Thus to solve $y^3 + py + q = 0$ when $\Delta > 0$, we find

$$\theta = \frac{1}{3}\arccos\left(\frac{-q}{2}\left(\frac{3}{-p}\right)^{3/2}\right).$$

Then the roots are

$$y_k = 2\sqrt{\frac{-p}{3}}\cos\left(\theta + \frac{2k\pi}{3}\right),$$

for $k = 0, 1, 2$.

A.6.5. Solution of the General Quartic Equation (Ferrari)

The general quartic equation

$$x^4 + bx^3 + cx^2 + dx + e = 0, \qquad a \neq 0$$

becomes

$$y^4 + py^2 + qy + r = 0$$

after substituting $x = y - (b/4)$. Then $y^4 = -py^2 - qy - r$, and adding $y^2z + (z^2/4)$ to both sides of the equation gives

$$\left(y^2 + \frac{1}{2}z\right)^2 = (z - p)y^2 - qy + \left(\frac{1}{4}z^2 - r\right).$$

If the right-hand side can be put in the form $(my + k)^2$, then the roots of the reduced quartic are the roots of

$$y^2 + \frac{1}{2}z = my + k$$

and

$$y^2 + \frac{1}{2}z = -my - k.$$

Since the right-hand side is a quadratic in y, it can be put in the form $(my + k)^2$ if and only if its discriminant is zero. This occurs if and only if

$$q^2 - 4(z - p)\left(\frac{z^2}{4} - r\right) = 0 ,$$

or equivalently, if and only if

$$z^3 - pz^2 - 4rz + (4pr - q^2) = 0 .$$

This latter equation in z is called the **resolvent cubic equation**, and any real root can be used to solve the reduced quartic.

For example, for the equation

$$y^4 + 3y^2 - 2y + 3 = 0 ,$$

the resolvent cubic is

$$z^3 - 3z^2 - 12z + 32 = 0 ,$$

which has the root $z = 4$. Thus

$$(y^2 + 2)^2 = y^2 + 2y + 1 = (y + 1)^2 ,$$

so either $y^2 + 2 = y + 1$ and then

$$y = \frac{1}{2} \pm \frac{\sqrt{3}}{2}i ,$$

or else $y^2 + 2 = -y - 1$ and then

$$y = -\frac{1}{2} \pm \frac{\sqrt{11}}{2}i .$$

A.6.6. Solution of the General Quartic Equation (Descartes)

We will try to factor the reduced quartic

$$y^4 + py^2 + qy + r = 0$$

as

$$(y^2 + ky + m)(y^2 - ky + n) = y^4 + (m + n - k^2)y^2 + (kn - km)y + mn .$$

We must have $m + n - k^2 = p$, $k(n - m) = q$, and $mn = r$. Since $m + n = p + k^2$ and $n - m = q/k$, we have $2n = p + k^2 + (q/k)$ and $2m = p + k^2 - (q/k)$. Then

$$4r = 2n2m = \left(p + k^2 + \frac{q}{k}\right)\left(p + k^2 - \frac{q}{k}\right) ,$$

which leads to the equation

$$(k^2)^3 + 2p(k^2)^2 + (p^2 - 4r)k^2 - q^2 = 0.$$

Any root of this equation gives a factorization of the desired form, and then the roots can be found from the quadratic formula.

For example, the equation $y^4 - 3y^2 + 6y - 2 = 0$ leads to the equation

$$(k^2)^3 - 6(k^2)^2 + 17k^2 - 36 = 0,$$

which has the root $k^2 = 4$. Thus

$$y^4 - 3y^2 + 6y - 2 = (y^2 + 2y - 1)(y^2 - 2y + 2),$$

which gives the four roots $y_1 = -1 + \sqrt{2}$, $y_2 = -1 - \sqrt{2}$, $y_3 = 1 + i$, $y_4 = 1 - i$.

We have shown that any polynomial equation of degree less than 5 can be solved by radicals. It is possible to show that certain equations of degree 5 or higher cannot be solved by radicals. For example, the equation $2x^5 - 10x + 5 = 0$ cannot be solved by radicals. We can summarize this section with the following theorem.

A.6.7 Theorem. *Any polynomial equation of degree ≤ 4 with real coefficients is solvable by radicals.*

EXERCISES: SECTION A.6

1. Verify that $\cos 3\theta = 4 \cos^3 \theta - 3 \cos \theta$.

2. Show that $x^3 + ax + 2 = 0$ has three real roots if and only if $a \leq -3$.

3. Use the method in A.6.4 to find the solutions of the equation $x^3 - 3x + 1 = 0$.

4. In A.6.4 check that y_1, y_2, y_3 as given are solutions of $y^3 + py + q = 0$.

5. To show that the discriminant of $y^3 + py + q = 0$ is $-4p^3 - 27q^2$, we can use the relations $y_1 + y_2 + y_3 = 0$, $y_1 y_2 + y_1 y_3 + y_2 y_3 = p$, and $y_1 y_2 y_3 = -q$ for the roots y_1, y_2, y_3. Substituting $y_1 = -y_2 - y_3$ in the second equation gives $p = -y_2^2 - y_3^2 - y_2 y_3$ and in the third equation gives $q = y_2^2 y_3 + y_2 y_3^2$. Compute $-27q^2 - 4p^3$ and check that it is equal to $(y_1 - y_2)^2 (y_1 - y_3)^2 (y_2 - y_3)^2$ when the same substitution $y_1 = -y_2 - y_3$ is made.

6. Give another proof that the discriminant of $y^3 + py + q = 0$ is $-4p^3 - 27q^2$ by using the following form of the roots: $y_1 = z_1 + z_2$; $y_2 = \omega z_1 + \omega^2 z_2$; $y_3 = \omega^2 z_1 + \omega z_2$; where $z_1 z_2 = -p/3$.

    Hint: Since $1, \omega, \omega^2$ are cube roots of unity, we have $\omega^3 = 1$, $\omega^2 + \omega + 1 = 0$, and $(x - 1)(x - \omega)(x - \omega^2) = x^3 - 1$, for any x. Substituting $x = z_1/z_2$ in the latter formula gives $(z_1 - z_2)(z_1 - \omega z_2)(z_1 - \omega^2 z_2) = z_1^3 - z_2^3$. Write out $(y_1 - y_2)(y_1 - y_3)(y_2 - y_3) = (1 - \omega)(z_1 - \omega^2 z_2)\dots$, square both sides, and simplify.

A.7   Dimension of a Vector Space

We assume that the student has already taken an elementary course in linear algebra. It is quite possible that only real numbers were allowed as scalars, whereas we need to be able to use facts about vector spaces with scalars from any field. (See Definition 4.1.1 for the definition of a field.) To make this book reasonably self-contained, in this section we include additional definitions, together with proofs of some facts about the dimension of a vector space.

For our purposes, the main application of techniques from linear algebra occurs when we need to study the structure of an extension field F of a field K. We can consider F as a vector space with scalars from K, and this makes it possible to utilize the concept of the dimension of F. We can get useful information about other extensions of K by comparing their dimension with that of F.

A.7.1 Definition. *A **vector space** over the field F is set V with a binary operation $+$ defined for all $\mathbf{u}, \mathbf{v} \in V$ and a **scalar multiplication** $a \cdot \mathbf{v} \in V$ defined for all $a \in F$ and $\mathbf{v} \in V$ such that the following conditions hold:*

  (i) $\mathbf{u} + \mathbf{v} \in V$, *for all* $\mathbf{u}, \mathbf{v} \in V$;

  (ii) $(\mathbf{u} + \mathbf{v}) + \mathbf{w} = \mathbf{u} + (\mathbf{v} + \mathbf{w})$, *for all* $\mathbf{u}, \mathbf{v}, \mathbf{w} \in V$;

  (iii)   *V contains an element $\mathbf{0}$ such that $\mathbf{0} + \mathbf{v} = \mathbf{v}$ for all $\mathbf{v} \in V$;*

  (iv)   *for each $\mathbf{v} \in V$ there exists an element $-\mathbf{v}$ such that $-\mathbf{v} + \mathbf{v} = \mathbf{0}$;*

  (v) $\mathbf{u} + \mathbf{v} = \mathbf{v} + \mathbf{u}$, *for all* $\mathbf{u}, \mathbf{v} \in V$;

  (vi) $a \cdot \mathbf{v} \in V$, *for all $a \in F$ and all $\mathbf{v} \in V$;*

  (vii) $a(b \cdot \mathbf{v}) = (ab) \cdot \mathbf{v}$, *for all $a, b \in F$ and all $\mathbf{v} \in V$;*

  (viii) $(a + b) \cdot \mathbf{v} = a \cdot \mathbf{v} + b \cdot \mathbf{v}$, *for all $a, b \in F$ and all $\mathbf{v} \in V$;*

  (ix) $a \cdot (\mathbf{u} + \mathbf{v}) = a \cdot \mathbf{u} + a \cdot \mathbf{v}$, *for all $a \in F$ and all $\mathbf{u}, \mathbf{v} \in V$; and*

  (x) $1 \cdot \mathbf{v} = \mathbf{v}$, *for all $\mathbf{v} \in V$.*

In the language of group theory, a vector space is an additive abelian group together with a scalar multiplication that satisfies conditions (vi) – (x). Note that scalar multiplication is not a binary operation on V. It must be defined by a function from $F \times V$ into V, rather than from $V \times V$ into V. As usual, we will use the notation $a\mathbf{v}$ rather than $a \cdot \mathbf{v}$.

If we denote the additive identity of V by $\mathbf{0}$ and the additive inverse of $\mathbf{v} \in V$ by $-\mathbf{v}$, then we have the following results: $\mathbf{0} + \mathbf{v} = \mathbf{v}$, $a \cdot \mathbf{0} = \mathbf{0}$, and $(-a)\mathbf{v} = a(-\mathbf{v}) = -(a\mathbf{v})$, for all $a \in F$ and $\mathbf{v} \in V$. The proofs are similar to those for the same results in a field, and involve the distributive laws, which provide the only connection between addition and scalar multiplication.

A.7.2 Definition. *Let V be a vector space over the field F. A nonempty subset W of V is called a **subspace** of V if it is a vector space under the operations of vector addition and scalar multiplication in F.*

A.7.3 Definition. *Let V be a vector space over the field F, and let S be a set $\{\mathbf{v}_1, \mathbf{v}_2, \ldots, \mathbf{v}_n\}$ of vectors in V. Any vector of the form $\mathbf{v} = \sum_{i=1}^{n} a_i \mathbf{v}_i$, for scalars $a_i \in F$, is called a **linear combination** of the vectors in S. The set of all linear combinations of vectors in S is called the **span** of S, denoted by $\mathrm{Span}\,(S)$. The set S is said to **span** V if $\mathrm{Span}\,(S) = V$.*

A.7.4 Proposition. *Let V be a vector space over the field F, and let S be a set $\{\mathbf{v}_1, \mathbf{v}_2, \ldots, \mathbf{v}_n\}$ of vectors in V. Then $\mathrm{Span}\,(S)$ is a subspace of V.*

A.7.5 Definition. *Let V be a vector space over the field F, and let S be a set $\{\mathbf{v}_1, \mathbf{v}_2, \ldots, \mathbf{v}_n\}$ of vectors in V. The vectors in S are said to be **linearly dependent** if one of the vectors can be expressed as a linear combination of the others. If not, then S is said to be a **linearly independent set**.*

In the preceding definition, if S is linearly dependent and, for example, $\mathbf{v}_j$ can be written as a linear combination of the remaining vectors in S, then we can rewrite the resulting equation as $a_1 \mathbf{v}_1 + \ldots + 1 \cdot \mathbf{v}_j + \ldots + a_n \mathbf{v}_n = \mathbf{0}$ for some scalars $a_i \in F$. Thus there exists a nontrivial (i.e., at least one coefficient is nonzero) relation of the form $\sum_{i=1}^{n} a_i \mathbf{v}_i = \mathbf{0}$. Conversely, if such a relation exists, then at least one coefficient must be nonzero. If a_j is nonzero, then since the coefficients are from a field, we can divide through by a_j and shift $\mathbf{v}_j$ to the other side of the equation to obtain $\mathbf{v}_j$ as a linear combination of the remaining vectors. If j is the largest subscript for which $a_j \neq 0$, then we can express $\mathbf{v}_j$ as a linear combination of $\mathbf{v}_1, \ldots, \mathbf{v}_{j-1}$.

From this point of view, S is linearly independent if and only if $\sum_{i=1}^{n} a_i \mathbf{v}_i = \mathbf{0}$ implies $a_1 = a_2 = \ldots = a_n = 0$. This is the condition that is usually given as the definition of linear independence.

If $\mathbf{v}_1, \ldots, \mathbf{v}_n$ span V and $\mathbf{v}_j$ is a linear combination of the other vectors, then omitting $\mathbf{v}_j$ still gives a spanning set. On the other hand, if $\mathbf{v}_1, \ldots, \mathbf{v}_n$ are linearly independent and do *not* span V, then adjoining to this family any vector $\mathbf{w}$ that is not in the span of $\{\mathbf{v}_1, \mathbf{v}_2, \ldots, \mathbf{v}_n\}$ produces a linearly independent set. These two remarks prove the following proposition.

A.7.6 Proposition. *Let V be a vector space spanned by a finite set S of vectors. Then any linearly independent subset of S is finite and forms a subset of a basis for V, and any subset of S that spans V contains a subset that is a basis.*

A.7.7 Theorem. *Let V be a vector space, let $S = \{\mathbf{u}_1, \mathbf{u}_2, \ldots, \mathbf{u}_m\}$ be a subset of V with $\mathrm{Span}\,(S) = V$, and let $\mathcal{T} = \{\mathbf{v}_1, \mathbf{v}_2, \ldots, \mathbf{v}_n\}$ be a linearly independent set. Then $n \leq m$, and V can be spanned by a set of m vectors that contains $\mathcal{T}$ and is contained in $S \cup \mathcal{T}$.*

Proof. Given $m > 0$, the proof will use induction on n. If $n = 1$, then of course $n \leq m$. Furthermore, $\mathbf{v}_1$ can be written as a linear combination of vectors in S, so the set $S' = \{\mathbf{v}_1, \mathbf{u}_1, \ldots, \mathbf{u}_m\}$ is linearly dependent. One of the vectors in S' can be written as a linear combination of $\mathbf{v}_1$ and the previous $\mathbf{u}$'s, so Proposition A.7.6 implies that deleting it from $\mathcal{S}'$ leaves a set with m elements that still spans V and also contains $\mathbf{v}_1$.

Now assume that the result holds for any set of n linearly independent vectors, and assume that $\mathcal{T}$ has $n + 1$ vectors. The first n vectors of $\mathcal{T}$ are still linearly independent, so by the induction assumption we have $n \leq m$ and V is spanned by some set $\{\mathbf{v}_1, \ldots, \mathbf{v}_n, \mathbf{w}_{n+1}, \ldots, \mathbf{w}_m\}$. Then $\mathbf{v}_{n+1}$ can be written as a linear combination of $\{\mathbf{v}_1, \ldots, \mathbf{v}_n, \mathbf{w}_{n+1}, \ldots, \mathbf{w}_m\}$. If $n = m$, then this contradicts the assumption that the set $\mathcal{T}$ is linearly independent, so we must have $n + 1 \leq m$. Furthermore, the set $\mathcal{S}' = \{\mathbf{v}_1, \ldots, \mathbf{v}_n, \mathbf{v}_{n+1}, \mathbf{w}_{n+1}, \ldots, \mathbf{w}_m\}$ is linearly dependent, so as before we can express one of the vectors $\mathbf{v}$ in $\mathcal{S}'$ as a linear combination of the previous ones. But $\mathbf{v}$ cannot be one of the vectors in $\mathcal{T}$, and so we can omit one of the vectors $\{\mathbf{w}_{n+1}, \ldots, \mathbf{w}_m\}$, giving the desired set. $\square$

A.7.8 Corollary. *Any two finite subsets that both span V and are linearly independent must have the same number of elements.*

The above corollary justifies the second sentence of the following definition.

A.7.9 Definition. *A subset of the vector space V is called a **basis** for V if it spans V and is linearly independent. If V has a finite basis, then it is said to be **finite dimensional**, and the number of vectors in the basis is called the **dimension** of V.*

The proofs of the remaining results are left as exercises for the reader.

A.7.10 Corollary. *Let V be an n-dimensional vector space. Then any set of more than n vectors is linearly dependent, and no set of fewer than n vectors can span V.*

A.7.11 Corollary. *Let V be an n-dimensional vector space, and let $\mathcal{B}$ be a set of n vectors in V. Then $\mathcal{B}$ is a basis for V if it is either linearly independent or spans V.*

A.7.12 Corollary. *Let V be an n-dimensional vector space. If W is any subspace of V, then $W = V$ if and only if W has dimension n.*

EXERCISES: SECTION A.7

1. Let V be an abelian group, with its operation denoted by $+$, and let p be a prime number. For any integer n and any element $v \in V$, we have defined $n \cdot v$ in Section 3.1. Show that V is a vector space over $\mathbf{Z}_p$ if and only if each nontrivial element of V has order p.

2. Let V be a finite dimensional vector space over the field F. Show that if W_1 and W_2 are subspaces of V, then there exists a basis $\mathcal{B}$ of V which has subsets $\mathcal{B}_1$ and $\mathcal{B}_2$ that are bases of W_1 and W_2, respectively. Show that this result cannot be extended to the case of three subspaces.

BIBLIOGRAPHY

Abstract Algebra

Artin, M., *Algebra*, Englewood Cliffs, N. J.: Prentice-Hall, Inc., 1991.

Birkhoff, G., and Mac Lane, S., *A Survey of Modern Algebra* (4th ed.). New York: Macmillan Publishing Co., Inc., 1977.

Clark, A., *Elements of Abstract Algebra*. New York: Dover Publications, Inc., 1984.

Dummit, D., and Foote, R., *Abstract Algebra* (3rd ed.). New York: John Wiley & Sons, Inc., 2003.

Fraleigh, J., and Katz, V., *A First Course in Abstract Algebra* (7th ed.). Reading, Mass.: Addison-Wesley Publishing Co., 2002.

Goldstein, L. J., *Abstract Algebra: A First Course*. Englewood Cliffs, N. J.: Prentice-Hall, Inc., 1973.

Herstein, I. N., *Topics in Algebra* (2nd ed.). New York: John Wiley & Sons, Inc., 1975.

Hungerford, T., *Algebra*. New York: Springer-Verlag New York, Inc., 1974.

Jacobson, N., *Basic Algebra I* (2nd ed.). San Francisco: W. H. Freeman & Company Publishers, 1985.

Lang, S., *Algebra* (3rd ed.). Reading, Mass.: Addison-Wesley Publishing Co., Inc., 1993.

Shapiro, L. W., *Introduction to Abstract Algebra*. New York: McGraw-Hill Book Company, 1975.

Van der Waerden, B. L., *Algebra*. vol. 1. New York: Frederick Unger Publishing Co., Inc., 1970.

Linear Algebra

Halmos, P. R., *Finite-Dimensional Vector Spaces*. New York: Springer-Verlag New York, Inc., 1993.

Hoffman, K. and Kunze, R., *Linear Algebra* (2nd ed.). Englewood Cliffs, NJ: Prentice-Hall, Inc., 1971.

Number Theory

Hardy, G. H., and Wright, E. M., *The Theory of Numbers* (5th ed.). Oxford, England: Oxford University Press, 1979.

Ireland, K., and Rosen, M., *A Classical Introduction to Modern Number Theory* (2nd ed.). New York: Springer-Verlag New York, Inc., 1990.

Niven, I. *Irrational Numbers* (Carus Mathematical Monograph No. 11). Washington, D. C.: The Mathematical Association of America, 1967.

Niven, I., Zuckerman, H. S. and Montgomery, H. L., *An Introduction to the Theory of Numbers* (5th ed.). New York: John Wiley & Sons, Inc., 1991.

Serre, J.–P., *A Course in Arithmetic*. New York: Springer-Verlag New York, Inc., 1973.

Theory of Equations

Dickson, L. E., *New First Course in the Theory of Equations*. New York: John Wiley & Sons, Inc., 1939.
Uspensky, J. V., *Theory of Equations*. New York: McGraw-Hill Book Company, 1948.

Group Theory

Hall, M., *The Theory of Groups*. New York: Macmillan Publishing Co., Inc., 1959.
Rotman, J. J., *Introduction to the Theory of Groups* (4th ed.). New York: Springer-Verlag New York, Inc., 1995.

Ring Theory

Beachy, J. A., *Introductory Lectures on Rings and Modules*. Cambridge, U.K.: Cambridge Univ. Press, 1999.
Herstein, I. N., *Noncommutative Rings* (Carus Mathematical Monographs No. 15). Washington, D. C.: The Mathematical Association of America, 1968.
Sharp, R. Y., *Steps in Commutative Algebra* (2nd ed.). Cambridge, U.K.: Cambridge Univ. Press, 2001.

Field Theory

Artin, E., *Galois theory: Lectures Delivered at the University of Notre Dame*. New York: Dover Publications, Inc., 1997.
Stewart, I., *Galois Theory* (3rd ed.). Boca Raton, FL: CRC Press, 2003.

History

Van der Waerden, B. L., *A History of Algebra*. New York: Springer-Verlag New York, Inc., 1985.

Other References

Cohen, L. and G. Ehrlich, *The Structure of the Real Number System*. Princeton, NJ: D. Van Nostrand Company, 1963.
Conway, J. B., *Functions of One Complex Variable* (2nd ed.). New York: Springer-Verlag New York, Inc., 1978.
Fulton, W., *Algebraic Topology: A First Course*. New York: Springer-Verlag New York, Inc., 1995.
Landau, E., *Foundations of Analysis* (2nd ed.). New York: Chelsea, 1960.

ANSWERS TO SELECTED EXERCISES

Exercises for which a solution is given are marked in the text by the symbol †.

Chapter 1

Section 1.1

3. (a) $\gcd(35, 14) = 7$ (c) $\gcd(252, 180) = 36$ (e) $\gcd(7655, 1001) = 1$

5. (a) $7 = 1 \cdot 35 + (-2) \cdot 14$ (c) $36 = (-2) \cdot 252 + 3 \cdot 180$

   (e) $1 = (-397) \cdot 7655 + 3036 \cdot 1001$

22. The integer x must have remainder 5 when divided by 11.

Section 1.2

1. (a) $35 = 5^1 7^1$, $14 = 2^1 7^1$, $(35, 14) = 7$, $[35, 14] = 70$.

   (c) $252 = 2^2 3^2 7^1$, $180 = 2^2 3^2 5^1$, $(252, 180) = 36$, $[252, 180] = 1260$.

   (e) $6643 = 7^1 13^1 73^1$, $2873 = 13^2 17^1$, $(6643, 2873) = 13$, $[6643, 2873] = 1468103$.

3. for $a = 4$: $\{1, 3\}$; for $a = 6$: $\{1, 5\}$; ... for $a = 9$: $\{1, 2, 4, 5, 7, 8\}$; ...

   for $a = 15$: $\{1, 2, 4, 7, 8, 11, 13, 14\}$; etc.

5. Diagrams of divisors of 9, 20, and 100:

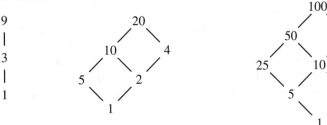

Section 1.3

1. (a) $x \equiv 2 \pmod 7$ (c) $x \equiv 13 \pmod{32}$

3. (a) $x \equiv 11 \pmod{21}$ (c) No solution

5. $x \equiv 9, 21, 33, 45, 57 \pmod{60}$

7. (a) 3   (c) 4

15. (a) $x \equiv 1, 7, 9, 15 \pmod{16}$ (c) $x \equiv 1, 3, 5, 7, 9, 11, 13, 15 \pmod{16}$

19. $x \equiv 43 \pmod{400}$

Section 1.4

1. Multiplication table for $\mathbf{Z}_{12}$:

·	0	1	2	3	4	5	6	7	8	9	10	11
0	0	0	0	0	0	0	0	0	0	0	0	0
1	0	1	2	3	4	5	6	7	8	9	10	11
2	0	2	4	6	8	10	0	2	4	6	8	10
3	0	3	6	9	0	3	6	9	0	3	6	9
4	0	4	8	0	4	8	0	4	8	0	4	8
5	0	5	10	3	8	1	6	11	4	9	2	7
6	0	6	0	6	0	6	0	6	0	6	0	6
7	0	7	2	9	4	11	6	1	8	3	10	5
8	0	8	4	0	8	4	0	8	4	0	8	4
9	0	9	6	3	0	9	6	3	0	9	6	3
10	0	10	8	6	4	2	0	10	8	6	4	2
11	0	11	10	9	8	7	6	5	4	3	2	1

3. (a) [14]   (c) Not invertible.

9. (a) The element [5] has multiplicative order 4, and [7] has multiplicative order 2.

11. There is no such congruence class in $\mathbf{Z}_8^\times$. In $\mathbf{Z}_7^\times$ each element is a power of [3] (or [5]).

13. (a) The idempotent elements of $\mathbf{Z}_6$ are [0], [1], [3], [4], and the idempotent elements of $\mathbf{Z}_{12}$ are [0], [1], [4], [9].

Chapter 2

Section 2.1

1. (a) The function f is one-to-one and onto.
   (c) The function f is one-to-one and onto if and only if $(m, n) = 1$.

3. (a) $f^{-1}(x) = x - 3$
   (c) If $(m, n) = 1$, and $km \equiv 1 \pmod{n}$, then $f^{-1}([x]_n) = [kx - kb]_n$, for all $[x]_n \in \mathbf{Z}_n$.

6. (a) There are 8 functions from S into T, and 9 from T into S.

8. The formula in (e) defines a function; the formulas in (a) and (c) do not.

10. (a) $f([8]_8) \neq f([0]_8)$ (c) $h([4]_4) \neq h([0]_4)$

Section 2.2

1. (a) We have $f(\mathbf{Z}) = \{1, i, -1, -i\}$, $\mathbf{Z}/f = \mathbf{Z}_4$, and the function $\bar{f} : \mathbf{Z}/f \to f(\mathbf{Z})$ is defined by $\bar{f}([n]_4) = i^n$.
   (c) We have $h(\mathbf{Z}_{12}) = \{[0]_{12}, [9]_{12}, [6]_{12}, [3]_{12}\}$ and $\mathbf{Z}_{12}/h = \{[[0]_{12}], [[1]_{12}], [[2]_{12}], [[3]_{12}]\}$, where $[[0]_{12}] = \{[0]_{12}, [4]_{12}, [8]_{12}\}$, $[[1]_{12}] = \{[1]_{12}, [5]_{12}, [9]_{12}\}$, $[[2]_{12}] = \{[2]_{12}, [6]_{12}, [10]_{12}\}$, $[[3]_{12}] = \{[3]_{12}, [7]_{12}, [11]_{12}\}$. The function $\bar{h} : \mathbf{Z}_{12}/h \to h(\mathbf{Z}_{12})$ is defined by $\bar{h}([[n]_{12}]) = h([n]_{12}) = [9n]_{12}$.

6. Define $\sim$ by $(x_1, y_1, z_1) \sim (x_2, y_2, z_2)$ if and only if $z_1 = z_2$.

8. (b) We have $[1] = \{\pm 1\}$ and $[6] = \{\pm 2^i 3^j \mid i \geq 1, j \geq 1\}$.

Section 2.3

1. (a) $\sigma\tau = \begin{pmatrix} 1 & 2 & 3 & 4 & 5 & 6 & 7 \\ 2 & 3 & 6 & 7 & 4 & 1 & 5 \end{pmatrix}$ (c) $\tau^2\sigma = \begin{pmatrix} 1 & 2 & 3 & 4 & 5 & 6 & 7 \\ 4 & 2 & 7 & 3 & 6 & 1 & 5 \end{pmatrix}$

(e) $\sigma\tau\sigma^{-1} = \begin{pmatrix} 1 & 2 & 3 & 4 & 5 & 6 & 7 \\ 1 & 3 & 2 & 7 & 6 & 4 & 5 \end{pmatrix}$

2. (a) $\sigma\tau = (1, 2, 3, 6)(4, 7, 5)$
   (c) $\tau^2\sigma = (1, 4, 3, 7, 5, 6)$
   (e) $\sigma\tau\sigma^{-1} = (2, 3)(4, 7, 5, 6)$

4. (a) $(1, 6)(2, 4, 3, 5)$ has order 4
   (c) $(1, 5, 3, 8)(2, 9)(4, 7, 6)$ has order 12

7. There is 1 cycle of length 1; there are 10 of length 2, 20 of length 3, 30 of length 4, and 24 of length 5. Elements can have order 1,2,3,4,5, or 6.

Chapter 3

Section 3.1

2. (a) If $a \neq \pm 1$, then there does not exist an inverse for a under $*$.
   (c) The operation is not associative, and has no identity element.
   (e) The operation defines a group structure on $\mathbf{R}+$.

7. Multiplication table for $\mathbf{Z}_7^\times$:

·	1	2	3	4	5	6
1	1	2	3	4	5	6
2	2	4	6	1	3	5
3	3	6	2	5	1	4
4	4	1	5	2	6	3
5	5	3	1	6	4	2
6	6	5	4	3	2	1

Section 3.2

1. The matrix $\begin{bmatrix} 1 & -1 \\ 1 & 0 \end{bmatrix}$ has order 6. The matrix $\begin{bmatrix} 1 & 1 \\ 0 & 1 \end{bmatrix}$ has infinite order.

5. (a) $\mathbf{Z}_6 = \langle 1 \rangle = \langle 5 \rangle; \langle 2 \rangle = \langle 4 \rangle = \{0, 2, 4\}; \langle 3 \rangle = \{0, 3\}; \langle 0 \rangle = \{0\}$.
   (c) $\mathbf{Z}_9^\times = \{1, 2, 4, 5, 7, 8\} = \langle 2 \rangle = \langle 5 \rangle; \langle 4 \rangle = \langle 7 \rangle = \{1, 4, 7\}; \langle 8 \rangle = \{1, 8\}; \langle 1 \rangle = \{1\}$.

12. (a) $\{\pm 1\}$ (b) The elements of finite order in $\mathbf{C}^\times$ are the complex roots of unity. (See Section A.5 of the appendix for a review of roots of unity.)

Section 3.3

1. $HK = \mathbf{Z}_{16}^\times$
3. For example, $H = \langle (1, 2) \rangle$ and $K = \langle (1, 3) \rangle$.
7. Let $G = \mathbf{Z}_2 \times S_3$.

Section 3.4

1. Define $\phi : \mathbf{Z}_4 \to \mathbf{Z}_{10}^\times$ by $\phi([n]_4) = [3]_{10}^n$.
3. Construct the group tables and find a one-to-one correspondence between the groups that preserves the entries of the tables.
5. No, because $\mathbf{C}^\times$ has an element of order 2 but $\mathbf{C}$ does not.
9. The groups $\mathbf{Z}_8$ and $\mathbf{Z}_2 \times \mathbf{Z}_4$ are not isomorphic.

Section 3.5

1. The element a^j has order $12/(j, 12)$.
5. $\left\langle \frac{\sqrt{2}}{2} + \frac{\sqrt{2}}{2}i \right\rangle = \{\pm 1, \pm i, \pm\frac{\sqrt{2}}{2} \pm \frac{\sqrt{2}}{2}i\}$

7. The groups $\mathbf{Z}_{18}^{\times}$ and $\mathbf{Z}_{27}^{\times}$ are cyclic.

9. $\langle([0]_4, [0]_2)\rangle = \{([0]_4, [0]_2)\}$ $\langle([0]_4, [1]_2)\rangle = \{([0]_4, [0]_2), ([0]_4, [1]_2)\}$

$\langle([2]_4, [0]_2)\rangle = \{([0]_4, [0]_2), ([2]_4, [0]_2)\}$ $\langle([2]_4, [1]_2)\rangle = \{([0]_4, [0]_2), ([2]_4, [1]_2)\}$

$\langle([1]_4, [0]_2)\rangle = \{([0]_4, [0]_2), ([1]_4, [0]_2), ([2]_4, [0]_2), ([3]_4, [0]_2)\} = \langle([3]_4, [0]_2)\rangle$

$\langle([1]_4, [1]_2)\rangle = \{([0]_4, [0]_2), ([1]_4, [1]_2), ([2]_4, [0]_2), ([3]_4, [1]_2)\} = \langle([3]_4, [1]_2)\rangle$

Section 3.6

1. (a) 4   (c) 4

8. 24

12. The maximum order in S_4 is 4; the maximum order in S_6 is 6; the maximum order in S_8 is 15.

14. In addition to the identity permutation, there are 15 products of two transpositions, 20 cycles of length 3, and 24 cycles of length 5.

Section 3.7

1. There are 3 different homomorphisms, given by the formulas $\phi_0([x]_6) = [0]_9$, $\phi_3([x]_6) = [3x]_9$, or $\phi_6([x]_6) = [6x]_9$, defined for all $[x]_6 \in \mathbf{Z}_6$.

5. For the given formula ϕ, we have $\phi(G) = \{1, 4\}$ and $\ker \phi = \{1, 4, 11, 14\}$.

7. The formulas in (a) and (c) define homomorphisms; the formula in (e) does not.

17. The normal subgroups of D_4 are: D_4; all three subgroups of order 4; $\{e, a^2\}$; and $\{e\}$.

Section 3.8

1. The cosets of $\langle[3]\rangle$ in $\mathbf{Z}_{24}$ are

$[0] + \langle[3]\rangle = \{[0], [3], [6], [9], [12], [15], [18], [21]\}$,

$[1] + \langle[3]\rangle = \{[1], [4], [7], [10], [13], [16], [19], [22]\}$, and

$[2] + \langle[3]\rangle = \{[2], [5], [8], [11], [14], [17], [20], [23]\}$.

3. The left cosets are $\{e, ab\}$, $\{a, a^2b\}$, and $\{a^2, b\}$; the right cosets are $\{e, ab\}$, $\{a, b\}$, and $\{a^2, a^2b\}$.

15. If N is a proper, nontrivial normal subgroup of D_4, then the factor group D_4/N is isomorphic to either $\mathbf{Z}_2$ or $\mathbf{Z}_2 \times \mathbf{Z}_2$.

17. $\mathbf{Z}_2 \times \mathbf{Z}_2$.

Chapter 4

Section 4.1

3. $f(x) = (2x^2 + 3x + 1)(x - 1) + 2$

5. (a) $f(x) = (2x^2 + 3x - 1)(x - 1) + 2$ (c) $f(x) = (x^2 + x + 1)(x - 1) + 2$

19. $f(x) = -4x^3 + 27x^2 - 41x + 3$

21. $(b_2 + 4b_1 + b_0)\frac{h}{3}$

Section 4.2

1. (a) $f(x) = g(x)(x^3 + 3x^2 - 2x + 1) + 9$ (c) $f(x) = g(x)(x^4 - x^3 + x^2 - x + 1)$

3. (a) $\gcd(f(x), f'(x)) = x - 1$ (c) $\gcd(f(x), f'(x)) = 1$

5. (a) $\gcd(x^4 + x^3 + x + 1, x^3 + x^2 + x + 1) = x^2 + 1$

(c) $\gcd(x^5 + 4x^4 + 6x^3 + 6x^2 + 5x + 2, x^4 + 3x^2 + 3x + 6) = x^3 + 4x^2 + 5x + 2$

7. (a) $x^2 + 1 = 1 \cdot (x^4 + x^3 + x + 1) + x \cdot (x^3 + x^2 + x + 1)$

(c) $x^3 + 4x^2 + 5x + 2 = 5 \cdot (x^5 + 4x^4 + 6x^3 + 6x^2 + 5x + 2) + (2x+1) \cdot (x^4 + 3x^2 + 3x + 6)$

12. As a partial answer, the irreducible polynomials of degree ≤ 4 are x, $x+1$, $x^2 + x + 1$, $x^3 + x^2 + 1$, $x^3 + x + 1$, $x^4 + x^3 + x^2 + x + 1$, $x^4 + x^3 + 1$, and $x^4 + x + 1$.

18. (a) $(a + bx)(c + dx) \equiv (ac - bd) + (ad + bc)x \pmod{x^2 + 1}$

(c) $(a + bx)(c + dx) \equiv (ac + bd) + (ad + bc + bd)x \pmod{x^2 + x + 1}$

20. (a) $q(x) = (a - bx)/(a^2 + b^2)$, if $a + bx \neq 0$ (c) $q(x) = (a+b) + bx$, if $a + bx \neq 0$

Section 4.3

7. The mapping $\phi : \mathbf{R}[x]/\langle x^2 + 1 \rangle \to \mathbf{C}$ defined by $\phi([a + bx]) = a + bi$ is an isomorphism.

16. If F is the field consisting of the 4 given matrices, then $\theta : \mathbf{Z}_2[x]/\langle x^2 + x + 1 \rangle \to F$ defined by $\theta \left(\begin{bmatrix} a & b \\ c & d \end{bmatrix} \right) = [a + bx]$ is an isomorphism.

17. One possible irreducible cubic over $\mathbf{Z}_2$ is $p(x) = x^3 + x + 1$. The elements of the ring $\mathbf{Z}_2[x]/\langle x^3 + x + 1 \rangle$ correspond to quadratic polynomials over $\mathbf{Z}_2$, and the identities necessary for multiplication are $[x][x^2] = [x + 1]$ and $[x^2][x^2] = [x^2 + x]$.

19. One possible irreducible cubic over $\mathbf{Z}_3$ is $p(x) = x^3 + 2x + 2$. The identities necessary for multiplication are $[x][x^2] = [x + 1]$ and $[x^2][x^2] = [x^2 + x]$.

21. (a) $[a + bx]^{-1} = [(a/c) - (b/c)x]$, for $c = a^2 + b^2$, if $[a + bx] \neq [0]$.

(c) $[x^2 - 2x + 1]^{-1} = [3x^2 + 4x + 5]$ (e) $[x]^{-1} = [4x + 4]$

23. The set of congruences $\mathbf{Z}_k[x]/\langle x^2 + 1 \rangle$ is a field for $k = 3, 7, 11$.

Section 4.4

3. (a) $-1, 2, -4, 5$. (c) $-12, -35$ (e) $8, 9$

5. (a) Use $p = 2$. (c) Substitute $x - 1$ and use $p = 2$.

15. We have the factorization $x^8 - 1 = (x - 1)(x + 1)(x^2 + 1)(x^4 + 1)$, and $x^4 + 1$ is irreducible over $\mathbf{Q}$ by Exercise 5 (a).

Chapter 5

Section 5.1

1. The set defined in (a) forms a subring, but the one defined in (c) does not.

3. The subsets defined in (a) and (c) form subrings; the one defined in (e) does not.

10. (a) Let $I = \{1, 2\}$, so that the associated ring R of all subsets of I consists of the elements I, $a = \{1\}$, $b = \{2\}$, and $\emptyset$. We have the following tables.

+	$\emptyset$	I	a	b
$\emptyset$	$\emptyset$	I	a	b
I	I	$\emptyset$	b	a
a	a	b	$\emptyset$	I
b	b	a	I	$\emptyset$

$\cdot$	$\emptyset$	I	a	b
$\emptyset$	$\emptyset$	$\emptyset$	$\emptyset$	$\emptyset$
I	$\emptyset$	I	a	b
a	$\emptyset$	a	a	$\emptyset$
b	$\emptyset$	b	$\emptyset$	b

19. (b) Units of $\mathbf{Z}_4 \oplus \mathbf{Z}_9$: $([1]_4, [1]_9)$, $([3]_4, [1]_9)$, $([1]_4, [2]_9)$, $([3]_4, [2]_9)$, $([1]_4, [4]_9)$, $([3]_4, [4]_9)$, $([1]_4, [5]_9)$, $([3]_4, [5]_9)$, $([1]_4, [7]_9)$, $([3]_4, [7]_9)$, $([1]_4, [8]_9)$, $([3]_4, [8]_9)$.

Section 5.2

13. The two possible ring homomorphisms π_1 and π_2 from $\mathbf{Z} \oplus \mathbf{Z}$ into $\mathbf{Z}$ are defined as follows, for all $(m, n) \in \mathbf{Z} \oplus \mathbf{Z}$: $\pi_1((m, n)) = m$ and $\pi_2((m, n)) = n$.

20. (a) *Hint*: With the notation of Exercise 10 of Section 5.1, define $\phi : \mathbf{Z}_2 \oplus \mathbf{Z}_2 \to R$ by $\phi(0, 0) = \emptyset, \phi(1, 0) = \{1\} = a, \phi(0, 1) = \{2\} = b,$ and $\phi(1, 1) = I$.

Section 5.3

1. We have $\mathbf{Z}_2[x]/\langle x^2 + 1\rangle = \{[0], [1], [x], [x + 1]\}$, where $[a] = a + \langle x^2 + 1\rangle$.

·	[0]	[1]	[x]	[x + 1]
[0]	[0]	[0]	[0]	[0]
[1]	[0]	[1]	[x]	[x + 1]
[x]	[0]	[x]	[1]	[x + 1]
[x + 1]	[0]	[x + 1]	[x + 1]	[0]

3. $[a_1x^2+b_1x+c_1][a_2x^2+b_2x+c_2] = [(5a_1a_2-2a_1b_2+a_1c_2-2b_1a_2+b_1b_2+c_1a_2)x^2+ (a_1a_2 + a_1b_2 + b_1a_2 + b_1c_2 + c_1b_2)x + (-6a_1a_2 + 3a_1b_2 + 3b_1a_2 + c_1c_2)]$

13. (b) $n\mathbf{Z} + m\mathbf{Z} = d\mathbf{Z}$, where $d = \gcd(n, m)$

16. (b) The factor ring R/I is isomorphic to $\mathbf{Z}_2$.

Section 5.4

7. $Q(D) \cong \mathbf{Q}(\sqrt{2})$

9. $Q(D) \cong \mathbf{Q}(i)$

Chapter 6

Section 6.1

1. (a) The number $\sqrt{2}$ satisfies the polynomial $f(x) = x^2 - 2$.
   (c) The number $\sqrt{3} + \sqrt{5}$ satisfies $f(x) = x^4 - 16x^2 + 4$.
   (e) The number $(-1 + \sqrt{3}i)/2$ satisfies $f(x) = x^2 + x + 1$.

5. (b) We have $u^{-1} = -1 - \frac{1}{3}u^2$, and $(1 + u)^{-1} = 4 - u + u^2$.

Section 6.2

1. (a) The number $\sqrt{3}$ has degree 2 over $\mathbf{Q}$. The set $\{1, \sqrt{3}\}$ is a basis for $\mathbf{Q}(\sqrt{3})$ over $\mathbf{Q}$.
   (c) We have $[\mathbf{Q}(\sqrt{3} + \sqrt{7}) : \mathbf{Q}] = 4$, and $\{1, u, u^2, u^3\}$ is a basis for $\mathbf{Q}(\sqrt{3} + \sqrt{7})$ over $\mathbf{Q}$, where $u = \sqrt{3} + \sqrt{7}$.
   (e) The degree of $\sqrt{2} + \sqrt[3]{2}$ over $\mathbf{Q}$ is 6, and $\{1, \sqrt[3]{2}, \sqrt[3]{4}, \sqrt{2}, \sqrt{2}\sqrt[3]{2}, \sqrt{2}\sqrt[3]{4}\}$ is a basis over $\mathbf{Q}$.

8. We have $[\mathbf{Q}(\sqrt{n}) : \mathbf{Q}] = 2$, unless n is a square, in which case $[\mathbf{Q}(\sqrt{n}) : \mathbf{Q}] = 1$.

Section 6.4

1. (a) The splitting field for $x^2 - 2$ over $\mathbf{Q}$ is $\mathbf{Q}(\sqrt{2})$.
   (c) The splitting field for $x^4 + x^2 - 6$ over $\mathbf{Q}$ is $\mathbf{Q}(\sqrt{2}, \sqrt{3}i)$.

3. (a) The splitting field for $x^2 + x + 1$ over $\mathbf{Z}_2$ is $\mathbf{Z}_2(u)$, where $u = x + \langle x^2 + x + 1\rangle$.
   (c) The splitting field for $x^3 + x + 1$ over $\mathbf{Z}_2$ is $\mathbf{Z}_2(u)$, where $u = x + \langle x^3 + x + 1\rangle$.

5. The polynomial $x^p - x$ splits over $\mathbf{Z}_p$.

Section 6.5

3. The polynomial $x^4 + x + 1$ is irreducible over $\mathbf{Z}_2$, so $\mathbf{Z}_2[x]/\langle x^4 + x + 1 \rangle$ is isomorphic to $GF(2^4)$. The element $x + \langle x^4 + x + 1 \rangle$ has order 15, and so it is a generator for the multiplicative group $GF(2^4)^\times$.

9. If $d = \gcd(e, f)$, then $E \cap F = GF(p^d)$, the unique subfield of $GF(p^n)$ with p^d elements.

Section 6.7

2. (a) $\left(\dfrac{231}{997}\right) = 1$ (b) $\left(\dfrac{783}{997}\right) = -1$

4. (a) $\left(\dfrac{5}{p}\right) = \begin{cases} 1 & p \equiv \pm 1 \pmod 5 \\ -1 & p \equiv \pm 2 \pmod 5 \end{cases}$

   (c) $\left(\dfrac{7}{p}\right) = \begin{cases} 1 & p \equiv \pm 1, \pm 3, \pm 9 \pmod{28} \\ -1 & p \equiv \pm 5, \pm 11, \pm 13 \pmod{28} \end{cases}$

Chapter 7

Section 7.1

3. The automorphism group of $\mathbf{Z}_2 \times \mathbf{Z}_2$ is isomorphic to S_3.

5. The function $\phi : G \to G$ defined by $\phi(g) = g^{-1}$ for all $g \in G$ is an automorphism if and only if G is abelian.

7. We have $\mathrm{Aut}(S_3) \cong S_3$.

Section 7.2

8. If we represent D_5 as $\{e, a, a^2, a^3, a^4, b, ab, a^2b, a^3b, a^4b\}$, where $a^5 = b^2 = e$ and $ba = a^{-1}b$, then the conjugacy classes are $\{e\}$, $\{a, a^4\}$, $\{a^2, a^3\}$, $\{b, ab, a^2b, a^3b, a^4b\}$.

10. The conjugacy classes of A_4 are $\{(1)\}$, $\{(1, 2, 3), (2, 4, 3), (1, 3, 4), (1, 4, 2)\}$,
    $\{(1, 3, 2), (2, 3, 4), (1, 4, 3), (1, 2, 4)\}$, $\{(1, 2)(3, 4), (1, 4)(2, 3), (1, 3)(2, 4)\}$.

12. (a) The conjugacy class equation for A_4 is $12 = 1 + (4 + 4 + 3)$.

Section 7.4

3. The centralizer of $(1, 2)(3, 4)$ is a Sylow 2-subgroup, and the cyclic subgroup generated by $(1, 2, 3)$ is a Sylow 3-subgroup.

6. The Sylow 3-subgroups of S_4 are $H_1 = \langle (1, 2, 3) \rangle$, $H_2 = \langle (1, 2, 4) \rangle$, $H_3 = \langle (1, 3, 4) \rangle$, and $H_4 = \langle (2, 3, 4) \rangle$, with $(3, 4)H_1(3, 4) = H_2$, $(2, 4)H_1(2, 4) = H_3$, and $(1, 4)H_1(1, 4) = H_4$.

Section 7.5

4. (a) $\mathbf{Z}_{20}^\times \cong \mathbf{Z}_2 \times \mathbf{Z}_4$ (c) $\mathbf{Z}_{70}^\times \cong \mathbf{Z}_4 \times \mathbf{Z}_2 \times \mathbf{Z}_3$

Section 7.6

3. The groups $\mathbf{Z}_4$ and $\mathbf{Z}_2 \times \mathbf{Z}_2$ provide an example.

Section 7.7

2. If $n \geq 5$, then A_n is the only proper nontrivial normal subgroup of S_n.

11. $|\mathrm{GL}_n(F)| = (q^n - 1)(q^n - q) \cdots (q^n - q^{n-1})$

Chapter 8

Section 8.1

2. The field $GF(2^3)$ can be described as $\{0, 1, \alpha, \alpha + 1, \alpha^2, \alpha^2 + 1, \alpha^2 + \alpha, \alpha^2 + \alpha + 1\}$ with multiplication given by $\alpha^3 = \alpha + 1$. The set $\{1, \alpha, \alpha^2\}$ is a basis for $GF(2^3)$ over $GF(2)$. The Galois group $\mathrm{Gal}(GF(2^3)/GF(2))$ is cyclic of order 3, generated by θ, where $\theta(x) = x^2$ for all $x \in GF(2^3)$.

4. In Example 8.1.2, $\{x \in \mathbf{Q}(\sqrt{2} + \sqrt{3}) \mid \theta_2(x) = x\} = \mathbf{Q}(\sqrt{3})$.

Section 8.2

7. The element $\omega + \sqrt[3]{2}$ is a primitive element for the extension $\mathbf{Q}(\omega, \sqrt[3]{2})$ of $\mathbf{Q}$.

Section 8.3

4. We have $\mathrm{Gal}(\mathbf{Q}(\sqrt{3}, \sqrt{2}i)/\mathbf{Q}) \cong \mathbf{Z}_2 \times \mathbf{Z}_2$.

Section 8.5

1. (a) $\Phi_8 = x^4 + 1$ (c) $\Phi_{15} = x^8 - x^7 + x^5 - x^4 + x^3 - x + 1$

Section 8.6

3. The Galois group of $x^5 - x - 1$ over $\mathbf{Q}$ is S_5.

Chapter 9

Section 9.2

1. (a) The greatest common divisor is $xy + 1$.

Section 9.3

1. The primes 3, 7, 11, 19 are irreducible, since they are congruent to 3 modulo 4. We have $2 = (1+i)(1-i)$, $5 = (2+i)(2-i)$, $13 = (2+3i)(2-3i)$, $17 = (4+i)(4-i)$.

INDEX OF SYMBOLS

GREEK ALPHABET

alpha	α	A
beta	β	B
gamma	γ	Γ
delta	δ	Δ
epsilon	ϵ, ε	E
zeta	ζ	Z
eta	η	H
theta	θ	Θ
iota	ι	I
kappa	κ	K
lambda	λ	Λ
mu	μ	M
nu	ν	N
xi	ξ	Ξ
omicron	o	O
pi	π	Π
rho	ρ	P
sigma	σ	Σ
tau	τ	T
upsilon	υ	Υ
phi	ϕ, φ	Φ
chi	χ	X
psi	ψ	Ψ
omega	ω	Ω

INDEX